11th Biennial Conference on Classical and Quantum Relativistic Dynamics of Particles and Fields (IARD 2018)

Journal of Physics: Conference Series
Volume 1239

Merida, Mexico
4 - 7 June 2018

Editor:

L.P. Horwitz

ISBN: 978-1-5108-8821-0
ISSN: 1742-6588

Printed from e-media with permission by:

Curran Associates, Inc.
57 Morehouse Lane
Red Hook, NY 12571

proceedings.com

Some format issues inherent in the e-media version may also appear in this print version.

Additional copies of this publication are available from:

Curran Associates, Inc.
57 Morehouse Lane
Red Hook, NY 12571 USA
Phone: 845-758-0400
Fax: 845-758-2633
Email: curran@proceedings.com
Web: www.proceedings.com

Table of contents

Volume 1239

The 11th Biennial Conference on Classical and Quantum Relativistic Dynamics of Particles and Fields

4–7 June 2018, Mérida, Yucatán, Mexico

Accepted papers received: 26 April 2019
Published online: 20 May 2019

Preface

The 11th Biennial Conference on Classical and Quantum Relativistic Dynamics of Particles and Fields

Peer review statement

Papers

Statistical Mechanics and Radiation

Modified Vlasov Equation and Collective Oscillations of Relativistic Radiating Electron Plasma

G. Ares de Parga and J. F. García-Camacho.....1

A Further Study of the Mixing of Relativistic Ideal Gases with Relative Relativistic Velocities: The Hot Plasma in the Sun's Corona, the Type II Spicules and CMEs.

R E Gonzalez-Narvaez, E E Díaz-Figueroa and G Ares de Parga.....27

Bulk viscosity in 2D relativistic uids: the effects of temperature and modifications to the Rayleigh-Brillouin spectrum

A L García-Perciante, L Franco-Pérez and A R Méndez.....43

A non-relativistic magnetized vector boson gas at any temperature

LC. Suárez-González, G. Quintero Angulo, A. Pérez Martínez and H. Pérez Rojas.....51

Mass-Energy-Momentum Radiation in Stueckelberg-Horwitz-Piron (SHP) Electrodynamics

Martin Land.....59

General Relativity, galactic rotation curves, dark matter, stars and planets

The Effect of Retardation on Galactic Rotation Curves

Asher Yahalom.....83

Significance of Deformity on the Mass Quadrupole Moment of Non-Rotating Neutron Stars

Omair Zubairi and Fridolin Weber.....90

Relativistic Beaming as a Probe of Stellar and Planetary Masses

Ben Placek.....99

Radial Acceleration and Tully-Fisher Relations in Conformal Gravity

James G. O'Brien, Thomas L. Chiarelli, Philip D. Mannheim, Mark A. Falcone, Muhannad H. AlQurashi and Jordan Carter.....108

Gravitation in the surface tension model of spacetime

H A Perko.....117

Geometrization of Newtonian Dynamics

Yaakov Friedman and Tzvi Scarr.....134

How Quasars with Thick Accretion Disk Eject Relativistic Jets

N Ben-Amots.....153

Relative and Absolute Time

Tepper L. Gill.....172

An Elementary Canonical Classical and Quantum Dynamics for General Relativity

L. P. Horwitz.....185

Time dispersion in quantum mechanics

John Ashmead.....204

Quantum Field Theory

How vacuum fluctuations determine the properties of the vacuum

G. B. Mainland and Bernard Mulligan.....299

Critical Scaling and a Dynamical Higgs Boson

Philip D. Mannheim.....316

Spinorial Formulations

Quantum equations for massless particles of any spin in curved spacetime

Amnon Moalem and Alexander Gersten.....330

Solutions in the 1/2,0 + 0,1/2 Representation of the Lorentz Group

Valeriy V. Dvoeglazov.....341

Maxwell's Equations, Fields or Potentials?

Alexander Gersten and Amnon Moalem.....352

Measurement and observation

The Einstein-Podolsky-Rosen paradox and SU(2) relativity

Paul O'Hara.....366

On the motion of microscopic bodies and observation of position in quantum theory

Alexey A. Kryukov.....378

IOP Conf. Series: Journal of Physics: Conf. Series **1239** (2019) 011001 doi:10.1088/1742-6596/1239/1/011001

Preface

The International Association for Relativistic Dynamics was organized in February 1998 in Houston, Texas, with John R. Fanchi as president.

Although the subject of relativistic dynamics has been explored, from both classical and quantum mechanical points of view, since the work of Einstein and Dirac, its most striking development has been in the framework of quantum field theory. The very accurate calculations of spectral and scattering properties, for example, of the anomalous magnetic moment of the electron and the Lamb shift in quantum electrodynamics, and many qualitative features of the strong and electroweak interactions, demonstrate the very great power of description achieved in this framework. Yet, many fundamental questions remain to be clarified, such as the structure of classical relativistic dynamical theories on the level of Hamilton and Lagrange in Minkowski space as well as on the curved manifolds of general relativity. There, moreover, remained the important questions of the covariant classical description of systems at high energy for which particle production effects are not large, such as discussed in Synge's book, *The Relativistic Gas*, and in Balescu's book on relativistic statistical mechanics, and the development of a consistent single and many body relativistic quantum theory.

In recent years, the very high accuracy of telescopes and advanced facilities for computation have brought a high level of interest in cosmological problems such as the structure of galaxies (dark matter) and the apparently anomalous expansion of the universe (dark energy). Some of the papers reported here deal with these problems, as well as other fundamental related issues.

It was for this purpose, to bring together researchers from a wide variety of fields, such as particle physics, astrophysics, cosmology, foundations of relativity theory, and mathematical physics, with a common interest in relativistic dynamics, to investigate fundamental questions of this type, that this Association was founded.

The second meeting took place, in 2000, at Bar Ilan University in Ramat Gan, Israel, the third, in 2002, at Howard University in Washington, D.C., and the fourth, on June 12-19, 2004, in Saas Fee, Switzerland. In 2006, the fifth meeting took place at the University of Connecticut campus in Storrs, Connecticut, and the sixth meeting, in Thessaloniki, Greece, in 2008, with the significant guidance of Professor Ioannis Antoniou, the help of the Aristotle University of Thessaloniki, at the Teleglion Foundation. The seventh meeting took place in Hualien, Taiwan, 2010,and the eighth at the Galileo Galilei Institute for Theoretical Physics (GGI) in Florence (Firenze) , Italy, in 2012. The ninth meeting took place again at the University of Connecticut in Storrs, Connecticut (2014), and the tenth,in 2016, in Ljubljana, Slovenia, with the help and guidance of Professor Matej Pavšič. The eleventh meeting took place in Mérida, Yucatán, Mexico, under the sponsorship of the Instituto Politécnic Nacional with the help and guidance of Professor Gonzálo Ares de Parga. This meeting forms the basis for the Proceedings that are recorded in this issue of the Journal of Physics: Conference Series.

IOP Conf. Series: Journal of Physics: Conf. Series **1239** (2019) 011001 doi:10.1088/1742-6596/1239/1/011001

Along with the work of some of the founding and newer but already much engaged members of the Association, we were fortunate to have lecturers from application areas that provided strong challenges for further developments in quantum field theory, cosmological problems, and in the dynamics of systems subject to accelerations and the effects of general relativity.

Topics treated in this issue include studies in relativistic statistical mechanics and radiation, including mass radiation in an off shell theory. A new approach to understanding the observed rotation curves of galaxies is discussed, based on retardation, consistent with the recent observation of gravitational waves. There is also a discussion of an elementary approach, following Einstein's original development, of the embedding of the relativistic dynamics of Stueckelberg, Horwitz and Piron into a canonical (with a complete phase space) classical and quantum theory of a particle moving in a gravitational field subject to additional forces, extended to many body systems with electromagnetic interaction. A geometrization of Newtonian mechanics is also studied, and the effect of vacuum fluctuations on the properties of the vacuum are also discussed. An important and timely study is given of the effect of critical scaling on the Higgs phenomenon is given. The spinorial formulation of relativistic wave equations is discussed, and the notion of measurement is rigorously studied. The EPR problem is also rigorously analyzed.

We thank the Scientific Advisory Committee for their invaluable guidance and advice:

Stephen Adler (Institute for Advanced Study)
Itzhak Bars (University of Southern California)
Gordon Baym (University of Illinois)
Fred Cooper (Los Alamos National Laboratory)
Bei-Lok Hu (University of Maryland)
Werner Israel (University of Victoria)
E.V. Shuryak (Brookhaven National Laboratory)
L.S. Shulman (Clarkson University)
William Unruh (University of British Columbia)
Luca Lusanna (National Institute for Nuclear Physics, INFN)
Benoit Famaey (Observatoire Astronomique de Strasbourg, CNRS)

The organizers express their gratitude to the Instituto Politécnic Nacional for its support in arranging excellent facilities in the charming city of Mérida.

We thank the participants who contributed through their lectures, personal discussions, and these papers, to the advancement of the subject and our understanding.

For the Editors and Organizing Committee,

L. P. Horwitz, Editor-in-Chief
Gonzalo Ares de Parga, Conference Chair and IARD Secretary
Martin Land, IARD President
Tepper Gill, IARD Treasurer
James O'Brien, IARD Vice President

IOP Conf. Series: Journal of Physics: Conf. Series **1239** (2019) 011002 doi:10.1088/1742-6596/1239/1/011002

Peer review statement

All papers published in this volume of *Journal of Physics: Conference Series* have been peer reviewed through processes administered by the proceedings Editors. Reviews were conducted by expert referees to the professional and scientific standards expected of a proceedings journal published by IOP Publishing.

IOP Conf. Series: Journal of Physics: Conf. Series **1239** (2019) 012001 doi:10.1088/1742-6596/1239/1/012001

Modified Vlasov Equation and Collective Oscillations of Relativistic Radiating Electron Plasma

G. Ares de Parga[1] **and J. F. García-Camacho**[1]

[1]Dpto. de Física, Escuela Superior de Física y Matemáticas, Instituto Politécnico Nacional, U. P. Adolfo López Mateos, Zacatenco, C.P. 07738, México City, México.

E-mail: gadpau@hotmail.com

Abstract. A Covariant Statistical Mechanics is developed for a system of electrons within the limits of plasma physics. By using the Landau-Lifshitz equation of motion for point charged particles, the phase space reduces to be a seven dimensional space instead of the ten dimensional space proposed by Hakim and Mangeney [J. Math. Phys. 1968, 9, 1, 116]. A modified Vlasov equation is obtained which permits one to deduce a general expression for the dispersion relations. Unlike other works, neither the Special Relativity nor any of the properties of the gauge are used to discard some of the wave modes. Instead, the continuity equation is used as a constraint embedded in the definition of the 4−vector current. This permits to make a mathematical analysis for calculating and discarding the transverse and longitudinal modes for any temperature. An expression for the longitudinal wave modes is deduced for any temperature and wave number. Usually, certain approximations are made to obtain such an expression that matches our result. But in general, without making such approximations, the result differs by a factor. The Landau damping and the damping due to the radiation reaction force are obtained and contrasted at low and high temperature cases. The results are compared with the ones obtained within Magnetohydrodynamics. Moreover, a numerical method is applied in order to compare the dispersion relations obtained with the approximations made. Surprisingly, it is sampled numerically that the dispersion relations are bounded for each temperature and that there is a cut-off for each wave number.

1. Introduction

During the sixteens of the last century, Hakim developed a manifestly Covariant Statistical Mechanics [1], [2]. The theory studied the electromagnetic interactions and described a rigorous hierarchy of equations for suitable reduced densities. Later, Hakim and Mangeney [3], including the radiation reaction force calculated by Dirac [5], obtained a relativistic Vlasov equation which permits one to obtain wave dispersion relations and the conductivity tensor. They also analyzed the irreversibility due to the interaction between the charges and the radiation reaction force and many other properties of a relativistic plasma. However, the use of the Lorentz-Dirac equation obligates to consider the hyper-acceleration as an independent variable since it plays the same role than the position and the velocity. The phase space results to be a $10N-$dimensional space ($12N-2N$, due to the kinematic relations). In an approximation process, by using the Lorentz-Dirac equation at first order in the characteristic time of the charges τ_o and by considering absence of correlation between particles, they obtained a modified Vlasov equation at first order in $ne^2\tau_o$ where n, e and $\tau_o = 2e^2/3mc^3$ represent the density of particles, the charge of the electron and the characteristic time of the electron, respectively. Among others properties that

IOP Conf. Series: Journal of Physics: Conf. Series **1239** (2019) 012001 doi:10.1088/1742-6596/1239/1/012001

can be derived, as we mentioned above, it has to be highlighted that this modified Vlasov equation can be linearized in order to obtain dispersion relations for electromechanical waves. They obtained general expressions for the transverse and longitudinal collective modes. They gave a solution for the cases $T = 0$ and $T \to \infty$, discriminating certain results due to very vague physical reasoning. Indeed, for $T = 0$, they eliminated solutions for the longitudinal mode ($w^2 = w_p^2 + k^2$) arguing that this solution comes from Special Relativity. Moreover, for high temperature without giving any proof for the longitudinal mode, they mentioned "it seems that the extreme relativistic plasma behaves like the vacuum for both longitudinal and transverse electromagnetic waves". In a later article [4], they accepted the lack of severity in the choice of the solutions and they claimed that the dispersion relations do not take into account the fact that the 4−current vector must satisfy the continuity equation. Although for low temperatures ($T \simeq 0$) the usual expected results are obtained just for the neighborhood of $k \simeq w_p$, for the high temperature case ($T \to \infty$), in the first paper about collective oscillations [3] they just assured that $k^\mu k_\mu = 0$ without any proof and they argued that the damping effects predominate and a wave cannot propagate in a high temperature plasma. Nevertheless, these types of contradictions were corrected in a second article [4] where they made some approximations for high temperature showing that the plasma frequency is shifted. Indeed, this does not present a surprise since radiation emission becomes more effective when the temperature increases. Although, for Hakim and Mangeney [3] in extreme relativistic plasmas, their own expressions are not longer valid since the first order approximation in $ne^2\tau_o$ becomes meaningless if bigger order terms are considered and, consequently, the decay of the wave must be accentuated. However, the order of approximation is related to the value of $ne^2\tau_o$ and high temperature systems can be considered within their model as long as the parameter $ne^2\tau_o$ stays small (see figure 2). Although the Landau-Lifshitz equation [6] can be considered or not as an approximation to order τ_o of the Lorentz-Dirac equation [5], the reality is that the differences with the Lorentz-Dirac equation or with the Eliezer-Ford-O´Connell equation [12], [10], [11] are negligible except in the cases of self-accelerations and pre-accelerations[14], [15], [13], [16]. Therefore, the method may be used to deduce the behavior of the longitudinal modes for high temperatures. One of the purposes of this paper consists in demonstrating that the longitudinal modes disappear when the temperatures increase but accompanied by the trend in which the longitudinal waves are canceled; that is: a dispersion relation is proposed for high temperature such that for $T \to \infty$ the frequency w becomes imaginary. Then, a cut-off exists above a temperature and all the perturbations are damped. Particularly, a cut-off appears for high frequencies leaving just the existence of low frequencies that does not possess any physical important effect.

On the other hand, as we mentioned before, although the Landau-Lifshitz equation is considered as a first order approximation in τ_o of the Lorentz-Dirac equation [5], many authors refer to it as the exact equation of motion for a charge particle. Indeed, for Spohn [7] and Rohrlich [8], [9] the Lorentz-Dirac equation must be restricted to its critical surface yielding the Landau-Lifshitz equation and consequently this last one represents the correct equation of motion of a spinless classical point charge. Moreover, by considering a generalized quantum Langevin equation and by giving a structure to the electron with a factor form and a finite cut-off parameter, Ford and O'Connell [10], [11] derived a non relativistic equation of motion for charged particles. Generalizing this equation to Special Relativity [11] they obtained an equation that Eliezer [12] derived fifty years before using distinct arguments. Although, the Landau-Lifshitz equation [6] and the Eliezer-Ford equation are mathematically different, their implications are physically equivalent [14], [15], [13], [16] within the Shen´s zone [17] where quantum effects are negligible. Recently, Hammond [18], [19], [20], [21] has proposed another approach for deducing the reaction force which eliminates the apparent paradox of the vanishing reaction force when a constant electric field is applied to a charge. Although for many authors [23], [24], [25], [26], [27], [28], this apparent paradox is explained by noticing that the radiation exits at the infinity; that is, the energy radiated to infinity is taken from the attached fields (the Scott term or the acceleration energy) and consequently even if the total radiation term in the equation of motion vanishes, the radiation to the infinity (the irreversible emission of radiation) exists. Moreover, the regular Larmor formula does not describes the radiated energy and it is substituted by a new expression that coincides with it for low accelerations [24]. Although, Hammond theory is consistent, it possesses the inconvenience that the expression for the reaction term must be deduced for each kind of interaction making it unpractical to deal with it when a system of particles is considered. Apart of this paradox, both equations, Landau-Lifshitz and Eliezer-Ford-O'Connell, are second order differential equations which do not present unphysical solutions, as pre-accelerations and runaway solutions. Whether or not the Landau-Lifshitz equation represents an exact or approximate description of the motion of charged particle, it must be subjected to experimental tests. However, such experiments present a very large degree of difficulty for various reasons [13], [14], [15]. However, dispersion relations can be easier measured and comparing them with the ones deduced by using the modified Vlasov equation, obtained starting from the Landau-Lifshitz equation, will give validity or not to the last equation. Otherwise, theory such as Hammond's may have more physical acceptance because at high speeds (high temperatures) and high accelerations great differences are presented between the different equations [19].

Once the Landau-Lifshitz equation is chosen to describe the motion of the charges, it is possible to eliminate the degree of liberty represented by considering the hyper-acceleration as an independent variable in the deduction of the density function. Indeed, the hyper-acceleration does not appear anymore. Therefore, we can follow Hakim [3] in the deduction of the density just by considering the positions and the velocities of the particles as independent variables. The phase space results to be a $7N-$dimensional space ($8N-N$, due to the kinematic relations) The final result will coincide with Hakim's generalization of the Vlasov equation in Special Relativity [3]. Moreover, a more general equation will be deduced by including a constant external electric field. The wave dispersion relations will be deduced and the effects of the coupling between the constant external electric field and the interaction of the particle will not play an important role and the dispersion relations will not be affected. Moreover, Landau damping can be deduced and compared with the damping due to the radiation reaction force. When a numerical method is used in order to solve the general expression for the longitudinal modes, it turns out that although the effect is much more pronounced in the case of very high temperatures, the dispersion relations have no real solution above a wave number for each temperature.

The article is organized as follows: in section 2, an equivalent expression of the Landau-Lifshitz equation of motion will be described in order to be used in the deduction of the modified relativistic Vlasov equation in section 3. In section 3, a modified relativistic Vlasov equation is deduced without considering the hyper-acceleration as independent variable and including an external field. In section 4, a perturbation is made in order to obtain a perturbed current and a constant electric field is considered in order to prove that no changes appear in the wave dispersion relations. In section 5, the dispersion relations will be deduced for different temperatures. The low temperature case will be analyzed for transverse and longitudinal modes and we give the reason why the wave solution ($w^2 = w_p^2 + k^2$) should be discarded for longitudinal modes without resorting to arguments based on not considering solutions that come from relativity or the absurd theory that they depend on the chosen gauge. Nor do we resort to the continuity equation as an extra constraint since it must already be contained in the Vlasov equation due to the fact that the last is deduced by using the definition of point particle current or the Maxwell´s equations. Of course, Landau damping [4] is deduced for some values of k and temperatures. The damping due to the radiation reaction force is deduced for low temperatures and compared with the Landau damping. In this section, the high temperature case is analyzed obtaining the dispersion relations for longitudinal modes and explaining how the wave disappears as long as the temperature increases. The existence of a cut-off for the frequencies is shown for high temperatures. Some concluding remarks are done in section 7.

2. Landau-Lifshitz Equation

For a system of N charged particles with the same mass m and equal charge e, for each particle, the Lorentz-Dirac equation [5] is

$$ma_i^\mu = \frac{e}{c}F_{(i)}^{\mu\nu}v_{\nu i} + m\tau_o\left[\dot{a}_i^\mu + \frac{a_i^2}{c^2}v_i^\mu\right], \qquad i = 1, ..., N \tag{1}$$

where c, $F_{(i)}^{\mu\nu}$ and $\tau_o = 2e^2/3mc^3$ are the speed of light, the field strength tensor due to the action of the other particles and the external field on the $i-$particle and the characteristic time of the charge e, respectively. The subscript "i" represents the $i-$particle and v_i^μ, a_i^μ and $\dot{a}_i^\mu = da_i^\mu/d\tau_i$ are the 4−velocity, the 4−acceleration, the 4−hyper-acceleration of the $i-$particle being τ_i the proper time of the $i-$particle, respectively. The equation can expressed in a different way as:

$$ma_i^\mu = \frac{e}{c}F_{(i)}^{\mu\nu}v_{\nu i} + m\tau_o\Delta^{\mu\nu}(v_{\rho i})\dot{a}_{\nu i}, \qquad i = 1, ..., N, \tag{2}$$

where

$$\Delta^{\mu\nu}(v_{i\rho}) = n^{\mu\nu} - \frac{v_i^\mu v_i^\nu}{c^2}. \tag{3}$$

In order to incorporate Landau-Lifshitz proposal on equation (2), the hyper-acceleration must be substituted by

$$\dot{a}_{\nu i} \rightarrow \frac{e}{mc}\frac{d}{d\tau_i}F_{\nu\alpha}^{(i)}v_i^\alpha = \frac{e}{mc}\left[F_{\nu\alpha}^{(i)}a_i^\alpha + v_i^\rho v_i^\alpha\frac{\partial F_{\nu\alpha}^{(i)}}{\partial x_i^\rho}\right], \tag{4}$$

and in a second step, in the first term of the right side of equation (4), the acceleration a_i^α must be substituted by the Lorentz force; that is:

$$\dot{a}_{\nu i} \rightarrow \frac{e}{mc}\left[\frac{e}{mc}F_{\nu\alpha}^{(i)}F_{(i)}^{\alpha\beta}v_{\beta i} + v_i^\rho v_i^\alpha\frac{\partial F_{\nu\alpha}^{(i)}}{\partial x_i^\rho}\right]. \tag{5}$$

Therefore, the Landau-Lifshitz equation of motion for each particle is

$$\begin{aligned} ma_i^\mu &= \frac{e}{c}F_{(i)}^{\mu\nu}v_{\nu i} + m\tau_o\Delta^{\mu\nu}(v_{\rho i})\frac{e}{mc}\left[\frac{e}{mc}F_{\nu\alpha}^{(i)}F_{(i)}^{\alpha\beta}v_{\beta i} + v_i^\rho v_i^\alpha\frac{\partial F_{\nu\alpha}^{(i)}}{\partial x_i^\rho}\right], \\ i &= 1, ..., N. \end{aligned} \tag{6}$$

On the other hand, the force applied on each particle must be decomposed in two parts: the external field and the field produced by the other particles on the $i-$particle; that is:

$$F_{(i)}^{\mu\nu} = F_{ext}^{\mu\nu} + \sum_{j\neq i}F_{ji}^{\mu\nu}(x_\rho), \tag{7}$$

where

$$\begin{aligned} F_{ji}^{\mu\nu}(x_\rho) &= \frac{4\pi e}{c}\int\int\int dx'^4 dv'^4 d\tau'\left[v'^\nu\partial^\mu - v'^\mu\partial^\nu\right]D_{ret}(x_\rho - x'_\rho) \\ &\times\delta(x'_\rho - x_{\rho j}(\tau'))\delta(v'_\rho - v_{\rho j}(\tau')), \end{aligned} \tag{8}$$

IOP Conf. Series: Journal of Physics: Conf. Series **1239** (2019) 012001 doi:10.1088/1742-6596/1239/1/012001

where D_{ret} represents the retarded Green function [29].

Finally, it is important to noticed that the equation of motion described by equation (6) does not depends on the hyper-acceleration and consequently the phase space is described by $8N-$dimensional space with the regular N constraint, $v^{\mu}v_{\mu} = c^2$. This point represents one of the different approaches to obtain the relativistic Vlasov equation compared with the work done by Hakim and Mangeney [3].

3. Relativistic Vlasov Equation with External Field Including Radiation Reaction Force

Hakim [1], [2] developed a manifestly covariant Statistical Mechanics generating an equation of BBGKY hierarchy in order to obtain a kinetic equation [3]. The work was so important that many others relied on it to develop various ideas. For example, Horwitz [30] proposed a quantum mechanical derivation of a manifestly covariant relativistic Boltzmann equation. We will follow Hakim and Mangeney method [3] but using the Landau-Lifshitz equation.

The phase space in Special Relativity must be constructed by following the dynamics of the particles in the system. By looking at the equation of motion, equation (6), It is clear that is composed as:

$$\Gamma = \{x_{\mu}\}^N \times \{v_{\mu}\}^N , \tag{9}$$

accompanied by the constraint

$$v_{\mu}v^{\mu} = c^2, \tag{10}$$

which can be incorporated directly in the densities by using a delta function. Notice that unlike Hakim and Mangeney [3] the hyper-acceleration is not taken as a part of the phase space since the equation of motion to be considered is the Landau-Lifshitz equation which is a second order differential equation. Using the regular assumptions of the Gibbs ensemble [1], [2], it is possible to define a random one-particle density as

$$R_1(x_{\mu}, v_{\mu}; \tau) = \sum_{i=1}^{N} \delta(x_{\mu} - x_{\mu i}(\tau))\delta(v_{\mu} - v_{\mu i}(\tau)), \tag{11}$$

where τ represents a parameter which will describe the trajectory of each particle by means of $v_{\mu i}(\tau) = dx_{\mu i}(\tau)/d\tau$. Also, following Hakim and Mangeney [3], we define the proper time-dependent density,

$$D_1(x_{\mu}, v_{\mu}; \tau) = N \langle R_1(x_{\mu}, v_{\mu}; \tau)\rangle , \tag{12}$$

where the brackets $\langle R_1 \rangle$ represents the regular average pondered by a random initials conditions [1], [2]. Also, we will need

$$\begin{aligned} R_2(x_{\mu}, v_{\mu}, \tau; x'_{\mu}, v'_{\mu}, \tau') &= \sum_{i \neq j} \delta(x_{\mu} - x_{\mu i}(\tau)) \\ &\times \delta(v_{\mu} - v_{\mu i}(\tau))\delta(x'_{\mu} - x_{\mu i}(\tau'))\delta(v'_{\mu} - v_{\mu i}(\tau')), \end{aligned} \tag{13}$$

and

$$D_2 = (x_{\mu}, v_{\mu}, \tau; x'_{\mu}, v'_{\mu}, \tau') = N(N-1) \left\langle R_2(x_{\mu}, v_{\mu}, \tau; x'_{\mu}, v'_{\mu}, \tau') \right\rangle . \tag{14}$$

Finally, the relativistic one-particle density:

$$\mathcal{N}_1(x_{\mu}, v_{\mu}) = \int_{-\infty}^{\infty} D_1(x_{\mu}, v_{\mu}; \tau) d\tau. \tag{15}$$

Let us follow the Klimontovich method [31]: first, by using the continuity equation, we have

$$\frac{\partial R_1(x_\mu, v_\mu;\tau)}{\partial \tau} + v^\mu \frac{\partial R_1(x_\mu, v_\mu;\tau)}{\partial x^\mu} + a^\mu \frac{\partial R_1(x_\mu, v_\mu;\tau)}{\partial v^\mu} = 0, \tag{16}$$

which can be written as

$$\frac{\partial R_1(x_\mu, v_\mu;\tau)}{\partial \tau} + v^\mu \frac{\partial R_1(x_\mu, v_\mu;\tau)}{\partial x^\mu} + \frac{\partial a^\mu R_1(x_\mu, v_\mu;\tau)}{\partial v^\mu} = 0. \tag{17}$$

Notice that a difference appears with Hakim and Mangeney dependence of R_1 since they obtained $R_1(x_i^\mu, v_i^\mu, \dot{a}_{\nu i})$ and our result is that $R_1 = R_1(x^\mu, v^\mu)$. This is obviously due to the fact that the equation of motion is taken from the beginning as the equation (6) which does not depend on $\dot{a}_{\nu i}$ but just in the fields and velocities as we mention above. Therefore, by using equations (6), (7) and (8), we arrive at:

$$\begin{aligned}&\frac{\partial R_1(x_\mu, v_\mu;\tau)}{\partial \tau} + v^\mu \frac{\partial R_1(x_\mu, v_\mu;\tau)}{\partial x^\mu}\\ &+\frac{e}{mc}\frac{\partial}{\partial v^\mu}\left[\begin{array}{c} F^{\mu\nu}v_\nu \\ +\tau_o \Delta^{\mu\nu}(v_\rho)\left[\frac{e}{mc}F_{\nu\alpha}F^{\alpha\beta}v_\beta + v^\rho v^\alpha \frac{\partial F_{\nu\alpha}}{\partial x^\rho}\right] R_1\end{array}\right] = 0.\end{aligned} \tag{18}$$

Let us analyze the second line of equation (18). First, the strength tensor $F^{\mu\nu}$ must be decomposed in two terms corresponding to the external force and the internal force, see equations (7) and (8); second, the non radiation terms are represented in the second line of equation (19). The radiation terms are represented in the third line of equation (19).

$$\begin{aligned}&\frac{\partial R_1(x_\mu, v_\mu;\tau)}{\partial \tau} + v^\mu \frac{\partial R_1(x_\mu, v_\mu;\tau)}{\partial x^\mu}\\ &+\frac{e}{mc}\frac{\partial}{\partial v^\mu}\left[F^{\mu\nu}_{ext}v_\nu R_1\right] + \frac{e}{mc}\frac{\partial}{\partial v^\mu}\left[F^{\mu\nu}_{int}v_\nu R_1\right]\\ &+\frac{e}{mc}\tau_o\frac{\partial}{\partial v^\mu}\left[\Delta^{\mu\nu}(v_\rho)\left[\begin{array}{c}\frac{e}{mc}\left[F^{ext}_{\nu\alpha}+F^{int}_{\nu\alpha}\right]\\ \times\left[F^{\alpha\beta}_{ext}+F^{\alpha\beta}_{int}\right]v_\beta\\ +v^\rho v^\alpha \frac{\partial\left[F^{ext}_{\nu\alpha}+F^{int}_{\nu\alpha}\right]}{\partial x^\rho}\end{array}\right] R_1\right] = 0.\end{aligned} \tag{19}$$

By using equation (8), the first two lines of equation (19) are expressed in the first two lines of equation (20). The radiation term must be analyze as follows. Terms depending on bigger order of e^2 will be neglected. In this order of ideas, as we can notice in equation (8), the internal $F^{\alpha\beta}_{int}$ depends on e and consequently the product of the external $F^{\alpha\beta}_{ext}$ with the internal $F^{\alpha\beta}_{int}$ as well as the product of the internal $F^{\alpha\beta}_{ext}$ with the internal $F^{\alpha\beta}_{ext}$ will be neglected since there is a factor $e\tau_o/mc$ in the radiation reaction term. Finally, the conserved terms are described in the third and fourth lines of equation (20).

$$\begin{aligned}&\frac{\partial R_1(x_\mu, v_\mu;\tau)}{\partial \tau} + v^\mu \frac{\partial R_1(x_\mu, v_\mu;\tau)}{\partial x^\mu} + \frac{e}{mc}\frac{\partial}{\partial v^\mu}\left[F^{\mu\nu}_{ext}v_\nu R_1\right]\\ &+\frac{4\pi e^2}{mc^2}v_\nu\frac{\partial}{\partial v_\mu}\int d\tau' dx^{4\prime} dv^{4\prime}\left[\left\{v^{\nu\prime}\partial^\mu - v^{\mu\prime}\partial^\nu\right\}D_{ret}(x_\rho - x'_\rho)\right]\\ &\times R_2(x_\mu, v_\mu, \tau; x'_\mu, v'_\mu, \tau')\\ +\ &\frac{e}{mc}\tau_o\frac{\partial}{\partial v^\mu}\left[\Delta^{\mu\nu}(v_\rho)\left[\frac{e}{mc}\left[F^{ext}_{\nu\alpha}\right]\left[F^{\alpha\beta}_{ext}\right]v_\beta + v^\rho v^\alpha\frac{\partial F^{ext}_{\nu\alpha}}{\partial x^\rho}\right]R_1\right]\\ &+\tau_o\frac{\partial}{\partial v^\mu}\left\{\begin{array}{c}\Delta^{\mu\nu}(v_\rho)\frac{4\pi e^2}{mc^2}v^\rho v^\alpha\int d\tau' dx^{4\prime} dv^{4\prime}\\ \left[\left\{v'_\alpha\partial_\nu - v'_\nu\partial_\alpha\right\}\frac{\partial}{\partial x^\rho}D_{ret}(x_\rho - x'_\rho)\right]R_2(x_\mu, v_\mu, \tau; x'_\mu, v'_\mu, \tau')\end{array}\right\}\\ =\ &0((ne^2\tau_o)^2).\end{aligned} \tag{20}$$

In the following, we will consider the case of an external constant electric field in the $x-$axis. Therefore, due to the antisymmetry of $F^{\mu\nu}_{ext}$ and remembering that the 4−current which produces the external field are outside of the volume, we have

$$\frac{e}{mc}\frac{\partial}{\partial v^{\mu}}\left[F^{\mu\nu}_{ext}v_{\nu}R_1\right]=\frac{e}{mc}F^{\mu\nu}_{ext}v_{\nu}\frac{\partial}{\partial v^{\mu}}R_1. \tag{21}$$

On the other hand, for a constant electric field it is a well known fact that the radiation term vanishes as it has been proved by many authors [23], [24], [25], [26], [27]. Indeed, consider the part of the external radiation reaction term, since the electric field is constant, we have

$$\Delta^{\mu\nu}(v_{\rho})\left[\frac{e}{mc}\left[F^{ext}_{\nu\alpha}\right]\left[F^{\alpha\beta}_{ext}\right]v_{\beta}+v^{\rho}v^{\alpha}\frac{\partial F^{ext}_{\nu\alpha}}{\partial x^{\alpha}}\right]=\Delta^{\mu\nu}(v_{\rho})\left[\frac{e}{mc}\left[F^{ext}_{\nu\alpha}\right]\left[F^{\alpha\beta}_{ext}\right]v_{\beta}\right], \tag{22}$$

then,

$$\begin{aligned}\Delta^{\mu\nu}(v_{\rho})\left[\frac{e}{mc}\left[F^{ext}_{\nu\alpha}\right]\left[F^{\alpha\beta}_{ext}\right]v_{\beta}+v^{\rho}v^{\alpha}\frac{\partial F^{ext}_{\nu\alpha}}{\partial x^{\alpha}}\right] &= (n^{\mu\nu}-\frac{v^{\mu}v^{\nu}}{c^2})\\ &\quad\times\left[\frac{e}{mc}\left[F^{ext}_{\nu\alpha}\right]\left[F^{\alpha\beta}_{ext}\right]v_{\beta}\right]\\ &= E^2v^{\mu}\left(1-\frac{c^2}{c^2}\right)=0,\end{aligned} \tag{23}$$

where E represents the constant electric field. Notice that within Hammond theory [18], the radiation reaction force does not vanish for a constant electric field and this will predict different dispersion relations. We arrive at

$$\begin{aligned}&\frac{\partial R_1(x_{\mu},v_{\mu};\tau)}{\partial\tau}+v^{\mu}\frac{\partial R_1(x_{\mu},v_{\mu};\tau)}{\partial x^{\mu}}+\frac{e}{mc}F^{\mu\nu}_{ext}v_{\nu}\frac{\partial}{\partial v^{\mu}}R_1\\ &+\frac{4\pi e^2}{mc^2}v_{\nu}\frac{\partial}{\partial v_{\mu}}\int d\tau' dx^{4\prime}dv^{4\prime}\left[\left\{v^{\nu\prime}\partial^{\mu}-v^{\mu\prime}\partial^{\nu}\right\}D_{ret}(x_{\rho}-x'_{\rho})\right]\\ &\times R_2(x_{\mu},v_{\mu},\tau;x'_{\mu},v'_{\mu},\tau')\\ &+\tau_o\frac{\partial}{\partial v^{\mu}}\left\{\begin{array}{c}\Delta^{\mu\nu}(v_{\rho})\frac{4\pi e^2}{mc^2}v^{\rho}v^{\alpha}\int d\tau' dx^{4\prime}dv^{4\prime}\\ \times\left[\left\{v'_{\alpha}\partial_{\nu}-v'_{\nu}\partial_{\alpha}\right\}\frac{\partial}{\partial x^{\rho}}D_{ret}(x_{\rho}-x'_{\rho})\right]R_2(x_{\mu},v_{\mu},\tau;x'_{\mu},v'_{\mu},\tau')\end{array}\right\}\\ =&\ 0((ne^2\tau_o)^2).\end{aligned} \tag{24}$$

Considering that the correlation function vanishes or is of order $(ne^2\tau_o)^2$; that is:

$$D_2(x_{\mu},v_{\mu},\tau;x'_{\mu},v'_{\mu},\tau')=D_1(x_{\mu},v_{\mu};\tau)D_1(x'_{\mu},v'_{\mu};\tau'). \tag{25}$$

By calculating the average of equation (12) and taking $N-1\simeq N$, we arrive at

$$\begin{aligned}&\frac{\partial D_1(x_{\mu},v_{\mu};\tau)}{\partial\tau}+v^{\mu}\frac{\partial D_1(x_{\mu},v_{\mu};\tau)}{\partial x^{\mu}}+\frac{e}{mc}F^{\mu\nu}_{ext}v_{\nu}\frac{\partial}{\partial v^{\mu}}D_1\\ &+\frac{4\pi e^2}{mc^2}Nv_{\nu}\frac{\partial}{\partial v_{\mu}}\int d\tau' dx^{4\prime}dv^{4\prime}\left[\left\{v^{\nu\prime}\partial^{\mu}-v^{\mu\prime}\partial^{\nu}\right\}D_{ret}(x_{\rho}-x'_{\rho})\right]\\ &\times D_1(x_{\mu},v_{\mu};\tau)D_1(x'_{\mu},v'_{\mu};\tau')\\ &+\tau_o\frac{4\pi e^2}{mc^2}N\frac{\partial}{\partial v^{\mu}}\left\{\begin{array}{c}\Delta^{\mu\nu}(v_{\rho})v^{\rho}v^{\alpha}\int d\tau' dx^{4\prime}dv^{4\prime}\\ \left[\left\{v'_{\alpha}\partial_{\nu}-v'_{\nu}\partial_{\alpha}\right\}\frac{\partial}{\partial x^{\rho}}D_{ret}(x_{\rho}-x'_{\rho})\right]\\ \times D_1(x_{\mu},v_{\mu};\tau)D_1(x'_{\mu},v'_{\mu};\tau')\end{array}\right\}\\ =&\ 0\end{aligned} \tag{26}$$

Considering that

$$D_{1 \atop \lim \tau \to \pm\infty} (x_\mu, v_\mu; \tau) = 0, \tag{27}$$

and integrating over the proper time τ, we obtain

$$\begin{aligned} & v^\mu \frac{\partial \mathcal{N}_1(x_\mu, v_\mu)}{\partial x^\mu} + \frac{e}{mc} F^{\mu\nu}_{ext} v_\nu \frac{\partial}{\partial v^\mu} \mathcal{N}_1(x_\mu, v_\mu) \\ & + \frac{4\pi e^2}{mc^2} N v_\nu \frac{\partial}{\partial v_\mu} \mathcal{N}_1(x_\mu, v_\mu) \\ & \times \int dx^{4\prime} dv^{4\prime} \left[\{ v^{\nu\prime} \partial^\mu - v^{\mu\prime} \partial^\nu \} D_{ret}(x_\rho - x'_\rho) \right] \mathcal{N}_1(x'_\mu, v'_\mu) \\ & + \tau_o \frac{4\pi e^2}{mc^2} N \frac{\partial}{\partial v^\mu} \left\{ \begin{array}{c} \Delta^{\mu\nu}(v_\rho) v^\rho v^\alpha \int dx^{4\prime} dv^{4\prime} \\ \left[\{ v'_\alpha \partial_\nu - v'_\nu \partial_\alpha \} \frac{\partial}{\partial x^\rho} D_{ret}(x_\rho - x'_\rho) \right] \\ \times \mathcal{N}_1(x_\mu, v_\mu) \mathcal{N}_1(x'_\mu, v'_\mu) \end{array} \right\} \\ = \; & 0, \end{aligned} \tag{28}$$

which represents a modified Vlasov equation with an external constant electric field up to the order $ne^2\tau_o$. It has to be highlighted that no collision term is obtained because all the interaction is electromagnetic and there are no collisions or friction forces. Nevertheless, the radiation reaction term can be taken as a collision term because without it we recover the usual Vlasov equation with no collisions.

4. Perturbation

4.1. Distribution function and the shielded potential

We assume now that the plasma is in an equilibrium state described by the Jüttner distribution function [32], [33], [34], [35] without an electric field. However, when a constant electric field is present the distribution function maybe pondered by a an exponential factor which takes into account the potential due to the electric constant field in the $x-$axis; that is:

$$\phi = \frac{\exp \frac{x}{\lambda} - \exp - \frac{x}{\lambda}}{\exp \frac{R}{\lambda}} \Rightarrow \mathcal{N}_{eq} = \mathcal{N}_J \exp - \frac{\phi}{\kappa T} \simeq \mathcal{N}_J, \tag{29}$$

where λ, R, κ, $\mathcal{N}_J$ and T correspond to the Debye length of the plasma in equilibrium, the length between the plates of the capacitor which generates the electric constant field, the Boltzmann constant, the Jüttner distribution function and the temperature, respectively. Therefore, in an equilibrium plasma the electric field is shielded [36]. Therefore, the Jüttner distribution function maybe considered good enough to describe an equilibrium plasma and the electric field represented by $F^{\mu\nu}_{ext} v_\nu$ is much smaller than it would exist if there were no plasma due to the shielding.

4.2. First order perturbation

Accordingly, a small perturbation can be described

$$\mathcal{N} = \mathcal{N}_{eq} + \mathcal{N}_p \simeq \mathcal{N}_J + \mathcal{N}_p. \tag{30}$$

Therefore, applying this to equation (28), considering that $\mathcal{N}_{eq}$ is solution of the unperturbed case which does not depend on x_μ, that is $\mathcal{N}_{eq} \neq \mathcal{N}_{eq}(x_\mu)$, and taking into account that both the derivatives of the perturbation and the product of perturbation can be neglected, we arrive at

$$\begin{aligned}
&v^{\mu}\frac{\partial \mathcal{N}_p(x_\mu, v_\mu)}{\partial x^\mu} \\
&+\frac{4\pi e^2}{mc^2} N v_\nu \frac{\partial}{\partial v_\mu} \mathcal{N}_{eq}(v_\mu) \int dx^{4\prime} dv^{4\prime} \\
&\times \left[\{v^{\nu\prime}\partial^\mu - v^{\mu\prime}\partial^\nu\} D_{ret}(x_\rho - x'_\rho)\right] \mathcal{N}_p(x'_\lambda, v'_\lambda) \\
&= -\tau_o \frac{4\pi e^2}{mc^2}\frac{\partial}{\partial v^\mu}\left\{\begin{array}{c} \Delta^{\mu\nu}(v_\rho)v^\rho v^\alpha \int dx^{4\prime} dv^{4\prime} \\ \left[\{v'_\alpha\partial_\nu - v'_\nu\partial_\alpha\} \frac{\partial}{\partial x^\rho} D_{ret}(x_\rho - x'_\rho)\right] \\ \times \mathcal{N}_{eq}(v_\lambda)\mathcal{N}_p(x'_\sigma, v'_\sigma) \end{array}\right\}. \qquad (31)
\end{aligned}$$

Notice that the term

$$\frac{e}{mc} F^{\mu\nu}_{ext} v_\nu \frac{\partial}{\partial v^\mu} \mathcal{N}_p(x_\mu, v_\mu), \qquad (32)$$

has been withdrawn because for high volume as it is the analyzed case the electric field is shielded and it represents the product of two small quantities. This means that the perturbed plasma will not be affected by an external constant electric field. Let us use the Fourier transformation in order to obtain the Fourier 4−current. We have

$$\widetilde{\mathcal{N}}_p(k_\mu, v_\mu) = \frac{1}{(2\pi)^2}\int dx^4 \exp\left[-ik^\mu x_\mu\right] \mathcal{N}_p(k_\rho, v'_\mu). \qquad (33)$$

By using equations (31) and (33), we obtain

$$\begin{aligned}
&ik^\mu v_\mu \widetilde{\mathcal{N}}_p(k_\mu, v_\mu) \\
&+i\left(\frac{4\pi e^2}{mc^2}\right) v_\nu \frac{\partial}{\partial v_\mu}\mathcal{N}_{eq}(v_\mu)\int dv^{4\prime} \frac{k^\mu v^{\nu\prime} - k^\nu v^{\mu\prime}}{-k^\lambda k_\lambda} \mathcal{N}_p(k_\rho, v'_\rho) \\
=\ &\left(\frac{4\pi e^2}{mc^2}\tau_o\right)\frac{\partial}{\partial v^\mu}\left\{\Delta^{\mu\nu}(v_\rho)v^\rho v^\alpha \mathcal{N}_{eq}(v_\lambda)\right\} \\
&\times \int dv^{4\prime}\frac{v'_\alpha k_\nu - v'_v k_\alpha}{-k^\lambda k_\lambda} k_\rho \mathcal{N}_p(k_\sigma, v'_\sigma). \qquad (34)
\end{aligned}$$

Finally, we arrive at

$$\begin{aligned}
\widetilde{\mathcal{N}}_p(k_\mu, v_\mu) \ =\ & \frac{\omega_p^2}{n_o} v_\nu \frac{\partial}{\partial v_\mu}\mathcal{N}_{eq}(v_\mu) \\
& \times \frac{1}{k^\mu v_\mu}\int dv^{4\prime}\frac{k^\mu v^{\nu\prime} - k^\nu v^{\mu\prime}}{k^\lambda k_\lambda}\mathcal{N}_p(x_\rho, v'_\rho) \\
& +i\left(\frac{\omega_p^2}{n_o}\tau_o\right)\frac{\partial}{\partial v^\mu}\left\{\Delta^{\mu\nu}(v_\rho)v^\rho v^\alpha \mathcal{N}_{eq}(v_\lambda)\right\} \\
& \times \frac{1}{k^\mu v_\mu}\int dv^{4\prime}\frac{v'_\alpha k_\nu - v'_v k_\alpha}{k^\lambda k_\lambda} k_\rho \mathcal{N}_p(k_\sigma, v'_\sigma), \qquad (35)
\end{aligned}$$

where ω_p represents the regular plasma frequency. If we want to calculate the Fourier component of the 4−current vector $J^\beta(k_v)$, we have

$$J^\beta(k_v) = \int dv^4 v^\beta \mathcal{N}_p(k_\nu, v_\sigma). \qquad (36)$$

Before continuing with the development, it is necessary to notice that some properties of the 4−vector current. Actually, we define the 4−vector point particles current for a system of point particles,

$$J^{\alpha}(x_{\rho}) = \int d\tau \left\{ \sum_{i=1}^{n} e_i \delta^4(x_{\rho} - x_{i\rho}(\tau)) \frac{dx_i^{\alpha}(\tau)}{d\tau} \right\}, \tag{37}$$

which represents a well defined 4−vector. It can be decomposed and written as:

$$\begin{aligned} \vec{J}(x,t) &= \sum_{i=1}^{n} e_i \delta^3(\vec{x} - \vec{x}_i(t)) \frac{d\vec{x}_i(t)}{dt} \\ &\text{and} \\ \rho(x,t) &= \sum_{i=1}^{n} e_i \delta^3(\vec{x} - \vec{x}_i(t)). \end{aligned} \tag{38}$$

It is easy to show that

$$\frac{\partial \rho(x,t)}{\partial t} + \nabla \cdot \vec{J}(x,t) = 0. \tag{39}$$

That is, the continuity equation is satisfied. If we take the Fourier transform of the 4−vector point particles current, the continuity equation must be satisfied. If we want to pass to a continuous system of particles with the same charge e, we put

$$\frac{\partial e\mathcal{N}}{\partial t} + \nabla \cdot \vec{J}(x,t) = 0. \tag{40}$$

where $\mathcal{N}$ represents a continuous particle density as the one introduced in equation (28). Of course, the continuity equation must be accomplished if we use the Maxwell equations. However, when a density of point particles is transformed to a continuous density, some properties can be lost and we must impose the continuity equation. Then, in equation (36) the continuity equation must be considered as a constraint embedded in the theory; that is:

$$k^{\mu} J_{\mu} = wJ^0(k_v) - \vec{k} \cdot \vec{J}(k_v) = 0. \tag{41}$$

This is equivalent of the procedure used in equations (9) and (10). By using equation (35), we arrive at

$$J^{\beta}(k_v) = \frac{w_p^2}{k^{\alpha}k_{\alpha}} \left\{ \begin{array}{c} -I_{\lambda}k^{\beta} - k^{\alpha}k_{\alpha}I^{\beta}_{\ \lambda} + k^{\alpha}I_{\alpha}g^{\beta}_{\ \lambda} \\ +i\tau_o \left(k^{\alpha}k_{\alpha}J^{\beta}_{\ \lambda} + \xi^{\alpha}k_{\alpha}\xi^{-1}g^{\beta}_{\ \lambda} - k^{\beta}\xi_{\lambda}\xi^{-1} \right) \end{array} \right\} J^{\lambda}(k_v) \tag{42}$$

where

$$I_{\nu}(k_{\rho}, \xi_{\rho}) = -\frac{1}{n_o} \int dv^4 \frac{v_v}{k^{\lambda}u_{\lambda}} \mathcal{N}_{eq}(v_{\mu}), \tag{43}$$

and

$$I_{\mu}^{\ v}(k_{\rho}, \xi_{\rho}) = -\frac{1}{n_o} \int dv^4 \frac{v_{\mu}v^{\nu}}{(k^{\lambda}u_{\lambda})^2} \mathcal{N}_{eq}(v_{\rho}), \tag{44}$$

and

$$J_{\mu}^{\ \nu}(k_{\rho}, \xi_{\rho}) = -\frac{1}{n_o} \int dv^4 \frac{v_{\mu}v^{\nu}}{k^{\lambda}u_{\lambda}} \mathcal{N}_{eq}(v_{\rho}), \tag{45}$$

4.3. The Jüttner distribution function

The Jüttner distribution function is expressed by Hakim and Mangeney [3] as

$$N_J = \frac{n_o m\xi}{4\pi K_2(m\xi)}\left[\exp -m\xi^\mu v_\mu\right] 2\Theta(v^0)\delta(v^\mu v_\mu - 1), \tag{46}$$

with

$$\xi = \frac{1}{\kappa T_o} \qquad \text{and} \qquad \xi^\mu = \xi \bar{v}^\mu, \tag{47}$$

where κ is the Boltzmann constant, K_2 represents the Kelvin function and $\gamma(\overrightarrow{u}) = (1-u^2)^{-1/2}$ (Normally, the Hankel function of second order is used in order to describes the Jüttner function [33], [35] but some factors are changed) and Θ represents the Heaviside function. Noticed that T_o is defined in the comoving frame of the space where the particles are moving with a velocity $\overrightarrow{u}$ where $\overrightarrow{u} = d\overrightarrow{r}/dt$ and $\overrightarrow{v} = \gamma(\overrightarrow{u})\overrightarrow{u}$ with $\gamma(\overrightarrow{u}) = (1-u^2)^{-1/2}$; that is: the rest frame K_o where the system is at rest and the temperature is T_o. The existence of such rest frame is assured because in the free particle system there is no interaction and a rest frame can be defined. However, when there is interaction between the particles it is necessary to define a volume in a particular frame and in such a frame the instantaneity is considered [37], [38], [39], [34], [35]. This means that Jüttner is defining the temperature in the rest frame K_o and a volume V_o where $\overrightarrow{u} = \overrightarrow{0}$. Therefore, Jüttner takes into account the effects of Relativistic Thermodynamics.

5. Dispersion Relations

Once we obtain the expression for the 4−vector current, equation (42), we need just to calculate the determinant and equals it to zero; that is:

$$\det G^\beta_\lambda = \det\left\{ \begin{array}{c} g^\beta{}_\lambda\left(1 - w_p^2 \frac{k^\alpha I_\alpha + i\tau_o \xi^\alpha k_\alpha \xi^{-1}}{k^\alpha k_\alpha}\right) + w_p^2\left(I^\beta{}_\lambda - i\tau_o J^\beta{}_\lambda\right) \\ + w_p^2 k^\beta \frac{I_\lambda + i\tau_o \xi_\lambda \xi^{-1}}{k^\alpha k_\alpha} \end{array} \right\} = 0 \tag{48}$$

In a reference frame (the lab frame) where $\xi^\mu = (\xi, \overrightarrow{0})$ and choosing the $z-$axis as the direction of propagation of the perturbation mode, we have

$$k^\mu = (k^0, 0, 0, k^3) = (w, 0, 0, k). \tag{49}$$

For the longitudinal mode, obtains

$$\begin{aligned} &\left\{\left[1 - \frac{w_p^2}{k^\alpha k_\alpha} G\right] + w_p^2\left\{I^0{}_0 - i\tau_o J^0{}_0\right\} + \frac{w_p^2 k^0}{k^\alpha k_\alpha}(I_0 + i\tau_o)\right\} \\ &\times \left\{\left[1 - \frac{w_p^2}{k^\alpha k_\alpha} G\right] + w_p^2\left\{I^3{}_3 - i\tau_o J^3{}_3\right\} + \frac{w_p^2 k^3}{k^\alpha k_\alpha}(I_3)\right\} \\ = \ & w_p^4 \left\{\frac{k^3}{k^\alpha k_\alpha}(I_0 + i\tau_o) + I^3{}_0 - i\tau_o J^3{}_0\right\} \times \left\{I^0{}_3 - i\tau_o J^0{}_3 + \frac{k^0 I_3}{k^\alpha k_\alpha}\right\}. \end{aligned} \tag{50}$$

where

$$G = k^\nu I_\nu + i\tau_o k^\nu \xi_\nu \xi^{-1}. \tag{51}$$

For the transverse mode, obtains

$$\left\{1 - \frac{w_p^2}{k^\alpha k_\alpha} G\right\} + w_p^2\left\{I^1{}_1 + i\tau_o J^1{}_1\right\} = 0, \tag{52}$$

5.1. Dispersion relations at $T = 0$

5.1.1. Real part of the dispersion relation for the longitudinal mode at $T = 0$ Let us consider equation (50) without taking into account the imaginary part which are multiplied by τ_o. As a first approximation since τ_o is too small, we will obtain a real dispersion relation; that is:

$$\begin{aligned} &\left\{\left[1 - \frac{w_p^2}{k^\alpha k_\alpha} G\right] + w_p^2 I^0{}_0 + \frac{w_p^2 k^0 I_0}{k^\alpha k_\alpha}\right\} \\ &\times \left\{\left[1 - \frac{w_p^2}{k^\alpha k_\alpha} G\right] + w_p^2 I^3{}_3 + \frac{w_p^2 k^3 I_3}{k^\alpha k_\alpha}\right\} \\ = \; & w_p^4 \left\{\frac{k^3 I_0}{k^\alpha k_\alpha} + I^3{}_0\right\} \left\{I^0{}_3 + \frac{k^0 I_3}{k^\alpha k_\alpha}\right\}. \end{aligned} \tag{53}$$

For a temperature very close to zero, we can suppose that $N_{eq}(v_\rho)$,

$$N_{eq}(v_\rho) = n_o \delta\left(v_\rho - \overline{v}_\rho\right) 2\Theta\left(\overline{v}^0\right) \delta\left(\overline{v}^\rho \overline{v}_\rho - 1\right). \tag{54}$$

It has to be noted that in our case since the lab frame is at rest with the plasma,

$$\overline{v}^\rho = (\overline{v}^0, \overrightarrow{0}) \qquad \text{with} \qquad \overline{v}^0 > 1 \qquad \text{and} \qquad \overline{v}^\rho \overline{v}_\rho = 1. \tag{55}$$

Therefore, by using equation (54) in equations (43), (44) and (45), we arrive at

$$I_\nu(k_\rho, \xi_\rho) = n_o^{-1} \int d^4 v \frac{v_\nu n_o \delta\left(\overline{v}_\rho - v_\rho\right)}{k^\alpha v_\alpha} = \int dv_0 d^3 v \frac{v_\nu \delta\left(\overline{v}_0 - v_0\right) \delta\left(\overrightarrow{\overline{v}} - \overrightarrow{v}\right)}{w v_o - \overrightarrow{k} \cdot \overrightarrow{v}}, \tag{56}$$

and

$$\begin{aligned} I^\nu{}_\mu(k_\rho, \xi_\rho) &= -n_o^{-1} \int d^4 v \frac{v_\mu v^\nu n_o \delta\left(\overline{v}_\rho - v_\rho\right)}{\left(k^\alpha v_\alpha\right)^2} \\ &= -\int dv_0 d^3 v \frac{v_\mu v^\nu \delta\left(\overline{v}_0 - v_0\right) \delta\left(\overrightarrow{\overline{v}} - \overrightarrow{v}\right)}{\left(w v_o - \overrightarrow{k} \cdot \overrightarrow{v}\right)^2}. \end{aligned} \tag{57}$$

By substituting equation (55) in equations (56) and (57), we obtain

$$I_0 = \frac{1}{w} \qquad \text{and} \qquad I_i = 0 \qquad \text{for} \qquad i = 1, 2, 3, \tag{58}$$

and

$$I^0{}_0 = -\frac{1}{w^2} \qquad \text{and} \qquad I^\nu_\mu = 0 \qquad \text{otherwise} \tag{59}$$

By using equations (58) and (59) in equation (50), we obtain

$$\left\{\left[1 - \frac{w_p^2}{k^\alpha k_\alpha} G\right] - \frac{w_p^2}{w^2} + \frac{w_p^2 k^0}{k^\alpha k_\alpha w}\right\} \left\{\left[1 - \frac{w_p^2}{k^\alpha k_\alpha} G\right]\right\} = 0. \tag{60}$$

Since we are not considering terms proportional to τ_o, from equation (51) G can be considered as

$$G = k^\nu I_\nu = w \frac{1}{w} = 1. \tag{61}$$

IOP Conf. Series: Journal of Physics: Conf. Series **1239** (2019) 012001 doi:10.1088/1742-6596/1239/1/012001

Therefore, equation (60) can be expressed as

$$(1-\frac{w_p^2}{w^2})(1-\frac{w_p^2}{k^\alpha k_\alpha})=0. \tag{62}$$

Hence, we have two solutions, the first one is

$$1-\frac{w_p^2}{k^\alpha k_\alpha}=0\Rightarrow w^2=w_p^2+k^2, \tag{63}$$

and the second is

$$(1-\frac{w_p^2}{w^2})=0\Rightarrow w^2=w_p^2. \tag{64}$$

We can conclude that for longitudinal mode, the dispersion relations are

$$w_{l1}^2=w_p^2+k^2 \qquad \text{and} \qquad w_{l2}^2=w_p^2. \tag{65}$$

The second corresponds to the regular non-propagating longitudinal mode which is also found in the non-relativistic case. However, the first dispersion relation corresponds to a mode which does not satisfy the continuity equation described in equation (41).

5.1.2. Real part of the dispersion relation for the transverse mode at $T=0$ Let us consider equation (52) which corresponds to the transverse modes in general. If we apply the same treatment than above using equations (59), (61) and not considering the imaginary part which multiply the characteristic time τ_o, we arrive at

$$1-\frac{w_p^2}{k^\alpha k_\alpha}=0, \tag{66}$$

which corresponds to

$$w_T^2=w_p^2+k^2. \tag{67}$$

5.1.3. Eigenvectors and the continuity equation In order to understand the reason why we have to discard one of the longitudinal mode, it is necessary to remember that equation (41) must be included in the deduction of the modes. The eigenvectors of equation (42) are

$$A_\alpha^\beta J^\alpha(k_v)=\lambda J^\beta(k_v) \qquad \text{with} \qquad \lambda=1, \tag{68}$$

and for $T=0$,

$$A_\alpha^\beta=\begin{pmatrix} \frac{w_p^2}{w^2} & 0 & 0 & 0 \\ 0 & \frac{w_p^2}{k^\mu k_\mu} & 0 & 0 \\ 0 & 0 & \frac{w_p^2}{k^\mu k_\mu} & 0 \\ -\frac{w_p^2 k}{w k^\mu k_\mu} & 0 & 0 & \frac{w_p^2}{k^\mu k_\mu} \end{pmatrix}. \tag{69}$$

The eigenvalues and eigenvectors correspond to:

Transverse modes

$$\lambda=\frac{w_p^2}{k^\mu k_\mu}=1\Rightarrow\begin{pmatrix}0\\J_1\\0\\0\end{pmatrix} \qquad \text{and} \qquad \begin{pmatrix}0\\0\\J_2\\0\end{pmatrix} \tag{70}$$

The eigenvectors correspond to the dispersion relations of equation (63).

Longitudinal modes

$$\lambda = \frac{w_p^2}{w^2} = 1 \Rightarrow \begin{pmatrix} J^0 \\ 0 \\ 0 \\ J^3 \end{pmatrix} = \begin{pmatrix} J^0 \\ 0 \\ 0 \\ \frac{w}{k} J^0 \end{pmatrix} \quad \text{and} \quad \lambda = \frac{w_p^2}{k^\mu k_\mu} = 1 \Rightarrow \begin{pmatrix} 0 \\ 0 \\ 0 \\ J^3 \end{pmatrix} \tag{71}$$

The first eigenvalue of equation (71) corresponds to the second dispersion relation of equation (65) and the second eigenvalue corresponds to the first dispersion relation of equation (65). However, in our case with $k^\mu = (w, 0, 0, k)$, the continuity equation turns to be

$$wJ^0 - kJ^3 = 0. \tag{72}$$

It is clear that the two transverse modes and the first transverse mode satisfy such equation. Nevertheless, the second longitudinal mode does not satisfy the continuity equation and must be discarded. This is why this mode does not exist and the reasons claimed by Hakim and Mangeney [3] of arguing that it was due to the choice of the gauge or because it was a solution coming from Special Relativity, are wrong. It is true that few years later, Hakim and Mangeney [4] used the continuity equation to discard this mode. However, the equation was introduced as an extra constraint of the theory and not as embedded in the theory itself as we explain in subsection 5.2. For non vanishing temperatures we will directly incorporate the continuity equation in the deduction of the normal modes.

5.1.4. Damping due to the radiation reaction force By neglecting the radiation term in equation (50), we obtain the dispersion relations from equation (60) at zero temperature ($T = 0$) but if we want to know the damping effect due to the radiation effect it is necessary to consider the term containing $i\tau_o$ in equation (50). By using equations (58) and (59) in subsection 5.2 and some identities of the Appendix of Hakim and Mangeney [3], for the values

$$J^0_{\ 0} = 0, J^3_{\ 3} = 0, J^0_{\ 3} = 0, J^3_{\ 0} = 0 \text{ and } J^1_{\ 1} = 0. \tag{73}$$

We have

For the transverse mode

$$\left\{w^2 - k^2 - w_p^2\right\} + i\tau_o \left\{w^2 - k^2 - w_p^2\right\} w_p^2 J^1_{\ 1} = 0. \tag{74}$$

Since $J^1_{\ 1} = 0$ (see equation (73)), we arrive at

$$\left\{w^2 - k^2 - w_p^2\right\} = 0, \tag{75}$$

which corresponds to the normal transverse mode which does not suffer any damping effect due to the radiation reaction force.

For the longitudinal mode

$$1 - \frac{w_p^2}{k^\alpha k_\alpha}(1 + \tau_o w) + w_p^2 \left(-\frac{1}{w^2} - i\frac{\tau_o}{w}\right) + \frac{w_p^2 w}{k^\alpha k_\alpha}\left(\frac{1}{w} + \tau_o\right) = 0. \tag{76}$$

Then,

$$w^2 k^\alpha k_\alpha - w_p^2 k^\alpha k_\alpha (1 + i\tau_o w) = 0. \tag{77}$$

IOP Conf. Series: Journal of Physics: Conf. Series **1239** (2019) 012001 doi:10.1088/1742-6596/1239/1/012001

Finally, we arrive at

$$w^2 = w_p^2(1 + i\tau_o w). \tag{78}$$

In order to find the damping term, we need to put

$$w_s = w_o + i\gamma, \tag{79}$$

where w_o represents the dispersion relation at zero order that in this case is $w_o^2 = w_p^2$. Therefore,

$$(w_o + i\gamma)^2 = w_p^2(1 + i\tau_o (w_o + i\gamma)). \tag{80}$$

We obtain

$$w_o^2 + 2iw_o\gamma - \gamma^2 = w_p^2 (1 - \tau_o\gamma) + i\tau_o w_p^2 w_o \tag{81}$$

Since $\gamma \ll w_p \simeq w_o$, we can conclude that

$$\gamma = \tau_o \frac{w_p^2}{2} \Rightarrow w_s = w_o + i\tau_o \frac{w_p^2}{2}. \tag{82}$$

We conclude that the damping effect corresponds to a constant damping close to zero temperature. This result will be verified when the damping effect is calculated for small temperatures in subsection 6.2.

5.2. Dispersion relation for the longitudinal mode at small temperature with $T \neq 0$ with damping effect (Landau and radiation reaction dampings)

5.2.1. The continuity equation and the imaginary integral Let us put on a side the transverse mode and put our attention just in the longitudinal modes. We have assumed that I_0 and $I^0_{\ 0}$ are real and consequently we obtained real dispersion relations. However, if we look at equations (56) and (57), imaginary parts can be deduced. Indeed, returning to equation (53) and putting

$$I_o = \mathrm{Re}I_0 + i\mathrm{Im}I_0 \qquad \text{and} \qquad I^0_{\ 0} = \mathrm{Re}I_0^0 + i\mathrm{Im}I_0^0 \tag{83}$$

and note that the frequency must have a damping term; that is:

$$w = w' + i\gamma, \tag{84}$$

where

$$\gamma \ll w'. \tag{85}$$

We will now incorporate the continuity equation in the calculation of the dispersion relations. Indeed, if we consider equations (42), (48), (38)and (72), we have a system of equations

$$\left\{ \begin{array}{c} G_\lambda^\beta J_\lambda = 0 \\ \det G_\lambda^\beta = 0 \\ k^\mu J_\mu = wJ^0 - kJ^3 = 0 \end{array} \right\}, \tag{86}$$

for which (see equation (48))

$$G_\lambda^\beta = \begin{pmatrix} G_0^0 & G_3^0 \\ G_0^3 & G_3^3 \end{pmatrix}. \tag{87}$$

We obtain

$$wG_3^0 + kG_0^0 = 0 \qquad \text{and} \qquad wG_3^0 + kG_0^0 = 0. \tag{88}$$

However, this two equations are equivalent as we can see easily. Therefore, we will just use the first of the two equations. In this case, and after a long way and using the relations obtained in the Appendix by Hakim and Mangeney [4], we arrive at

$$\begin{aligned}\frac{k^2 + w_p^2 k^\alpha I_\alpha}{k^\alpha k_\alpha} &= \frac{2ww_p^2 I_o}{k^\alpha k_\alpha} + w_p^2 \frac{\partial I_o}{\partial w} \\ &+ i\tau_o \frac{w_p^2}{k^\alpha k_\alpha}\left(w + k^\alpha k_\alpha \frac{\partial I_o}{\partial w} - k^\alpha k_\alpha \left(k^\alpha I_\alpha + \frac{3}{\alpha}\right) I_o\right), \qquad (89)\end{aligned}$$

where

$$\alpha = m\xi = \frac{m}{\kappa T_o}.$$

This result differs from Hakim and Mangeney [2] by an extra factor w/k in the imaginary part. However, they always calculated the dispersion relations for $w \simeq k$ and the results are equivalent.

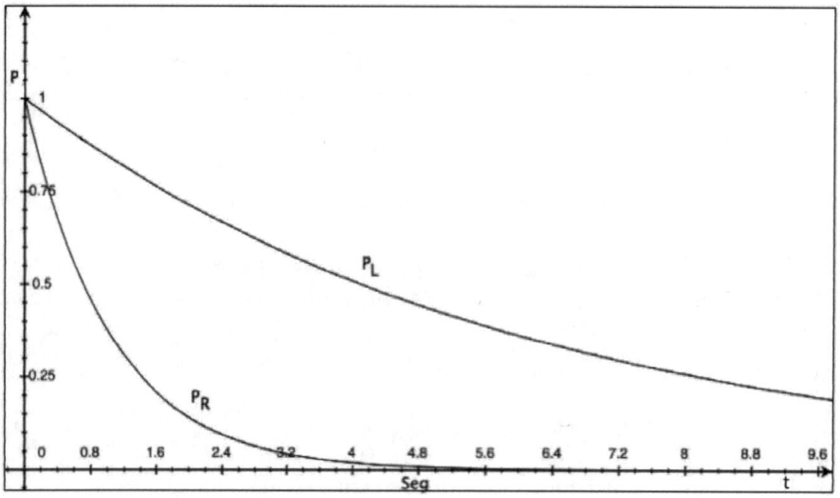

Figure 1. In a typical Tokamak, P_L is greater than P_R which means that the losses due to the radiation reaction damping are bigger than the one due to the Landau damping.

5.2.2. The suprathermic case the suprathermic case normally is described by using the relation

$$v_T < v_\phi < 1, \qquad (90)$$

where v_T and v_ϕ represent the thermic velocity $\sqrt{m\xi} = \sqrt{m/\kappa T_o} = \sqrt{\alpha}$ (for non relativistic case) and the phase velocity w/k, respectively. However, to obtain analytical expressions, it is necessary to add other constraints as

$$\begin{aligned} 1/m^2\xi^2(1 - v_\phi^2) &\ll 1 \qquad (91) \\ (1 - v_\phi^2) &\ll m^2\xi^2, \end{aligned}$$

which means that

$$w \simeq k \qquad \text{and} \qquad w \simeq w_p \qquad (92)$$

Plasma	Density $N_{0e}(m^{-3})$	Electron´s temperature T(Kelvin)	α	$N_{0e}e^2\tau_0$
Solar core	10^{32}	10^{7}	593.56	$1.6*10^{-29}$
Tokamak	10^{20}	10^{8}	59.35	$1.6*10^{-41}$
Glow discharge	10^{16}	10^{4}	593565.05	$1.6*10^{-45}$
Solar wind	10^{6}	10^{5}	59356.50	$1.6*10^{-55}$
Ionosphere	10^{12}	10^{3}	5935650.50	$1.6*10^{-58}$
Sun´s crown	$5.5*10^{14}$	10^{7}	593.56	$8.8*10^{-47}$
Chromosphere	10^{15}	10^{4}	593565.05	$1.6*10^{-46}$

Figure 2. Some characteristic plasmas

and accompanied by

$$m\xi > 1. \tag{93}$$

That is: the temperature is not small. Accordingly, we can now develop around a new parameter

$$\eta^2 = \alpha^{-2}\left(1 - v_\phi^2\right). \tag{94}$$

Without considering the imaginary quantities (Landau and radiation reaction damping terms are neglected in first instance), equation (84) can be written as

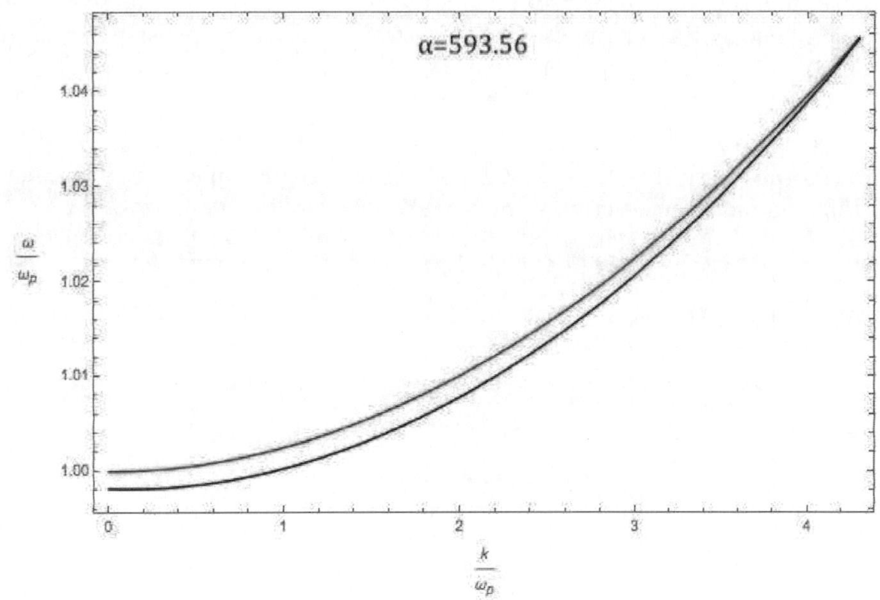

Figure 3. For α bigger than 500, the approximation (red line) coincides with the numerical method (blue line) as it is expected.

$$k^2 + w_p^2 k^\alpha I_\alpha = 2w_p^2 w I_o + w_p^2\left(w^2 - k^2\right)\frac{\partial I_o}{\partial w}. \tag{95}$$

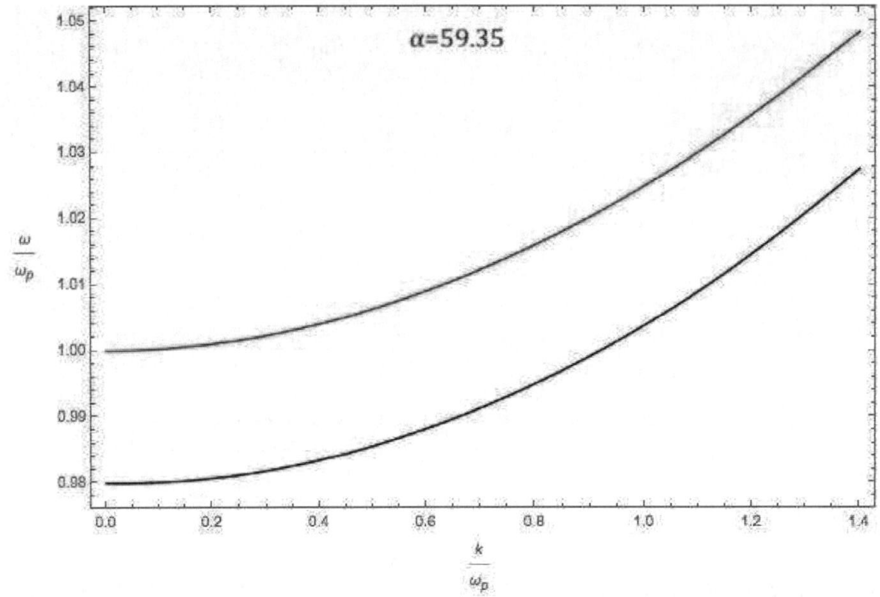

Figure 4. For a typical Tokamak, the approximation is still valid. But it has to be noticed that the numerical method (blue line) predicts that the value when $k \to 0$, $w = 0.98w_p$

In order to continuous with real part of equation (96), just $\mathrm{Re}I_o$is considered, we follow the same calculation done by Hakim and Mangeney [2] which make an expansion in η^2, we arrive at

$$w^2 = w_p^2 + \frac{3w_p^4}{k^2}\alpha^{-1} = w_p^2 + 3\left(kv_T\right)^2, \tag{96}$$

which represents the classical result. Now if we consider, the complex terms in equation (97), (remember that the radiation reaction term are not considered) we obtain

$$k^\alpha k_\alpha - w_p^2 + k^\alpha k_\alpha w_p^2\left(\mathrm{Re}I_0^0 + i\mathrm{Im}I_0^0\right) + w_p^2 k^0\left(\mathrm{Re}I_0 + i\mathrm{Im}I_0\right) = 0. \tag{97}$$

Equation (97) can be written as

$$k^\alpha k_\alpha - w_p^2 + k^\alpha k_\alpha w_p^2\mathrm{Re}I_0^0 + w_p^2 k^0\mathrm{Re}I_0 + i\left(k^\alpha k_\alpha w_p^2\mathrm{Im}I_0^0 + w_p^2 k^0\mathrm{Im}I_0\right) = 0. \tag{98}$$

By substituting the values of the real part, see equation (64), we have

$$\left(w^2 - w_p^2\right)k^\alpha k_\alpha + iw^2 w_p^2\left(k^\alpha k_\alpha\mathrm{Im}I_0^0 + k^0\mathrm{Im}I_0\right) = 0. \tag{99}$$

Putting

$$\varepsilon = k^\alpha k_\alpha\mathrm{Im}I_0^0 + k^0\mathrm{Im}I_0, \tag{100}$$

we obtain

$$\left(w^2 - w_p^2\right)k^\alpha k_\alpha + iw^2 w_p^2\varepsilon = 0. \tag{101}$$

This is equivalent to

$$\left(w^2 - w_p^2\right)\left(w^2 - k^2\right) + iw^2 w_p^2\varepsilon = 0. \tag{102}$$

By using equation (84), we arrive at

$$\left[\left(w' + i\gamma\right)^2 - w_p^2\right]\left[\left(w' + i\gamma\right)^2 - k^2\right] + i\left(w' + i\gamma\right)^2 w_p^2\varepsilon = 0. \tag{103}$$

Finally,

$$\left(w'^2 + 2i\gamma w' - \gamma^2 - w_p^2\right)\left(w'^2 + 2i\gamma w' - \gamma^2 - k^2\right) + i\left(w' + i\gamma\right)^2 w_p^2 \varepsilon = 0. \tag{104}$$

By neglecting the term γ^2, $w' = w_p$ and for the damping part, we obtain

$$\begin{aligned} 2\left(\gamma w_p\right)\left(w_p^2 - k^2\right) + w_p^4 \varepsilon &= 0 \\ \gamma &= -\frac{w_p^3 \varepsilon}{2(w_p^2 - k^2)}. \end{aligned} \tag{105}$$

Obtaining the value of ε in this approximation, we arrive to

$$\gamma = \frac{1}{2}\sqrt{\frac{\pi}{2}}\frac{w_p^4}{k^3}\left(m\xi\right)^{3/2} \exp -(m\xi v_\phi^2)/2, \tag{106}$$

which represents the Landau damping term. It has to be noticed that within this approximation, the radiation reaction damping is equal to the one calculated for zero temperature. We can affirm that the intensity of the perturbation P is damped as

$$P = \exp -\gamma t \exp -\tau_o \frac{w_p^2}{2} t. \tag{107}$$

Therefore, If we consider a typical Tokamak case with

$$w_p = 5.6 \times 10^{11} Hertz \qquad \text{and} \qquad \beta = \frac{v_T}{c} = \alpha^{-1/2} = 0.12, \tag{108}$$

we can compare graphically both dampings (see figure 1)

$$P_L = \exp -\gamma t = \exp -0.16 t \qquad \text{and} \qquad P_{\tau_o} = \exp -\tau_o \frac{w_p^2}{2} t = \exp -0.98 t, \tag{109}$$

which shows that in this case the Bremsstrahlung generates a more important damping.

5.3. The superluminic and relativistic cases (longitudinal modes)

Let us now analyze the results which can be obtained when

$$v_\phi \gg 1 \tag{110}$$

In this case, let us rewrite equation (89) in another form without any approximation

$$\begin{aligned} \Omega_p^2 + k^2 &= 2w_p^2 w I_o + w_p^2(w^2 - k^2)\frac{\partial I_o}{\partial w} \\ &+ i\tau_o w_p^2\left[w + (w^2 - k^2)\frac{\partial I_o}{\partial \alpha} + (w^2 - k^2)\left(G + \frac{3}{\alpha}\right) I_o\right], \end{aligned} \tag{111}$$

where

$$\Omega_p^2 = w_p^2\left[K_1(\alpha)/K_2(\alpha)\right] = w_p^2 k^\nu I_\nu = w_p^2 \mathrm{Re} G = w_p^2 \Im. \tag{112}$$

We can make an expansion in $(k/w)^2 = v_\phi^{-2} \ll 1$; we arrive to

$$\begin{aligned}\Omega_p^2 + k^2 &= 2w_p^2 \left[\sum_{l=0}^{\infty} \left(\frac{k}{w}\right)^{2l} \phi_l(\alpha)\right] \\ &-w_p^2 \left(1 - \frac{k^2}{w^2}\right) \left[\sum_{l=0}^{\infty} \left(\frac{k}{w}\right)^{2l} (2l+1)\, \phi_l(\alpha)\right] \\ &+i\tau_o w_p^2 \frac{w^2}{k} \left\{ \begin{array}{c} 1 + (1 - \frac{k^2}{w^2}) \left[\sum_{l=0}^{\infty} \left(\frac{k}{w}\right)^{2l} \frac{d\phi_l(\alpha)}{d\alpha}\right] \\ + \left(1 - \frac{k^2}{w^2}\right) (\Im + \frac{3}{\alpha}) \left[\sum_{l=0}^{\infty} \left(\frac{k}{w}\right)^{2l} \phi_l(\alpha)\right] \end{array} \right\}, \end{aligned} \tag{113}$$

where $\phi_l(\alpha)$ are defined in the Appendix by Hakim and Mangeney [3]. It has to be remembered that

$$\tau_o \frac{w^2}{k} \ll 1 \qquad \text{and} \qquad k \ll w. \tag{114}$$

If we do not consider the imaginary part, we arrive at

$$w_o^2 = (\Im - \phi_1)\, w_p^2 + 3k^2 \left[(\phi_1 - \phi_2) / (\Im - \phi_1)\right]. \tag{115}$$

5.3.1. Low temperature For low temperatures, we obtain

$$w_o^2 = w_p^2 + 3\,(\kappa T/m)\, k^2. \tag{116}$$

which coincides with the suprathermic case and suspects that this result is valid for any value of k.

5.3.2. High temperature For high temperatures, and just for the first order relativistic expansion, we need to continuous the expansion in the ϕ_l's and it can be deduced that (we recover the value of the speed of light c)

$$w_o^2 = w_p^2 + 3k^2\,(\kappa T/m) - \frac{5}{2} w_p^2 (\kappa T/mc^2) - \frac{33}{2}\left(\frac{k^2}{c^2}\right)(\kappa T/m)^2, \tag{117}$$

which means that when $k \to 0$, the w_o does not go to w_p but to

$$w_o^2 = w_p^2 \left(1 - \frac{5}{2}(\kappa T/mc^2)\right). \tag{118}$$

This relation shows that there exists a shift of the plasma frequency by a factor $(1 - (5/2)\,(\kappa T/mc^2))$. This approximation is valid for not such high temperature since when $T \to \infty$, $w_o^2 \to -\frac{5}{2}(\kappa T/mc^2)$ and w_o will be imaginary which corresponds to a damping. Nevertheless, high expansion in the ϕ_l's will turn on a similar effect as we will see in the ultra-relativistic case. Considering the imaginary part (the radiation reaction term), we have

$$w = w_o + i\gamma. \tag{119}$$

We arrive at

$$\gamma = \tau_o w_p^2 (w_o^3/k^3)\left[\Im\,(\Im + 3/\alpha)\right]. \tag{120}$$

which represents a tremendous damping effect. The Landau damping vanishes for high temperature because the imaginary part of the integrals can be neglected.

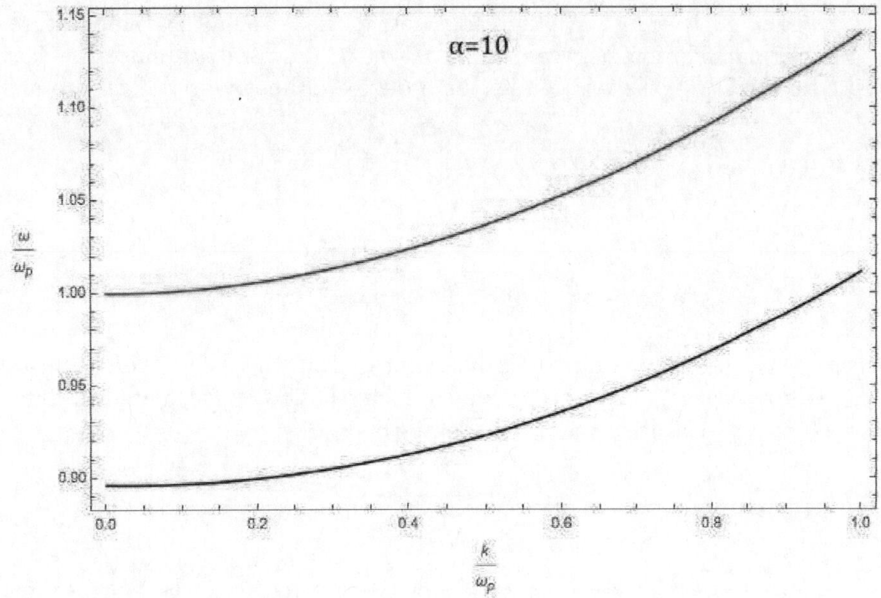

Figure 5. The approximation is good but the numerical method (blue line) shows that as α decays the approximation is less accurate (red line).

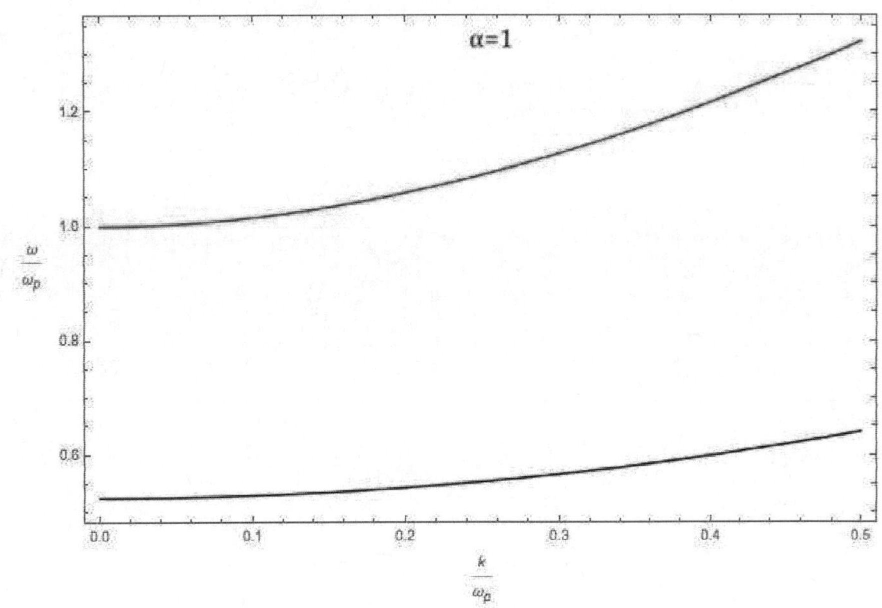

Figure 6. For $\alpha \simeq 1$, equation (116) is not anymore valid (red line) and it has to be substituted by an equation close to equation (117) or by the numerical solution (blue line)

5.3.3. The ultra relativistic case

IOP Conf. Series: Journal of Physics: Conf. Series **1239** (2019) 012001 doi:10.1088/1742-6596/1239/1/012001

The relation between α and v_T We have to notice that when the temperature is ultra-relativistic, the normal relation between the $\alpha = m\xi = \frac{m}{\kappa T_o}$ and v_T must be changed. For non relativistic temperature, the relation is (the 3 degree of liberty are taken into account)

$$v_T^2 = 3/\alpha \qquad \text{with} \qquad \alpha^{-1} = \frac{\kappa T_o}{m}, \tag{121}$$

with the particularity that

$$\Im = 1 - \frac{3}{2\alpha} + 0(\alpha^{-2}) \qquad \text{for} \qquad \alpha \gg 1. \tag{122}$$

However, when the temperature is ultra-relativistic, it is clear that $v_T = 1$ and α must constraint to a minimum value. Nevertheless, the thermal velocity has to be calculated in different form. Indeed, following the Special Relativity, the average energy,

$$\langle E \rangle = \frac{m}{\sqrt{1 - v_T^2}}, \tag{123}$$

represents an appropriate relativistic definition of the thermal velocity. On the other hand, Synge [40] has proved that

$$\langle E \rangle = m\,(\Im + 3/\alpha)\,. \tag{124}$$

Noticing that

$$\Im = \frac{1}{2}\alpha + 0(\alpha^2) \qquad \text{for} \qquad \alpha \ll 1 \tag{125}$$

which gives

$$v_T^2 \simeq 1 - \frac{1}{9}\alpha^2. \tag{126}$$

Therefore, $\alpha \to 0$, $v_T^2 \to 1$ as it was expected.

Dispersion relation for the ultra-relativistic case Now, we can continuous analyzing the ultra-relativistic case. Taking the real part of equation (89), we arrive to ($\alpha \to 0$)

$$k^2 + w_p^2 k^\alpha I_\alpha = 2ww_p^2 \left[\frac{\alpha}{4k}\log\left(\frac{w-k}{w+k}\right)\right] + w_p^2 k^\alpha k_\alpha \frac{\alpha}{2k^\alpha k_\alpha}. \tag{127}$$

In this case,

$$k^\alpha I_\alpha \simeq \frac{\alpha}{2}. \tag{128}$$

We obtain

$$k^2 = ww_p^2 \left[\frac{\alpha}{2k}\ln\left(\frac{w-k}{w+k}\right)\right]. \tag{129}$$

In the superlumic case, $k \to 0$, we have

$$k^2 = -ww_p^2\alpha \left[\frac{\ln(w+k) - \ln(w+k)}{2k}\right], \tag{130}$$

that is:

$$k^2 = -ww_p^2\alpha \frac{d\ln w}{dw} = -ww_p^2 \frac{1}{w} = -w_p^2. \tag{131}$$

which means that for $k \to 0$, there is no perturbation as we will show in section 6 (Numerical Results of the Dispersion Relations).

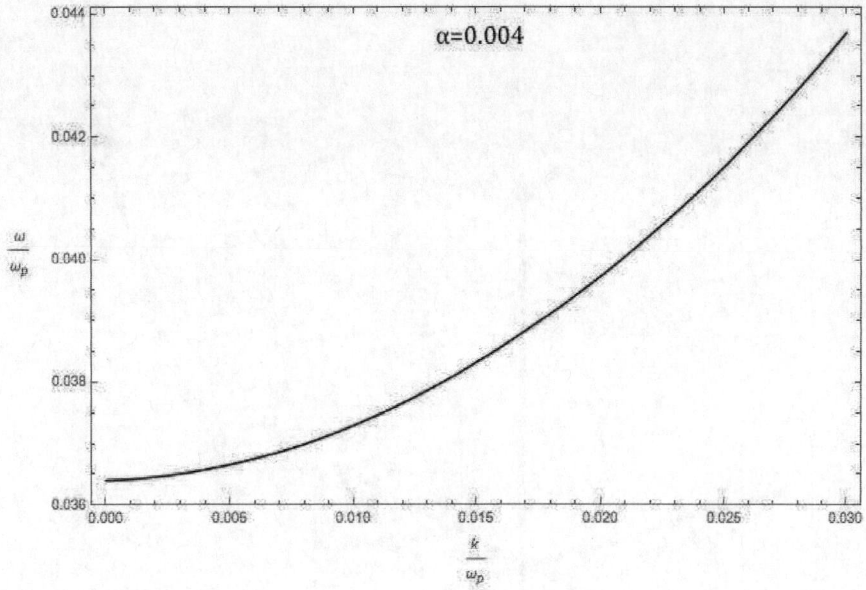

Figure 7. For very low α, there is not exist a good approximation and just the numerical method is valid. It has to be highlighted that when $k \to 0$, $w = 0.37w_p$

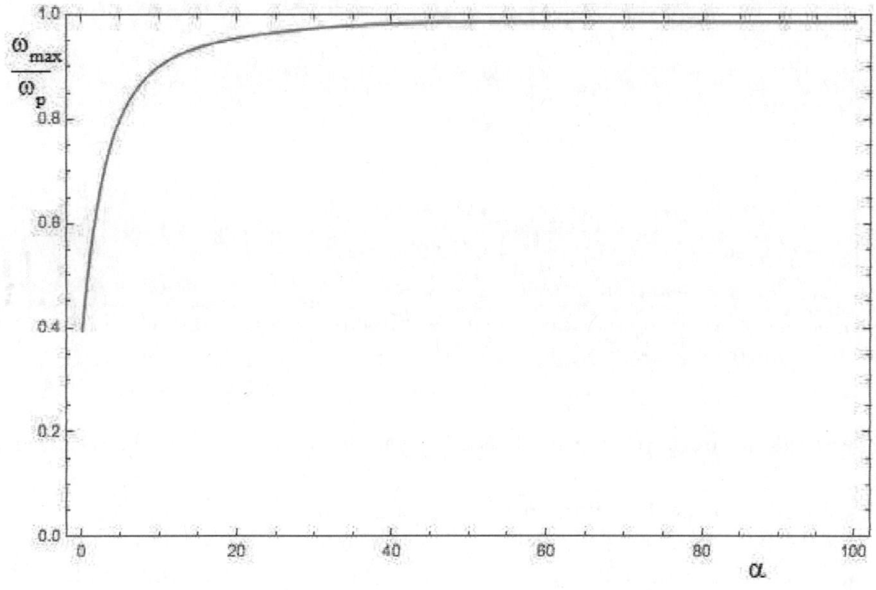

Figure 8. The value of w when $k \to 0$ as a function of α. It has to be highlighted that the value of $w \to 3$ as $\alpha \to 0$.

6. Numerical Results of the Dispersion Relations

Until now we have given dispersion relations in many ranges of validity but no for any α neither k in general. Due to difficulty to solve in general equation (89), it is necessary to use numerical methods to describe the dispersion relations. Let us present some important results. In figure 2,

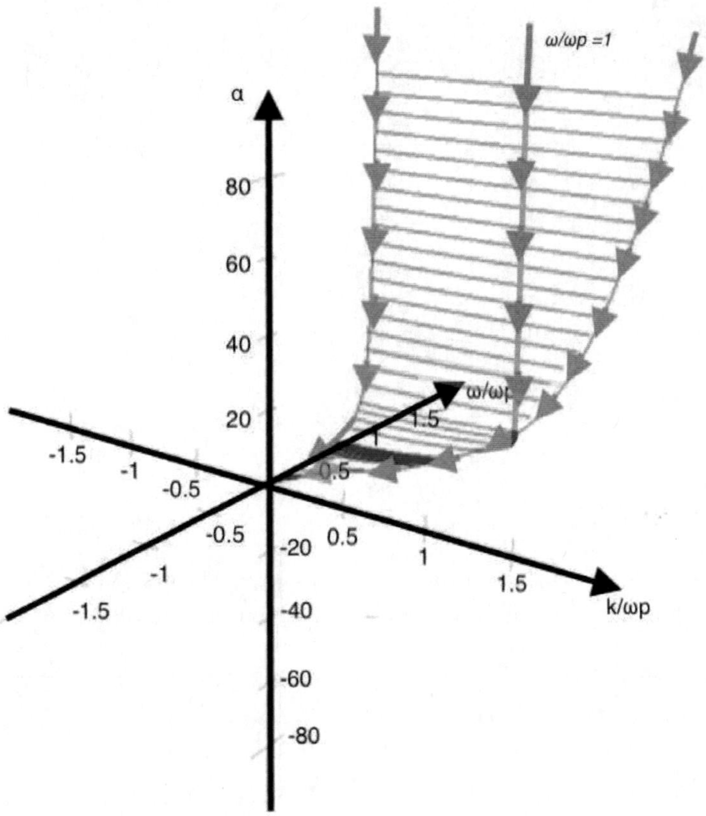

Figure 9. Dispersion relations surface in the 3−dimension space k, w, α. The red arrow line which crosses the surface, represent $(w/w_p) = 1$

some characteristic plasmas are described showing that all of then satisfy the required condition $ne^2\tau_o \ll 1$.

Some dispersion relations for different α's are shown from figure 3 to figure 7. A comparison is made between the dispersion relations obtained by making some approximations, equation (116) and the numerical method for different values of α. Notice that the relation described in equation (118) is much accurate for decreasing α. The red lines in the figures just corresponds to equation (116) which does not describe the shift in the plasma frequency.

In figure (8), it is numerically showed that there exists a limit for the value of w when $k \to 0$ for ultra-relativistic temperatures. However, the most interesting fact consists on noticing that when α is decreasing the dispersion relation has a cut-off for the wave number k (see figure 9); that is: for each α there is a maximum wave number k_α such that above it the wave no longer exists. Moreover, below a certain value of α, it can be said that there is not anymore a dispersion relation but unique values of w and k; that is

$$\alpha < \alpha_p \to w = w(\alpha) \qquad \text{and} \qquad k = k(\alpha). \tag{132}$$

7. Conclusion

We follow Hakim [3] in the deduction of the density of particles just by considering the positions and the velocities of the particles as independent variables without taking into account the hyper-acceleration as an independent variable. This is due to the fact that Landau-Lifshitz equation of motion for charged particles is used to describe the motion of the charges. The final result coincides with Hakim generalization of the Vlasov equation in Special Relativity [3]. Moreover, a more general equation is deduced by including a constant external electric field. The wave dispersion relations are obtained and the effect of the coupling between the constant external electric field and the interaction of the particle does not play an important role since the dispersion relations are not affected.

We recover well-known dispersion relations for different range of values and approximations comparing their with numerical solutions. The original results are threefold:

A- There exist a cut-off wave number k_α for each temperature for which above this wave number there is no wave.

B- For temperatures above $T_{\alpha=1}$ (for $\alpha < 1$), the value of w at $k \to 0$ decreases below $0.5w_p$

C- For temperatures above T_p ($\alpha < \alpha_p$), the dispersion relation does not exist and we have a unique pair, $w = w(\alpha)$ and $k = k(\alpha)$.

It remains to do a similar study using the Hammond model. Indeed, although Hammond model does not give an expression for the radiation reaction force, we will make an attempt in order to deduce an effective Hammond radiation reaction force for the case of an external constant electric field in a plasma. The radiation reaction force must not vanish as it happens for the other equations and consequently different dispersion relations will be obtained to test the viability of Hammond model. It has to be noted that Hammond proposed an experiment to check the difference between the different equations by measuring the gain of kinetic energy under a high laser intensities ($10^{22} W cm^{-2}$) [19], [22]. We think that dispersion relations are easier to be measured.

Acknowledgement : This work was partially supported by COFAA, EDI SIP - IPN and SNI CONACYT.

8. References

[1] Hakim R 1967 Remarks on Relativistic Statistical Mechanics. I *Journal of Mathematical Physics* **8** 1315
[2] Hakim R 1967 Remarks on Relativistic Statistical Mechanics. II. Hierarchies for the Reduced Densities *Journal of Mathematical Physics* **18** 1379
[3] Hakim R and Mangeney A 1968 Relativistic Kinetic Equations Including Radiations Effects. I Vlasov Approximation *Journal of Mathematical Physics* **9** 1 116
[4] Hakim R and Mangeney A 1971 Collective Oscillations of a Relativistic Radiating Electron Plasma *Physics of Fluids* **14** 2751
[5] Dirac P A M 1938 Classical theory of radiating electrons *Proceedings of the Royal Society of London. Series A. Mathematical and Physical Sciences* **167** 148–169
[6] Landau L D and Lifshitz E M 1962 *The Classical Theory of Fields* (Pergamon, London 2nd. edn.) §76
[7] Spohn H 2000 The critical manifold of the Lorentz-Dirac equation *Europhysics Letters* **50** 287
[8] Rohrlich F 2000 The self-force and radiation reaction *American Journal of Physics* **68** 1109
[9] Rohrlich F 2001 The correct equation of motion of a classical point charge *Physics Letters A* **283** 5–6 276
[10] Ford G W and O'Connell R F 1991 Radiation reaction in electrodynamics and the elimination of runaway solutions *Physics Letters A* **157** 4–5 217
[11] Ford G W and O'Connell R F 1993 Relativistic form of radiation reaction *Physics Letters A* **174** 3 182
[12] Eliezer C J 1948 On the classical theory of particles *Proceedings of the Royal Society of London. Series A. Mathematical and Physical Sciences* **194** 543
[13] Kravets Y, Noble A and Jaroszynski D 2013 Radiation reaction effects on the interaction of an electron with an intense laser pulse *Physical Review E* **88** 011201(R)
[14] Di Piazza A 2008 Exact Solution of the Landau-Lifshitz Equation in a Plane Wave *Letters in Mathematical Physics* **83** 3 305

[15] Di Piazza A, Müller C, Hatsagortsyan K Z and Keitel C H 2012 Extremely high-intensity laser interactions with fundamental quantum systems *Review of Modern Physics* **84** 1177

[16] García-Camacho J F, Salinas E, Avalos-Vargas A and Ares de Parga G 2015 Mathematical differences and physical similarities between Eliezer-Ford-Oconnell equation and Landau-Lifshitz equation *Revista Mexicana de Física* **61** 363–371

[17] Shen C S 1972 Magnetic Bremsstrahlung in an Intense Magnetic Field *Physical Review D* **6** 2736

[18] Hammond R T 2010 Relativistic Particle Motion and Radiation Reaction in Electrodynamics *Electronic Journal of Theoretical Physics* **7** 221–258

[19] Hammond 2010 R T Radiation reaction at ultrahigh intensities *Physical Review A* **81** 062104

[20] Hammond R T 2011 Charged particle in a constant, uniform electric field with radiation reaction *Advanced Studies in Theoretical Physics* **5** 275

[21] Hammond R T 2009 New approach to radiation reaction in classical electrodynamics *arXiv:0902.4231v2* [physics.class-ph] **21 Jul**

[22] Yanovsky V, Chvykov V, Kalinchenko G, Rousseau P, Planchon T, Matsuoka T, Maksimchuk A, Nees J, Cheriaux G, Mourou G and Krushelnick K 2008 Ultra-high intensity- 300-TW laser at 0.1 Hz repetition rate *Optics Express* **16** 2109

[23] Ares de Parga G, Mares R and Domínguez S 2005 Landau-Lifshitz equation of motion for a charged particle revisited *Annales de la Fondation Louis de Broglie* **30** 283

[24] Ares de Parga G 2006 A Physical Deduction of an Equivalent LandauLifshitz Equation of Motion in Classical Electrodynamics. A New Expression for the Large Distance Radiation Rate of Energy *Foundations of Physics* **36** 10 1474–1510

[25] Medina R 2006 Radiation reaction of a classical quasi-rigid extended particle *Journal of Physics A: Mathematical and General* **39** 3801

[26] Rohrlich F 2007 *Classical Charged Particles* (World Scientific, New Jersey) Supplement **257**

[27] Ares de Parga G, Ortiz-Domínguez M and Mares R 2008 Particle Motion in Static Homogeneous Electric and Magnetic Fields and its Consequences in a Typical Confinement Situation in Plasma Physics *Indian Journal of Physics* **82** 9 927-937

[28] DeWitt B S and Brehme R W 1960 Radiation damping in a gravitational field *Annals of Physics* **9** 2 220

[29] Jackson J D 1975 *Classical Electrodynamics* (Wiley, New York, 2nd edn) Chaps **12** 12 and **14**

[30] Horwitz L P, Shashoua S and Schieve W C 1989 A manifestly covariant relativistic Boltzmann equation for the evolution of a system of events *Physica A: Statistical Mechanics and its Applications* **161** 2 300

[31] Klimontovich Yu L 1960 Relativistic Transport Equations for a Plasma. I. *Soviet Physics JETP* **37** 10 524

[32] Jüttner F 1911 Das Maxwellsche Gesetz der Geschwindigkeitsverteilung in der Relativtheorie *Annalen der Physik* **34** 856

[33] Cubero D, Casado-Pascual J, Dunkel J, Talkner P and Hänggi P 2007 Thermal equilibrium and statistical thermometers in special relativity *Physical Review Letters* **99** 170601

[34] López-Carrera B and Ares de Parga G 2008 Relativistic transformation of the canonical distribution function in relativistic statistical mechanics *Physica A* **387** 1099–1109

[35] Ares de Parga G and López-Carrera B 2011 Relativistic Statistical Mechanics vs Relativistic Thermodynamics *Entropy* **13** 9 1664-1693

[36] Krall N A and Trivelpiece A W 1973 *Principles of Plasma Physics* (McGraw-Hill, New York)

[37] Gamba A 1967 Physical quantities in different reference systems according to relativity *American Journal of Physics* **35** 83–88

[38] Nakamura T K 2006 Relativistic energy-momentum of a body with a finite volume *Space Science Reviews* **122** 271–278

[39] Ares de Parga G, López-Carrera B and Angulo-Brown F 2005 A proposal for relativistic transformations in thermodynamics *Journal of Physics A: Mathematical and General* **38** 2821–2834

[40] Synge J L 1957 *The Relativistic Gas* (North-Holland, Amsterdand)

A Further Study of the Mixing of Relativistic Ideal Gases with Relative Relativistic Velocities: The Hot Plasma in the Sun's Corona, the Type II Spicules and CMEs.

***R E Gonzalez-Narvaez, E E Díaz-Figueroa, G Ares de Parga**

Dpto. de Física, Escuela Superior de Física y Matemáticas, Instituto Politécnico Nacional, U. P. Adolfo López Mateos, Zacatenco, C. P. 07738, CDMX, México.

E-mail: * justicia.12@hotmail.com

Abstract. The Redefined Relativistic Thermodynamics and the conservation of the 4-vector energy–momentum predict a tremendous increment of the temperature after mixing two gases with relative sub-relativistic velocity. This phenomenon can be used to describe the heating of a cold clump with shocked jets material and to predict an improving of the ignition in a Tokamak by injecting a plasma with a sub-relativistic velocity. First, by using the same effect, the mixing of the type II spicule and Coronal Mass Ejections with the plasma of a cold Sun's corona will explain the increase temperature of the Sun's corona. Second, considering, in agreement with the observations, that a part of the type II spicule, fall off into the Chromosphere and that the rest of the type II spicule that shocks with the Sun's corona possesses a higher average relative velocity, the mixing of such a part and the plasma in the Sun's corona will maintain the high temperature of the Sun's corona. Moreover, the shock of Coronal Mass Ejections with the Sun's corona it is a energy source to maintain the high temperature of the Sun's corona too.

1. Introduction

One of the biggest problems in Modern Astrophysics is the origin of the Sun's corona temperatures ($1 \times 10^6 K - 2 \times 10^6 K$). There have been proposed many theories that have given an interesting advance in the understanding of the Sun's atmosphere mechanism [1, 2].
On the development of a theory of the upper solar atmosphere heating, there have to be included realistic and theoretical models about the energy release process and the plasma response to this warming. These models range from heating by accretion [1], heating by acoustic waves [3], heating by Alfvén waves, energy deposition through the damping of magnetohydrodynamics waves [4], nanoflares (arise when magnetic field is stressed, vía reconnection) [5], velocity filtration [6] and spicules [7]. Moreover, most efforts have focused in attempts to explain and to find the causes for the coronal energy balance.
Indeed, it is necesary to identify and understand why the upper solar atmosphere temperature is so much higher even when the lower layers are cooler. This strange event was discovered in 1940, when Grotrian [8] and Edlén [2] realized that the emission lines during a total solar eclipse did not belong to a new hypothetical element named coronio and on the contrary, the emission lines belonged to known elements which were in high ionization states, for example, Fe XIV.

The scientific community assumed that this kind of events were induced by magnetic reconection processes or by shock waves locally heating the plasma. However, these observations were questioned [4].
De Pontieu *et al* [9] were the first ones who made known chromospheric plasma jets originated in the solar quiet areas, with the following characteristics:

- Diameter:$\sim 200 \quad Km$
- Ejection velocities: $50Km/s - 150Km/s$
- Life time:$10s - 150s$.

The chromospheric plasma jets with the features just mentioned were named type II spicules and they are different from type I spicule, because the first ones are faster, thinner, and they appear more frequently . Other important detail is that a part of the type II spicule ($\sim 10\%$) reaches $1 \times 10^6 K - 2 \times 10^6 K$ temperatures while it is going up into the solar corona, and the coldest rest part in going back to the Chromosphere [10]. This phenomenon gave way to the heating process understanding in the upper solar atmosphere and type II spicules came to be the favourite candidate to the solar energy supply and solar mass supply. Before the discovery of the type II spicules, the type I spicules were proposed as being the mainly resposible for the solar material supply [11]. However, type I spicules are not frequent and energetic enough to meet the requirements.
De Pontieu's [7,9] observations are not enough to know the physics of the heating. Nevertheless Klimchuk [10] proposed a classical thermodynamic model that reveals that type II spicules are not adequately energetic to heat solar corona. However Martínez-Sykora *et al* [12,13] think as a possibility, that the study of the spicules could help to explain the temperatures in solar corona.
It is also known that the Coronal Mass Ejections [CMEs] were discovered in the early 1970s when spaceborn coronographs revealed that eruptions of plasma are ejected from the Sun. Eruptive flares, filament eruptions, CMEs and failed eruptions represent different manifestations of a disruption and reconfiguration of the coronal magnetic field, which ultimately leads to an ejection of magnetized plasma that may, or may not, propagate into the Heliosphere.
Coronal mass ejections are the sudden release of $10^{15}g - 10^{16}g$ of chromospheric and coronal plasma to the Heliosphere. Their speed ranges are $100Km/s < v < 10\ 000Km/s$ [14]. A total of 6907 CMEs were observed from January 1996 to December 2002 [15]. Then, an average value of ~ 3 CMEs per day occur on the Sun.
In the theory of relativistic transformation laws of the thermodynamic quantities, many authors enriched the topic with different proposals. In regard to the various controversies, in 1998, Sieniutycz mentioned that the three more common and acceptable relativistic transformations of the temperature still survive as $T = \gamma^d T_0$, where $\gamma = \left(1 - v^2/c^2\right)^{-1/2}$, where $d = 1,\ -1$ or 0 depending on an Ott-like proposal, an Einstein-Plank-like proposal and on a Landsberg like-proposal, respectively.
On the other hand, Ares de Parga *et al* [17] redefined some thermodynamic quantities in order to conserve the laws of Thermodynamics in a moving frame and hence they proved the possibility of the existence of a relativistic transformation of the temperature. Then, based on Fermi's ideas [18] about the definition of 4−vectors in Special Relativity, Gamba [19] was able to understand the meaning of physical quantities in different reference systems. With this background, Nakamura [20, 21] developed a technique in order to covariantly define some mechanic and thermodynamic quantities. Some years later, Ares de Parga *et al* [17,22,23] were able to match Nakamura's ideas and the redefinition of some thermodynamic quantities in order to obtain a covariant theory of the relativistic laws of the transformation of the thermodynamic quantities and they called it the Redefined Relativistic Thermodynamics [RRT] [24]. This proposal which is based on the concept of simultaneity, permits matching it with a Relativistic Statistical Mechanics. Then Gonzalez-Narvaez *et al* [25] computed the final temperature of a

IOP Conf. Series: Journal of Physics: Conf. Series **1239** (2019) 012002 doi:10.1088/1742-6596/1239/1/012002

two ideal gases mixture with relative velocities. They considered different velocity cases, even relative relativistic velocities. Therefore, they computed the final temperature of a two ideal gases mixture with relative velocities based on the RRT and they compared the non-relativistic and the relativistic results. The differences between the non-relativistic and the relativistic results are negligible for sub-relativistic velocities.
In the present work, from the mixing of two ideal gases with relative velocities, the values of the final temperature, of the heating of a cold clump with shocked jets material and the ignition in a Tokamak by injecting a plasma, reported in Gonzalez-Narvaez *et al* [25] are computed more accurately. Moreover, this phenomenon can be used to describe the sharp increase of temperature of the Sun's corona by the mixing of the type II spicules and CMEs. In this work, even when only relative sub-relativistic velocities are considered, the final temperature is computed based on RRT to preserve the form.

2. Mixture of two ideal gases

2.1. The Redefined Relativistic Thermodynamics

The theory is based on being aware that in a physical system which is composed of particles or a perfect fluid constrained in a volume, the definition of a 4−vector energy-momentum is subject to instantaneity in a particular reference frame. Such property may be described by noticing that to an infinite volume a normal volume 4−vector w^μ can be associated; that is, each event x^μ which belongs to the 3−hyper-surface V is such that

$$\left(x^\mu - x^\mu_V\right) w_\mu = 0, \tag{1}$$

being x^μ_V any event in the volume V. In the reference frame where the volume is at rest, the time is the same for all the points satisfying equation (1) (for example, if the system properties are measured in the lab frame K, the instantaneity in K will be measured in a volume V and the normal volume 4−vector is $w^\mu = (1,\ 0)$, measured in K.) Then, covariantly, the volume is expresed as

$$dV^\mu = \frac{w^\mu dV_0}{u^\lambda w_\lambda}, \tag{2}$$

with u^μ is the corresponding 4−vector of the velocity u between K and K_0 (the rest frame of the moving system, the comoving frame) and V_0 is the volume of the system in K_0.
Now, the goal consists of defining the thermodynamic quantities in a covariant form. Before, some quantities that will permit us to recover the first, the second and the third law of Thermodynamics, are needed.
For a perfect fluid , the energy-momentum tensor is

$$T^{\alpha\beta} = (p+\rho)\, u^\alpha u^\beta - p n^{\alpha\beta} \tag{3}$$

where ρ and p represents the energy density and the pressure in the rest frame, respectively, and $n^{\alpha\beta}$ the Minkowsky metric. The 4−vector energy momentum is defined, with instantaneity in K (the lab frame) as

$$\begin{aligned} P^\alpha &= \int_V T^{\alpha\beta} dV_\beta \\ &= \int_V \left[(\rho+p)\, u^\alpha - \frac{p w^\alpha}{w_\lambda u^\lambda}\right] dV_0. \end{aligned} \tag{4}$$

Therefore, the bulk energy is

$$\begin{aligned} G^\mu &= \left[(\rho+p)\, u^\mu u^{nu} - (\rho+p)\, n^{\mu\nu}\right] \frac{w_\nu}{w_\lambda u^\lambda} V_0 \\ &= (E_0 + pV_0)\, u^\mu - (E_0 + pV_0)\, \frac{w_\nu}{w_\lambda u^\lambda}, \end{aligned} \tag{5}$$

IOP Conf. Series: Journal of Physics: Conf. Series **1239** (2019) 012002 doi:10.1088/1742-6596/1239/1/012002

where E_0 represents the total internal energy in the rest frame.
Then, a 4−vector redefined energy-momentum ξ^μ may be expressed as

$$\xi^\mu = P^\mu - G^\mu = \frac{w^\mu}{w_\lambda u^\lambda} E_0. \tag{6}$$

It is easy to show that with the redefined energy-momentum,the result is

$$d\xi^\mu = đ\Theta^\mu + dW^\mu, \tag{7}$$

where

$$đ\Theta^\mu = w^\mu \frac{đ\Theta_0}{w_\lambda u^\lambda} \qquad and \qquad dW^\mu = w^\mu \frac{pdV_0}{w_\lambda u^\lambda}. \tag{8}$$

Equation (7) represents the first law of Thermodynamics but in a covariant form. On the other hand, the 4−vector temperature and inverse 4−vector temperature are

$$T^\mu = w^\mu \frac{T_0}{w_\lambda u^\lambda} \qquad and \qquad \beta_\mu = \frac{u_\mu}{kT_0}, \tag{9}$$

where k represents the Boltzman constant. Moreover, from the definition of the entropy, which is an invariant, the second law is obtained

$$\begin{aligned} dS &= \beta_\mu \frac{w^\mu V_0}{w_\lambda u^\lambda} T^{\mu\nu} + \beta_\mu p dV_0 \\ &= \frac{1}{kT_0} dE_0 + \frac{p}{kT_0} dV_0. \end{aligned} \tag{10}$$

The equations of state and the laws of Thermodynamics are invariant as long as some thermodynamic quantities transform as necessary; that is, some quantities transform as the temperature, equation (9) and some parameters are invariant but the rest transform to get invariant equations.

2.2. Mixture of two ideal gases
The problem solved by Gonzalez-Narvaez *et al* [25] consists of deducing the final temperature and final velocity of the mixture of two relativistic ideal gases with different temperatures and distinct velocities.
Then, based on RRT , if the instantaneity is in K, the 4−velocity and the normal volume 4−vector are

$$u^\alpha = (\gamma,\, \gamma u) \qquad and \qquad w^\alpha = (1,\, 0)\,, \tag{11}$$

the 4−vector energy momentum is

$$P^\alpha = \left(\gamma \left(E_0 + \frac{u^2}{c^2} pV_0\right),\, \gamma \frac{u}{c}\left(E_0 + pV_0\right)\right), \tag{12}$$

which represents the Einstein-Planck-Tolman energy-momentum. Therefore, to calculate the final state when two isolated identical ideal gases with different temperatures and relative motion are mixed, the process is divided into two parts (the process may be divided into two parts because the computation is simplified).

- First, the two systems are put in contact by using an adiabatic wall. Then the wall is replaced by a diathermic wall allowing a flux of energy, obtaining an equilibrium state with temperature which will be called the equilibrium temperature T_{1eq} measured in K_1, the rest

frame of one of the gases which will be considered as the lab frame (The relative motion in this process is not varied).
Hence, taking into account the first law

$$d\xi^0 = đ\Theta^0 + dW^0, \tag{13}$$

the conservation of energy implies that

$$d\xi^0_{11} = dQ_{11} = -dQ_{12} = -d\xi^0_{12}, \tag{14}$$

since no work is done. The first subscript represents the reference frame where the instantaneity is achieved and the second subscript describes the considered ideal gas. Therefore, by using the heat capacities of each gas

$$Q_{11} = -Q_{12} \quad \Rightarrow \quad C_{V1}\left(T_{1eq} - T_{11}\right) = C_{V2}\left(T_{12} - T_{1eq}\right), \tag{15}$$

where T_{11} is the initial temperature measured in the lab frame K_1 of the gas which is at rest in K_1 and T_{12} is the initial temperature measured in the lab frame K_1 of the gas which is at rest in K_2. C_{V1} and C_{V2} are the corresponding heat capacities (which are invariant). In our case, since the particles are identical for both gases with mass m and each gas has particle numbers N_1 and N_2, respectively,

$$C_{Vi} = mN_i c_V, \tag{16}$$

with $i = 1,\ 2$. Then, equation (15) may be written as

$$N_1\left(T_{1eq} - T_{11}\right) = N_2\left(T_{12} - T_{1eq}\right). \tag{17}$$

Thereby, the equilibrium temperature is

$$T_{1eq} = \frac{N_2 T_{12} + N_1 T_{11}}{N_1 + N_2}. \tag{18}$$

- Second, much of the diathermic wall is eliminated to mix both gases obtaining an ideal gas with a final temperature T_{1f} and a certain final velocity u_f respect to the lab frame K_1(The action of the walls of the volumes is neglected in such a way that a final volume is obtained with the same final velocity than the final gas in K_1).
Recalling that an isolated system was the first hypothesis, the 4−vector energy-momentum must be conserved

$$P^{\mu}_{in1tot} = P^{\mu}_{eq1tot} = P^{\mu}_{eq11} + P^{\mu}_{eq12} = P^{\mu}_{f1tot}, \tag{19}$$

where the subscript f represents the final state of the mixture of the ideal gases. Taking the temporary component of the last two terms of equation (19) and substituting the 4−vector energy momentum, equation (12), one finds

$$a = P^0_{eq1tot} = E_{eq11} + \gamma\left(E_{eq11} + \gamma N_2 k T_{1eq}\frac{u^2}{c^2}\right) = \gamma_f\left(E_{ff} + \gamma_f N k T_{1f}\frac{u_f^2}{c^2}\right) \tag{20}$$

where the value pV was replaced by NkT, $T_{2eq} = \gamma^{-1} T_{1eq}$, and $N = N_1 + N_2$. For the spatial component

$$b = P^1_{eq1tot} = \gamma\left(E_{eq22} + N_2 k T_{1eq}\gamma\right)\frac{u}{c} = \gamma_f\left(E_{ff} + NkT_{1f}\gamma_f\right)\frac{u_f}{c}. \tag{21}$$

The quantities a and b are known because the initial and intermediate conditions are set. Then by calculating $au_f - b$, the final temperature may be expressed as

$$T_{1f} = \frac{b - au_f}{Nku_f}. \tag{22}$$

3. Computation improvements of final temperature T_{1f}

In order to deduce the final temperature T_{1f}, the final velocity u_f must be obtained, or vice versa. For this purpose, it is necessary to express the internal energy E_{ff} as a function of the temperature. The relativistic internal energy of the final gas [27] in its own rest frame is

$$E_{ff} = Nm\left(\frac{K_1(\beta_{ff}m)}{K_2(\beta_{ff}m)} + \frac{3}{\beta_{ff}m}\right), \tag{23}$$

where $\beta_{ff} = 1/kT_{ff}($ This expression for the internal energy considers the total energy of the particles; that is, there is no subtraction of the kinetic energy of the rest mass).
Gonzalez-Narvaez *et al* [25] obtained the final velocity u_f, and then they computed the final temperature. However, as shown below, a small variation in the final velocity u_f, produces a great change in the final temperature T_{1f}.
Using the equation (22), the temperature derivative is

$$dT_{1f} = \frac{-b}{Nku_f^2}du_f, \tag{24}$$

so that,

$$\triangle T_{1f} = \frac{b}{Nku_f^2} \triangle u_f = \alpha \triangle u_f. \tag{25}$$

In order to know the magnitude of α, the data and the final velocity u_f used and computed, respectively, in Gonzalez-Narvaez *et al* [25] are take into account in such a way that the result is

$$\alpha > 10^5,$$

that is, a small variation in the final velocity u_f, produces a great change in the final temperature T_{1f}, and then a big mistake in final temperature T_{1f} could be made if the final velocity u_f is first computed.
Accordingly, it is more convenient to compute the final temperature T_{1f} first and then to determinate the final velocity u_f.
From equation (22) the final velocity u_f is a function of the final temperature, that is,

$$u_f = \frac{bc}{a + NkT_{1f}}, \tag{26}$$

then

$$\gamma_f = \frac{a + NkT_{1f}}{\sqrt{(a + NkT_{1f})^2 - b^2}}. \tag{27}$$

On the other hand, from the last equality of equation (21), the total momentum is a function of the final velocity u_f and the final temperature T_{1f}

$$b = \gamma_f\left(E_{ff} + NkT_{1f}\gamma_f\right)\frac{u_f}{c}. \tag{28}$$

Then, equation (23), is substituted in the last equation

$$\begin{aligned}\frac{bc}{\gamma_f u_f} &= Nmc^2\left[\frac{K_1\left(\frac{mc^2}{\gamma_f kT_{1f}}\right)}{K_2\left(\frac{mc^2}{\gamma_f kT_{1f}}\right)} + \frac{3\gamma_f kT_{1f}}{mc^2}\right] + \gamma_f NkT_{1f} \\ &= Nmc^2\frac{K_1\left(\frac{mc^2}{\gamma_f kT_{1f}}\right)}{K_2\left(\frac{mc^2}{\gamma_f kT_{1f}}\right)} + 4\gamma_f NkT_{1f}.\end{aligned}$$

From equations (26) and (27), $bc/\gamma_f u_f = \sqrt{(a + NkT_{1f})^2 - b^2}$, then

$$
\begin{aligned}
1 &= Nmc^2 \frac{K_1\left(\frac{mc^2}{\gamma_f kT_{1f}}\right)}{K_2\left(\frac{mc^2}{\gamma_f kT_{1f}}\right)} \frac{1}{\sqrt{(a + NkT_{1f})^2 - b^2}} + \frac{4\gamma_f NkT_{1f}}{\sqrt{(a + NkT_{1f})^2 - b^2}}, \\
0 &= Nmc^2 \frac{K_1\left(\frac{mc^2}{\gamma_f kT_{1f}}\right)}{K_2\left(\frac{mc^2}{\gamma_f kT_{1f}}\right)} \frac{1}{\sqrt{(a + NkT_{1f})^2 - b^2}} + \frac{4\gamma_f NkT_{1f}}{\sqrt{(a + NkT_{1f})^2 - b^2}} - 1. \qquad (29)
\end{aligned}
$$

The last equation will be called $ET\,(T_{1f})$. Finally, γ_f is substituted in (29),

$$
\begin{aligned}
ET &= \frac{K_1\left(\frac{mc^2\sqrt{(a+NkT_{1f})^2-b^2}}{kT_{1f}(a+NkT_{1f})}\right)}{K_2\left(\frac{mc^2\sqrt{(a+NkT_{1f})^2-b^2}}{kT_{1f}(a+NkT_{1f})}\right)} \frac{Nmc^2}{\sqrt{(a + NkT_{1f})^2 - b^2}} + \frac{4NkT_{1f}\,(a + NkT_{1f})}{(a + NkT_{1f})^2 - b^2} - 1 \\
&= 0 \qquad (30)
\end{aligned}
$$

So, the final temperature T_{1f} is such that $ET\,(T_{1f}) = 0$.
Now, final temperatures and final velocities are computed more accurately for two cases considered by Gonzalez-Narvaez *et al* [25].

- Jet-clump
  Gonzalez-Narvaez *et al* [25] considered one of the two opposite jets that are regularly launched simultaneously, such that mass deposition rate is in the z direction and shocks a dense-cool clump. The jet [26] is initially launched with a sub-relativistic velocity equal to $u_{jet} = 9.5 \times 10^6 m/s$, the collision jet temperature is $T_{1j} = 4 \times 10^7 K$ and the jet particle number per unit length is $N_j = 1.9440 \times 10^{44}$. Let us consider that the clump is at rest before the collision with the initial clump temperature, $T_{1clump} = 2 \times 10^7 K$, and the clump particle number per unit length is $N_{clump} = 8.5355 \times 10^{43}$.
  Because the jet-clump distance is $20Kpc$, it is assumed the jet loses energy on the way, so that the initial collision jet velocity u, is such that $u < u_{jet}$.
  * Hillel and Soker [26] obtained that the final temperature are $T_{1f} \sim 4.9 \times 10^7 K$ and the final velocity is of the order $10^6 m/s$.
  * Gonzalez-Narvaez *et al* [25] obtained that for an initial collision jet velocity, $u = 8 \times 10^5 m/s$, the final temperature is $T_{1f} = 4.5459 \times 10^7 K$.
  * In this work (computing the final temperature T_f first and then determining the final velocity u_f) for a initial collision jet velocity $u = 1.35 \times 10^6 m/s < u_{jet}$, a drift velocity of the same order and a final temperature, $T_{1f} = 4.975 \times 10^7 K$, are predicted.

  The result exposed by Gonzalez-Narvaez *et al* [25] was not so accurate, namely, the final temperature reported by them is smaller by $5.991 \times 10^6 K$, with an initial jet velocity which is smaller too, because in Gonzalez-Narvaez *et al* [25] the final drift velocity was the first quantity computed.
- Tokamak
  Gonzalez-Narvaez *et al* [25] proposed too a process for increasing the ignition Tokamak temperature. An ideal gas confined in a Tokamak at room temperature is considered, and suddenly and ideal gas at room temperature too is injected with a relative velocity to attain the ignition temperature $T_{1f} \sim 10^8 K$, considering ITER tokamak conditions, that is, if equilibrium temperature is $T_{1eq} = 300K$ and particle numbers are $N_1 = 3 \times 10^{23}$ and $N_2 = 4 \times 10^{23}$.

* Gonzalez-Narvaez *et al* [25] considered an initial velocity condition $u = 3 \times 10^6 m/s$ to get the ignition temperature order, that is, $T_{1f} = 1.9530 \times 10^8 K$.
* In this work (computing the final temperature T_f first and then to determinate the final velocity u_f) for an initial velocity $u = 4 \times 10^6 m/s$, the final temperature is $T_{1f} = 1.608416 \times 10^8 K$. Then, with initial velocities $u > 4 \times 10^6 m/s$, it is possible to get the desired order of temperatures, $T_{ITER} = 25KeV = 3 \times 10^8 K$.

As it can be seen, Gonzalez-Narvaez *et al* [25] results were not so accurate because they computed final drift velocity first.

4. Type II Spicules and CMEs

In this section type II spicules and CMEs are analized as a possible energy sources to heat and maintain the current Sun's corona temperature, considering the mixture of two ideal gases exposed in section 2. That is, because the conservation of the 4−vector energy-momentum, the mixture of ideal gases on Sun's regions, as Chromosphere (type II spicules and CMEs), Transition Region [TR] and solar corona, the solar corona is heated and maintained to temperatures of order $> 10^6 K$.

4.1. Mixture type II spicules ideal gas and Sun's corona ideal gas

Type II spicules are chromospheric plasma jets with velocities of the order $u \sim 10^5 m/s$ and they occur thousands of times a day. Martínez-Sykora *et al* [12, 13] showed that the interactions between magnetic field and solar atmosphere partially ionized gas, with charged and neutral particles, generate spicules. Moreover, they think as a possibility, that the study of the spicules could help to explain the temperatures in solar corona (see figure 1 in Martínez-Sykora *et al* [12]). Then, when these plasma jets are ejected, they have to go through two layers which are the TR and solar corona. Therefore, the model proposed in this work, to understand the temperature increment and the current temperature of the solar corona, two subprocess must be considered, namely, the mixture of a type II spicule ideal gas with the TR ideal gas and the mixture of the resulting ideal gas of the first subprocess and the solar corona ideal gas (quantities of the second subprocess; that is, in the solar corona, are primed). Moreover, there are considered columns of ideal gas such that the bases have a radius of $2 \times 10^5 m$, and the heights depend on the size of the shock object and the size of the layer.
The average height of a type II spicule is $h_{IIS} = 10^6 m$ (which is considered equal to the corona column height $h_{IIS} = h_{SC}$), the height of the TR is $h_{TR} = 3 \times 10^5 m$, the particle density of Chromosphere and TR are considered equal, that is, $n_{TR} = n_{IIS} = 10^{15} m^{-3}$, and the particle density of solar corona is $n_{SC} = 5.5 \times 10^{14} m^{-3}$. It is known, from the observations that the plasma ejected velocities going into the solar corona are in ranges $5 \times 10^4 m/s - 1.5 \times 10^5 m/s$. Accordingly, the initial velocities of spicules are proposed in order to consider the observed velocity ranges as the final velocities of this first subprocess.

- Subprocess 1. Mixture of type II spicule ideal gas with the TR ideal gas.
  The TR ideal gas with $N_{TR} = N_1$ particles and initial temperature $T_{1TR} = T_{11}$, is at rest in the TR reference frame(or lab frame), $K_{TR} = K_1$, while the spicule ideal gas with $N_{IIS} = N_2$ particles and initial temperature $T_{1IIS} = T_{12}$, is at rest in the spicule reference frame $K_{IIS} = K_2$ and K_2 moving with initial relative velocity u respect to reference frame K_1.
  When spicule ideal gas shocks and come into contact with the TR ideal gas (assuming that this happen in a negligible time period), there is only a energy flow (heat), so the system spicule-TR reaches an intermediate thermal equilibrium, with an equilibrium temperature, T_{1eq}, equation (18). Then, ideal gases get mixed, and the final ideal gas has a final tem-

IOP Conf. Series: Journal of Physics: Conf. Series **1239** (2019) 012002 doi:10.1088/1742-6596/1239/1/012002

perature T_{1f}, equation (22), and final velocity u_f, with respect to the lab frame, K_1. The results are shown in Table 1.

$u(\times 10^4 m/s)$	Process	$u_f(\times 10^4 m/s)$	$T_{1f}(\times 10^4 K)$
6.5	NR	5	7.0783
	R	5	7.0783
9.75	NR	7.5	10.9263
	R	7.5	10.9263
13	NR	10	16.3134
	R	10	16.3134
16.25	NR	12.5	23.2398
	R	12.5	23.2398
19.5	NR	15	31.7053
	R	15	31.7053

Table 1. Subprocess 1. Mixture of type II spicule ideal gas with the TR ideal gas, where $T_{1TR} = 4 \times 10^4 K$, $T_{1IISC} = 4 \times 10^4 K$ $T_{eq} = 4 \times 10^4 K$, $N_1 = 9.424777961 \times 10^{30}$ and $N_2 = 3.141592641 \times 10^{31}$. There is no difference between nonrelativistic (NR) and the relativistic (R) computation because relative velocities are sub-relativistic.

The mixing of this subprocess, is called total spicule for convenience.

- Subprocess 2. Mixture total spicule ideal gas and the solar corona ideal gas.
The solar corona ideal gas with $N_{SC} = N_1'$ particles and initial temperature $T_{1SC} = T_{11}'$, is at rest in the solar corona reference frame (or lab frame too), $K_{SC} = K_1' = K_1$, while the total spicule ideal gas with $N_{IITS} = N_2'$ particles and initial temperature $T_{1IITS} = T_{12}'$, is at rest in the total spicule reference frame $K_{IITS} = K_2'$ and K_2' moving with initial relative velocity, $u' = u_f$, respect to reference frame K_1.
When total spicule ideal gas shocks and come into contact with the solar corona ideal gas, the system total spicule-solar corona reaches an intermediate thermal equilibrium, with an equilibrium temperature, T_{1eq}', equation (18). Then, ideal gases get mixed, and the final ideal gas has a final temperature T_{1f}', equation (22), and final velocity u_f', respect to the lab frame, K_1. The final temperature, T_{1f}', is the the quantity that has to have the same order of temperature of the current solar corona.
In Table 2, solar corona temperature $T_{1SC} = 2 \times 10^4 K$ is considered; that is: it is assumed that in the beginning of time the solar corona was cool, and it got to higher temperatures with time.

Now, because the shock of the total spicule and the solar corona results in an increment in the solar corona temperature, it is assumed the fastest total spicule shocks the solar corona once again in order to look for the temperature reached. This process is repeated several times with the purpose of knowing the maximum temperature reached by the solar corona. The results are shown in Table 3.

Then, the maximum temperature reached with consecutive total spicule shocks is $\sim 0.59 \times 10^6 K$, which is more than a half of the current solar corona temperature.

On the other hand, the current temperature contribution of spicules to solar corona is also analyzed. If the solar corona temperature is $T_{1SC} = T_{11}' = 1 \times 10^6 K$, final temperatures

$u(\times 10^4 m/s)$	$T'_{eq}(\times 10^4 K)$	Process	$u'_f(\times 10^4 m/s)$	$T'_{1f}(\times 10^4 K)$
5	5.5685	NR	3.5135	7.7122
		R	3.5135	7.7122
75	8.2725	NR	5.2702	13.0958
		R	5.2702	13.0958
10	12.0581	NR	7.0270	20.6328
		R	7.0270	20.6328
12.5	16.9252	NR	8.7837	30.3233
		R	8.7837	30.3233
15	22.8740	NR	10.5405	42.1672
		R	10.5405	42.1672

Table 2. Subprocess 2. Mixture total spicule ideal gas and the solar corona ideal gas where $T_{1SC} = T'_{11} = 2 \times 10^4 K$, $T_{1f} = T'_{12} = T_{1IITS}$, $N'_1 = 1.727875859 \times 10^{31}$ and $N'_2 = 4.084070437 \times 10^{31}$. There is no difference between nonrelativistic (NR) and the relativistic (R) computation because relative velocities are sub-relativistic.

n−shock	$T'_{eq}(\times 10^5 K)$	$u'_f(\times 10^5 m/s)$	$T'_{1f}(\times 10^5 K)$
2	3.48156	1.05405	5.41088
3	3.83658	1.05405	5.76590
4	3.94213	1.05405	5.87145
5	3.97351	1.05405	5.90283

Table 3. Subprocess 2. Mixture total spicule ideal gas and the solar corona ideal gas where T'_{11} is the previous shock considered temperature , $T_{1f} = T'_{12} = T_{1IITS}$, $N'_1 = 1.727875859 \times 10^{31}$, $N'_2 = 4.084070437 \times 10^{31}$ and $u' = 150000.0035625011 m/s$. In this table are showed five consecutive shocks.

are shown in Table 4. Final temperatures are lower than $1 \times 10^6 K$.

$u(\times 10^4 m/s)$	$T'_{eq}(\times 10^5 K)$	Process	$u'_f(\times 10^4 m/s)$	$T'_{1f}(\times 10^5 K)$
5	3.47037	NR	3.5135	3.68474
		R	3.5135	3.68474
7.5	3.74076	NR	5.2702	4.22310
		R	5.2702	4.22310
10	4.11932	NR	7.0270	4.97680
		R	7.0270	4.97680
12.5	4.60604	NR	8.7837	5.94584
		R	8.7837	5.94584
15	5.20091	NR	10.5405	7.13023
		R	10.5405	7.13023

Table 4. Subprocess 2. Mixture total spicule ideal gas and the solar corona ideal gas where $T_{1SC} = T'_{11} = 1 \times 10^6 K$, $T_{1f} = T'_{12} = T_{1IITS}$, $N'_1 = 1.727875859 \times 10^{31}$ and $N'_2 = 4.084070437 \times 10^{31}$. There is no difference between nonrelativistic (NR) and the relativistic (R) computation because relative velocities are sub-relativistic.

Moreover, if solar corona temperature is $T_{1SC} = T'_{11} = 2 \times 10^6 K$, final temperatures are shown in Table 5. Final temperatures are lower than $2 \times 10^6 K$. Therefore, the actual spicule contribution to solar corona temperature consists of coolling a little the solar corona.

$u(\times 10^4 m/s)$	$T'_{eq}(\times 10^5 K)$	Process	$u'_f(\times 10^4 m/s)$	$T'_{1f}(\times 10^5 K)$
5	6.44334	NR	3.5135	6.65771
		R	3.5135	6.65771
7.5	6.71374	NR	5.2702	7.19607
		R	5.2702	7.19607
10	7.09229	NR	7.0270	7.94977
		R	7.0270	7.94977
12.5	7.57901	NR	8.7837	8.91882
		R	8.7837	8.91882
15	8.17388	NR	10.5405	10.10321
		R	10.5405	10.10321

Table 5. Subprocess 2. Mixture total spicule ideal gas and the solar corona ideal gas where $T_{1SC} = T'_{11} = 2 \times 10^6 K$, $T_{1f} = T'_{12} = T_{1IITS}$, $N'_1 = 1.727875859 \times 10^{31}$ and $N'_2 = 4.084070437 \times 10^{31}$.There is no difference between nonrelativistic (NR) and the relativistic (R) computation because relative velocities are sub-relativistic.

There is no difference between nonrelativistic and the relativistic computation because relative velocities are sub-relativistic.
But, if an hypothetical relativistic type II spicule is considered, namely, its initial velocity is $u = 2.6 \times 10^8 m/s$, the discrepancy between no-relativistic and relativistic results is awesome as shown in Table 6 and Table 7. As it can be seen, there is a big difference between no-relativistic and relativistic results.

$u(\times 10^8 m/s)$	Process	$u_f(\times 10^8 m/s)$	$T_{1f}(\times 10^{11} K)$
2.6	NR	2	4.9253
	R	2.1770	7.6252

Table 6. Subprocess 1. Mixture of type II spicule ideal gas with the TR ideal gas, where $T_{1TR} = 4 \times 10^4 K$, $T_{1IISC} = 4 \times 10^4 K$ $T_{eq} = 4 \times 10^4 K$, $N_1 = 9.424777961 \times 10^{30}$ and $N_2 = 3.141592641 \times 10^{31}$.

$u(\times 10^8 m/s)$	$T'_{eq}(\times 10^{11} K)$	Process	$u'_f(\times 10^8 m/s)$	$T'_{1f}(\times 10^{11} K)$
2.1770	5.3582	NR	1.5298	9.4222
		R	1.6457	11.5607

Table 7. Subprocess 2. Mixture total spicule ideal gas and the solar corona ideal gas where $T_{1SC} = T'_{11} = 2 \times 10^4 K$, $T_{1f} = T'_{12} = T_{1IITS}$, $N'_1 = 1.727875859 \times 10^{31}$ and $N'_2 = 4.084070437 \times 10^{31}$.

However, the idea of a total spicule material with relativistic velocities is not so out of place. Klimchuk [10] observed that while spicule go up and go through the solar corona, a spicule fraction ($\sim 10\%$) is heated to temperatures $1 - 2MK$. Moreover, the rest and cooler spicule go back to the Chromosphere.
On the other hand there exists magnetic fields, therefore taking into account these two facts, it is achievable to think of in the existence of an image apparatus in the upper spicule.

4.1.1. Image Apparatus. In astrophysical and thermonuclear applications (terrestrial and stellar magnetic fields) it is of considerable interest to know how particles behave in magnetic

fields that vary in space. Often the variation are such that a perturbation solution for the motion, that is, the distance over wich $\vec{B}$ changes appreciably in magnitud or direction is large compared to the gyration radius of the particle.
Let us consider a simple situation in which a static magnetic field $\vec{B}$ acts mainly in the z direction, but has a small positive gradient in that direction. In addition to the z component of field there is a small radial component due to the curvature of the lines of force [28]. This happens in the image apparatus.
The image apparatus configuration is as follows: the central part magnetic field $\vec{B_1}$ is weaker than extreme part magnetic field $\vec{B_2}$. A charged particle in a magnetic field satisfactorily follows

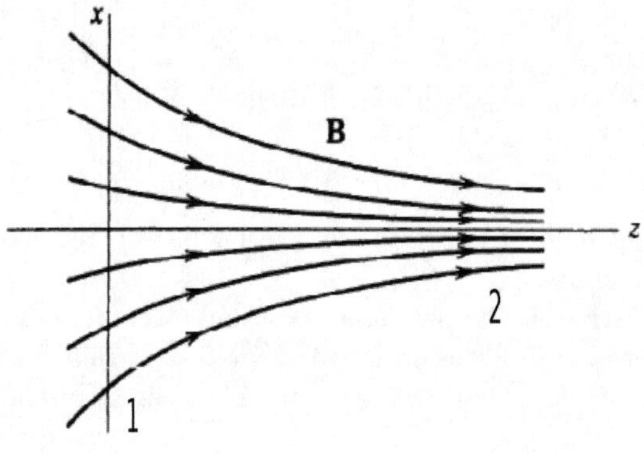

Figure 1. Lines represent lines magnetic field.

the magnetic field lines if

$$r_L \frac{|\nabla B|}{B} \ll 1, \tag{31}$$

where r_L is the Larmor radius. This condition means the particle spin radii are much smaller than the magnetic field spin radios ($|\nabla B|/B = 1/R$).
From the charged particle movement equation, an adiabatic invariant is obtained

$$\frac{V_\perp^2}{B} = cte \quad \Rightarrow \quad \mu = \frac{\gamma m V_\perp^2}{2B}. \tag{32}$$

Considering this adiabatic invariant, as in the image apparatus, one obtains

$$\frac{V_\perp^2}{B} = cte \quad \Rightarrow \quad \frac{V_{\perp 2}^2}{V_{\perp 1}^2} = \frac{B_2}{B_1} \tag{33}$$

and

$$V_\perp^2 + V_z^2 = cte. \tag{34}$$

From equations (33) and (34) is obtained that all the particles which satisfied the condition

$$V_{\perp 1}^2 \left(\frac{B_2}{B_1} - 1 \right) \leq V_z^2 \tag{35}$$

will escape from region 2 (see Figure 1) as it happens in the image apparatus.
Then, if there exists an image apparatus effect in the upper spicule, it plays a selecting mechanism role for the spicule particles which satisfy equation (35); that is: all these particles with such

velocities, may be the $\sim 10\%$ that remains in the solar corona. Moreover, this particular particles are much more energetic, so their mean kinetic energy is bigger and the temperature of this set is greater than the remaining spicule. Furthermore, because these particles impact with the solar corona, there will be an amazing solar corona temperature increment.
Accordingly, if these particles have a bigger mean kinetic energy, they are faster and it is not irrational to think that particles have relativistic velocities and hence the Redefined Relativistic Thermodynamics has to be considered for the system analysis.

4.2. Mixture CME ideal gas and Sun's corona ideal gas
CMEs are coronal mass ejections, that is, an ejection of magnetized chromospheric plasma that may or may not propagate into the Heliosphere. Therefore, it is valid to considerer the mechanism exposed in section 2 as a possible cause of the temperature order in the solar corona.
The model proposed to understand the temperature increment and the current temperature of solar corona, unlike how it was done with spicules, is just one process, because the TR height is negligible respect to the CMEs height [16].
There are considered columns of ideal gas such that the bases have a radius of $2 \times 10^5 m$, and the heights depend on the size of the shock object.
The range height of CMEs is $h_{CME} > R_s$ (which is considered equal to the corona column height $h_{CME} = h_{SC}$, and $R_s = 6.957 \times 10^8 m$), the particle density of the Chromosphere is $n_{TR} = n_{IIS} = 10^{15} m^{-3}$, and the particle density of the solar corona is $n_{SC} = 5.5 \times 10^{14} m^{-3}$. It is known, from the observations that the plasma ejected velocities going into the solar corona are in ranges $> 4 \times 10^5 m/s$ [16].
The CME ideal gas with $N_{SC} = N_1$ particles and initial temperature $T_{1CS} = T_{11}$, is at rest in the solar corona reference frame(or lab frame), $K_{SC} = K_1$, while CME ideal gas with $N_{CME} = N_2$ particles and initial temperature $T_{1CME} = T_{12}$, is at rest in CME reference frame $K_{CME} = K_2$ and K_2 moving with initial relative velocity u respect to reference frame K_1.
When the CME ideal gas shocks and comes into contact with the solar corona ideal gas (assuming that this happen in a negligible time period), there is only an energy flow (heat), so the system CME-solar corona reaches an intermediate thermal equilibrium, with an equilibrium temperature, T_{1eq}, equation (18). Then, ideal gases get mixed, and equation (22), and final velocity u_f, respect to the lab frame, K_1.
When the solar corona is cool, $T_{SC} = 2 \times 10^4 K$ (that is, it is assumed that in the beginning of the time solar corona was at the same Chromosphere temperature) solar corona final temperatures resulting from the shock of a CME and solar corona are shown in Table 8.

On the other hand, when solar corona is at the current temperature, that is, $T_{SC} = 1 \times 10^6 K$, solar corona final temperatures resulting from the shock of a CME and solar corona are shown in Table 9.

Finally, solar corona final temperatures, even when the solar corona is cool, are much bigger than $1 \times 10^6 K - 2 \times 10^6 K$, because a CME shock. In fact, there are final temperatures of order $\sim 10^7 K$.

As an example, if an hypothetical relativistic CME is considered, that is, e.g., for initial velocity $u = 2.6 \times 10^8 m/s$, the discrepancy between nonrelativistic and the relativistic results is awesome as shown below

5. Energy balance
Considering the results presented in the last sections, there has to be an energy balance between spicules, CMEs, and the radiation due to the temperature to get the current solar corona tem-

$h_{CME}(R_s)$	$u(\times 10^5 m/s)$	$N_{SC}(\times 10^{34})$	$N_{CME}(\times 10^{34})$	$T_{eq}(K)(\times 10^4)$	Process	$u_f(\times 10^5 m/s)$	$T_{1f}(\times 10^6 K)$
3.33	8.62	4.002937406	7.27806801	3.2903	NR	5.56129	7.014803
					R	5.56129	7.014825
4.21	11.97	5.060770714	9.201401298	3.2903	NR	7.72258	13.496075
					R	7.72259	13.496158
5.98	21.32	7.188458164	13.06992393	3.2903	NR	13.75483	42.743288
					R	13.75490	42.744128
6.06	7.61	7.284624828	13.24477242	3.2903	NR	4.90967	5.474525
					R	4.90968	5.474593
11.7	14	14.06437467	25.57159031	3.2903	NR	9.03225	18.449745
					R	9.032277	18.449901
14.89	20.67	17.89902041	32.54367348	3.2903	NR	13.33548	40.178696
					R	13.33554	40.179438

Table 8. Mixture CME ideal gas and the solar corona ideal gas where $T_{SC} = T_{11} = 2 \times 10^4 K$ and $T_{CME} = T_{12} = 4 \times 10^4 K$. There is no difference between nonrelativistic (NR) and the relativistic (R) computation because relative velocities are sub-relativistic.

$h_{CME}(R_s)$	$u(\times 10^5 m/s)$	$N_{SC}(\times 10^{34})$	$N_{CME}(\times 10^{34})$	$T_{eq}(K)(\times 10^5)$	Process	$u_f(\times 10^5 m/s)$	$T_{1f}(\times 10^6 K)$
3.33	8.62	4.002937406	7.27806801	3.8064	NR	5.5612	7.3625
					R	5.5612	7.3625
4.21	11.97	5.060770714	9.201401298	3.8064	NR	7.7225	13.8438
					R	7.7225	13.8439
5.98	21.32	7.188458164	13.06992393	3.8064	NR	13.7548	43.0910
					R	13.7549	43.0918
6.06	7.61	7.284624828	13.24477242	3.8064	NR	4.9096	5.8222
					R	4.9096	5.8222
11.7	14	14.06437467	25.57159031	3.8064	NR	9.0322	18.7974
					R	9.0322	18.7976
14.89	20.67	17.89902041	32.54367348	3.8064	NR	13.3354	40.0526
					R	13.3355	40.0527

Table 9. Mixture CME ideal gas and the solar corona ideal gas where $T_{SC} = T_{11} = 1 \times 10^6 K$and $T_{CME} = T_{12} = 4 \times 10^4 K$. There is no difference between nonrelativistic (NR) and the relativistic (R) computation because relative velocities are sub-relativistic.

$h_{CME}(R_s)$	$u(\times 10^8 m/s)$	$N_{SC}(\times 10^{34})$	$N_{CME}(\times 10^{34})$	$T_{eq}(K)(\times 10^4)$	Process	$u_f(\times 10^8 m/s)$	$T_{1f}(\times 10^{11} K)$
14.89	2.6	17.89902041	32.54367348	3.2903	NR	1.6774	6.3519
					R	1.9293	10.4938

Table 10. Mixture CME ideal gas and the solar corona ideal gas where $T_{SC} = T_{11} = 2 \times 10^4 K$ and $T_{CME} = T_{12} = 4 \times 10^4 K$.

perature.

- Spicules.
  De Pontieu *et. al.* [7] exploited the detection of the disk counterpart of type II spicules: rapid blueshifted events (RBEs). Using an automated detection code, they found 2434 RBEs ocurring in an active-region plage footpoints of coronal loops during 1-hour-long time series on 25 April 2010. The chromospheric mass flux that spicules propel to coronal heights is estimated to be two orders of magnitude larger than the mass flux of the solar wind [29].
  Based in the ubiquity of these events and the observed coronal intensities, De Pontieu *et.*

al. [7] estimated that these events carry a mass flux density of $1.5 \times 10^{-9} g/cm^2/s$ and an energy flux density of $\sim 2 \times 10^6 erg/cm^2/s$ into the corona.
This is of the order that is required to sustain the energy lost from the active region corona [30].

- CMEs.
  The Solar and Heliospheric Observatory (SOHO) mission's white light coronographs have observed a total of 6907 CMEs, from January 1996 to December 2002, that is ~ 3 CMEs per day, [15].

- Radiation due to the temperature.
  The solar corona lose energy by the photon emission and the heat conduction to the active region of the base. The cooling rate varies from the $10^5 erg/cm^2/s$ in quiet regions to $10^7 erg/cm^2/s$ in active regions [5].
  Moreover, solar corona would get cool by radiation due to the temperature in just 10min.

Given the conservative nature of the De Pontieu *et. al.* [7] estimation, type II spicules are likely to play a substantial role in the coronal energy balance, because the contrast of the energy flux density of spicules is the same order of the cooling rates by the radiation due to the temperature. On the other hand, CMEs supply energy to the solar corona if some of the ejected mass is kept in the solar corona. If some CMES are so energetic that the total mass is ejected into the space, they will not contribute to the balance of energy because, in the limiting case, what enters the solar corona will come out without interfering. At best, some CMEs particles will collide with particles of the solar corona and the temperature of the latter will tend to grow.

6. Conclusions

The mixing of ideal gases with relative velocities results in an amazing increment in the mixture final temperature.
Type II Spicules can locally heat solar corona to the maximum temperature $\sim 0.59 \times 10^6 K$, which is more than a half of the current solar corona temperature, but it is not enough, while CMEs cause final temperatures of order $\sim 10^7 K$.
Furthermore, when the solar corona is at the current temperature $1 \times 10^6 K - 2 \times 10^6 K$, type II spicules cool the solar corona temperature whereas CMEs are still heating it and ejecting mass to the space.
The use of The Redefined Relativistic Thermodynamics is relevant when relativistic velocities appear, as in the ($\sim 10\%$) selected by the possible image apparatus in the upper spicule part.

References

[1] Pandey K, Narain U 2008 *On Solar Corona Heating Mechanisms*, (India: Astrophysics Research Group, Meerut College)
[2] Edlen B 1942 *Zeitschrift f. Astrophys* **22** 30.
[3] Alexander C E, 2012 *Phenomena in the Solar Corona: characteristic structure and dynamic evolution*, (England: Jeremiah Horrocks Institute for Astrophysics and Supercomputing, University of Central Lancashire)
[4] De Pontieu B, Hansteen V H, Carlsson M, Schruver C J, Tarbell T D, Title A M, Shine R A, Suematsu Y, Tsuneta S, Katsukawa Y, Ichimoto K, Shimizu T, Nagata S 2007 *Publ. Astron. Soc. Jpn.* **59** S655S662.
[5] Ashwanden M J 2012 *Physis of the Solar Corona: An Introduction* (Chichester, UK: Springer).
[6] Jewett J W Jr 2008 *Phys. Teach.* **46** 269.
[7] De Pontieu B, McIntosh S W, Carlsson M, Hansteen V H, Tarbell T D, Boerner P, Martínez-Sykora J, Schrijver C J, Title A M 2011 *Science* **331** 6013 55-58.
[8] Grotrian W 1939 *Die Naturwissenschaften* **27** 13 214.
[9] De Pontieu B, Hansteen V H, Rouppe Van der Voort L, Van Noort M, Carlsson M 2007 *Apj.* **655** 624-641.

[10] Klimchuk J A 2012 *JGR* **117** A12102.
[11] Athay R G 1984 *Apj.* **287** 412-417.
[12] Martínez-Sykora J,De Pontieu B, Hansteen V H, Rouppe van der Voort L, Carlsson M, Pereira T M D 2017 *Science* **356** 1269-1272.
[13] Martínez-Sykora J, De Pontieu B, De Moortel I, Hansteen V H, Carlsson M 2018 Impact of type II spicules in the corona: simulations and synthetic observables *Preprint* arXiv: 1805.06475v1
[14] Green L M, Török T, Vršnak B, Manchester W, Veronig A 2018 *Space Sci. Rev.* **214** 46 1-52.
[15] Yashiro S, Gopalswamy N, Michalek G, Cyr O C St, Plunkett S P, Rich N B, Howard R A 2004 *JGR* **109** A07105.
[16] Kim R S, Gopalswamy N, Moon Y J, Cho K S, Yashiro S 2012 *Ap. J.* **746** 118 1-8.
[17] Ares de Parga G, Lopez-Carrera B, Angulo-Brown F 2005 *J. Phys. A* **38** 2821–2834.
[18] Segre E, Fermi E 1962 *Note e Memorie-Collected Papers* vol. 1 (Chicago Il. USA:University of Chicago Press).
[19] Gamba A 1967 *Amer. J. Phys.* **35** 83–89.
[20] Nakamura T K 2006 *Space Sci. Rev.* **122** 271–278.
[21] Nakamura T K 2012 *Progr. Theoret. Phys* **128** 463–475.
[22] López-Carrera B, Granados V, Ares de Parga G 2008 *Rev. Mex. Phys.* **54** 15–19.
[23] López-Carrera B, Rosales M A, Ares de Parga G 2010 *Chin. Phys.* B **19** 4.
[24] Ares de Parga G, Lopez-Carrera B 2009 *Phys.* A **388** 4345–4356.
[25] Gonzalez-Narvaez R E, Ares de Parga A M, Ares de Parga G 2017 *Annals of Physics* **376** 391-411.
[26] Hillel S, Soker N 2014 *Mon. Not. R. Astron. Soc.* **445** 34161-4174.
[27] Greiner L, Neiser L, Stocker L 1995 *Thermodynamics and Statistical Mechanics* (New York:Springer-Verlag).
[28] Jackson J D 1999 *Classsical electrodynamics* (USA:John Wiley and Sons, Inc.).
[29] Withbroe G L 1983 *Astrophys. J.* **267** 825.
[30] Priest E R 1982 *Solar Magnetohydrodynamics* (Dordrecht, Netherlands: D Reidel Publishing).

IOP Conf. Series: Journal of Physics: Conf. Series **1239** (2019) 012003 doi:10.1088/1742-6596/1239/1/012003

Bulk viscosity in 2D relativistic fluids: the effects of temperature and modifications to the Rayleigh-Brillouin spectrum

A L García- Perciante, L Franco-Pérez and A R Méndez
Universidad Autónoma Metropolitana - Cuajimalpa, Departamento de Matemáticas Aplicadas y Sistemas, 05348, Mexico City, México.
E-mail: algarcia@correo.cua.uam.mx, lfranco@correo.cua.uam.mc, amendez@correo.cua.uam.mx

Abstract. The dependence of the bulk viscosity with the relativistic parameter $z = kT/mc^2$, obtained through the complete Boltzmann equation [1, 2], is thoroughly analyzed. A complete and rigorous examination of the relevant non-relativistic and ultra-relativistic limits is carried out in the case of a hard disk model and compared with the results obtained in the relaxation time approximation. The modifications of the non-vanishing bulk viscosity to the Rayleigh-Brillouin is briefly discussed.

1. Introduction

The study of relativistic bidimensional fluids is a relevant and interesting topic which is still in need of extensive analysis. Additional to its importance in the modelling of axisymmetric systems, for example a charged fluid in the presence of a magnetic field, orbiting gases, or accretion in gravitational potentials, the interest in the study of bidimensional systems in relativistic scenarios has seen a substantial increase due to the new generation of thin materials and their several applications. Also, relativistic fluids still pose a challenge, plagued with unaswered questions and topics of intense debate even in very fundamental issues, numerical simulations in low dimensionality have been one of the most valuable assets in order to corroborate theoretical predictions.

The establishment of relativistic transport coefficients in a special relativistic framework from the complete Boltzmann equation has only been addressed recently in the 2D scenario. Indeed, in a separate publication (see Ref. [3]) the constitutive equations are established and the transport coefficients expressed in terms of collision integrals which, for a hard disks model, can be numerically evaluated. In particular, the bulk viscosity is expressed in terms of the integral

$$\mathcal{I}(z) = \int_{\frac{2}{z}}^{\infty} e^{-\left(x-\frac{2}{z}\right)} \left(\frac{1}{x} + \frac{3}{x^2} + \frac{3}{x^3}\right) \left(z^2x^2 - 4\right)^{5/2} dx \qquad (1)$$

whose dependence on z is not trivial. Moreover, its convergence in the non-relativistic ($z \to 0$) and ultrarelativistic ($z \to \infty$) limits is not completely justified. This is precisely part of the goal of the present work, the formal establishment of the behaviour of Eq. (1) in the non-relativistic and ultrarelativistic limits. Also, in order to address the effect of a non-zero bulk viscosity of the gas for the complete range of z we study the corresponding modifications on the Rayleigh-Brillouin spectrum.

In order to accomplish such a task, the rest of the work is organized as follows. In sections 2 and 3 we briefly outline the procedure carried out in Ref. [3] in order to establish the analytical expression for the bulk viscosity in the specific case of a hard disks model. Section 4 is devoted to the analysis of the integral defined in Eq. (1) in the non relativistic and ultrarelativistic limits, while section 5 addresses the modification that a non-vanishing bulk viscosity has on the Brillouin peaks for a light scattering spectrum. The discussion of the results and final remarks are included in section 6.

2. Boltzmann equation and Chapman-Enskog approximation

In this section, we provide a brief description of the procedure leading to the function $\mathcal{I}(z)$ in Eq. (1), whose relevant limits are explored in the main part of this work. The complete calculation will be published elsewhere (a preprint can be found in Ref. [3]) together with the calculation of the rest of the relevant coefficients, to which the reader is refered to for further details.

The starting point is the relativistic Boltzmann equation [2] in a flat spacetime with a $(+--)$ metric which reads

$$v^{\alpha}\frac{\partial}{\partial x^{\alpha}}f\left(x^{\nu},v^{\nu}\right)=\int\int\left(\tilde{f}\tilde{f}_1-ff_1\right)F\sigma d\chi dv_1^*,\tag{2}$$

where $f\left(x^{\nu},v^{\nu}\right)$ is the one particle distribution function, F is the invariant flux, $\gamma=u^{\mu}v_{\mu}/c^2$, with v^{μ} being the molecular 3-velocity measured in an arbitrary frame, u^{μ} is the fluid's 3-velocity and σ and χ are the corresponding scattering cross section and solid angle in 2D. The local equilibrium solution to Eq. (2) is given by a bidimensional Maxwell-Jüttner distribution, that is

$$f^{(0)}\left(v^{\nu}\right)=\frac{ne^{\frac{1}{z}}}{2\pi c^2 z\left(1+z\right)}e^{-\frac{u^{\alpha}v_{\alpha}}{zc^2}}\tag{3}$$

where u^{α} corresponds to the fluid's 3-velocity. With such distribution one can establish statistical definition for the state variables: n (number density), u^{ν} (hydrodynamic 3-velocity) and ε (internal energy). Considering the particles frame (Eckart's frame), one has

$$n=\int f^{(0)}\left(v^{\nu}\right)\gamma dv^*,\tag{4}$$

$$nu^{\nu}=\int f^{(0)}\left(v^{\nu}\right)v^{\nu}dv^*,\tag{5}$$

$$n\varepsilon=mc^2\int f^{(0)}\left(v^{\nu}\right)\gamma^2 dv^*,\tag{6}$$

from which one can obtain

$$n\varepsilon=nmc^2 g\left(z\right)\qquad\text{with}\qquad g\left(z\right)=\frac{2z^2+2z+1}{z\left(z+1\right)}.\tag{7}$$

These quantities, in the absence of dissipation, follow Euler's equations. In order to establish the first order in the gradients distribution, the Chapman-Enkog solution to Eq. (2) reads

$$f\left(v^{\nu}\right)=f^{(0)}\left(v^{\nu}\right)\left(1+\phi\left(v^{\nu}\right)\right),\tag{8}$$

where the first order correction to the local equilibrium distribution function $\phi\left(v^{\nu}\right)$ is given by the solution of the linearized Boltzmann equation.

$$f^{(0)} \int\int f^{(0)}(v_1)\left(\tilde{\phi}_1 + \tilde{\phi} - \phi_1 - \phi\right) F\sigma d\chi dv_1^* = v^\alpha \frac{\partial f^{(0)}(v^\nu)}{\partial x^\alpha}. \tag{9}$$

The dissipative fluxes, which arise from such deviation from equilibrium, are found as moments of $f^{(0)}(v)\,\phi(v^\nu)$. In particular, for the energy momentum tensor, which is the focus of the present work, one has

$$\pi^{\mu\nu} = mh^{\mu\alpha}h^{\nu\beta}\int f^{(0)}(v)\,\phi(v^\nu)\,v_\alpha v_\beta d^*v. \tag{10}$$

Here $h^{\alpha\beta} = \eta^{\alpha\beta} - u^\alpha u^\beta/c^2$ is the spatial projector corresponding to the (2+1) representation where u^α corresponds to the temporal direction in the comoving frame.

The details of the calculation that follows can be found in Ref.[3]. However the authors consider it worthwhile to point out here the particular step to which the occurrence of a finite bulk viscosity in the relativistic regime, opposed to the zero value obtained for non-relativistic gases, can be traced down to. Equation (9) is a linear integral equation, whose solution is a superposition of the homogeneous and a particular solutions. It is from the latter that the driving terms for the deviation will appear as the gradients of the state variables. In such a term one finds, for the part that depends on the velocity gradient and will thus lead to viscous dissipation,

$$\left(v^\alpha f^{(0)}_{,\alpha}\right)_{visc.} = f^{(0)}\left\{\left(\frac{v^\beta u_\beta}{c^2}\right)\left(\frac{1}{n}u^\alpha n_{,\alpha} + \frac{1}{T}\left(\frac{\gamma}{z} - g(z)\right)u^\alpha T_{,\alpha}\right) + \frac{v_\mu v^\beta h^\alpha_\beta}{zc^2}u^\mu_{,\alpha}\right\} \tag{11}$$

where we have introduced the usual decomposition $v^\alpha = v^\beta h^\alpha_\beta + \left(v^\beta u_\beta/c^2\right)u^\alpha$, in this case in a (2+1) framework. Notice that the proper time derivatives of the number density and temperature are coupled with the velocity gradient through Euler's equations

$$u^\alpha n_{,\alpha} = -nu^\alpha_{,\alpha} \qquad \text{and} \qquad u^\alpha T_{,\alpha} = -k_p(z)\,Tu^\alpha_{,\alpha} \tag{12}$$

where $k_p(z) = k/C_n = (z+1)/(z(g(z) + 2(z+1)))$. Also, the term $u^\alpha u^\mu_{,\alpha}$ is not included in Eq. (11) since the momentum equation in the local equilibrium case couples it only with the pressure gradient.

Introducing Eqs. (12) in Eq. (11) and splitting the velocity gradient in antisymmetric, symmetric traceless parts and the trace times an identity tensor one can write

$$\left(v^\alpha f^{(0)}_{,\alpha}\right)_{visc.} = \left[\left(\frac{1}{2} - k_p(z)\right)\frac{\gamma_k^2}{z} + \left(k_p(z)\,g(z) - 1\right)\gamma_k - \frac{1}{2z}\right]u^\nu_{,\nu} - \frac{v^\mu v^\alpha}{zc^2}\mathring{\sigma}_{\mu\alpha}, \tag{13}$$

where $\mathring{\sigma}_{\mu\alpha}$ is the symmetric traceless component of $u_{\mu,\alpha}$. Notice that, in the non-relativistic case, the term in brackets vanishes since in such a limit $z \to 0$, $\gamma \to 1$ and thus $g(z) \sim z^{-1} + 1$ and $k_p(z) \sim 1 - 2z$. Due to this fact, the bulk viscosity is zero for the monoatomic ideal gas at low temperatures. However, as will be verified in the next section, is non-zero for $z \neq 0$ and increases rapidly for $0 < z < 1$.

Once the driving term given by Eq. (13) is substituted in Eq. (9), a constitutive equation for the scalar part of the Navier tensor can be found. Such a relation is written as

$$\pi^\mu_\mu = 2\mu u^\alpha_{,\alpha}. \tag{14}$$

Following the standard procedure (see for example Refs. [1, 4]) one can obtain a first approximation for μ. The steps are carefully detailed in Ref. [3], here we only quote the result:

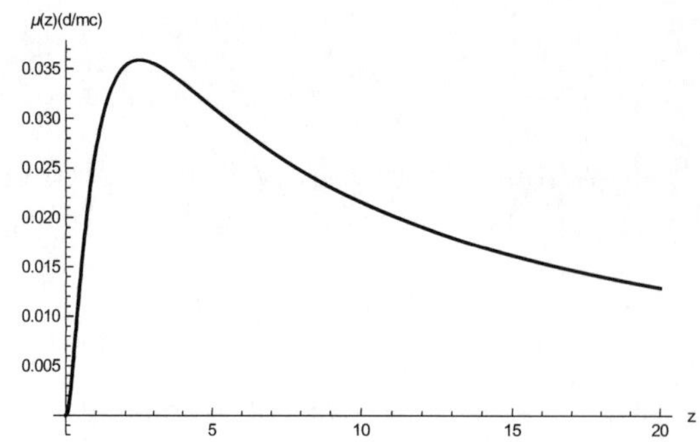

Figure 1. The dimensionless bulk viscosity as a function of the parameter z in the relativistic scenario.

$$\mu = \frac{4z^7}{(1+z)^2} \frac{mc^2}{(2z^2+4z+1)^2} \left[\gamma_k^2, \gamma_k^2\right]^{-1}, \tag{15}$$

where

$$[H, G] = -\frac{1}{n^2} \int \mathcal{C}(H)\, G f^{(0)} d^* v \tag{16}$$

is the collision bracket, which satisfies

$$[H, G] = -\frac{1}{4n^2} \int \left(H' + H_1' - H - H_1\right)\left(G' + G_1' - G - G_1\right) f^{(0)} f_*^{(0)} F \sigma(\chi) d\chi dv_1^* dv^* d^* v. \tag{17}$$

In the next section, the behavior of μ as a function of z for a hard disks model will be described.

3. The temperature dependence of the viscosity for a hard disk gas

The so-called collision integrals defined in Eq. (16) depend on a molecular interaction model for the system. The simplest case in a bidimensional scenario consists on a hard disk model for which the scattering cross section is given by

$$\sigma(\chi) = \frac{d}{2} \left|\sin\left(\frac{\chi}{2}\right)\right|.$$

In such a case, as is shown in Appendix C of Ref. [3], the relevant collision integral is given by

$$\left[\gamma^2, \gamma^2\right] = \frac{2cz^3 d}{15} \frac{1}{(z+z^2)^2} \mathcal{I}(z), \tag{18}$$

where the integral $\mathcal{I}(z)$ is given in Eq. (1). Thus, one can write the bulk viscosity for the system as

$$\mu(z) = \frac{30mc}{d} \frac{z^6}{(2z^2+4z+1)^2} \mathcal{I}(z)^{-1}. \tag{19}$$

In the next section, the non-relativistic limit of this expression will be carefully addressed. However, it can be seen at this point, by inspection of Fig. 1 that μ vanishes at $z = 0$. It is important to notice that for z finite, $\mu \neq 0$ and moreover, it increases very rapidly with temperature.

Bulk viscosity for the monoatomic ideal gas is thus non-zero as long as $T \neq 0$ and becomes relevant for some range of values of z. The fact that it reaches a maximum value, which can be numerically obtained as $\mu_{max} \sim \mu(2.4886) \sim 0.0359$, and then decreases for higher temperatures is also found in the three dimensional case and deserves a closer analysis. One can then conclude that for a limited range of temperature, for each gas, viscosity is enhanced. This could lead to faster damping/enhancing of instabilities.

4. Non-relativistic and ultrarelativistic limits of $\mu(z)$

In this section, a formal proof of the non-relativistic and ultrarelativistic limits of the expression for μ obtained in Ref. [3] and quoted in the previous section is detailed. As mentioned above, inspection of Fig. 1 points towards the bulk viscosity approaching zero in the non-relativistic and ultra-relativistic limits. This behavior is proven separately for each case. In particular, the non-relativistic limit of this and other transport coefficients is far from trivial since the lower integration limit tends to infinity and thus the usual techniques cannot be applied.

We thus begin by addressing the low temperature, non-relativistic, case. Let us start defining the function

$$h(x,z) = e^{-\left(x-\frac{2}{z}\right)}\left(\frac{1}{x}+\frac{3}{x^2}+\frac{3}{x^3}\right)\left(z^2x^2-4\right)^{5/2}, \tag{20}$$

which corresponds to the integrand in $\mathcal{I}(z)$ and an auxiliary function

$$k(x,z) = e^{-\left(x-\frac{2}{z}\right)}\left(\frac{1}{x^2}\right)\left(z^2x^2-4\right)^{3}, \tag{21}$$

both well defined on $A = (x,z) \in \mathbb{R}^2, \mid xz \geq 2, z > 0$.

Proposition 1. *For all $(x,z) \in A$ and $z < 1$, it is verified $0 < k(x,z) \leq h(x,z)$.*

Proof. The equality holds for $x = 2/z$. For $x > 2/z$ we claim

$$e^{-\left(x-\frac{2}{z}\right)}\left(\frac{1}{x^2}\right)\left(z^2x^2-4\right)^{3} < e^{-\left(x-\frac{2}{z}\right)}\left(\frac{1}{x}+\frac{3}{x^2}+\frac{1}{x^3}\right)\left(z^2x^2-4\right)^{5/2}$$

which is equivalent to

$$\left(z^2x^2-4\right)^{1/2} < \left(x+3+\frac{3}{x}\right).$$

If we define $x = 2\alpha/z$ for $\alpha \in (1,\infty)$, last inequality can be written as

$$2z\left(\alpha^2-1\right)^{1/2} < 2\alpha + 3z + \frac{3z^2}{2\alpha}$$

which holds for any $\alpha > 1$ and for $z \in (0,1)$. □

Theorem 1. *$\mu(z) \to 0^+$ as $z \to 0^+$.*

Proof. Since $\mu \geq 0$ for $z > 0$, we can establish that the limit vanishes by simply upper-bounding the function by an auxiliary real-valued analytic function that tends to zero in such a limit.

From Proposition 1 we have

$$0 < \int_{\frac{2}{z}}^{\infty} k(x,z)\,dx \leq \int_{\frac{2}{z}}^{\infty} h(x,z)\,dx, \tag{22}$$

where the integral on the left hand side can be computed as

$$\begin{aligned} \int_{\frac{2}{z}}^{\infty} k(x,z)\,dx &= 8\left(-4z+2z^2-2z^3+3z^4+6z^5+3z^6\right)+64e^{2/z}\int_{2/z}^{\infty}\frac{e^{-t}}{t}dt, \\ &= 96z^5+\mathcal{O}(z^6) \end{aligned} \tag{23}$$

for $z \in (0, \epsilon)$, where $0 < \epsilon < 1$ is such that the last expression becomes true, namely there exists an $M > 0$ and

$$\left| \int_{\frac{2}{z}}^{\infty} k(x,z)\, dx - 96z^5 \right| \leq M \left| z^7 \right|$$

for $0 < z < \epsilon$.

Thus we have

$$0 < \mu(z) < \frac{30mc}{d} \frac{z^6}{(2z^2+4z+1)^2} \left(96z^5 + \mathcal{O}(z^6)\right)^{-1},$$

which formally shows that $\mu \to 0^+$ as $z \to 0^+$ in the non relativistic limit. □

The ultrarelativistic case is established in the following.

Theorem 2. *$\mu(z) \to 0^+$ as $z \to +\infty$.*

Proof. We realize that $\frac{2}{z} \to 0$, therefore

$$\begin{aligned} \lim_{z\to\infty} \mathcal{I}(z) &= \left(\lim_{z\to\infty} z^5\right) \int_0^{\infty} e^{-x} x^2 \left(x^2 + 3x + 3\right) dx \\ &= 48 \lim_{z\to\infty} z^5 . \end{aligned}$$

Thus

$$\lim_{z\to\infty} \mu(z) = \frac{5}{8} \frac{mc}{d} \lim_{z\to\infty} \frac{z}{(2z^2+4z+1)^2} = 0.$$

as is claimed □

Theorem 1 and 2 claim that the bulk viscosity becomes negligible in the limits. This implies that, whatever effect it has on the dynamics of the fluids, it should only be relevant in a finite interval of z. In order to assess the possible effect of bulk viscosity dissipation and provide an example of an experiment that may lead to the measure of such effect, we carry out a linear analysis of a free relativistic gas to first order in statistical density fluctuations.

5. Modification to the Brillouin peaks due to bulk viscosity

The transport equation for the relativistic gas can be readily established by multiplying Eq. (2) by the collisional invariants and integrating in velocity space. The procedure is the standard one, and leads to two conservation equations

$$N^{\nu}_{,\nu} = 0, \qquad T^{\mu\nu}_{,\nu} = 0 \tag{24}$$

where the particle flux is given by

$$N^{\nu} = \int f(v^{\nu})\, v^{\nu} dv^{*} \tag{25}$$

and the energy momentum tensor is

$$T^{\mu\nu} = m \int f(v^{\nu})\, v^{\mu} v^{\nu} dv^{*}. \tag{26}$$

The relation of such moments with the state variables is given, in Eckart's frame and using the (2+1) decomposition by $N^{\nu} = nu^{\nu}$ and $T^{\mu\nu} = n\varepsilon u^{\mu}u^{\nu}/c^2 + ph^{\mu\nu} + \pi^{\mu\nu} + q^{\mu}u^{\nu}/c^2 + u^{\mu}q^{\nu}/c^2$. Such expressions are then introduced in the balance equations (Eq. (24, 25)) and the state variables

IOP Conf. Series: Journal of Physics: Conf. Series **1239** (2019) 012003 doi:10.1088/1742-6596/1239/1/012003

are assumed to be given by an equilibrium value plus a small fluctuation: $X = X_0 + \delta X$. The resulting system of equations, to first order in fluctuations (δn, δu^μ and δT) can be then transformed to Fourier-Laplace space, in which the corresponding dispersion relation is given by

$$s^3 + a_1 q^2 s^2 + \left(a_2 q^2 + a_3\right) q^2 s + a_4 q^4 = 0 \tag{27}$$

where

$$a_1 = -\frac{1}{\rho}\left(A - \frac{\rho}{p} k_p(z) L_T + \frac{1}{c^2}\left(L_n + k_p(z) L_T\right)\right),$$
$$a_2 = -\frac{k_p(z)}{\rho} A L_T,$$
$$a_3 = \frac{p}{\rho}\left(1 + k_p(z)\right),$$
$$a_4 = \frac{k_p(z)}{\rho}\left(L_T - L_n\right),$$

here bulk viscosity enters in Eq. (27) through the relation $A = 4\eta/3 + \mu$ where η is the shear viscosity, and L_T and L_n are the transport coefficients appearing in the relativistic heat flux constitutive equation [3, 7]. Equation (27) has the same structure as the dispersion relation in the non-relativistic case which can be analyzed using Mountain's method [8]. Following such approximation, one can identify a purely decaying mode, corresponding to a real root given by $s_1 = -a_4 q^2/a_3$. The remaining two roots correspond to a conjugate pair

$$s_{2,3} = -\frac{1}{2}\left(a_1 + \frac{a_4}{a_3}\right) \pm iq\sqrt{a_3} \tag{28}$$

which leads to decaying, oscillating modes. Thus, the corrections due to the relativistic nature of the molecular dynamics of the disks could be measured in a light scattering experiment where a Rayleigh-Brillouin spectrum can be obtained. In particular, since the focus of this work is the effect of the bulk viscosity, we are only concerned with the Brillouin doublet width, which is given by the real part of the complex roots (Eq. (28)). The width of the central peak as well as the location of the lateral ones is not affected by the presence of μ. Let's call

$$W(\mu(z)) = -\frac{1}{2}\left(a_1 + \frac{a_4}{a_3}\right), \tag{29}$$

the width of the Brillouin peaks. Also, in order to understand better the effect of μ, in Fig. 2 we show the ratio $\left(W(\mu) - W(\mu = 0)\right)/W(\mu = 0)$ as a function of z. Clearly $W(\mu \neq 0)$ and $W(\mu = 0)$ are equal at $z = 0$, the ratio reaches a maximum for an intermediate value of z and finally tends to zero for large values of the temperature.

6. Discussion and final remarks

The integral in Eq. (1) that results in a complete identification of the dependence of μ with the temperature in the two-dimensional relativistic system is far of being simple. In this work we showed that in both the non-relativistic and ultrarelativistic limits the bulk viscosity of a 2D relativistic system vanishes and the formal proof was presented.

Dissipation in single component fluids in the absence of external forces is composed of two effects: viscous and thermal. While thermal dissipation in relativistic systems modifies in a somewhat dramatical fashion the structure of the transport equations, the viscous tensor effects remain the same in form however the coefficients are significantly altered. In particular, the Rayleigh-Brillouin spectrum is modified by the presence of a non-vanishing bulk viscosity in

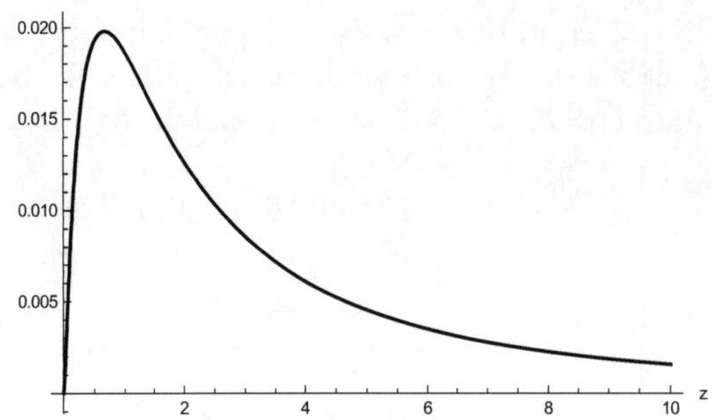

Figure 2. The ratio of the Brillouin widths $(W(\mu) - W(\mu = 0))/W(\mu = 0)$ as a function of z.

the relativistic scenario. The dynamics of density fluctuations clearly depend on all dissipative contributions. In general, the shape of the Brillouin peaks is given in terms of all the transport coefficients (see Eq. (29)) but the presence of a non zero bulk coefficient in the intermediate temperature regime leads to a slight modification of the spectrum, in particular the width of the lateral peaks is altered, as is shown in Fig. 2 and Eq. (29). The analysis of the extreme limits shown in this paper for all the transport coefficients in the (2+1) case is important in the context of the development of new two-dimensional materials as graphene and will be addressed elsewhere.

References

[1] S. Chapman, T. G. Cowling, *The Mathematical Theory of Non-Uniform Gases*, Cambridge Univ. Press (1970).

[2] C. Cercignani, G. Medeiros Kremer, *The Relativistic Boltzmann Equation: Theory and Applications*, 3rd edn. Cambridge University press, Cambridge (1991).

[3] García-Perciante A. L. and Méndez A. R., *Dissipative properties of relativistic two-dimensional gases* (arxiv:1810.04342), (2018).

[4] R. Courant and D. Hilbert, *Methods of Mathematical Physics*, Interscience Publishers, Inc., New York, (1937).

[5] García-Perciante A. L., Sandoval-Villalbazo A., García-Colín L. S, *On the microscopic nature of dissipative effects in special relativistic kinetic theory*, Jour. Non-Equilib. Thermodyn. **37**, 43 (2012)

[6] M. Mendoza, I. Karlin, S. Succi and H. J. Herrmann, Ultrarelativistic transport coefficients in two dimensions, J. Stat. Mech. (2013) P02036.

[7] W. Israel. *Relativistic kinetic theory of a simple gas*, J. Math. Phys., **4**, 1163, (1963).

[8] R. D. Mountain, *Spectral Distribution of Scattered Light in a Simple Fluid*, Rev. Mod. Phys. **38**, 1 (1966).

IOP Conf. Series: Journal of Physics: Conf. Series **1239** (2019) 012004 doi:10.1088/1742-6596/1239/1/012004

A non-relativistic magnetized vector boson gas at any temperature

LC. Suárez-González[1], G. Quintero Angulo[2], A. Pérez Martínez[1] and H. Pérez Rojas[1]

[1] Instituto de Cibernética Matemática y Física (ICIMAF), Calle E esq 15 No. 309 Vedado, La Habana, CP 10400, Cuba

[2] Facultad de Física, Universidad de La Habana,San Lázaro y L, CP 10400, Cuba

E-mail: `lismary@icimaf.cu`

Abstract. We study the thermodynamic properties of a neutral vector boson gas in presence of a constant magnetic field, by means of a semi-classical approach that allows to introduce the spin in the non-relativistic spectrum of the bosons. Bose-Einstein condensation is obtained and it turns out to depend on all the parameters involved in the problem: temperature, particle density and magnetic field. A spontaneous magnetization appears at low temperature as a consequence of the condensed state. The axial symmetry imposed in the system by the magnetic field presence, splits the pressure in two components, one along and another perpendicular to the magnetic axis. Under certain conditions, the perpendicular pressure becomes negative signaling that the system undergoes a transversal magnetic collapse. The spontaneous magnetization might be useful to model magnetic field production inside compact stars, while the negative pressures imposes certain limits to the temperatures and densities needed inside these objects to support a given magnetic field.

1. Introduction

Neutron stars (NS) are the smaller and denser objects known in our Universe so far [1]. Although they have been widely studied, there are no consensus about their internal composition, mainly because matter at such extreme conditions cannot been obtained yet at lab. Nevertheless, a lot of exotic particles and phases have been conjecture to exist in NS interiors [2]. In particular, it has been proposed that, at some stage of the NS evolution, it might contain certain amount of bosons formed up by the pairing of neutrons and protons in the crust and core, and electron and positrons in the magnetosphere [2, 3, 4]. Given the high magnetic fields present in most NS, the magnetic properties of these particles are expected to be relevant for the structure, composition and other physical phenomena related to these compact objects.

Our main purpose is to study the thermodynamic properties of a magnetized gas of neutral spin-1 particles (neutron-neutron or electron-positron spin parallel paired), with the aim of providing equations of state that allows more accurate descriptions of magnetic neutron stars. However, note that the occurrence of Bose-Einstein condensation, one of the most outstanding properties of bosonic systems, in presence of magnetic fields is of interest in others areas, such as condensed matter and particle physics [5, 6], in which this work might be also applied.

The thermodynamics of relativistic magnetized boson gases has been tackled in [7, 8, 9, 10, 11] for the low temperature regime ($T << m$). In this sense, our work can be seen as an extension of the former papers, because Bose-Einstein condensation (BEC), magnetization and the anisotropic equations of state (EoS) are obtained at any temperature, although so far we are taken the bosons as non relativistic. The extension of the study at any temperature is convenient mainly for two reasons. First, it provides an easy way to check the computations consistency since in the high temperature region the well known classical behaviour has to be recovered. And second, this extension eliminates the restriction that T has to be much less than the particle mass, that depending on the kind of bosons might be in contradiction with the temperatures expected inside the NS.

On the other hand, a non-relativistic magnetized spin-1 boson gas has been investigated in [12]. In what concerns BEC and magnetic properties, our results are in general agreement with those shown in the above mentioned paper. However, in [12] the breaking of the SO3 rotational symmetry produced by the magnetic field is never taken into account, in spite of this is a very relevant feature that splits the pressure in two components, perpendicular and parallel to the magnetic field direction, and may cause instabilities in the system [13]. In order to provide a complete description of the non-relativistic magnetized spin-1 gas, we devote a section to the analysis of the stability and the anisotropic pressures.

The paper is organized as follows. In Section 2, the thermodynamical potential of the non-relativistic spin-1 gas under the action of an external magnetic field is computed. Section 3 is devoted to Bose-Einstein condensation and magnetic properties. Anisotropic EoS and magnetic instabilities are discussed in Section 4. Finally, in Section 5 concluding remarks are given.

2. Thermodynamical potential of a non-relativistic vector boson gas interacting with a magnetic field

In this section we compute the thermodynamical potential of an ideal non-relativistic neutral vector boson gas interacting with a constant and uniform magnetic field $\vec{B} = (0, 0, B)$. The spectrum of the bosons is $\varepsilon(p, S_z) = \vec{p}^2/2m - S_z\kappa B$, being m the mass of the particles, κ their magnetic moment, $\vec{p}$ the momentum and $S_z = -1, 0, +1$ the projection of the spin in the z direction. For such a gas, the density of states $g(\epsilon)$ reads

$$\begin{aligned} g(\epsilon) &= \frac{4\pi V}{(2\pi\hbar)^3} \sum_{S_z=-1,0,1} \int_0^\infty dp\, p^2 \delta\left(\epsilon - \varepsilon(p, S_z)\right) \\ &= \frac{4\pi V}{(2\pi\hbar)^3} \sum_{S_z=-1,0,1} \int_0^\infty dp\, p^2 \delta\left(\epsilon - \frac{\vec{p}^2}{2m} + S_z\kappa B\right) \\ &= \frac{4\pi V}{(2\pi\hbar)^3} \left[\int_0^\infty dp\, p^2 \delta\left(\epsilon - \frac{\vec{p}^2}{2m} - \kappa B\right) + \int_0^\infty dp\, p^2 \delta\left(\epsilon - \frac{\vec{p}^2}{2m}\right) \right. \\ &+ \left. \int_0^\infty dp\, p^2 \delta\left(\epsilon - \frac{\vec{p}^2}{2m} + \kappa B\right) \right]. \end{aligned} \tag{1}$$

where ϵ is the energy. Note that Eq. (1) can be separated in three terms, each one corresponding to a specific spin state. After doing the integration $g(\epsilon)$ becomes

$$g(\epsilon) = \frac{4\pi m V}{(2\pi\hbar)^3} \left[\sqrt{2m(\epsilon - \kappa B)} + \sqrt{2m\epsilon} + \sqrt{2m(\epsilon + \kappa B)} \right]. \tag{2}$$

Using Eq. (2) the thermodynamical potential $\Omega(\mu, T, B)$ might be written as

$$\Omega(\mu, T, B) = \frac{T}{V} \int_0^\infty d\epsilon g(\epsilon)\; ln(f_{BE}(\epsilon, \mu)), \;\; \forall\, \mu < \epsilon, \tag{3}$$

with $f_{BE}(\epsilon,\mu) = \left[1 - e^{\beta(\mu-\epsilon)}\right]^{-1}$ being the Bose-Einstein distribution function, μ the chemical potential, T the absolute temperature and $\beta = 1/T$. As the density of states, $\Omega(\mu,T,B)$ can be divided in three terms $\Omega(\mu,T,B) = \Omega_{-}(\mu,T,B) + \Omega_0(\mu,T,B) + \Omega_{+}(\mu,T,B)$, where $\Omega_{-}(\mu,T,B)$, $\Omega_0(\mu,T,B)$ and $\Omega_{+}(\mu,T,B)$ corresponds to the states with $S_z = -1$, $S_z = 0$ and $S_z = 1$ respectively. Integrating over the energy in Eq. (3) one gets

$$\Omega_{-}(\mu,T,B) = -\frac{T}{\lambda^3} g_{5/2}(z_{-}), \tag{4}$$

$$\Omega_0(\mu,T,B) = -\frac{T}{\lambda^3} g_{5/2}(z), \tag{5}$$

$$\Omega_{+}(\mu,T,B) = -\frac{T}{\lambda^3} g_{5/2}(z_{+}), \tag{6}$$

where $g_{5/2}(x) = \sum_{l=1}^{\infty} x^l/l^{5/2}$ is the polylogarithmic function of order 5/2, $\lambda = \sqrt{2\pi/mT}$ is the thermal wavelength, $z = e^{\mu/T}$ is the fugacity and $z_\sigma = ze^{\sigma\frac{\kappa B}{T}}$ where $\sigma = -,+$. With the use of Eqs. (4)-(6), all the thermodynamic magnitudes -particle density, magnetization, pressures, etc.-, and the contributions to them of each spin state, can be determined.

3. Bose-Einstein condensation and magnetic properties

We start the study of Bose-Einstein condensation (BEC) by calculating the particle number density ρ

$$\rho = \rho_{gs}(T,B) - \frac{\partial\Omega(\mu,T,B)}{\partial\mu}, \tag{7}$$

where ρ_{gs} stands for the particles in the condensate, i.e. in the ground state, and is such that $\rho_{gs}(T,B) = 0$ if $T \geq T_c$ and $\rho_{gs}(T,B) > 0$ if $T < T_c$, being T_c the BEC critical temperature. After doing the derivative in Eq. (7), the particle density takes the form

$$\rho = \rho_{gs}(T,B) + \rho_{-}(\mu,T,B) + \rho_0(\mu,T,B) + \rho_{+}(\mu,T,B), \tag{8}$$

$$\rho = \rho_{gs}(T,B) + \frac{g_{3/2}(z_{-})}{\lambda^3} + \frac{g_{3/2}(z)}{\lambda^3} + \frac{g_{3/2}(z_{+})}{\lambda^3}. \tag{9}$$

Using Eq. (9) we compute T_c numerically as a function of ρ and B. To do so, it is enough to recall that when the transition to the BEC begins, $\rho_{gs} = 0$ and the chemical potential μ equals the particle's rest energy $\varepsilon_{min} = -\kappa B$. The result is depicted in left panel of Fig.1, where T_c is plotted as a function of the magnetic field for several fixed values of the particle density. The results are given for bosons of mass $m = 2m_e$ and $\kappa = 2\mu_B$, where m_e is the electron mass and μ_B the Bohr magneton. $B_c = m/2\kappa$ is the magnetic field value at which the magnetic energy becomes comparable to the mass of the particles.

The left panel of Fig.1 shows how the BEC critical temperature grows starting from its $B = 0$ value

$$T_c(0) = \frac{2\pi}{m}\left(\frac{\rho}{3g_{3/2}(1)}\right)^{2/3}, \tag{10}$$

until saturation when $B \to \infty$ for wich

$$T_c(\infty) = \frac{2\pi}{m}\left(\frac{\rho}{g_{3/2}(1)}\right)^{2/3}. \tag{11}$$

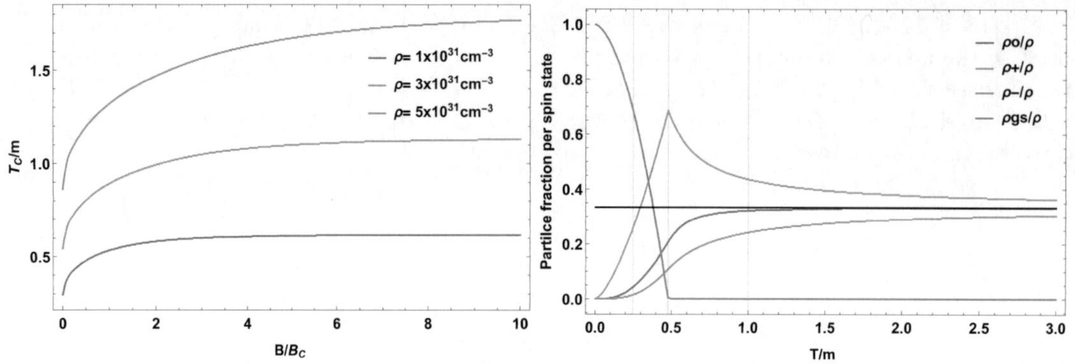

Figure 1. Left panel, BEC critical temperature as a function of the magnetic field for several values of the particle density. Right panel, particle density per spin state as a function of temperature for $\rho = 10^{31}\text{cm}^{-3}$ and $B = 0.5B_c$.

These extreme values of the BEC critical temperature are in agreement with the ones obtained in [12]. On the other hand, the asymptotic behavior of T_c with growing B constitutes a major difference between this case and the relativistic one, in which the BEC critical temperature always increases with B and diverges when $B \to B_c$ [11].

The saturation of T_c in the strong magnetic field region give us valuable information about the influence of B on the BEC of non-relativistic particles. For magnetic field values before saturation, increasing B increases T_c in a noticeable way, driving the system to condensation. But when the magnetic field reaches the saturated region, further changes barely affects T_c, in spite of what we should not forget that $T_c(\infty)$ is bigger than $T_c(0)$. Therefore, one can conclude that in general, for a fixed particle density, the magnetic field effect on the BEC is to increase the critical temperature, and the weaker the field, the more sensitive is the system to change on its values. Note also that, although the BEC critical temperature increases with the particle density, the ratio between the two extreme values $T_c(\infty)/T_c(0) = \sqrt[3]{9}$ is constant.

Furthermore, it is of interest to look at the behavior of the particle fraction in the BEC ρ_{gs}/ρ and per spin state ρ_σ/ρ, $\sigma = -, 0, +$ as a function of temperature. This is shown in right panel of Fig. 1. In the high temperature region, $T \gg m$, $\rho_{gs} = 0$ and $\rho_\sigma/\rho \to 1/3$ for all σ, because this is a temperature dominating region in which thermal disorder rules. When T decreases, the magnetic field ordering effect begins to be noticed and the fraction of particles with spin aligned to the field ρ_+/ρ becomes the higher one. This behavior continues through the low temperature region $T \ll m$ being the next appreciable change when $T = T_c$. At this point the fraction of particles in the BEC becomes non zero and increases with decreasing temperature until it equals 1 at $T = 0$, where $\rho_\sigma/\rho = 0$ for all σ. Nevertheless, since the condensate in the ground state is determined by $S_z = 1$, it is expected that, for a magnetic boson gas, a non zero magnetization exists even in the absence of magnetic field. To check on this, we computed the gas magnetization

$$M = \kappa\rho_{gs} - \left(\frac{\partial\Omega}{\partial B}\right) = \kappa(\rho_{gs} + \rho_+) - \kappa\rho_-, \tag{12}$$

and plotted it in Fig. 2 for $\rho = 10^{31}\text{cm}^{-3}$ and several values of B, including $B = 0$. The curves on this graph are in accordance with the ones in Fig. 1 (right panel): $M \to 0$ for high T, while $M \to \kappa\rho$ when $T \to 0$. The outstanding feature of Fig. 2 is that this happens also for $B = 0$, i.e. the gas shows a spontaneous magnetization that in this case is not due to an interaction between the spin of the particles but to the BEC. Note that, for $B = 0$, $M \neq 0$ only when $T < T_c$. This

IOP Conf. Series: Journal of Physics: Conf. Series **1239** (2019) 012004 doi:10.1088/1742-6596/1239/1/012004

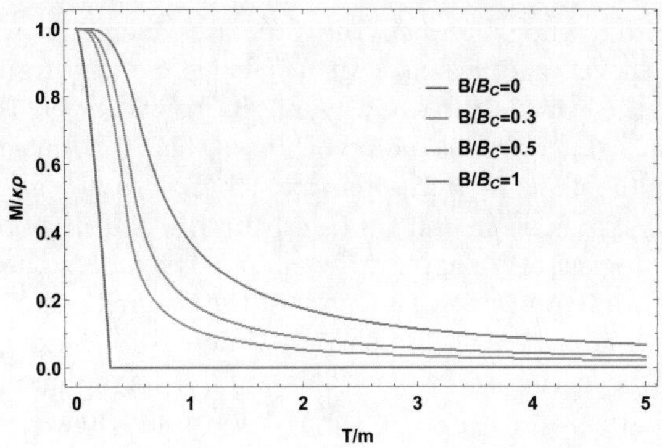

Figure 2. Magnetization as a function of temperature for $\rho = 10^{31}\text{cm}^{-3}$ and several values of the magnetic field.

can be also obtained by direct substitution of $B = 0$ in Eq. (12), which gives $M(B = 0) = \kappa\rho_{gs}$. This result acquires astrophysical relevance since the spontaneous magnetization might provide a magnetic field source for the interior of compact objects.

4. Equations of state

To compute the EoS, the breaking of the SO3 rotational symmetry produced by the magnetic field in the system has to be taken into account [13]. As a consequence of that, the energy-momentum tensor of the particles becomes anisotropic and the pressure splits in two components, one parallel, $P_{\parallel}$, and the other perpendicular, $P_{\perp}$, to the magnetic axis. Therefore, for a magnetized quantum gas of either neutral or charged, bosonic or fermionic, particles the EoS read

$$
\begin{aligned}
E &= \Omega + \mu\rho - TS, && (13)\\
P_{\parallel} &= -\Omega, && (14)\\
P_{\perp} &= -\Omega - \mathcal{M}B, && (15)
\end{aligned}
$$

where $S = -\left(\partial\Omega_B/\partial T\right)_{T,V}$ is the entropy of the system, that in the present case is

$$
S = -\frac{5\,\Omega}{2\,T} - \mu\frac{\rho - \rho_{gs}}{T} - \kappa B\frac{\rho_{+} - \rho_{-}}{T}. \tag{16}
$$

Combining Eq. (16) with Eqs. (13) and (4), we get for the energy density

$$
E = \frac{3}{2}P_{\parallel} - \kappa B(\rho_{gs} + \rho_{+} - \rho_{-}). \tag{17}
$$

While from Eqs. (14) and (15) one can see that the pressures are expressed in terms of quantities Ω and M computed above. The EoS, Eqs. (14), (15) and (17), can be used to model magnetized compact objects composed partially of entirely of magnetized bosons. However, care must be taken, since, as shows Fig. 3, in dependence of the temperature, the magnetic field and the particle density, the perpendicular pressure might be negative.

Left panel of Fig. 3 shows the pressures as a function of the magnetic field for $\rho = 10^{31}\text{cm}^{-3}$ and several values of the temperature. At zero magnetic field, $P_{\parallel} = P_{\perp}$ and the system is isotropic. If $B \neq 0$ the difference between the pressures increases when decreasing the temperature or augmenting the magnetic field. In fact, it is evident from the plot that the magnetic field barely affects the parallel pressure, while its contribution is very important to the perpendicular one. From a microscopic point of view, this is because the magnetic field diminishes the perpendicular momentum of the particles, but does not affect the parallel one. Macroscopically, this effect is expressed in the subtractive term $-MB$ (note that $M > 0$) that appears in the perpendicular pressure. That is why $P_\perp$ is always less than $P_\parallel$ and decreases until becoming zero or even negative with the increase of the magnetic field. Since the effect of a negative perpendicular pressure is to push the particles inwards to the magnetic axis, this can be interpreted as the system becoming unstable. This kind of instability has been previously observed in other magnetized quantum gases and it is known as transverse magnetic collapse [13, 14, 15, 16, 17].

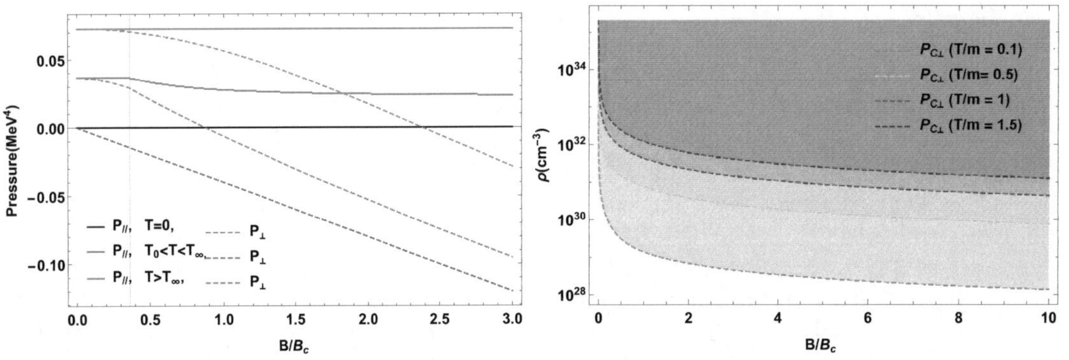

Figure 3. Left plot, the pressures as a function of the magnetic field for several values of the temperature and $\rho = 10^{31}\mathrm{cm}^{-3}$. Right plot, phase diagram for the transverse magnetic collapse in the particle density-magnetic field plane for several values of the temperature; the dashed lines stands for the solution of $P_\perp(T, B, \rho) = 0$ at a fixed temperature; for each temperature, the instability region has been shadowed.

In right panel of Fig. 3 a phase diagram for the collapsed/non collapsed gas is drawn in the particle density-magnetic field plane, for several values of the temperature. The dashed lines stand for the solution of $P_\perp(T, B, \rho) = 0$ at a fixed temperature. The gas is stable in the region under the lines and unstable over them. At $T \neq 0$ and $B = 0$, an infinite number of particles is needed for the system to become unstable, an expected result since the magnetic field is the cause of the instability. When $T = 0$ and $B \neq 0$, $P_\parallel = 0$ and $P_\perp = -MB$ is always negative, i.e. the pure condensed state is unstable for any non-zero magnetic field value. In the case that both the temperature and the magnetic field are different from zero decreasing the temperature favors the collapse, as well as increasing the magnetic field. In addition, Fig. 3 also shows that the collapse is favored by particle density augmentation, because the denser the gas, the higher are M and T_c. In consequence, for a fixed temperature and magnetic field increasing ρ decreases the thermal pressure $(-\Omega(\mu, T, B))$ and increases the magnetic pressure $(-MB)$ driving $P_\perp$ to negative values. In this sense, the collapse imposes an upper limit to the particle densities that can exist inside compact stars with a given magnetic field. Nevertheless, we would like to remark that since a compact star has an heterogeneous composition, the occurrence or not of the collapse will depend on the pressures, and therefore, on the magnetic response, of all the present species.

5. Conclusions

We have studied the properties of the BEC, magnetization and EoS of a magnetized vector boson gas at any temperature. The bosons are considered non-relativistic and the spin is included under a semi-classical approach.

The phase transition to the BEC depends on the temperature, the particle density and the magnetic field, in a way that increasing the particle density or the magnetic field, as well as decreasing the temperature, drives the system to the condensed state. For a fixed particle density, the critical temperature grows with the magnetic field from its value at $B = 0$, $T_c(0)$ to the limiting value $T_c(\infty)$, being the ratio between this extreme values independent of density.

Below the BEC critical temperature, the gas shows a spontaneous magnetization. This magnetization is not due to a spin coupling between the particles, but to the fact that bosons in the condensed phase are in the state of lowest energy. For a system of magnetic bosons this state is such as all the particles have the same spin projection. The ability of this system to be spontaneously magnetized might be connected to magnetic field production in astrophysical objects.

As happens for other magnetized quantum gases, the magnetic field presence imposes its axial symmetry to the system and separates the pressures in two components, one along and the other perpendicular to the magnetic axis. Under certain conditions, the perpendicular pressure might vanish or be negative and the system becomes unstable. As was shown, this instability is caused by the magnetic field, while the temperature opposes it. In consequence, the cooler the matter, the more susceptible it is to undergo a transverse magnetic collapse. On the other hand, increasing the particle density also destabilizes the gas. This imposes an upper limit to the boson densities allowed in astronomical objects for a given temperature and magnetic field.

Acknowledgments

This work have been supported by the grant No. 500.03401 of PNCB-MES, Cuba. A.P.M acknowledges the support of the organizing committee of IARD 2018 who allowed her participate in the nice IARD conference. She also thanks the Caribbean network of OEA-ICTP for supporting her visit to CINVESTAV, Mexico, where this work was finished. G.Q.A. express gratitude to the Service de Coopération et d'Action Culturelle (SCAC) of the Embassy of France in Cuba and to the Federated Institutes Programme of the ICTP, Italy, for support during the final stage of this work. L.C.S.G. and G.Q.A. acknowledges D. Alvear Terrero for her help in some calculations and in the elaboration of the figures. The authors would also like to express their gratitude to the referees for their comments and useful suggestions.

References

[1] Camenzind M 2007 *Compact objects in astrophysics : white dwarfs, neutron stars, and black holes*
[2] Page D, Prakash M, Lattimer J M and Steiner A W 2011 *PoS* **XXXIVBWNP** 005 (*Preprint* `1110.5116`)
[3] Shternin P S, Yakovlev D G, Heinke C O, Ho W C G and Patnaude D J 2011 *mnras* **412** L108–L112 (*Preprint* `1012.0045`)
[4] Chavanis P H and Harko T 2012 *Phys. Rev.* **D86** 064011 (*Preprint* `1108.3986`)
[5] Berman O L, Kezerashvili R Y, Lozovik Y E and Snoke D W 2010 *Philosophical Transactions of the Royal Society of London A: Mathematical, Physical and Engineering Sciences* **368** 5459–5482
[6] Higbie J M, Sadler L E, Inouye S, Chikkatur A P, Leslie S R, Moore K L, Savalli V and Stamper-Kurn D M 2005 *Physical Review Letters* **95** 050401 (*Preprint* `cond-mat/0502517`)
[7] Rojas H P 1996 *Physics Letters B* **379** 148–152 (*Preprint* `hep-th/9510191`)
[8] Khalilov V, Ho C L and Yang C 1997 *Modern Physics Letters A* **12** 1973–1981
[9] Khalilov V R and Ho C L 1999 *Phys. Rev. D* **60**(3) 033003 URL `https://link.aps.org/doi/10.1103/PhysRevD.60.033003`
[10] Perez Rojas H and Villegas-Lelovski L 2000 *Brazilian Journal of Physics* **30** 410 – 418 ISSN 0103-9733
[11] Angulo G Q, Martínez A P and Rojas H P 2017 *Physical Review C* **96** 045810
[12] Yamada K 1982 *Progress of Theoretical Physics* **67** 443–453

IOP Conf. Series: Journal of Physics: Conf. Series **1239** (2019) 012004 doi:10.1088/1742-6596/1239/1/012004

[13] Chaichian M, Masood S S, Montonen C, Perez Martinez A and Perez Rojas H 2000 *Phys. Rev. Lett.* **84** 5261–5264 (*Preprint* hep-ph/9911218)

[14] Pérez Martínez A, Pérez Rojas H and Mosquera Cuesta H J 2003 *European Physical Journal C* **29** 111–123 (*Preprint* astro-ph/0303213)

[15] Felipe R G, Mosquera Cuesta H J, Perez Martinez A and Perez Rojas H 2005 *Chin. J. Astron. Astrophys.* **5** 399 (*Preprint* astro-ph/0207150)

[16] Pérez Rojas H, Rodríguez Querts E and Pérez Mart'inez A 2017 *International Journal of Modern Physics: Conference Series* **45** 1760045 (*Preprint* https://doi.org/10.1142/S201019451760045X) URL https://doi.org/10.1142/S201019451760045X

[17] Quintero Angulo G, Pérez Martínez A and Pérez Rojas H 2017 *Astron. Nachr./AN.* **338** 1127–1130 (*Preprint* 1804.00641)

IOP Conf. Series: Journal of Physics: Conf. Series **1239** (2019) 012005 doi:10.1088/1742-6596/1239/1/012005

Mass-Energy-Momentum Radiation in Stueckelberg-Horwitz-Piron (SHP) Electrodynamics

Martin Land
Department of Computer Science, Hadassah College, 37 HaNeviim Street, Jerusalem
E-mail: martin@hac.ac.il

Abstract. In Stueckelberg-Horwitz-Piron (SHP) electrodynamics individual particles are not restricted to fixed mass shells. Particles and fields may exchange mass along with energy-momentum, while total mass-energy-momentum is conserved for the system. By extension, interacting particles may exchange mass through the fields, and much as a radio transmitter produces electromagnetic waves that carry power to a remote radio receiver, the corresponding SHP apparatus can, in principle, transport mass across spacetime. The conserved Noether current contains energy density and energy flux density into space (Poynting 3-vector) as in Maxwell theory, as well as two additional features of the electromagnetic field: the mass density and mass flux density into spacetime (Poynting 4-vector). In this paper we consider the radiation from a simple dipole antenna in SHP, and use the mass-energy-momentum tensor to compare the results with the standard treatment in Maxwell theory. We find that mass radiation can only take place when the net charge on the entire antenna oscillates along with the current. We show that much as an ideal amplifier drives current into the antenna as the source of the radiated energy, the oscillating net charge drives spacetime events into the antenna as the source of the radiated mass. This process illuminates the event-oriented nature of mass dynamics in SHP theory.

1. Introduction

As a classical field theory, Maxwell electrodynamics describes particles, which interact relativistically through U(1) gauge fields, and fields, which are induced by currents associated with particle worldlines in spacetime. In a slightly different manner, Stueckelberg-Horwitz-Piron (SHP) electrodynamics [1–6] describes individual spacetime events interacting relativistically through U(1) gauge fields, and the fields induced by currents associated with the construction of particle worldlines by events evolving through spacetime. As a canonical evolution theory, events in 4D spacetime are defined on an unconstrained 8D phase space

$$\left(x^\mu(\tau), \dot{x}^\mu(\tau)\right) \qquad \text{where} \qquad \dot{x}^\mu = \frac{dx^\mu}{d\tau} \qquad \text{and} \qquad \mu, \nu = 0,1,2,3 \tag{1}$$

such that each worldline is traced out as the coordinate-independent evolution parameter τ advances monotonically. The absence of constraints among the four-velocities means that the system is not generally reparametrization invariant and so particle trajectories evolving dynamically are not *a priori* restricted to fixed mass shells. In particular, events and gauge fields may exchange mass, although the total mass of the particles and fields is conserved. A self-interaction that tends to restore particle masses to their on-shell values has been described [7] and Horwitz [8] has shown that a particle constructed as a statistical ensemble of events will return to its equilibrium mass following collisions. Still, preliminary work indicates that a Gaussian wavepacket representing a free particle in SHP quantum mechanics may radiate mass, so considerable work remains in understanding these mass dynamics and the analytical techniques used to describe them.

In this paper we explore mass radiation by generalizing to SHP the standard treatment of the power radiated from a simple dipole antenna. Just as the Maxwell antenna radiates energy provided by the amplifier driving the oscillating current density, we find that the SHP antenna will generally radiate both mass and energy provided by the amplifier. However, when the oscillating current density is such that the total charge on the antenna remains fixed, mass radiation vanishes and power radiation reduces to the standard Maxwell expression. This result will be shown to illustrate the relationship between mass radiation and event creation.

Section 2 presents a brief overview of Stueckelberg-Horwitz-Piron electrodynamics, including the field equations and mass-energy-momentum tensor. In section 3 we briefly discuss some preliminary examples that are useful in understanding the mass-energy-momentum tensor: sourceless plane wave solutions of the pre-Maxwell field equations, the absence of radiation from a constant current in Maxwell theory, and the corresponding result in pre-Maxwell theory. Section 4 provides a brief review of the standard treatment in Maxwell theory of energy radiation from a simple dipole antenna. In section 5 we generalize this treatment to SHP theory and in section 6 we discuss the results.

2. Stueckelberg-Horwitz-Piron (SHP) electrodynamics

SHP may be seen as formalizing the distinction between two aspects of time, the chronological world time τ that determines monotonic ordering of events, and the coordinate time x^0 that locates each event on the laboratory clock [9]. Given this distinction, the theory is obtained by generalizing nonrelativistic mechanics to relativistic form through the substitution of τ for the Newtonian time t, and Poincaré symmetry for Galilean symmetry. The free particle Lagrangian is then $L = \frac{1}{2}M\dot{x}^\mu \dot{x}_\mu \implies M\ddot{x}^\mu = 0$ so that defining $p^\mu = \partial L/\partial \dot{x}_\mu$ leads to $E = p^0 c = Mc^2(dt/d\tau)$. This aspect of the formalism permits a unified description of particles (positive energy, future directed events with $\dot{x}^0 > 0$) and antiparticles (negative energy, past directed events with $\dot{x}^0 < 0$), with smoothly evolving pair processes possible under appropriate interactions in classical mechanics.

In analogy to $x^0 = ct$, we introduce a notation for "fifth" scalar quantities

$$x^5 = c_5\tau \qquad \dot{x}^5 = c_5 \qquad \partial_5 = \frac{1}{c_5}\frac{\partial}{\partial\tau} \qquad \begin{array}{c}\alpha,\beta,\gamma,\delta,\epsilon = 0,1,2,3,5 \\ \lambda,\mu,\nu,\rho = 0,1,2,3\end{array} \tag{2}$$

and note that the action

$$\begin{aligned} S &= \int d\tau \left\{ \frac{1}{2}M\dot{x}^\mu \dot{x}_\mu + \frac{e}{c}\left[\dot{x}^\mu a_\mu(x,\tau) + c_5 a_5(x,\tau)\right] \right\} \\ &= \int d\tau \left\{ \frac{1}{2}M\dot{x}^\mu \dot{x}_\mu + \frac{e}{c}\dot{x}^\alpha a_\alpha(x,\tau) \right\} \end{aligned} \tag{3}$$

is invariant under the 5D gauge transformation

$$a_\alpha(x,\tau) \longrightarrow a'_\alpha(x,\tau) = a_\alpha(x,\tau) + \partial_\alpha \Lambda(x,\tau)\,. \tag{4}$$

We rewrite the interaction term as

$$\dot{x}^\alpha a_\alpha(x,\tau) \longrightarrow \frac{1}{c}\int d^4x\, j^\alpha(x,\tau) a_\alpha(x,\tau) \tag{5}$$

IOP Conf. Series: Journal of Physics: Conf. Series **1239** (2019) 012005 doi:10.1088/1742-6596/1239/1/012005

where the event current

$$j^\alpha(x,\tau) = c\dot{x}^\alpha(\tau)\delta^4\left(x - x(\tau)\right) \tag{6}$$

is defined at each τ, with support restricted to the spacetime location of the event at $x = x(\tau)$. This current is related to the standard divergenceless Maxwell current by

$$J^\mu(x) = \int d\tau\, j^\mu(x,\tau) = c\int d\tau\, \dot{x}^\mu(\tau)\delta^4\left(x - x(\tau)\right), \tag{7}$$

an integration called concatenation [10], which can be seen as the sum at x of all contributions to the current by the events that trace out the worldline represented by $J^\mu(x)$. Although the interaction term (5) exhibits a formal 5D symmetry — possibly O(3,2) or O(4,1), which contain O(3,1) as a subgroup — the physical symmetry is understood to be a vector+scalar representation of O(3,1).

We specify the dynamical action with electromagnetism [5,11] as

$$\begin{aligned} S = &\int d\tau\, \frac{1}{2}M\dot{x}^\mu\dot{x}_\mu \\ &+ \int d\tau d^4x\, \left\{\frac{e}{c^2}\, j^\alpha(x,\tau)a_\alpha(x,\tau) - \int \frac{ds}{\lambda}\frac{1}{4c}\left[f^{\alpha\beta}(x,\tau)\Phi(\tau - s)f_{\alpha\beta}(x,s)\right]\right\} \end{aligned} \tag{8}$$

where we have defined the field intensities

$$f_{\alpha\beta} = \partial_\alpha a_\beta(x,\tau) - \partial_\beta a_\alpha(x,\tau) \tag{9}$$

which are invariant under the 5D gauge transformation (4). Expanding the expression

$$f^{\alpha\beta}f_{\alpha\beta} = f^{\mu\nu}f_{\mu\nu} + 2f^{5\mu}f_{5\mu} = f^{\mu\nu}f_{\mu\nu} + 2g^{55}f_5^{\ \mu}f_{5\mu} \tag{10}$$

we may avoid interpreting g^{55} as an element of a 5D metric, and rather see its role as equivalent to the choice of sign for the vector contribution $f_5^{\ \mu}f_{5\mu}$ to the field energy. The field interaction kernel is

$$\Phi(\tau) = \delta(\tau) - (\xi\lambda)^2\delta''(\tau) = \int \frac{d\kappa}{2\pi}\left[1 + (\xi\lambda\kappa)^2\right]e^{-i\kappa\tau} \tag{11}$$

where λ is a parameter with dimensions of time and

$$\xi = \frac{1}{2}\left[1 + \left(\frac{c_5}{c}\right)^2\right] \tag{12}$$

is chosen so that the low energy Lorentz force agrees with Coulomb's law. Writing

$$\varphi(\tau) = \lambda\int \frac{d\kappa}{2\pi}\frac{e^{-i\kappa\tau}}{1 + (\xi\lambda\kappa)^2} = \frac{1}{2\xi}e^{-|\tau|/\xi\lambda} \tag{13}$$

we see that

$$\int \frac{d\tau}{\lambda}\,\varphi(\tau) = 1 \qquad\qquad \int \frac{ds}{\lambda}\,\varphi(\tau - s)\,\Phi(s) = \delta(\tau) \tag{14}$$

permitting the convolution in (8) to be inverted. Integrating by parts the nonlocal term $\delta''(\tau)$ in (8), puts the kinetic part for the fields into the form

$$\begin{aligned} \int d\tau\, ds\, f^{\alpha\beta}(x,\tau)\Phi(\tau - s)f_{\alpha\beta}(x,s) = &\int d\tau\, f^{\alpha\beta}(x,\tau)f_{\alpha\beta}(x,\tau) \\ &+ \int d\tau\, (\xi\lambda)^2\left(\partial_\tau f^{\alpha\beta}(x,\tau)\right)\partial_\tau f_{\alpha\beta}(x,\tau) \end{aligned} \tag{15}$$

which breaks any 5D symmetry $(x,\tau) \to (x',\tau')$ to O(3,1).

Varying the action with respect to x^μ leads to the Lorentz force [12]

$$M\ddot{x}^\mu(\tau) = \frac{e}{c} f^\mu{}_\alpha(x,\tau)\dot{x}^\alpha(\tau) = \frac{e}{c}\left[f^\mu{}_\nu(x,\tau)\dot{x}^\nu(\tau) + c_5 f^\mu{}_5(x,\tau)\right] \tag{16}$$

$$\frac{d}{d\tau}(-\tfrac{1}{2}M\dot{x}^2) = -M\dot{x}^\mu\ddot{x}_\mu = \frac{ec_5}{c} f_{5\mu}\dot{x}^\mu \tag{17}$$

where (17) exhibits the possibility of mass exchange between events and fields. Varying the action with respect to $a_\alpha(x,\tau)$ produces the field equations [4,11]

$$\partial_\beta f^{\alpha\beta}(x,\tau) = \frac{e}{c} j^\alpha_\varphi(x,\tau) \qquad \partial_\alpha f_{\beta\gamma} + \partial_\gamma f_{\alpha\beta} + \partial_\beta f_{\gamma\alpha} = 0 \tag{18}$$

whose source

$$j^\alpha_\varphi(x,s) = \int ds\ \varphi(\tau - s)\ j^\alpha(x,s) \tag{19}$$

where we used (14) to replace $\Phi(\tau)$ in the field term on LHS with $\varphi(\tau)$ in the current term on the RHS. These field equations are formally similar to Maxwell's equations in 5D, and are called pre-Maxwell equations. Rewriting the field equations in 4-vector and scalar components, they take the form

$$\partial_\mu f^{5\mu} = \frac{e}{c} j^5_\varphi = \frac{ec_5}{c}\rho_\varphi \qquad \partial_\mu f_{\nu\rho} + \partial_\nu f_{\rho\mu} + \partial_\rho f_{\mu\nu} = 0 \tag{20}$$

$$\partial_\nu f^{\mu\nu} - \frac{1}{c_5}\frac{\partial}{\partial\tau} f^{5\mu} = \frac{e}{c} j^\mu_\varphi \qquad \partial_\nu f_{5\mu} - \partial_\mu f_{5\nu} + \frac{1}{c_5}\frac{\partial}{\partial\tau} f_{\mu\nu} = 0 \tag{21}$$

which may be compared with the 3-vector form of Maxwell equations

$$\nabla\cdot\mathbf{E} = \frac{e}{c}J^0 = e\rho \qquad \nabla\cdot\mathbf{B} = 0 \tag{22}$$

$$\nabla\times\mathbf{B} - \frac{1}{c}\frac{\partial}{\partial t}\mathbf{E} = \frac{e}{c}\mathbf{J} \qquad \nabla\times\mathbf{E} + \frac{1}{c}\frac{\partial}{\partial t}\mathbf{B} = 0 \tag{23}$$

with $f^{5\mu}$ playing the role of the vector electric field and $f^{\mu\nu}$ playing the role of the magnetic field. For our purposes it is convenient to have the 3-vector and scalar forms

$$(\mathbf{e})^i = f^{0i} \qquad (\mathbf{b})_i = \varepsilon_{ijk} f^{jk} \qquad (\boldsymbol{\epsilon})^i = f^{5i} \qquad \epsilon^0 = f^{50} \tag{24}$$

for which the field equations split into four generalizations of the 3-vector Maxwell equations

$$\begin{aligned} &\nabla\cdot\mathbf{e} - \frac{1}{c_5}\frac{\partial}{\partial\tau}\epsilon^0 = \frac{e}{c}j^0_\varphi = e\rho^0_\varphi && \nabla\cdot\mathbf{b} = 0 \\ &\nabla\times\mathbf{b} - \frac{1}{c}\frac{\partial}{\partial t}\mathbf{e} - \frac{1}{c_5}\frac{\partial}{\partial\tau}\boldsymbol{\epsilon} = \frac{e}{c}\mathbf{j}_\varphi && \nabla\times\mathbf{e} + \frac{1}{c}\frac{\partial}{\partial t}\mathbf{b} = 0 \end{aligned} \tag{25}$$

and three new equations for the fields $\boldsymbol{\epsilon}$ and ϵ^0

$$\begin{aligned} &\nabla\cdot\boldsymbol{\epsilon} + \frac{1}{c}\frac{\partial}{\partial t}\epsilon^0 = \frac{e}{c}j^5_\varphi = \frac{ec_5}{c}\rho_\varphi \qquad \nabla\times\boldsymbol{\epsilon} - g^{55}\frac{1}{c_5}\frac{\partial}{\partial\tau}\mathbf{b} = 0 \\ &\nabla\epsilon^0 + \frac{1}{c}\frac{\partial}{\partial t}\boldsymbol{\epsilon} + g^{55}\frac{1}{c_5}\frac{\partial}{\partial\tau}\mathbf{e} = 0. \end{aligned} \tag{26}$$

The field equations (18) can be combined to find the wave equation

$$\left(\partial^\mu\partial_\mu + g_{55}\frac{1}{c_5^2}\frac{\partial^2}{\partial\tau^2}\right) f^{\alpha\beta} = \partial^\alpha j^\beta - \partial^\beta j^\alpha \tag{27}$$

for which the principal part Green's function [13]

$$G_P(x,\tau) = -\frac{1}{2\pi}\delta(x^2)\delta(\tau) - \frac{c_5}{2\pi^2}\frac{\partial}{\partial x^2}\theta(-g_{55}g_{\alpha\beta}x^\alpha x^\beta)\frac{1}{\sqrt{-g_{55}g_{\alpha\beta}x^\alpha x^\beta}} \tag{28}$$

includes a Maxwell term with support on the lightcone at simultaneous τ, and a term with timelike or spacelike support, depending on g^{55}. In most cases, field contributions from the second term in (28) can be ignored in the far field compared to contributions from the first term. It follows from (27) that this approximation is equivalent to dropping terms on the order $\delta''(\tau)/c_5^2$ and this exclusion should be maintained consistently.

The connection between SHP electrodynamics and standard Maxwell theory can be established through the concatenation integral defined for the current in (7). Under the equilibrium boundary conditions

$$\rho_\varphi(x,\tau) \xrightarrow[\tau\to\pm\infty]{} 0 \qquad\qquad f^{5\mu}(x,\tau) \xrightarrow[\tau\to\pm\infty]{} 0 \tag{29}$$

integration over worldline permits the identification

$$\left.\begin{matrix} \partial_\nu f^{\mu\nu} - \dfrac{1}{c_5}\dfrac{\partial}{\partial\tau}f^{5\mu} = \dfrac{e}{c}j^\mu_\varphi \\[2ex] \partial_\mu f_{\nu\rho} + \partial_\nu f_{\rho\mu} + \partial_\rho f_{\mu\nu} = 0 \\[2ex] \partial_\alpha j^\alpha = 0 \end{matrix}\right\} \xrightarrow[\int \frac{d\tau}{\lambda}]{} \left\{\begin{matrix} \partial_\nu F^{\mu\nu}(x) = \dfrac{e}{c}J^\mu(x) \\[2ex] \partial_\mu F_{\nu\rho} + \partial_\nu F_{\rho\mu} + \partial_\rho F_{\mu\nu} = 0 \\[2ex] \partial_\mu J^\mu(x) = 0 \end{matrix}\right. \tag{30}$$

where

$$A^\mu(x) = \int\frac{d\tau}{\lambda}\,a^\mu(x,\tau) \qquad F^{\mu\nu}(x) = \int\frac{d\tau}{\lambda}\,f^{\mu\nu}(x,\tau) \qquad J^\mu(x) = \int\frac{d\tau}{\lambda}\,j^\mu(x,\tau). \tag{31}$$

Similarly, concatenation of (28) produces

$$\int d\tau\; G_P(x,\tau) = -\frac{1}{2\pi}\delta(x^2) \tag{32}$$

recovering the Maxwell Green's function.

The translation invariance of the action leads to a conserved Noether current [4] in the form

$$\partial_\alpha T^{\alpha\beta}_\Phi = -\frac{e}{c^2}f^{\beta\alpha}j_\alpha \qquad\qquad T^{\alpha\beta}_\Phi = \frac{1}{c}\left(f^{\alpha\gamma}_\Phi f^\beta{}_\gamma - \frac{1}{4}f^{\delta\varepsilon}_\Phi f_{\delta\varepsilon}g^{\alpha\beta}\right) \tag{33}$$

where $T^{\alpha\beta}_\Phi$ is interpreted as the mass-energy-momentum tensor and

$$f^{\alpha\beta}_\Phi(x,\tau) = \int\frac{ds}{\lambda}\,\Phi(\tau-s)f^{\alpha\beta}(x,s) \tag{34}$$

is the convolved field strength appearing in the action (8). In terms of the 3-vector fields, the mass-energy-momentum tensor components are

$$\begin{aligned} T_\Phi^{00} &= \frac{1}{2c}\left[\mathbf{e}\cdot\mathbf{e}_\Phi + \mathbf{b}\cdot\mathbf{b}_\Phi + g_{55}\left(\boldsymbol{\epsilon}\cdot\boldsymbol{\epsilon}_\Phi + \epsilon^0\epsilon_\Phi^0\right)\right] \\ T_\Phi^{0i} &= \frac{1}{c}\left(\mathbf{e}\times\mathbf{b}_\Phi + g_{55}\epsilon^0\boldsymbol{\epsilon}_\Phi\right)^i \\ T_\Phi^{50} &= \frac{1}{c}\mathbf{e}\cdot\boldsymbol{\epsilon}_\Phi \\ T_\Phi^{5i} &= \frac{1}{c}\left(\boldsymbol{\epsilon}\times\mathbf{b}_\Phi + \epsilon^0\mathbf{e}_\Phi\right)^i \\ T_\Phi^{55} &= \frac{1}{2c}\left[\boldsymbol{\epsilon}\cdot\boldsymbol{\epsilon}_\Phi - \epsilon^0\epsilon_\Phi^0 + g^{55}\left(\mathbf{e}\cdot\mathbf{e}_\Phi - \mathbf{b}\cdot\mathbf{b}_\Phi\right)\right]. \end{aligned} \tag{35}$$

Integrating the $\beta = 5$ component of the conservation law (33) over spacetime and using (6) for the event current leads to

$$\frac{1}{c_5}\frac{d}{d\tau}\int d^4x\, T_\Phi^{55} = -\frac{e}{c}f^{5\mu}\left(x(\tau),\tau\right)\dot{x}_\mu(\tau) \tag{36}$$

for the event trajectory $x(\tau)$. Using the Lorentz force (17) on the RHS this becomes

$$\frac{d}{d\tau}\left[\int d^4x\, T_\Phi^{55} + g_{55}\left(-\frac{1}{2}M\dot{x}^2\right)\right] = 0 \tag{37}$$

expressing conservation of the combined mass of particles and fields [12], with T^{55} interpreted as the spacetime density of mass associated with the fields, in analogy to T^{00} as energy density. It is precisely this interpretation that will be studied in the following sections.

3. Preliminary examples

Before studying the classical antenna we examine the interpretation of $T^{\alpha\beta}$ by considering plane waves and the fields from a uniform current.

3.1. Sourceless fields

From the wave equation (27) for $j^\alpha(x,\tau) = 0$ we may write the field in terms of the Fourier transform [12]

$$f^{\alpha\beta}(x,\tau) = \frac{1}{(2\pi)^5}\int d^5k\, e^{ik\cdot x} f^{\alpha\beta}(k) = \frac{1}{(2\pi)^5}\int d^4k\, d\kappa\, e^{i(\mathbf{k}\cdot\mathbf{x} - k_0x^0 + g_{55}c_5\kappa\tau)} f^{\alpha\beta}(k,\kappa) \tag{38}$$

where

$$\kappa = k^5 = g^{55}k_5 \tag{39}$$

is understood to represent the mass carried by the plane wave, much as k^0 and $\mathbf{k}$ represent energy and 3-momentum. This interpretation is supported by the wave equation which imposes the 5D constraint

$$k^\alpha k_\alpha = \mathbf{k}^2 - (k^0)^2 + g_{55}\kappa^2 = 0 \quad\Longrightarrow\quad g_{55}\kappa^2 = (k^0)^2 - \mathbf{k}^2 \tag{40}$$

expressing κ in terms of the difference between energy and momentum. Under concatenation, the field becomes

$$F^{\alpha\beta}(x) = \int\frac{d\tau}{\lambda}\, f^{\alpha\beta}(x,\tau) = \int\frac{d^4k}{(2\pi)^4}\, e^{ik_\mu x^\mu}\left(\frac{1}{\lambda c_5}f^{\alpha\beta}(k,0)\right) = \int\frac{d^4k}{(2\pi)^4}\, e^{ik_\mu x^\mu} F^{\alpha\beta}(k) \tag{41}$$

IOP Conf. Series: Journal of Physics: Conf. Series **1239** (2019) 012005 doi:10.1088/1742-6596/1239/1/012005

and recovers the 4D mass-shell constraint $k^\mu k_\mu = 0$ for the Maxwell field. In the transform domain, the sourceless pre-Maxwell equations take the form

$$\begin{array}{lll} \mathbf{k}\cdot\mathbf{e} - g_{55}\kappa\epsilon^0 = 0 & \mathbf{k}\cdot\mathbf{b} = 0 & \mathbf{k}\cdot\boldsymbol{\epsilon} - k^0\epsilon^0 = 0 \\ \mathbf{k}\times\mathbf{e} - k^0\mathbf{b} = 0 & & \mathbf{k}\times\mathbf{b} + k^0\mathbf{e} - g_{55}\kappa\boldsymbol{\epsilon} = 0 \\ \mathbf{k}\times\boldsymbol{\epsilon} - \kappa\mathbf{b} = 0 & & -\kappa\mathbf{e} + k^0\boldsymbol{\epsilon} - \mathbf{k}\epsilon^0 = 0 \end{array} \tag{42}$$

which can be solved by taking $\boldsymbol{\epsilon}_\parallel$ and $\mathbf{e}_\perp$ as independent 3-vector polarizations, and writing

$$\mathbf{e}_\parallel = g_{55}\frac{\kappa}{k^0}\boldsymbol{\epsilon}_\parallel \qquad \boldsymbol{\epsilon}_\perp = \frac{\kappa}{k^0}\mathbf{e}_\perp \qquad \epsilon^0 = \frac{1}{k^0}\mathbf{k}\cdot\boldsymbol{\epsilon}_\parallel \qquad \mathbf{b} = \frac{1}{k^0}\mathbf{k}\times\mathbf{e}_\perp \tag{43}$$

for the remaining fields. Unlike Maxwell plane waves, for which $\mathbf{E}$, $\mathbf{B}$, and $\mathbf{k}$ are mutually orthogonal, the pre-Maxwell electric fields $\mathbf{e}$ and $\boldsymbol{\epsilon}$ have both transverse and longitudinal components. When $\kappa \to 0$, we find that $\mathbf{e}$, $\mathbf{b}$, and $\mathbf{k}$ become mutually orthogonal and $\boldsymbol{\epsilon}$ becomes a decoupled longitudinal polarization parallel to $\mathbf{k}$.

We use (11) to write the convolved field as

$$f_\Phi^{\alpha\beta}(x,\tau) = \int \frac{ds}{\lambda}\,\Phi(\tau - s) f^{\alpha\beta}(x,s) = \frac{1}{(2\pi)^5}\int d^4k\, d\kappa\, e^{i(\mathbf{k}\cdot\mathbf{x} - k_0 x^0 + g_{55}c_5\kappa\tau)} f_\Phi^{\alpha\beta}(k,\kappa) \tag{44}$$

where

$$f_\Phi^{\alpha\beta}(k,\kappa) = \frac{1 + (\xi\lambda c_5\kappa)^2}{\lambda} f^{\alpha\beta}(k,\kappa) \tag{45}$$

introduces a multiplicative factor that will appear once in each field bilinear of $T_\Phi^{\alpha\beta}$. Using (35), the energy density is

$$T_\Phi^{00} = \frac{1}{2c}\left[\mathbf{e}\cdot\mathbf{e}_\Phi + \mathbf{b}\cdot\mathbf{b}_\Phi + g_{55}\left(\boldsymbol{\epsilon}\cdot\boldsymbol{\epsilon}_\Phi + \epsilon^0\epsilon^0_\Phi\right)\right] = \frac{1}{c}\left(\mathbf{e}_\perp^2 + g_{55}\boldsymbol{\epsilon}_\parallel^2\right)\frac{1 + (\xi\lambda\kappa)^2}{\lambda} \tag{46}$$

which, since $\mathbf{e}_\perp^2 = \frac{1}{2}\left(\mathbf{e}_\perp^2 + \mathbf{b}^2\right)$, is equivalent to the energy density in Maxwell theory

$$\theta^{00} = \frac{1}{2c}\left(\mathbf{E}^2 + \mathbf{B}^2\right) \tag{47}$$

with the addition of the independent polarization $\boldsymbol{\epsilon}_\parallel$. The mass density is found to be

$$T_\Phi^{55} = \frac{\kappa^2}{ck_0^2}\left(\mathbf{e}_\perp^2 + g_{55}\boldsymbol{\epsilon}_\parallel^2\right)\frac{1 + (\xi\lambda\kappa)^2}{\lambda} = \frac{\kappa^2}{k_0^2}T_\Phi^{00} \tag{48}$$

expressing energy density scaled by the squared mass-to-energy ratio for the field. The energy flux — the standard Poynting 3-vector — is

$$T_\Phi^{0i} \longrightarrow \mathbf{T}_\Phi^0 = \frac{\mathbf{k}}{k^0}T_\Phi^{00} \tag{49}$$

expressing the energy density T_Φ^{00} flowing uniformly in the direction of the momentum normalized to energy. We note that for a free particle

$$\frac{\mathbf{k}}{k^0} = \frac{\mathbf{p}}{E/c} = \frac{1}{c}\frac{M\,d\mathbf{x}/d\tau}{M\,dt/d\tau} = \frac{\mathbf{v}}{c} \tag{50}$$

which will not generally be a unit vector unless $\kappa = 0$, as it must be for Maxwell plane waves. The mass flux vector — a second Poynting 3-vector — can be written

$$T_{\Phi}^{5i} \longrightarrow \mathbf{T}_{\Phi}^{5} = \frac{\mathbf{k}}{\kappa} T_{\Phi}^{55} \tag{51}$$

expressing the mass density T_{Φ}^{55} flowing uniformly in the direction of the momentum normalized to mass. Finally,

$$T_{\Phi}^{50} = \frac{k^0}{\kappa} T_{\Phi}^{55} = \frac{\kappa}{k^0} T_{\Phi}^{00} \tag{52}$$

so that $T_{\Phi}^{5\mu}$ can be written as

$$T_{\Phi}^{5\mu} = \frac{k^\mu}{\kappa} T_{\Phi}^{55} = \frac{\kappa k^\mu}{k_0^2} T_{\Phi}^{00} \tag{53}$$

expressing the mass density T_{Φ}^{55} flowing in the direction of the normalized 4-momentum. In this sense, T_{Φ}^{50} represents the flow of mass into the time direction. We notice that when $\kappa \longrightarrow 0$, as is the case for Maxwell plane waves, $\mathbf{k}/k^0$ becomes a unit vector and $T_{\Phi}^{5\alpha} = 0$, so that mass density and flow vanish. The interpretation of plane waves carrying energy and momentum (energy flux) uniformly to infinity is thus seen to generalize to mass flow, where mass is best understood through (40) as the non-identity of energy and momentum.

Suppose that a plane wave of this type impinges on a test particle in its rest frame, described by $x^\alpha(\tau) = (c\tau, \mathbf{0}, c_5\tau)$. Since $\dot{\mathbf{x}} = \mathbf{0}$, the wave will interact with the event through the Lorentz force (16) and (17) as

$$M\ddot{x}^\mu(\tau) = \frac{e}{c}\left[f^\mu{}_0(x,\tau)\dot{x}^0(\tau) + c_5 f^\mu{}_5(x,\tau)\right] \qquad \frac{d}{d\tau}(-\tfrac{1}{2}M\dot{x}^2) = -\frac{ec_5}{c} g_{55}\epsilon^0 \dot{x}^0 \tag{54}$$

which for $\epsilon_\parallel \neq 0 \Rightarrow \mathbf{k}\cdot\epsilon_\parallel \neq 0$ becomes

$$\ddot{t} = -g_{55}e\frac{c_5}{Mc^2}\frac{1}{k^0}\mathbf{k}\cdot\epsilon_\parallel \qquad \ddot{\mathbf{x}} = \frac{e}{M}\left[\mathbf{e}_\perp\left(1 + \frac{c_5}{c}\frac{\kappa}{k^0}\right) + \epsilon_\parallel\left(\frac{c_5}{c} + g_{55}\frac{\kappa}{k^0}\right)\right] \tag{55}$$

$$\frac{d}{d\tau}(-\tfrac{1}{2}M\dot{x}^2) = -g_{55}ec_5\frac{1}{k^0}\mathbf{k}\cdot\epsilon_\parallel \tag{56}$$

showing that the incident plane wave will initially accelerate the test event in such a way as to transfer mass. If the plane wave is a far field approximation to the radiation field of an accelerating charge, then the resulting picture describes the transfer of mass by the radiation field between charged events. As the test event accelerates, the equations of motion become more complex

$$\ddot{t} = \frac{e}{Mc^2}\left[\left(g_{55}\frac{\kappa}{k^0}\epsilon_\parallel + \mathbf{e}_\perp\right)\cdot\dot{\mathbf{x}} - g_{55}c_5\frac{1}{k^0}\mathbf{k}\cdot\epsilon_\parallel\right] \tag{57}$$

$$\ddot{\mathbf{x}} = \frac{e}{M}\left[\mathbf{e}_\perp\left(\dot{t} + \frac{c_5}{c}\frac{\kappa}{k^0}\right) - \frac{1}{k^0 c}\left[(\dot{\mathbf{x}}\cdot\mathbf{e}_\perp)\,\mathbf{k} - (\dot{\mathbf{x}}\cdot\mathbf{k})\,\mathbf{e}_\perp\right] + \epsilon_\parallel\left(\frac{c_5}{c} + g_{55}\frac{\kappa}{k^0}\dot{t}\right)\right] \tag{58}$$

$$\frac{d}{d\tau}(-\tfrac{1}{2}M\dot{x}^2) = -ec_5 g_{55}\left[\frac{1}{k^0}\mathbf{k}\cdot\epsilon_\parallel\dot{t} + \frac{1}{c}\left(\epsilon_\parallel + \frac{\kappa}{k^0}\mathbf{e}_\perp\right)\cdot\dot{\mathbf{x}}\right] \tag{59}$$

and depend on the relative orientations of the fields $\mathbf{e}_\perp$ and $\epsilon_\parallel$, and the phase

$$\exp\{i(\mathbf{k}\cdot\mathbf{x} - k_0 x^0 + g_{55}c_5\kappa\tau)\}. \tag{60}$$

IOP Conf. Series: Journal of Physics: Conf. Series **1239** (2019) 012005 doi:10.1088/1742-6596/1239/1/012005

3.2. *Energy flux from a uniformly moving charge in Maxwell electrodynamics*

In the rest frame of a Maxwell charge, the potentials are

$$A^0(x) = \frac{e}{4\pi R} \qquad \mathbf{A} = 0 \tag{61}$$

leading to fields

$$\mathbf{E}(x) = -\nabla A^0 = \frac{e}{4\pi R^2}\hat{R} \qquad \mathbf{B}(x) = \nabla \times \mathbf{A} = 0 \tag{62}$$

so the energy density and Poynting vector are

$$T^{00} = \frac{1}{2}\left(\mathbf{E}^2 + \mathbf{B}^2\right) = \frac{1}{2}\left(\frac{e}{4\pi R^2}\right)^2 \qquad \mathbf{T} = \mathbf{E} \times \mathbf{B} = 0. \tag{63}$$

The vanishing Poynting vector is understood to mean that a static charge does not transfer energy, and in particular does not radiate. To sharpen this point, we step outside the rest frame to consider a uniformly moving charge along the $\hat{\mathbf{x}}$-axis with τ-dependent position

$$r = u\tau = (\gamma c\tau, \gamma\beta c\tau, 0, 0) = \gamma c\tau\hat{\mathbf{t}} + \gamma\beta c\tau\hat{\mathbf{x}} \tag{64}$$

where

$$\gamma = \frac{1}{\sqrt{1-\beta^2}} \tag{65}$$

as usual. From an observation point

$$x = (0, a, b, 0) = a\hat{\mathbf{x}} + b\hat{\mathbf{y}} \tag{66}$$

the field strengths are easily found to be

$$\mathbf{E} = -e\gamma\frac{a\hat{\mathbf{x}} + b\hat{\mathbf{y}}}{4\pi\left(a^2\gamma^2 + b^2\right)^{3/2}} \qquad \mathbf{B} = -e\gamma\beta\frac{b\,\hat{\mathbf{x}} \times \hat{\mathbf{y}}}{4\pi\left(a^2\gamma^2 + b^2\right)^{3/2}} \tag{67}$$

so that the nonzero Poynting vector

$$\mathbf{T} = \mathbf{E} \times \mathbf{B} = \frac{(e\gamma)^2\, b\beta}{(4\pi)^2\left(a^2\gamma^2 + b^2\right)^3}\left(b\hat{\mathbf{x}} - a\hat{\mathbf{y}}\right) \tag{68}$$

represents energy transfer along the path of the moving charge. To see that the uniformly moving charge does not radiate — transfer energy that is lost to the charge — we enclose the path along the $\hat{\mathbf{x}}$-axis in an infinitely long cylinder of radius ρ with surface area element

$$d\mathbf{S} = \rho\left(\cos\phi\hat{\mathbf{y}} + \sin\phi\hat{\mathbf{z}}\right) d\phi\; da. \tag{69}$$

Summing the total energy flux through the cylinder leads to

$$\int (\mathbf{E} \times \mathbf{B}) \cdot d\mathbf{S} = -r\int_0^{2\pi} \cos\phi\; d\phi \int_{-\infty}^{\infty} da\; \frac{ab\,(e\gamma)^2\,\beta}{(4\pi)^2\left(a^2\gamma^2 + b^2\right)^3} = 0 \tag{70}$$

showing that although the field varies in time at any fixed observation point, with nonzero Poynting vector representing energy flux at that point, the symmetry (odd parity) of the field configuration (both radially and along the direction of the current) means that no net energy is radiated away by the uniformly moving charge.

3.3. Energy flux from a "static" charge in SHP electrodynamics

In SHP theory, an event in its rest frame is in uniform motion in the time direction, and so the position is given by the 4-vector and scalar

$$X(\tau) = (c\tau, \mathbf{0}) \qquad X^5(\tau) = c_5\tau \tag{71}$$

with current components are

$$j_\varphi(x,\tau) = \left(c\delta^3(\mathbf{x})\,\varphi(t-\tau), \mathbf{0}\right) \qquad j^5(x,\tau) = \frac{c_5}{c} j^5(x,\tau). \tag{72}$$

Using the leading term of Green's function (28) leads to the potential

$$a^0(x,\tau) = \frac{e}{4\pi R}\varphi(\tau-\tau_R) \qquad \mathbf{a}(x,\tau) = 0 \qquad a^5(x,\tau) = \frac{c_5}{c}a^0(x,\tau) \tag{73}$$

where $\tau_R = t - R/c$ is the retarded time at an observation point $(ct, \mathbf{x})$ and $R = |\mathbf{x}|$. The field strengths are easily found to be

$$\begin{aligned} \mathbf{e} &= -\nabla a^0 = \frac{e}{4\pi R^2}\left(\varphi(\tau-\tau_R) - \frac{R}{c}\varphi'(\tau-\tau_R)\right)\hat{\mathbf{R}} \qquad && \mathbf{b} = \nabla\times\mathbf{a} = 0 \\ \epsilon^0 &= g_{55}\frac{1}{c_5}\frac{\partial}{\partial\tau}a^0 + \frac{1}{c}\frac{\partial}{\partial t}a^5 = -\frac{e}{4\pi R}\left(g_{55} - \frac{c_5^2}{c^2}\right)\frac{\varphi'}{c_5} \qquad && \boldsymbol{\epsilon} = -\nabla a^5 = \frac{c_5}{c}\mathbf{e}. \end{aligned} \tag{74}$$

Some care is required in treating the convolved fields $f_\Phi^{\alpha\beta}$. The factor

$$\varphi(\tau-\tau_R) - \frac{R}{c}\varphi'(\tau-\tau_R) \tag{75}$$

contained in $\mathbf{e}$ will be replaced under convolution with $\Phi(\tau)$ by

$$\delta(\tau-\tau_R) - \frac{R}{c}\delta'(\tau-\tau_R) \tag{76}$$

so that the bilinear $\mathbf{e}\cdot\mathbf{e}_\Phi$ contains the product

$$\left[\varphi(\tau-\tau_R) - (R/c)\varphi'(\tau-\tau_R)\right]\left[\delta(\tau-\tau_R) - (R/c)\delta'(\tau-\tau_R)\right] = \frac{1}{2\xi}\delta(\tau-\tau_R) \tag{77}$$

where we use

$$\varphi(\tau) = \frac{1}{2\xi}e^{-|\tau|/\xi\lambda} \quad\Longrightarrow\quad \varphi(0) = \frac{1}{2\xi} \ , \quad \varphi'(0) = 0 \tag{78}$$

and discard terms containing $\varphi''(\tau)$ under the approximation discussed in connection with the Green's function (28). The remaining δ-function in (77) can be handled by integrating over x^0, so that the energy and mass densities are found to be

$$\int dx^0\, T^{00} = \left(1 + \frac{c_5^2}{c^2}g^{55}\right)\frac{1}{2}\left(\frac{e}{4\pi R^2}\right)^2 \tag{79}$$

$$\int dx^0\, T^{55} = g^{55}\left(1 + \frac{c_5^2}{c^2}g^{55}\right)\frac{1}{2}\left(\frac{e}{4\pi R^2}\right)^2. \tag{80}$$

As in the Maxwell case, the Poynting 3-vectors contain cross products with $\mathbf{b} = 0$, but also terms containing ϵ^0, which vanishes because $\varphi'(0) = 0$. Therefore

$$\mathbf{T}^0 = \frac{1}{c} g_{55} \epsilon^0 \boldsymbol{\epsilon}_\Phi = g_{55} \frac{c_5}{c} \frac{1}{c} \epsilon^0 \mathbf{e}_\Phi = g_{55} \frac{c_5}{c} \mathbf{T}^5 = 0 \tag{81}$$

so that the mass and energy flux into space vanishes and there is no radiation. Nevertheless, the mass flux into the time direction is

$$\int dx^0 \, T^{50} = \frac{c_5}{c} \left(\frac{e}{4\pi R^2} \right)^2 \tag{82}$$

representing the flow of the event, and the corresponding mass density, from the past to the future, along the path of the event.

4. Radiation from a simple dipole antenna in Maxwell theory

The radiation from an oscillating current density is both a practical model in communications engineering and a textbook example of radiation from an accelerating charge. In order to compare the treatments of this problem in Maxwell theory and SHP, we begin with a brief review of the standard derivation [14] of radiation from a dipole antenna, which assumes a separable current density oscillating according to

$$\mathbf{J}(\mathbf{x}, t) = \mathbf{J}(\mathbf{x}) \, e^{i\omega t} \qquad \nabla \cdot \mathbf{J}(\mathbf{x}) + i\omega\rho = 0 \tag{83}$$

where the second expresses represents current conservation. As an aside, we examine possible considerations leading to separability. In a microscopic approach to the antenna problem, we would pose a collection of oscillating charges with position 4-vectors

$$X_n(\tau) = \left(ct_n(\tau), \mathbf{a}_n e^{i\omega\tau} \right) \tag{84}$$

for which we may assume nonrelativistic motion so that $t_n(\tau) = t = \tau$ for each particle. The Maxwell current for this collection is

$$\begin{aligned} \mathbf{J}(\mathbf{x}, t) &= \sum_n \int d\tau \; c\dot{\mathbf{X}}_n(\tau) \, \delta^4(x - X_n(\tau)) \\ &= \sum_n \int d\tau \; i\omega \; c\mathbf{a}_n e^{i\omega\tau} \delta(ct - c\tau) \, \delta^3\left(\mathbf{x} - \mathbf{a}_n e^{i\omega\tau}\right) \\ &= \left[\sum_n i\omega \; \mathbf{a}_n \delta^3 \left(\mathbf{x} - \mathbf{a}_n e^{i\omega t}\right) \right] e^{i\omega t}. \end{aligned} \tag{85}$$

We now replace the term in square brackets with its time average over one cycle of oscillation $T = 2\pi/\omega$ and write

$$\mathbf{J}(\mathbf{x}, t) \simeq \left[\frac{1}{T} \int_0^T dt \; \sum_n i\omega \; \mathbf{a}_n \delta^3 \left(\mathbf{x} - \mathbf{a}_n e^{i\omega t}\right) \right] e^{i\omega t} = \mathbf{J}(\mathbf{x}) \, e^{i\omega t}. \tag{86}$$

Thus, $\mathbf{J}(\mathbf{x})$ approximates the time-dependent current density by a time averaged static configuration in space, rendering the antenna problem tractable.

Using the current (83) and the 4D Green's function $D(x) = -\delta\left(x^2\right)/2\pi$, the vector potential is

$$\begin{aligned}\mathbf{A}(\mathbf{x},t) &= \frac{1}{4\pi}\int d^3x'dt' \, \frac{1}{|\mathbf{x}-\mathbf{x}'|}\delta\left(t-t'+\frac{|\mathbf{x}-\mathbf{x}'|}{c}\right)\mathbf{J}\left(\mathbf{x}'\right)e^{i\omega t'} \\ &= \frac{1}{4\pi}\int d^3x' \, \frac{e^{-ik|\mathbf{x}-\mathbf{x}'|}}{|\mathbf{x}-\mathbf{x}'|}\mathbf{J}\left(\mathbf{x}'\right)\, e^{i\omega t} \\ &= \mathbf{A}(\mathbf{x})\, e^{i\omega t} \end{aligned} \tag{87}$$

where $k = \omega/c = 2\pi/\lambda$. Now, writing $\mathbf{x} = r\hat{\mathbf{r}}$, we make the far field approximation

$$R = \left|\mathbf{x}-\mathbf{x}'\right| = \left(r^2 + \left(\mathbf{x}'\right)^2 - 2r\hat{\mathbf{r}}\cdot\mathbf{x}'\right)^{1/2} \simeq r - \hat{\mathbf{r}}\cdot\mathbf{x}' \tag{88}$$

and the dipole approximation

$$\left|k\hat{\mathbf{r}}\cdot\mathbf{x}'\right| < kd = \frac{2\pi d}{\lambda} \ll 1 \Rightarrow e^{ik\hat{\mathbf{r}}\cdot\mathbf{x}'} \simeq 1 \qquad \frac{r-\hat{\mathbf{r}}\cdot\mathbf{x}'}{c} \simeq \frac{r}{c}\left(1-\hat{\mathbf{r}}\cdot\hat{\mathbf{x}}'\,\frac{d}{r}\right) \simeq \frac{r}{c} \tag{89}$$

leading to

$$\mathbf{A}(\mathbf{x}) = \frac{1}{4\pi}\int d^3x' \, \frac{e^{-ik|\mathbf{x}-\mathbf{x}'|}}{|\mathbf{x}-\mathbf{x}'|}\mathbf{J}\left(\mathbf{x}'\right) \simeq \frac{e^{-ikr}}{4\pi r}\int d^3x' \, \mathbf{J}\left(\mathbf{x}'\right). \tag{90}$$

Integrating by parts and using current conservation, this becomes

$$\mathbf{A}(\mathbf{x}) = -\frac{e^{-ikr}}{4\pi r}\int d^3x' \, \mathbf{x}' \, \nabla\cdot\mathbf{J} = \frac{e^{-ikr}}{4\pi r}\overbrace{i\omega\int d^3x' \, \mathbf{x}'\rho\left(\mathbf{x}'\right)}^{i\omega\mathbf{p} \,=\, Id\hat{\mathbf{d}}} \tag{91}$$

where the second integral is the dipole moment $\mathbf{p}$ of the charge distribution $\rho(\mathbf{x})$ and $i\omega\mathbf{p}$ can be described as a constant current I along a dipole of length d in the direction $\hat{\mathbf{d}}$.

The magnetic field is found from the potential as

$$\mathbf{B} = \nabla\times\mathbf{A} = \frac{Id}{4\pi}(-ik)\frac{e^{-ikr}}{r}\left(1+\frac{1}{ikr}\right)\hat{\mathbf{r}}\times\hat{\mathbf{d}} \simeq \frac{\omega k}{4\pi}\frac{e^{-ikr}}{r}\hat{\mathbf{r}}\times\hat{\mathbf{p}} \tag{92}$$

where we used $\lambda \ll r \Rightarrow 1/kr \ll 1$. The electric field follows from Maxwell's equations as

$$\mathbf{E} = \frac{1}{i\omega}\nabla\times\mathbf{B} = \frac{1}{4\pi}\left(k^2\frac{e^{-ikr}}{r}(\hat{\mathbf{r}}\times\mathbf{p})\times\hat{\mathbf{r}} + [3\hat{\mathbf{r}}(\hat{\mathbf{r}}\cdot\mathbf{p})-\mathbf{p}]\left(\frac{1}{r^3}+\frac{ik}{r^2}e^{-ikr}\right)\right) \tag{93}$$

so that in the far field

$$\mathbf{E} \simeq \frac{k^2}{4\pi}\frac{e^{-ikr}}{r}(\hat{\mathbf{r}}\times\mathbf{p})\times\hat{\mathbf{r}} = \frac{k}{\omega}\,\mathbf{B}\times\hat{\mathbf{r}}. \tag{94}$$

Expressing the direction of oscillation in radial coordinates as

$$\hat{\mathbf{d}} = \hat{\mathbf{z}} = \cos\theta\,\hat{\mathbf{r}} - \sin\theta\,\hat{\boldsymbol{\theta}} \tag{95}$$

leads to expressions for the fields

$$\mathbf{B} = \frac{\omega k}{4\pi}\frac{e^{-ikr}}{r}\sin\theta\,\hat{\boldsymbol{\phi}} \qquad\qquad \mathbf{E} = \frac{\omega k}{4\pi}\frac{e^{-ikr}}{r}\sin\theta\hat{\boldsymbol{\theta}} \tag{96}$$

and the time-averaged average Poynting vector

$$\bar{\boldsymbol{\theta}} = \frac{1}{2}\mathbf{E} \times \mathbf{B}^* = \frac{1}{2}\left(\frac{k}{4\pi}\right)^2 \frac{(Id)^2}{r^2} \sin^2\theta \, \hat{\mathbf{r}} \tag{97}$$

from which the total radiated power is found by integrating over the surface of a sphere of radius r

$$\bar{P} = \frac{1}{2}\int d\Omega \, r^2 \, \hat{\mathbf{r}} \cdot \left[\left(\frac{k}{\omega}\mathbf{B} \times \hat{\mathbf{r}}\right) \times \mathbf{B}\right] = \frac{1}{2}\int_0^{2\pi} d\phi \int_0^{\pi} d\theta \left(\frac{\omega k}{4\pi}|\mathbf{p}|\right)^2 \sin^3\theta = \frac{k^2 (Id)^2}{12\pi}. \tag{98}$$

This nonzero integral is understood to describe the power invested by the amplifier in accelerating the charges to produce a radiation field (identifible by its $1/r$ dependence) and transferred away through any closed surface enclosing the antenna.

5. The oscillating dipole in SHP theory

In this section we follow the essential steps taken in the previous section in deriving the radiated power, adapting them as necessary to the microscopic dynamics of SHP.

5.1. Modeling the oscillating current

Carrying over the approximations used to write the Maxwell current (83) from the previous section requires some care, because the microscopic dynamics in SHP impose certain additional restrictions on the 5D pre-Maxwell current.

First, the time averaging (86) that renders the Maxwell problem tractable does not produce a clear separation into a space-dependent $\mathbf{J}(\mathbf{x})$ and time-dependent oscillation for the pre-Maxwell event current. The microscopic SHP current for a collection of nonrelativistic charged events is

$$\mathbf{j}(\mathbf{x}, t, \tau) = c\sum_n \dot{\mathbf{X}}_n(\tau)\, \delta^4(x - X_n(\tau)) = \sum_n i\omega \, \mathbf{a}_n e^{i\omega\tau} \delta(t-\tau)\, \delta^3\left(\mathbf{x} - \mathbf{a}_n e^{i\omega\tau}\right) \tag{99}$$

from which the exact potential takes the complicated form

$$\mathbf{a}_\varphi(x, \tau) = \frac{i\omega}{c}\sum_n \mathbf{a}_n e^{i\omega\tau} \frac{e}{4\pi\left|\mathbf{x} - \mathbf{a}_n e^{i\omega\tau}\right|}\, \varphi\left(t - \frac{\left|\mathbf{x} - \mathbf{a}_n e^{i\omega\tau}\right|}{c} - \tau\right). \tag{100}$$

But the presence of $\delta(t-\tau)$ in the current (99) as an unintegrated distribution renders time averaging meaningless. Proceeding (twice) with an averaging procedure would lead to

$$\mathbf{j}(\mathbf{x}, t, \tau) \simeq \begin{cases} \left[\frac{1}{T}\sum_n i\omega \, \mathbf{a}_n \delta^3\left(\mathbf{x} - \mathbf{a}_n e^{i\omega\tau}\right)\right] e^{i\omega\tau} \simeq \mathbf{J}(\mathbf{x})\, e^{i\omega\tau} & , \ t \text{ average} \\ \left[\frac{1}{T}\sum_n i\omega \, \mathbf{a}_n \delta^3\left(\mathbf{x} - \mathbf{a}_n e^{i\omega t}\right)\right] e^{i\omega t} \simeq \mathbf{J}(\mathbf{x})\, e^{i\omega t} & , \ \tau \text{ average} \end{cases} \tag{101}$$

which should be equivalent because $t = \tau$. However, integrating the Green's function (28) with the τ-independent current $\mathbf{J}(\mathbf{x})\, e^{i\omega t}$ recovers the wavelike Maxwell potential found in (87), while integrating with the t-independent current $\mathbf{J}(\mathbf{x})\, e^{i\omega\tau}$ leads to an oscillating Coulomb potential with no wavelike behavior. Since equivalent currents should not induce qualitatively different potentials, we conclude that time averaging is not a useful approach here.

Second, the component $j^5(x,\tau)$ represents an event density in spacetime and must be non-negative. For example, given a collection of events as in (84), the event density is

$$j^5(x,\tau) = c\sum_{n=1}^{N} \dot{X}_n^5(\tau)\,\delta^4(x - X_n(\tau)) = cc_5 \sum_{n=1}^{N} \delta^4(x - X_n(\tau)) \geq 0 \tag{102}$$

and the total number of events is

$$\frac{1}{cc_5}\int d^4x\, j^5(x,\tau) = N. \tag{103}$$

In the general case, conservation of the 5-current

$$\partial_\beta f^{\alpha\beta}(x,\tau) = \frac{e}{c}\, j^\alpha(x,\tau) \quad \Longrightarrow \quad \partial_\alpha j^\alpha(x,\tau) = 0 \tag{104}$$

implies continuity in the form

$$\frac{1}{c_5}\frac{d}{d\tau}\int d^4x\, j^5(x,\tau) = c\frac{dN}{d\tau} = -\int d^4x\, \partial_\mu j^\mu(x,\tau) \tag{105}$$

so that total event number is conserved unless the four-current $j^\mu(x,\tau)$ includes an external contribution — such as an ideal amplifier driving an antenna — whose divergence does not vanish at the boundary of spacetime.

Third, the component $j^0(x,\tau)$ must remain non-negative in order to eliminate pair processes. For a single event, $j^0(x,\tau) < 0$ is associated with $\dot{t} < 0$ through

$$j^0(x,\tau) = c^2\dot{t}(\tau)\,\delta^4(x - X(\tau)) \tag{106}$$

and is characteristic of an antiparticle. We therefore take the oscillation as a perturbation $\rho(\mathbf{x})$ to a static background current density $\rho_0(\mathbf{x})$ that preserves these requirements on the event current.

In analogy to the approximation used in the Maxwell approach, and in light of the relationship between j^0 and j^5 seen in (72), we pose a current of the form

$$j^0(x,\tau) = c\left[\rho_0(\mathbf{x}) + \rho(\mathbf{x})\,e^{i\omega t}\right]\phi(\tau - t) \tag{107}$$

$$\mathbf{j}(x,\tau) = \mathbf{J}(\mathbf{x})\,e^{i\omega t}\phi(\tau - t) \tag{108}$$

$$j^5(x,\tau) = \frac{c_5}{c}j^0(x,\tau) = c_5\left[\rho_0(\mathbf{x}) + \rho(\mathbf{x})\,e^{i\omega t}\right]\phi(\tau - t) \tag{109}$$

where $\phi(\tau - t)$ expresses a correlation between t and τ, inserted by hand instead of using a time averaging procedure. In this sense, the replacement

$$\mathbf{j}(\mathbf{x},t,\tau) \quad \longrightarrow \quad \mathbf{J}(\mathbf{x})\,e^{i\omega t}\phi(\tau - t) \tag{110}$$

may be less precise than the comparable approximation in Maxwell theory, and we must be attentive to artifacts introduced by the model. In analogy to (13) we choose

$$\phi(\tau - t) = \frac{1}{2\sigma}e^{-|\tau - t|/\sigma} \tag{111}$$

which imposes a correlation $\tau - t \sim \sigma$ through

$$\phi(\tau - t) \longrightarrow \begin{cases} \text{strong correlation:} & \sigma \to 0 \Rightarrow \phi(\tau - t) \to \delta(\tau - t) \Rightarrow t = \tau \\ \text{weak correlation:} & \sigma \to \text{large} \Rightarrow t - \tau \text{ evenly distributed.} \end{cases} \tag{112}$$

Notice that in the strong correlation limit, the potential found from the Green's function (28) is

$$\begin{aligned} \mathbf{a}(x,\tau) &= \frac{e}{2\pi}\int d^3x' d(ct')\, \delta\left((\mathbf{x}-\mathbf{x}')^2 - c^2(t-t')^2\right) \mathbf{J}(\mathbf{x}')\, e^{i\omega t'} \delta(\tau - t') \\ &= \frac{e}{4\pi c} e^{i\omega\tau} \int d^3x' \frac{1}{|\mathbf{x}-\mathbf{x}'|}\, \delta\left(\tau - t + \frac{|\mathbf{x}-\mathbf{x}'|}{c}\right) \mathbf{J}(\mathbf{x}') \end{aligned} \tag{113}$$

describing a Coulomb-like potential that oscillates in τ simultaneously across spacetime, rather than a wave propagating with phase $(kr - \omega t)$. This suppression of the expected wavelike behavior can be characterized by the dimensionless parameter

$$\frac{1}{\omega\sigma} = \frac{T}{2\pi\sigma} = \frac{\text{antenna period}}{\text{correlation time}} \tag{114}$$

which we take to be small but greater than zero. We note the useful properties

$$\phi(-\tau) = \phi(\tau) \qquad \int d\tau\, \phi(\tau) = 1 \qquad \phi'(\tau) = -\frac{\varepsilon(\tau)}{\sigma}\phi(\tau) \tag{115}$$

and the Fourier transform

$$\varphi(\tau - t) = \int \frac{d\omega}{2\pi} \Phi(\omega)\, e^{i\omega(\tau - t)} \qquad \Phi(\omega) = \frac{1}{1 + (\sigma\omega)^2}\ . \tag{116}$$

Since the background density $\rho_0(\mathbf{x})$ is independent of t and τ, conservation of the 5D current becomes

$$\begin{aligned} 0 = \frac{1}{c}\frac{\partial}{\partial t} j^0 + \nabla \cdot \mathbf{j} + \frac{1}{c_5}\frac{\partial}{\partial \tau} j^5 &= \rho(\mathbf{x}) \frac{\partial}{\partial t}\left[\left(e^{i\omega t}\right)\phi(\tau - t)\right] + \nabla \cdot \mathbf{J}(\mathbf{x})\, e^{i\omega t}\phi(\tau - t) \\ &\quad + \rho(\mathbf{x})\, e^{i\omega t} \frac{\partial}{\partial \tau}\phi(\tau - t) \\ &= i\omega\rho(\mathbf{x})\, e^{i\omega t}\phi(\tau - t) - \underline{\rho(\mathbf{x})\, e^{i\omega t}\phi'(\tau - t)} \\ &\quad + \nabla \cdot \mathbf{J}(\mathbf{x})\, e^{i\omega t}\phi(\tau - t) + \underline{\rho(\mathbf{x})\, e^{i\omega t}\phi'(\tau - t)} \\ &= \left[i\omega\rho(\mathbf{x}) + \nabla \cdot \mathbf{J}(\mathbf{x})\right] e^{i\omega t}\phi(\tau - t) \end{aligned} \tag{117}$$

so that $\rho(\mathbf{x})$ and $\mathbf{J}(\mathbf{x})$ are related as in the Maxwell problem through

$$\rho(\mathbf{x}) = -\frac{1}{i\omega}\nabla \cdot \mathbf{J}(\mathbf{x}) \quad \longrightarrow \quad e\int d^3x\, \mathbf{J}(\mathbf{x}) = -e\int d^3x\, \mathbf{x}\, \nabla \cdot \mathbf{J} = \overbrace{i\omega \int d^3x\, \mathbf{x} e\rho(\mathbf{x})}^{i\omega\mathbf{p} = Id\hat{\mathbf{d}}}\,. \tag{118}$$

The total charge at time τ is found from the spacetime integral

$$\begin{aligned} Q(\tau) &= \frac{e}{c}\int d^4x\, j^0(x,\tau) = \frac{e}{c}\int d^4x\, c\left[\rho_0(\mathbf{x}) + \rho(\mathbf{x})\, e^{i\omega t}\right]\phi(\tau - t) \\ &= \int d^3x\, e\rho_0(\mathbf{x}) \int dt\, \phi(\tau - t) + \int d^3x\, e\rho(\mathbf{x}) \int dt\, e^{i\omega t}\phi(\tau - t) \\ &= \int d^3x\, e\rho_0(\mathbf{x}) + \int d^3x\, e\rho(\mathbf{x})\ \Phi(\omega)\, e^{i\omega\tau} \\ &= Q_0 + Q\, \Phi(\omega)\, e^{i\omega\tau} \end{aligned} \tag{119}$$

where we write

$$Q_0 = \int d^3x \, e\rho_0 (\mathbf{x}) \qquad Q = \int d^3x \, e\rho (\mathbf{x}) \tag{120}$$

and used (115) and (116). The total charge in spacetime takes the form of a background charge with an oscillating (externally driven) perturbation. Similarly, the total current density is

$$\mathbf{J}(\tau) = \frac{e}{c} \int d^4x \, \mathbf{J}(\mathbf{x}) \, e^{i\omega t} \phi(\tau - t) = \Phi(\omega) \, e^{i\omega\tau} \int d^3x \, e\mathbf{J}(\mathbf{x}) = \Phi(\omega) \, i\omega \mathbf{p} e^{i\omega\tau} \tag{121}$$

representing an oscillating dipole.

5.2. *Induced potential and field strengths*

Using the leading term in the Green's function (28)

$$G_{Maxwell} = -\frac{1}{2\pi}\delta(x^2)\delta(\tau) \tag{122}$$

the induced potential is

$$a^\alpha(x,\tau) = \frac{e}{c} \int d^3x' \, \frac{1}{4\pi|\mathbf{x}-\mathbf{x}'|} j^\alpha \left(c\left(t - \frac{|\mathbf{x}-\mathbf{x}'|}{c}\right), \mathbf{x}', \tau \right) \tag{123}$$

with components

$$\begin{aligned} a^0(x,\tau) =& \, e \int d^3x' \, \frac{\rho_0(\mathbf{x}')}{4\pi|\mathbf{x}-\mathbf{x}'|} \phi\left(\tau - t + \frac{|\mathbf{x}-\mathbf{x}'|}{c}\right) \\ &+ e \int d^3x' \, \frac{\rho(\mathbf{x}')}{4\pi|\mathbf{x}-\mathbf{x}'|} e^{i\omega\left(t - \frac{|\mathbf{x}-\mathbf{x}'|}{c}\right)} \phi\left(\tau - t + \frac{|\mathbf{x}-\mathbf{x}'|}{c}\right) \end{aligned} \tag{124}$$

$$\mathbf{a}(x,\tau) = \frac{e}{c} \int d^3x' \, \frac{\mathbf{J}(\mathbf{x}')}{4\pi|\mathbf{x}-\mathbf{x}'|} e^{i\omega\left(t - \frac{|\mathbf{x}-\mathbf{x}'|}{c}\right)} \phi\left(\tau - t + \frac{|\mathbf{x}-\mathbf{x}'|}{c}\right) \tag{125}$$

$$a^5(x,\tau) = \frac{c_5}{c} a^0(x,\tau). \tag{126}$$

Applying the far field and dipole approximations these potentials can be evaluated as

$$\begin{aligned} a^0(x,\tau) \simeq& \frac{1}{4\pi r}\phi\left(\tau - t + \frac{r}{c}\right) \underbrace{\int d^3x' \, e\rho_0(\mathbf{x}')}_{Q_0} + \frac{1}{4\pi r} e^{i\omega t} e^{-ikr} \phi\left(\tau - t + \frac{r}{c}\right) \underbrace{\int d^3x' \, e\rho(\mathbf{x}')}_{Q} \\ =& \frac{Q_0}{4\pi r} + \frac{Q}{4\pi r} \, e^{-i(kr-\omega t)} \, \phi\left(\tau - t + \frac{r}{c}\right) \end{aligned} \tag{127}$$

$$\begin{aligned} \mathbf{a}(x,\tau) \simeq& \frac{1}{4\pi r} e^{-ikr} e^{i\omega t} \underbrace{\frac{e}{c} \int d^3x' \, \mathbf{J}(\mathbf{x}')}_{ik\mathbf{p}} \phi\left(\tau - t + \frac{r}{c}\right) \\ =& \, \mathbf{p} \, \frac{ik}{4\pi r} \, e^{-i(kr-\omega t)} \, \phi\left(\tau - t + \frac{r}{c}\right) \end{aligned} \tag{128}$$

$$a^5(x,\tau) = \frac{c_5}{c} a^0(x,\tau) \tag{129}$$

from which the components of $f^{\alpha\beta}(x,\tau)$ are found by taking appropriate derivatives. We define the spherical wave factor

$$\chi(x,\tau) = \frac{e^{-i(kr-\omega t)}}{4\pi r}\phi\left(\tau - t + \frac{r}{c}\right) \tag{130}$$

and split the field strengths into a spacetime factor and polarization factors, as

$$\begin{aligned}
&\mathbf{b} = \nabla\times\mathbf{a} = \widehat{\mathbf{b}}\,\chi && \widehat{\mathbf{b}} = -ikId\left(1+\frac{\varepsilon_R}{i\omega\sigma}\right)\hat{\mathbf{r}}\times\hat{\mathbf{d}} \\
&\mathbf{e} = -\frac{1}{c}\frac{\partial}{\partial t}\mathbf{a} - \nabla a^0 = \frac{Q_0}{4\pi r^2}\hat{\mathbf{r}} + \widehat{\mathbf{e}}\,\chi && \widehat{\mathbf{e}} = ik\left(1+\frac{\varepsilon_R}{i\omega\sigma}\right)\left(Q\hat{\mathbf{r}} - Id\hat{\mathbf{d}}\right) \\
&\epsilon = g_{55}\frac{1}{c_5}\frac{\partial}{\partial\tau}\mathbf{a} - \frac{c_5}{c}\nabla a^0 = \frac{c_5}{c}\frac{Q_0}{4\pi r^2}\hat{\mathbf{r}} + \widehat{\epsilon}\,\chi && \widehat{\epsilon} = ik\left[\frac{c_5}{c}\left(1+\frac{\varepsilon_R}{i\omega\sigma}\right)Q\hat{\mathbf{r}} - g_{55}\frac{c}{c_5}\frac{\varepsilon_R}{i\omega\sigma}Id\hat{\mathbf{d}}\right] \\
&\epsilon^0 = g_{55}\frac{1}{c_5}\frac{\partial}{\partial\tau}a^0 - \frac{1}{c}\frac{\partial}{\partial t}a^5 = \widehat{\epsilon}^{\,0}\chi && \widehat{\epsilon}^{\,0} = ik\left[\frac{c_5}{c}\left(1+\frac{\varepsilon_R}{i\omega\sigma}\right) - g_{55}\frac{c}{c_5}\frac{\varepsilon_R}{i\omega\sigma}\right]Q
\end{aligned} \tag{131}$$

where we used

$$\phi'(\tau) = -\frac{\varepsilon(\tau)}{\sigma}\,\phi(\tau) \qquad \frac{1}{kr} \ll 1 \qquad \mathbf{p} = \frac{1}{i\omega}Id\hat{\mathbf{d}} \tag{132}$$

and write $\varepsilon_R = \varepsilon(\tau - t + R/c)$. We drop the static Coulomb terms produced by Q_0, as these do not contribute to radiation. Since we saw in (112) and (113) that $\omega\sigma \sim$ small tends to suppress wavelike behavior, we take

$$1 + \frac{\varepsilon_R}{i\omega\sigma} \simeq 1 \tag{133}$$

but note [5] that since $c_5/c \ll 1$ we leave factors

$$\frac{c}{c_5}\frac{\varepsilon_R}{i\omega\sigma} \tag{134}$$

unchanged. Taking the orientation of the antenna to be $\hat{\mathbf{d}} = \hat{\mathbf{z}}$ the polarizations then simplify to

$$\begin{aligned}
&\widehat{\mathbf{e}} \simeq ik\left(Q\hat{\mathbf{r}} - Id\hat{\mathbf{z}}\right) && \widehat{\mathbf{b}} \simeq -ikId\,\hat{\mathbf{r}}\times\hat{\mathbf{z}} \\
&\widehat{\epsilon}^{\,0} \simeq ik\left[\frac{c_5}{c} - g_{55}\frac{c}{c_5}\frac{\varepsilon_R}{i\omega\sigma}\right]Q && \widehat{\epsilon} \simeq ik\left[\frac{c_5}{c}Q\hat{\mathbf{r}} - g_{55}\frac{c}{c_5}\frac{\varepsilon_R}{i\omega\sigma}Id\hat{\mathbf{z}}\right]
\end{aligned} \tag{135}$$

and we notice that terms containing $1/\omega\sigma$ appear only in the components of $\widehat{\epsilon}^{\mu}$. Such terms are artifacts of modeling the time correlation by $\phi(\tau - t)$, and can be understood as the contribution to the fields required to impose this correlation across spacetime. As was seen for plane waves, these fields will accelerate a test event initially at rest through the Lorentz force

$$\begin{aligned}
\ddot{t} &= \frac{e}{Mc^2}\left(\mathbf{e}\cdot\dot{\mathbf{x}} - g_{55}c_5\epsilon^0\right) = -g_{55}\frac{ec_5}{Mc^2}\epsilon^0 \\
\ddot{\mathbf{x}} &= \frac{e}{M}\left[\mathbf{e}\dot{t} + \frac{1}{c}\mathbf{b}\times\dot{\mathbf{x}} + \frac{c_5}{c}\epsilon\right] = \frac{e}{M}\left[\mathbf{e} + \frac{c_5}{c}\epsilon\right] \\
\frac{d}{d\tau}\left(-\tfrac{1}{2}M\dot{x}^2\right) &= -eg_{55}c_5\left(\epsilon^0\dot{t} - \epsilon\cdot\dot{\mathbf{x}}(\tau)\right) = -eg_{55}c_5\epsilon^0
\end{aligned} \tag{136}$$

in such a way as to transfer mass to the event.

IOP Conf. Series: Journal of Physics: Conf. Series **1239** (2019) 012005 doi:10.1088/1742-6596/1239/1/012005

5.3. *Mass-energy-momentum tensor*

Examining the polarizations (135), we see that the mass-energy-momentum tensor will contain bilinear field combinations of the type

$$T^{\alpha\beta} = \frac{1}{c}\left(f_{\Phi}^{\alpha\gamma} f^{\beta}{}_{\gamma} - \frac{1}{4} f_{\Phi}^{\delta\varepsilon} f_{\delta\varepsilon} g^{\alpha\beta}\right) \quad \longrightarrow \quad \operatorname{Re}\left[\left(\mathbf{A}^{\alpha} + i\,\mathbf{B}^{\alpha}\right)\chi\right] \cdot \operatorname{Re}\left[\left(\mathbf{C}^{\beta} + i\,\mathbf{D}^{\beta}\right)\chi\right] \tag{137}$$

where the common spacetime factor (130) contains the spherical wave and the time correlation function ϕ. It is convenient to separate the resulting products as $T^{\alpha\beta} = T_0^{\alpha\beta} + T_\sigma^{\alpha\beta}$, where $T_\sigma^{\alpha\beta}$ includes all terms containing $1/\omega\sigma$. We will justify this separation below. To facilitate writing the real parts of these expressions, we designate

$$\begin{aligned} S(x,\tau) &= k^2 \left(\frac{\phi\left(\tau - t + \frac{r}{c}\right)}{4\pi r}\right)^2 \sin^2(kr - \omega t) \\ C(x,\tau) &= k^2 \left(\frac{\phi\left(\tau - t + \frac{r}{c}\right)}{4\pi r}\right)^2 \cos^2(kr - \omega t) \\ X(x,\tau) &= k^2 \left(\frac{\phi\left(\tau - t + \frac{r}{c}\right)}{4\pi r}\right)^2 2\sin(kr - \omega t)\cos(kr - \omega t) \end{aligned} \tag{138}$$

and note that these functions drop off as $1/r^2$ and so will produce nonzero surface integrals at large r, as is characteristic of radiation fields. Using these functions, the components of $T_0^{\alpha\beta}$ are

$$\begin{aligned} T_0^{00} &= \frac{1}{2}\left[(Q + Id\,\cos\theta)^2 + 2(Id)^2\left(1 - (\cos\theta)^2\right) + 2g_{55}\left(\frac{c_5}{c}\right)^2 Q^2\right] S(x,\tau) \\ \mathbf{T}_0^{0} &= \left[Id\,(Q - Id\,\cos\theta)\,\hat{\mathbf{z}} + \left(Id\,(Id - Q\,\cos\theta) + g_{55}\left(Q\frac{c_5}{c}\right)^2\right)\hat{\mathbf{r}}\right] S(x,\tau) \\ T_0^{50} &= Q\frac{c_5}{c}\,(Q - Id\,\cos\theta)\,S(x,\tau) \\ \mathbf{T}_0^{5} &= Q\frac{c_5}{c}\,(Q - Id\,\cos\theta)\,S(x,\tau)\,\hat{\mathbf{r}} = T_0^{50}\,\hat{\mathbf{r}} \\ T_0^{55} &= \frac{1}{2}g^{55}\,(Q - Id\,\cos\theta)^2\,S(x,\tau) \end{aligned} \tag{139}$$

all of which have spacetime dependence $S(x,\tau)$. The components of $T_\sigma^{\alpha\beta}$ are

$$\begin{aligned} T_\sigma^{00} &= \frac{1}{2}\left[\left(\frac{c}{c_5}\frac{1}{\omega\sigma}\right)^2\left[(Id)^2 + Q^2\right] C(x,\tau) - \frac{\varepsilon_R}{\omega\sigma} Q\,[Q + Id\,\cos\theta]\,X(x,\tau)\right] \\ \mathbf{T}_\sigma^{0} &= -Q\frac{\varepsilon_R}{\omega\sigma}\left[Id\left(X(x,\tau) + g_{55}\frac{c}{c_5}C(x,\tau)\right)\hat{\mathbf{z}} + Q\,X(x,\tau)\,\hat{\mathbf{r}}\right] \\ T_\sigma^{50} &= g_{55}\frac{c}{c_5}\frac{\varepsilon_R}{\omega\sigma} Id\,(Id - Q\,\cos\theta)\,X(x,\tau) \\ \mathbf{T}_\sigma^{5} &= \frac{c}{c_5}g_{55}\left[\frac{\varepsilon_R}{\omega\sigma} Id\,[Q - Id\,\cos\theta]\,\hat{\mathbf{z}} + \frac{\varepsilon_R}{\omega\sigma}\left((Id)^2 - Q^2\right)\hat{\mathbf{r}}\right] X(x,\tau) \\ T_\sigma^{55} &= \frac{1}{2}\left[\left(\frac{c}{c_5}\frac{1}{\omega\sigma}\right)^2\left((Id)^2 - Q^2\right) C(x,\tau) - g_{55}\frac{\varepsilon_R}{\omega\sigma} Q\,(Q - Id\,\cos\theta)\,X(x,\tau)\right] \end{aligned} \tag{140}$$

whose spacetime dependence is determined by $C(x,\tau)$ and $X(x,\tau)$ and is thus out of phase with the $T_0^{\alpha\beta}$.

As expected from the transfer of mass made possible by the fields ϵ^μ in (135) and seen explicitly in (136), and we find a nonzero mass density T^{55} and mass flux T^0 and $\mathbf{T}^5$ into time and space. Moreover, integrating over a sphere of radius r, the net mass flux into space will be of the form

$$\begin{aligned} P &= \int d\Omega\, r^2\, \hat{\mathbf{r}} \cdot \mathbf{T}_0^5 \\ &= \int d\Omega\, r^2\, \hat{\mathbf{r}} \cdot \left[Q\frac{c_5}{c}\left(Q - Id\left(\cos\theta\right)\right) S\left(x,\tau\right)\hat{\mathbf{r}}\right] \\ &= Q\frac{c_5}{c}r^2\, k^2 \left(\frac{\phi\left(\tau - t + \frac{r}{c}\right)}{4\pi r}\right)^2 \sin^2\left(kr - \omega t\right) \int d\Omega\ \left[Q - Id\cos\theta\right] \\ &= \frac{k^2 c_5}{4\pi c} Q^2 \left(\phi\left(\tau - t + \frac{r}{c}\right)\right)^2 \sin^2\left(kr - \omega t\right) \end{aligned} \tag{141}$$

and thus nonzero wherever $\tau \simeq t - r/c$. Just as the energy radiated by a Maxwell dipole antenna must be provided by the amplifier that drives the oscillating current density, the mass radiated by an SHP antenna is continuously provided by an amplifier that creates events and drives them into the antenna.

5.4. *Neutral antenna*

We consider two scenarios for generating a current density in a split dipole antenna. In the first scenario, shown in Figure 1a, the current source simultaneously drives charged events into the left and right antenna segments, so that the total charge in each segment fluctuates around its neutral value by $\pm Q$ over one cycle of oscillation. In the second scenario, shown in Figure 1b, the alternating current source removes charge Q from one segment and drives charge Q into the other segment, so that the antenna remains neutral at all times.

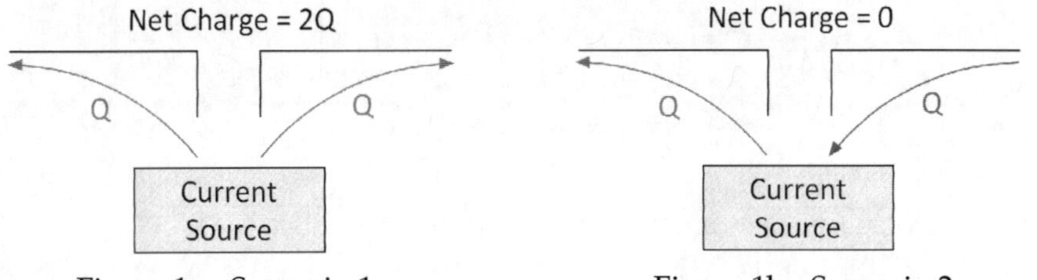

Figure 1a: Scenario 1 Figure 1b: Scenario 2

Taking the current along the z-axis in an antenna of length d, the charge density is described by

$$\rho\left(\mathbf{x}\right) = \begin{cases} \delta\left(x\right)\delta\left(y\right)\rho_z\left(z\right) & , \ -\frac{d}{2} \leq z \leq \frac{d}{2} \\ 0 & , \ \text{otherwise} \end{cases} \tag{142}$$

where

$$\rho_z\left(-z\right) = \begin{cases} +\rho_z\left(z\right) & , \ \text{Scenario 1} \\ -\rho_z\left(z\right) & , \ \text{Scenario 2} \end{cases} \tag{143}$$

and so the current density is related through

$$\rho\left(\mathbf{x}\right) = -\frac{1}{i\omega}\nabla \cdot \mathbf{J}\left(\mathbf{x}\right) \tag{144}$$

and has opposite parity. Evaluating the dipole moment, we find

$$\begin{aligned} Id\,\hat{\mathbf{d}} &= i\omega e \int d^3x\ \mathbf{x}\rho\left(\mathbf{x}\right) = ie\omega\hat{\mathbf{z}} \int_0^{\frac{d}{2}} dz\ z\ \left(\rho_z\left(z\right) - \rho_z\left(-z\right)\right) \\ &= \begin{cases} 0 & ,\ \text{Scenario 1} \\ 2ie\omega\hat{\mathbf{z}} \displaystyle\int_0^{\frac{d}{2}} dz\ z\ \rho_z\left(z\right) & ,\ \text{Scenario 2} \end{cases} \end{aligned} \tag{145}$$

restating the inherent condition that a nonzero dipole moment requires an asymmetric charge distribution. Discounting Scenario 1, we find that in Scenario 2 the total charge on the antenna

$$Q_{\text{total}} = \int d^3x\ e\rho_0\left(\mathbf{x}\right) + \int d^3x\ e\rho\left(\mathbf{x}\right) = Q_0 + \underbrace{e\int_{-\frac{d}{2}}^{\frac{d}{2}} dz\ \rho_z\left(z\right)}_{Q=0} = Q_0 \tag{146}$$

is just the background charge, and the total number of events in the antenna

$$\frac{1}{cc_5}N = \frac{1}{cc_5}\int d^4x\ j^5\left(x,\tau\right) = \int d^3x dt\ \left[\rho_0\left(\mathbf{x}\right) + \rho\left(\mathbf{x}\right)e^{i\omega t}\right]\phi\left(\tau - t\right) = \frac{Q_0}{e} \tag{147}$$

is also given by the constant background charge.

The system simplifies considerably in Scenario 2 with $Q = 0$. The fields reduce to

$$\begin{aligned} \hat{\mathbf{e}} &= -ikId\hat{\mathbf{z}} & \widehat{\mathbf{b}} &= -ikId\ \hat{\mathbf{r}} \times \hat{\mathbf{z}} \\ \widehat{\epsilon}^{\,0} &= 0 & \widehat{\boldsymbol{\epsilon}} &= -ikg_{55}\frac{c}{c_5}\frac{\varepsilon_R}{i\omega\sigma}Id\hat{\mathbf{z}} \end{aligned} \tag{148}$$

so that the effect of the waves on a test event at rest reduce to

$$\ddot{t} = -g_{55}\frac{ec_5}{Mc^2}\epsilon^0 = 0 \qquad \ddot{\mathbf{x}} = -\frac{e}{M}ikId\hat{\mathbf{z}}\left[1 + g_{55}\frac{c}{c_5}\frac{\varepsilon_R}{i\omega\sigma}\right] \tag{149}$$

$$\frac{d}{d\tau}(-\tfrac{1}{2}M\dot{x}^2) = -eg_{55}c_5\epsilon^0 = 0 \tag{150}$$

which does not involve transfer of mass. The components of become $T_0^{\alpha\beta}$

$$\begin{aligned} T_0^{00} &= (Id)^2\left(1 - \frac{1}{2}\cos^2\theta\right)S\left(x,\tau\right) \\ \mathbf{T}_0^{0} &= (Id)^2\left(-\cos\theta\ \hat{\mathbf{z}} + \hat{\mathbf{r}}\ \right)S\left(x,\tau\right) \\ T_0^{50} &= 0 \\ \mathbf{T}_0^{5} &= 0 \\ T_0^{55} &= \frac{1}{2}g^{55}\left(Id\right)^2\cos^2\theta\ S\left(x,\tau\right) \end{aligned} \tag{151}$$

describing no transfer of mass into space or time.

The components of $T_\sigma^{\alpha\beta}$ also simplify to

$$\begin{aligned}
T_\sigma^{00} &= \frac{1}{2}\left(\frac{c}{c_5}\frac{1}{\omega\sigma}\right)^2 (Id)^2\, C(x,\tau) = \frac{1}{2}\left(\frac{c}{\omega c_5}\frac{\phi'}{\phi}\right)^2 (Id)^2\, C(x,\tau) \\
\mathbf{T}_\sigma^0 &= 0 \\
T_\sigma^{50} &= -\frac{c}{c_5} g_{55}\frac{\varepsilon_R}{\omega\sigma}(Id)^2\; X(x,\tau) = \frac{c}{\omega c_5} g_{55}\frac{\phi'}{\phi}(Id)^2\; X(x,\tau) \\
\mathbf{T}_\sigma^5 &= T^{50}\,(\hat{\mathbf{z}} - \hat{\mathbf{r}}) \\
T_\sigma^{55} &= \frac{1}{2}\left(\frac{c}{c_5}\frac{1}{\omega\sigma}\right)^2 (Id)^2\, C(x,\tau) = \frac{1}{2}\left(\frac{c}{\omega c_5}\frac{\phi'}{\phi}\right)^2 (Id)^2\, C(x,\tau)
\end{aligned} \tag{152}$$

where we used (116) to replace $-\varepsilon_R/\sigma$ with ϕ'/ϕ. These expressions involve no transfer of energy but do describe nonzero transfer of mass into space and time directions. Once again, we understand this transfer as an artifact of the time correlation model that enters through the derivative of $\phi(\tau - t)$, rather than an inherent feature of radiation from an oscillating charge. In particular, all of the nonzero terms in the expression for mass conservation contain ϕ', so that these terms are separately conserved among themselves. To see this we write the mass component of (33) as

$$\partial_\alpha T^{\alpha 5} = -\frac{e}{c} f^{5\alpha} j_\alpha \quad\longrightarrow\quad \frac{1}{c}\frac{\partial}{\partial t}T^{50} + \nabla\cdot\mathbf{T}^5 + \frac{1}{c_5}\frac{\partial}{\partial\tau}T^{55} = -\frac{e}{c}\epsilon^\mu j_\mu \tag{153}$$

which becomes

$$\frac{1}{c}\frac{\partial}{\partial t}T_\sigma^{50} + \nabla\cdot\mathbf{T}_\sigma^5 + \frac{1}{c_5}\frac{\partial}{\partial\tau}\left(T_0^{55} + T_\sigma^{55}\right) = -\frac{e}{c}\epsilon_\sigma^\mu j_\mu \tag{154}$$

because $T_0^{50} = \mathbf{T}_0^5 = 0$. We also write the field as ϵ_σ^μ because it is seen in (148) to contain the factor $\varepsilon_R/\omega\sigma$. Finally, we note that because T_0^{55} depends on τ only through the factor of ϕ^2 in $S(x,\tau)$, the derivative $\partial_\tau T_0^{55}$ must similarly contain ϕ'. Thus, each term in (154) enters through the derivative of the time correlation model, and these terms are conserved among themselves with no corresponding energy transfer.

Integrating the energy Poynting vector over the surface of a sphere of radius r we must evaluate

$$\begin{aligned}
\hat{\mathbf{r}}\cdot\mathbf{T}^0 &= (Id)^2\,\hat{\mathbf{r}}\cdot\left(-\cos\theta\,\hat{\mathbf{z}} + \hat{\mathbf{r}}\right) S(x,\tau) \\
&= (Id)^2\left(-\cos^2\theta + 1\right) S(x,\tau) \\
&= (Id)^2 \sin^2\theta\; S(x,\tau)
\end{aligned} \tag{155}$$

to find the instantaneous radiated power

$$\begin{aligned}
P &= \int d\Omega\, r^2\;(Id)^2\, S(x,\tau)\sin^2\theta \\
&= (Id)^2\, r^2\, k^2\left(\frac{\phi\left(\tau - t + \frac{r}{c}\right)}{4\pi r}\right)^2 \sin^2(kr - \omega t)\int_0^{2\pi} d\phi\int_0^{\pi} d\theta\;\sin^3\theta \\
&= (Id)^2\; k^2\left(\frac{\phi\left(\tau - t + \frac{r}{c}\right)}{4\pi}\right)^2 \sin^2(kr - \omega t)\,\frac{8\pi}{3} \\
&= \frac{k^2 (Id)^2}{6\pi}\left(\phi\left(\tau - t + \frac{r}{c}\right)\sin(kr - \omega t)\right)^2 .
\end{aligned} \tag{156}$$

IOP Conf. Series: Journal of Physics: Conf. Series **1239** (2019) 012005 doi:10.1088/1742-6596/1239/1/012005

Since we have assumed that $1/\omega\sigma$ is small, we may take $\phi\left(\tau - t + \frac{r}{c}\right)$ as effectively constant over one cycle of the wave, so that the average radiated power over one cycle is

$$\bar{P} \simeq \frac{k^2 (Id)^2}{6\pi} \left(\phi\left(\tau - t + \frac{r}{c}\right)\right)^2 \frac{1}{T} \int_0^T dt\ \left(\sin\left(kr - \omega t\right)\right)^2 = \frac{k^2 (Id)^2}{12\pi} \left(\phi\left(\tau - t + \frac{r}{c}\right)\right)^2 \tag{157}$$

which agrees with (98) up to the factor of ϕ^2. The neutral antenna radiates energy in agreement with the Maxwell result and radiates no mass (leaving aside the derivatives of the arbitatrarily chosen function ϕ).

6. Discussion

As a canonical classical field theory, Stueckelberg-Horwitz-Piron (SHP) defines events in 4D spacetime on an unconstrained 8D phase space. It is therefore unsurprising, from an abstract point of view, that interactions may involve the transfer of both four-vector quantities and scalar quantities formed from four-vectors. Nevertheless, the concrete statement that electromagnetic fields and particles may exchange both energy-momentum and mass seems to demand further elaboration.

In standard formulations of relativistic mechanics [15] the *a priori* constraint associated with parameterization by proper time

$$-c^2 ds^2 = \eta_{\mu\nu} dx^\mu dx^\nu = \eta_{\mu\nu} \dot{x}^\mu\left(s\right) \dot{x}^\nu\left(s\right) ds^2 \quad \Longrightarrow \quad \dot{x}^2 = -c^2 \tag{158}$$

restricts the degrees of freedom carried by the massless electromagnetic field in such a way that its squared energy and squared momentum are equal. This restriction is reflected in the fact that only three components of the Lorentz force, usually taken as 3-velocity, are independent, while the fourth component, (3-velocity)^2 $\sim$ energy, is dependent on the other three. In SHP theory, however, the proper time interval

$$-c^2 ds^2(\tau) = \eta_{\mu\nu} dx^\mu dx^\nu = \eta_{\mu\nu} \dot{x}^\mu\left(\tau\right) \dot{x}^\nu\left(\tau\right) d\tau^2 \tag{159}$$

is a dynamical scalar quantity dependent on the four unconstrained components of velocity. Therefore, four components of the Lorentz force, usually taken as 4-velocity, are independent, and while the fifth component, (4-velocity)^2 $\sim$ mass, is dependent, determined by whatever the difference between squared energy and squared momentum turns out to be in any given interaction. As a result, the mass of an individual particle (a "fifth", scalar quantity) is only conserved under physical conditions for which certain other "fifth" quantities vanish: a^5, $\partial_5 a^\mu$, and $\partial_5 j^5 = -\partial_\mu j^\mu$. From a general potential found from the Green's function (28)

$$a^\alpha\left(x,\tau\right) = -\frac{e}{c} \int d^4x' d\tau'\ G_P\left(x - x', \tau - \tau'\right) j^\alpha_\varphi\left(x', \tau'\right) \tag{160}$$

we may conclude that $a^5(x,\tau) = 0$ requires $j^5(x,\tau) = 0$, and that $\partial_5 a^\mu(x,\tau) = 0$ requires $\partial_5 j^\mu(x,\tau) = 0$. These requirements, together with $\int d\tau\ G_P\left(x,\tau\right) = -\delta\left(x^2\right)/2\pi$, imply that mass conservation constrains $j^\mu(x,\tau) = j^\mu(x)$ to be a divergenceless Maxwell current.

In the absence of these restrictions, a simple plane wave impinging on a test particle may transfer momentum, energy, and mass, where the total 4-vector and scalar quantities are conserved for the particle and field, and the independence of four momentum components is expressed in the 5D mass shell condition (40). As seen in (56), for any fixed value of $\mathbf{k}$, the mass transfer becomes more pronounced as κ^2, the difference between k_0^2 and $\mathbf{k}^2$, grows, supporting the interpretation of κ as representing the mass of the electromagnetic field. For these plane waves, the mass-energy-momentum tensor is composed of energy density T^{00}, mass density

T^{55}, a Poynting 3-vector $T^{0i} = (k^i/k^0)T^{00}$ representing energy flux into space, and a Poynting 4-vector $T^{5\mu} = (k^\mu/\kappa)T^{55}$ representing mass flux into spacetime.

A radiating particle may lose both energy and mass, whether or not the radiation is absorbed through subsequent interactions. Electromagnetic fields, in Maxwell and SHP theory alike, can be split into a retarded part, associated with uniform inertial motion, that falls off as $1/R^2$, and a radiation part, associated with acceleration, that falls off as $1/R$. The corresponding flux densities therefore split into retarded and radiative parts with $1/R^4$ and $1/R^2$ dependence, respectively. Since a closed surface in 3D has area element $R^2 d\Omega$, the total flux from the retarded part decreases rapidly at long distances, while the radiation flux can propagate to infinity. Examples of these dependencies were seen in the Poynting vectors (68) for uniform motion and (97) for accelerated motion. But the symmetries of the field configurations are equally significant. For uniform motion, the Poynting vector is antisymmetric with respect to the path of the particle, so that the total flux vanishes identically at any distance, preventing energy loss by radiation regardless of the R-dependence. But the Poynting vector for the oscillating charge is directed radially outward at any observation point, so that the radiative energy flux will be nonzero at any finite distance, even if the flux were to fall off more strongly than $1/R^2$.

In the case of the simple SHP dipole antenna, the model we used to simplify the oscillating current density presents two opposite scenarios. In Scenario 1, the amplifier alternates between injecting and withdrawing charge Q symmetrically into the antenna, with current $I = 0$ across the antenna. Under these conditions, we see from (139) that the field possesses both mass and energy density, each producing radiative flux. In Scenario 2, the antenna remains electrically neutral with $Q = 0$, while the amplifier shifts charge back and forth across the antenna segments characterized by an alternating current with magnitude I. Under these conditions, expressions (151) indicates that the field possesses both mass and energy density, producing radiation of energy equivalent to the Maxwell result, but no radiation of mass.

The absence of mass radiation in Scenario 2 can be attributed to the absence of event injection into the antenna. As was seen in (147), when the net injected charge Q vanishes, the total number of events in the antenna is given by the background charge, with no contribution from the oscillating current. This association of event injection with mass radiation can also be understood through the continuity equation, which relates the event number N to the divergence of the 4-current. Rewriting (105) in the form

$$\Delta N = N(\infty) - N(-\infty) = -\frac{1}{c}\int d\tau \int d^4x \, \partial_\mu j^\mu(x,\tau) \tag{161}$$

we generally interpret the spacetime integral of the total divergence as the current crossing the outer boundary of spacetime at infinity. We expect this integral to vanish on qualitative grounds unless the current includes an external component, such as an ideal amplifier injecting charged events into the system. Indeed, changing the order of integration and using the boundary conditions (29) we find

$$\Delta N = -\frac{1}{c}\int d^4x \, \partial_\mu \int d\tau \, j^\mu(x,\tau) = -\frac{1}{c}\int d^4x \, \partial_\mu J^\mu(x) = 0 \tag{162}$$

expressing the net zero change in event number in Maxwell theory as the equilibrium limit of SHP. However, for the current defined in (107) and (108), and recalling (117), equation (105) becomes

$$\begin{aligned}\frac{dN}{d\tau} &= -\frac{1}{c}\int d^4x\, \partial_\mu j^\mu(x,\tau)\\ &= -\frac{1}{c}\int d^4x\, \left\{ i\omega\rho\left(\mathbf{x}\right)e^{i\omega t}\phi\left(\tau - t\right) - \rho\left(\mathbf{x}\right)e^{i\omega t}\phi'\left(\tau - t\right) + \nabla\cdot\mathbf{J}\left(\mathbf{x}\right)e^{i\omega t}\phi\left(\tau - t\right)\right\}\\ &= \int d^3x\, \rho\left(\mathbf{x}\right)\int dt\, e^{i\omega t}\phi'\left(\tau - t\right)\\ &= \frac{Q}{e}\, i\omega\Phi\left(\omega\right)e^{i\omega\tau}\\ &= \frac{1}{e}\,\frac{d}{d\tau}Q\left(\tau\right)\end{aligned} \tag{163}$$

where $Q(\tau)$ is the total spacetime charge given in (119). As expected, integrating over τ takes $e^{i\omega\tau} \longrightarrow \delta(\omega)$ so that $\Delta N = 0$. However, at any finite τ the injection of charge $Q \neq 0$ into the antenna is equivalent to a 4-current with non-vanishing divergence, and is thus equivalent to an oscillating event injection, leading to mass radiation.

References

[1] Stueckelberg E 1941 *Helv. Phys. Acta* **14** 321–322 (In French)
[2] Stueckelberg E 1941 *Helv. Phys. Acta* **14** 588–594 (In French)
[3] Horwitz L and Piron C 1973 *Helv. Phys. Acta* **48** 316–326
[4] Saad D, Horwitz L and Arshansky R 1989 *Found. Phys.* **19** 11251149
[5] Land M 2017 *Journal of Physics: Conference Series* **845** 012024 URL `http://stacks.iop.org/1742-6596/845/i=1/a=012024`
[6] Horwitz L P 2015 *Relativistic Quantum Mechanics* (Dordrecht, Netherlands: Springer)
[7] Land M 2017 *Journal of Physics: Conference Series* **845** 012025 URL `http://stacks.iop.org/1742-6596/845/i=1/a=012025`
[8] Horwitz L P 2017 *Journal of Physics: Conference Series* **845** 012026 URL `http://stacks.iop.org/1742-6596/845/i=1/a=012026`
[9] Horwitz L, Arshansky R and Elitzur A 1988 *Found. Phys.* **18** 11591193
[10] Arshansky R, Horwitz L and Lavie Y 1983 *Found Phys* **13** 1167
[11] Land M 2003 *Found. of Phys.* **33** 1157
[12] Land M and Horwitz L 1991 *Found. Phys. Lett.* **4** 61
[13] Land M and Horwitz L 1991 *Found. Phys.* **21** 299–310
[14] Jackson J 1975 *Classical Electrodynamics* (New York, NY, USA: Wiley) chapter 9, p. 391ff
[15] Fock V 1937 *Phys. Z. Sowjetunion* **12** 404–425 URL `http://www.neo-classical-physics.info/uploads/3/4/3/6/34363841/fock_-_wkb_and_dirac.pdf`

IOP Conf. Series: Journal of Physics: Conf. Series **1239** (2019) 012006 doi:10.1088/1742-6596/1239/1/012006

The Effect of Retardation on Galactic Rotation Curves

Asher Yahalom
Ariel University, Ariel 40700, Israel
E-mail: asya@ariel.ac.i

Abstract.
Galaxies are huge physical systems having dimensions of many tens of thousands of light years. Thus any change at the galactic center will be noticed at the rim only tens of thousands of years later. Those retardation effects seems to be neglected in present day galactic modelling used to calculate rotational velocities of matter in the rims of the galaxy and surrounding gas. The significant differences between the predictions of Newtonian instantaneous action at a distance and observed velocities are usually explained by either assuming dark matter or by modifying the laws of gravity (MOND). In this paper we will show that taking general relativity seriously without neglecting retardation effects one can explain the radial velocities of galactic matter without postulating dark matter.

1. Introduction

The general theory of relativity (GR) is verified by many observations. Nevertheless, some observations seems not to fit GR and observed matter. As soon as 1933 Fritz Zwicky realized that the velocities of the Galaxies within the Comma Cluster are way larger than those predicted by the virial theorem in Newtonian theory [1]. He remarked that the amount of matter needed to account for the velocities could be 400 times that of the visible matter. Which led to postulating an unseen form of matter permeating the cluster. Volders in 1959 remarked that stars in the periphery of the neighbor spiral galaxy M33 do not move as expected [2]. The virial theorem in Newtonian Gravity predicts that $MG/r \sim Mv^2$, that is to say, the rotation curve should increase and at some point bend down and the velocity should drop off as $1/\sqrt{r}$. In the seventies Rubin and Ford [3, 4] showed for a very large sample of spiral galaxies that this behavior is a general feature: velocities at the periphery of the galaxies do not bend down, attain a plateau at some velocity for each galaxy. In figure 1 we see a rotation curves for the M33 galaxy describing this situation. In what follows we will show that such effects can be deduced from GR if retardation effects are not neglected.

IOP Conf. Series: Journal of Physics: Conf. Series **1239** (2019) 012006 doi:10.1088/1742-6596/1239/1/012006

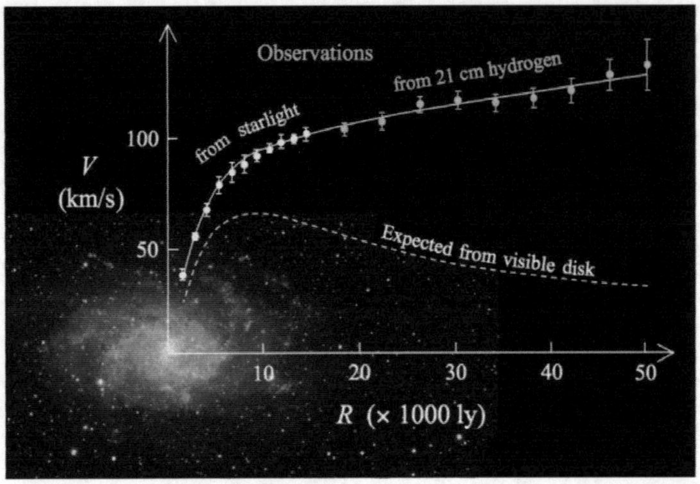

Figure 1. M33 rotation curve [5]

2. General Relativity

The general theory of relativity is based on two fundamental equations, Einstein equations [6, 8, 9, 7]:

$$G_{\mu\nu} = -\frac{8\pi G}{c^4} T_{\mu\nu} \tag{1}$$

in which $G_{\mu\nu}$ is the Einstein tensor, $T_{\mu\nu}$ is the stress-energy tensor, $G \simeq 6.67\ 10^{-11}\ m^3 kg^{-1} s^{-2}$ is the gravitational constant and $c \simeq 3\ 10^8\ ms^{-1}$ is the velocity of light in vacuum (Greek indices are in the range $0-3$). And the geodesic equation:

$$\frac{d^2x^\alpha}{ds^2} + \Gamma^\alpha_{\mu\nu}\frac{dx^\mu}{ds}\frac{dx^\nu}{ds} = \frac{du^\alpha}{ds} + \Gamma^\alpha_{\mu\nu} u^\mu u^\nu = 0 \tag{2}$$

in which $x^\alpha(s)$ are the four dimensional coordinates of the particle in space-time, s is a parameter along the trajectory that for massive particles can be the length of the trajectory, $u^\mu = \frac{dx^\mu}{ds}$ and $\Gamma^\alpha_{\mu\nu}$ is the affine connection (Einstein summation convention is assumed). The stress-energy tensor of matter is usually taken in the form:

$$T_{\mu\nu} = (p + \rho c^2) u_\mu u_\nu - p\ g_{\mu\nu} \tag{3}$$

In the above p is the pressure, ρ is the density. Lowering and raising indices is done through the metric $g_{\mu\nu}$ and the inverse metric $g^{\mu\nu}$, that is $u_\mu = g_{\mu\nu} u^\nu$. The same metric serves to calculate s:

$$ds^2 = g_{\mu\nu} dx^\mu dx^\nu, \tag{4}$$

and the affine connection:

$$\Gamma^\alpha_{\mu\nu} = \frac{1}{2} g^{\alpha\beta}\left(g_{\beta\mu,\nu} + g_{\beta\nu,\mu} - g_{\mu\nu,\beta}\right), \qquad g_{\beta\mu,\nu} \equiv \frac{\partial g_{\beta\mu}}{\partial x^\nu} \tag{5}$$

The affine connection serves to calculate the Riemann and Ricci tensors and the curvature scalar:

$$R^\mu_{\nu\alpha\beta} = \Gamma^\mu_{\nu\alpha,\beta} - \Gamma^\mu_{\nu\beta,\alpha} + \Gamma^\sigma_{\nu\alpha}\Gamma^\mu_{\sigma\beta} - \Gamma^\sigma_{\nu\beta}\Gamma^\mu_{\sigma\alpha}, \quad R_{\alpha\beta} = R^\mu_{\alpha\beta\mu}, \quad R = g^{\alpha\beta} R_{\alpha\beta} \tag{6}$$

which in turn serves to calculate the Einstein tensor:

$$G_{\alpha\beta} = R_{\alpha\beta} - \frac{1}{2} g_{\alpha\beta} R. \tag{7}$$

Hence matter distribution determines the metric through equation (1) and metric determines trajectories through equation (2) as is well known.

3. Linear Approximation of GR

Except for the extreme cases of compact objects (black holes and neutron stars) and the very early universe (big bang) one need not consider the full non-linear Einstein equation. In most other cases of astronomical interest (galactic dynamics included) one can linearize those equations around the flat Lorentz metric $\eta_{\mu\nu}$ such that[1]:

$$g_{\mu\nu} = \eta_{\mu\nu} + h_{\mu\nu}, \quad \eta_{\mu\nu} \equiv \text{diag } (1, -1, -1, -1), \quad |h_{\mu\nu}| \ll 1 \tag{8}$$

One than defines the quantity:

$$\bar{h}_{\mu\nu} \equiv h_{\mu\nu} - \frac{1}{2}\eta_{\mu\nu}h, \quad h = \eta^{\mu\nu}h_{\mu\nu}, \tag{9}$$

$\bar{h}_{\mu\nu} = h_{\mu\nu}$ for non diagonal terms. For diagonal terms:

$$\bar{h} = -h \Rightarrow h_{\mu\nu} = \bar{h}_{\mu\nu} - \frac{1}{2}\eta_{\mu\nu}\bar{h}. \tag{10}$$

It can be shown ([6] page 75 exercise 37, see also [7, 8, 9]), that one can choose a gauge such that the Einstein equations are:

$$\bar{h}_{\mu\nu,\alpha}{}^{\alpha} = -\frac{16\pi G}{c^4}T_{\mu\nu}, \qquad \bar{h}_{\mu\alpha,}{}^{\alpha} = 0. \tag{11}$$

Equation (11) can always be integrated to take the form [10][2]:

$$\bar{h}_{\mu\nu}(\vec{x}, t) = -\frac{4G}{c^4}\int \frac{T_{\mu\nu}(\vec{x}', t - \frac{R}{c})}{R} d^3x',$$
$$t \equiv \frac{x^0}{c}, \quad \vec{x} \equiv x^a \quad a, b \in [1, 2, 3], \quad \vec{R} \equiv \vec{x} - \vec{x}', \quad R = |\vec{R}|. \tag{12}$$

The factor before the integral is small: $\frac{4G}{c^4} \simeq 3.3 \; 10^{-44}$ hence in the above calculation one can take $T_{\mu\nu}$ which is zero order in $h_{\alpha\beta}$. Let us now calculate the affine connection in the linear approximation:

$$\Gamma^{\alpha}_{\mu\nu} = \frac{1}{2}\eta^{\alpha\beta}\left(h_{\beta\mu,\nu} + h_{\beta\nu,\mu} - h_{\mu\nu,\beta}\right). \tag{13}$$

The affine connection has only first order terms, hence for a first order approximation of $\Gamma^{\alpha}_{\mu\nu}u^{\mu}u^{\nu}$ appearing in the geodesic, $u^{\mu}u^{\nu}$ is zeroth order. In the zeroth order:

$$u^0 = \frac{1}{\sqrt{1 - \frac{v^2}{c^2}}}, \qquad u^a = \vec{u} = \frac{\frac{\vec{v}}{c}}{\sqrt{1 - \frac{v^2}{c^2}}}, \qquad \vec{v} \equiv \frac{d\vec{x}}{dt}, \quad v = |\vec{v}|. \tag{14}$$

For non relativistic velocities:

$$u^0 \simeq 1, \qquad \vec{u} \simeq \frac{\vec{v}}{c}, \qquad u^a \ll u^0 \qquad \text{for} \quad v \ll c. \tag{15}$$

Inserting equation (13) and equation (15) in the geodesic equation we arrive at the approximate form:

$$\frac{dv^a}{dt} \simeq -c^2\Gamma^{a}_{00} = -c^2\left(h^{a}_{0,0} - \frac{1}{2}h_{00,}{}^{a}\right) \tag{16}$$

[1] Private communication with the late Professor Donald Lynden-Bell

[2] For reasons why the symmetry between space and time is broken see [11, 12]

Let us now look at $T_{\mu\nu} = (p + \rho c^2)u_\mu u_\nu - p\ g_{\mu\nu}$. In the current case $\rho c^2 \gg p$, combining this with equation (15) we arrive at $T_{00} = \rho c^2$ while all other components of the tensor $T_{\mu\nu}$ are significantly smaller. This implies that $\bar{h}_{00}$ is significantly larger than other components of the tensor $\bar{h}_{\mu\nu}$. Of course one should be careful and not deduce from the different magnitudes of quantities that such a difference exist between their derivatives. In fact by the gauge condition in equation (11):

$$\bar{h}_{\alpha 0,}{}^{0} = -\bar{h}_{\alpha a,}{}^{a} \qquad \Rightarrow \bar{h}_{00,}{}^{0} = -\bar{h}_{0a,}{}^{a}, \quad \bar{h}_{b0,}{}^{0} = -\bar{h}_{ba,}{}^{a}. \tag{17}$$

Hence the zeroth derivative of $\bar{h}_{00}$ (contains a $\frac{1}{c}$ factor) is the same order as the spatial derivative of $\bar{h}_{0a}$ and like wise the zeroth derivative of $\bar{h}_{0a}$ (which appears implicitly in equation (16)) is the same order of the spatial derivative of $\bar{h}_{ab}$. However, it is safe to compare spatial derivatives of $\bar{h}_{00}$ and $\bar{h}_{ab}$ and conclude that the former is significantly larger than the later. Using equation (10) and taking the above consideration into account we write equation (16) as:

$$\frac{dv^a}{dt} \simeq \frac{c^2}{4}\bar{h}_{00,}{}^{a} \Rightarrow \frac{d\vec{v}}{dt} = -\vec{\nabla}\phi = \vec{F}, \qquad \phi \equiv \frac{c^2}{4}\bar{h}_{00} \tag{18}$$

Thus ϕ is a gravitational potential of the motion which can be calculated using equation (12):

$$\phi = \frac{c^2}{4}\bar{h}_{00} = -\frac{G}{c^2}\int \frac{T_{00}(\vec{x}', t - \frac{R}{c})}{R} d^3x' = -G\int \frac{\rho(\vec{x}', t - \frac{R}{c})}{R} d^3x' \tag{19}$$

and $\vec{F}$ is the force per unit mass. If ρ is static we are in the realm of the Newtonian instantaneous action at a distance theory. However, it is unlikely that ρ is static as a galaxy will attract mass from the intergalactic medium.

4. Beyond the Newtonian Approximation

The retardation time $\frac{R}{c}$ which may be a few tens of thousands of years is short with respect to the time that the galactic density changes significantly. This means that we can write a Taylor series for the density [15]:

$$\rho(\vec{x}', t - \frac{R}{c}) = \sum_{n=0}^{\infty} \frac{1}{n!}\rho^{(n)}(\vec{x}', t)(-\frac{R}{c})^n, \qquad \rho^{(n)} \equiv \frac{\partial^n \rho}{\partial t^n}. \tag{20}$$

Inserting equation (20) into equation (19) and keeping the first three terms we will obtain:

$$\phi = -G\int \frac{\rho(\vec{x}', t)}{R} d^3x' + \frac{G}{c}\int \rho^{(1)}(\vec{x}', t)d^3x' - \frac{G}{2c^2}\int R\rho^{(2)}(\vec{x}', t)d^3x' \tag{21}$$

The first term will provide the Newtonian potential, the second term does not contribute, the third term will result in the lower order correction to the Newtonian theory:

$$\phi_r = -\frac{G}{2c^2}\int R\rho^{(2)}(\vec{x}', t)d^3x' \tag{22}$$

The total force per unit mass:

$$\begin{aligned} \vec{F} &= \vec{F}_N + \vec{F}_r \\ \vec{F}_N &= -\vec{\nabla}\phi_N = -G\int \frac{\rho(\vec{x}', t)}{R^2}\hat{R}d^3x', \qquad \hat{R} \equiv \frac{\vec{R}}{R} \\ \vec{F}_r &\equiv -\vec{\nabla}\phi_r = \frac{G}{2c^2}\int \rho^{(2)}(\vec{x}', t)\hat{R}d^3x' \end{aligned} \tag{23}$$

While the Newtonian force $\vec{F}_N$ is always attractive the retardation force $\vec{F}_r$ can be either attractive or repulsive. Also notice that while the Newtonian force decreases as $\frac{1}{R^2}$, the retardation force is independent of distance as long as the Taylor approximation of equation (20) is valid. For short distances the Newtonian force is dominant but as the distances increase the retardation force becomes dominant. Newtonian force can be neglected for distances significantly larger than the retardation distance:

$$R \gg R_r \equiv c\Delta t \tag{24}$$

Δt is the typical duration in which the density ρ changes. Of course for $R \ll R_r$ the retardation effect can be neglected and only Newtonian forces should be considered. For large distances $r = |\vec{x}| \to \infty$ such that $\hat{R} \simeq \frac{\vec{x}}{|\vec{x}|} \equiv \hat{r}$ we obtain:

$$\vec{F}_r = \frac{G}{2c^2}\hat{r}\int \rho^{(2)}(\vec{x}', t)d^3x' = \frac{G}{2c^2}\hat{r}\ddot{M}, \qquad \ddot{M} \equiv \frac{d^2M}{dt^2}. \tag{25}$$

Now as the galaxy attracts intergalactic gas its mass increases thus $\dot{M} > 0$, however, as the intergalactic gas is depleted the rate at which the mass increases must decrease hence $\ddot{M} < 0$. Thus in the galactic case:

$$\vec{F}_r = -\frac{G}{2c^2}|\ddot{M}|\hat{r} \tag{26}$$

and the retardation force is attractive.

5. Dark Matter

In what circumstances can one confuse retardation with the effect of a non existent "dark matter"? Let us ignore retardation effects and suppose that radial velocities are a result of some mysterious dark matter. In this case we can write for a spherically symmetric mass distribution [13]:

$$-\frac{v_c^2}{r}\hat{r} = \vec{F}_d = -\frac{GM_d(r)}{r^2}\hat{r} \tag{27}$$

v_c is the speed of a test particle of constant radius r and $M_d(r)$ is the amount of dark matter inside the radius r. Comparing equation (27) and equation (26) we see that the "dark matter" mass can be calculated as follows:

$$M_d(r) = \frac{r^2|\ddot{M}|}{2c^2} \tag{28}$$

Now since:

$$M_d(r) = 4\pi\int_0^r r'^2\rho_d(r')dr', \qquad \frac{dM_d(r)}{dr} = 4\pi r^2\rho_d(r) \tag{29}$$

it follows:

$$\rho_d(r) = \frac{|\ddot{M}|}{4\pi c^2 r} \tag{30}$$

and the asymptotic rotation velocity is thus:

$$v_c = \sqrt{\frac{G}{2c^2}|\ddot{M}|r} \tag{31}$$

This is consistent with observational data of [5] who concluded that the "dark matter" density decreases as $r^{-1.3}$ for M33. We notice that the asymptotic rotation velocity can only be achieved for distances which are much larger than the galaxy itself and the exceed considerably the distances for which Newtonian contributions are significant and at the same time do not violate the second order Taylor expansion approximation which we consider.

IOP Conf. Series: Journal of Physics: Conf. Series **1239** (2019) 012006 doi:10.1088/1742-6596/1239/1/012006

6. MOND

Another approach to explaining galactic rotation curves is the claim that either the laws of dynamics (Newton's second law) or the laws of Gravitation (GR) should be modified. This approach championed by Milgorm is denoted "MOND" (Modified Newtonian dynamics) [14]. In one version of this approach Newton's law of gravity is modified:

$$\vec{F}_M = -\frac{GM}{\mu(\frac{a}{a_0})r^2}\hat{r} \tag{32}$$

In the above μ is the interpolation function that should be 1 for $a_0 \ll a$. Let us assume:

$$\mu(\frac{a}{a_0}) = \frac{1}{1 + (\frac{a_0}{a})^2} \tag{33}$$

If $a_0 \gg a$, $\mu \simeq (\frac{a}{a_0})^2$. A test particle revolving in a constant radius will have centrifugal acceleration $a = \frac{v^2}{r}$ and thus:

$$\vec{F}_M = -\frac{GMa_0^2}{v^4}\hat{r} \tag{34}$$

For v constant at a far away distance this expression is similar to the retardation force and thus:

$$|\ddot{M}| = \frac{2Ma_0^2c^2}{v^4}. \tag{35}$$

Milgrom found $a_0 = 1.2\ 10^{-10}ms^{-2}$ to be most fitting to the data. The mass of the M33 galaxy is $9.95\ 10^{40}kg$ and the velocity far away[3] from the galaxy is $179,000\ ms^{-1}$. We thus obtain $\ddot{M} \simeq 2.51\ 10^{17}kgs^{-2}$ and a ratio $\frac{|\ddot{M}|}{M} \simeq 2.52\ 10^{-24}\ s^{-2}$. This amounts to a typical accumulation acceleration time scale of $20,000$ years and retardation distance of $20,000$ light years which seems reasonable according to figure 1. We notice that such an accumulation acceleration cannot be maintained throughout the life of the galaxy and is connected with "recent" (in cosmological time scale) depletion of gas around the galaxy.

7. Conclusion

We show that "dark matter" and "MOND" effects are explained in the framework of standard GR as effects due to retardation without assuming any exotic matter or modifications of the theory of gravity.

What will happen if the mass out side the galaxy is not yet depleted or totally depleted? In this case $\ddot{M} = 0$ and retardation force should vanish. This was indeed reported recently [16] for the young galaxy NGC1052-DF2.

Retardation effects in electromagnetic theory were discussed in [17, 18, 19].

[1] F. Zwicky,"On a New Cluster of Nebulae in Pisces," In: Proc. Natl. Acad. Sci. U S A., May 1937, vol. 23(5), pp. 251256.

[2] L.M.J.S. Volders, "Neutral Hydrogen in M33 and M101," *Bull. astr. Inst. Netherl.*, vol. 14, 323, 1959. V.C. Rubin, W.K. Ford Jr., N. Thonnard, and M.S. Roberts, "Motion of the Galaxy and the Local Group Determined from the Velocity Anisotropy of Distant Sc I Galaxies. I. The Data and II. The Analysis for the Motion," *Astrophys. J.*, vol. 81, 687 and 719, 1976.

[3] V.C. Rubin, and W.K. Ford Jr., "Rotation of the Andromeda Nebula from a Spectroscopic Survey of Emission Regions," *Astrophys. J.*, vol. 159, 379, 1970.

[4] V.C. Rubin, W.K. Ford Jr., and N. Thonnard, "Rotational Properties of 21 Sc Galaxies with a Large Range of Luminosities and Radii from NGC 4605 (R=4kpc) to UGC 2885 (R=122kpc)," *Astrophysical Journal*, vol. 238, 471, 1980.

[3] This is only a crude estimation, of course the retardation force does not allow strictly constant circular velocity profiles

[5] E. Corbelli; P. Salucci (2000). "The extended rotation curve and the dark matter halo of M33". Monthly Notices of the Royal Astronomical Society. 311 (2): 441447. arXiv:astro-ph/9909252. doi:10.1046/j.1365-8711.2000.03075.x.
[6] Narlikar, J. V. (1993). Introduction to Cosmology, Second Edition. Cambridge University Press.
[7] A. S. Eddington, "The mathematical theory of relativity" Cambridge University Press (1923)
[8] S. Weinberg "Gravitation and Cosmology: Principles and Applications of the General Theory of Relativity" John Wiley & Sons, Inc. (1972)
[9] C. W. Misner, K.S. Thorne & J.A. Wheeler, "Gravitation" W.H. Freeman & Company (1973)
[10] J. D. Jackson, Classical Electrodynamics, Third Edition. Wiley: New York, (1999).
[11] Asher Yahalom "The Geometrical Meaning of Time" Foundations of Physics, http://dx.doi.org/10.1007/s10701-008-9215-3 Volume 38, Number 6, Pages 489-497 (June 2008).
[12] Yahalom, A. (2009). The Gravitational Origin of the Distinction between Space and Time, International Journal of Modern Physics D, Vol. 18, Issue: 14, pp. 2155-2158. Gravity Research Foundation - 2009 Honorable Mention.
[13] J. Binney & S. Tremaine, Galactic Dynamics, Princeton University Press (1987)
[14] Milgrom, M. (1983). "A modification of the Newtonian dynamics as a possible alternative to the hidden mass hypothesis". Astrophysical Journal. 270: 365370. doi:10.1086/161130. Milgrom, M. (1983). "A modification of the Newtonian dynamics - Implications for galaxies". Astrophysical Journal. 270: 371389. doi:10.1086/161131. Milgrom, M. (1983). "A modification of the Newtonian dynamics - Implications for galaxy systems". Astrophysical Journal. 270: 384. doi:10.1086/161132.
[15] A. Yahalom "Retardation Effects in Electromagnetism and Gravitation" Proceedings of the Material Technologies and Modeling the Tenth International Conference, Ariel University, Ariel, Israel, August 20 24, 2018. (arXiv:1507.02897v2)
[16] Pieter van Dokkum, Shany Danieli, Yotam Cohen, Allison Merritt, Aaron J. Romanowsky, Roberto Abraham, Jean Brodie, Charlie Conroy, Deborah Lokhorst, Lamiya Mowla, Ewan OSullivan & Jielai Zhang "A galaxy lacking dark matter" Nature volume 555, pages 629632 (29 March 2018) doi:10.1038/nature25767.
[17] M. Tuval & A. Yahalom *Newton's Third Law in the Framework of Special Relativity*, Eur. Phys. J. Plus **129**, 240 (2014). DOI: 10.1140/epjp/i2014-14240-x. (arXiv:1302.2537 [physics.gen-ph])
[18] M. Tuval and A. Yahalom, *Momentum Conservation in a Relativistic Engine*, Eur. Phys. J. Plus **131**, 374 (2016). DOI: 10.1140/epjp/i2016-16374-1
[19] Asher Yahalom "Retardation in Special Relativity and the Design of a Relativistic Motor". Acta Physica Polonica A, Vol. 131 (2017) No. 5, 1285-1288. DOI: 10.12693/APhysPolA.131.1285.

IOP Conf. Series: Journal of Physics: Conf. Series **1239** (2019) 012007 doi:10.1088/1742-6596/1239/1/012007

Significance of Deformity on the Mass Quadrupole Moment of Non-Rotating Neutron Stars

Omair Zubairi[1] and Fridolin Weber[2,3]

[1] Department of Sciences, Wentworth Institute of Technology, 550 Huntington Ave Boston, MA 02115, USA

[2] Department of Physics, San Diego State University, 5500 Campanile Drive, San Diego, CA 92182, USA

[3] Center for Astrophysics and Space Sciences, University of California, San Diego, La Jolla, CA 92093, USA

E-mail: `zubairio@wit.edu`[1], `fweber@sdsu.edu`[2,3]

Abstract. Generally, when modeling non-rotating neutron stars, assumptions of perfect spherical symmetry is made; however, this assumption is not correct if the interior composition of the star is described by an anisotropic equation of state. Due to extremely high magnetic fields, magnetars and/or neutron stars that contain color-superconducting quark matter cores can exhibit such anisotropies. Recent publications on models of the global structure of highly magnetized neutron stars indicate that these objects can be deformed making them either oblate or prolate spheroids. Due to these deformations, the gravitational quadrupole moment is expected to be non-zero resulting in a non-homogeneous mass distribution in either the equatorial or polar directions, thus resulting in different masses from the spherical case. In this work, we examine this inhomogeneity by calculating the gravitational mass quadrupole moment of non-rotating deformed neutron stars in the framework of general relativity and investigate any changes from traditional spherical models.

1. Introduction

Non-rotating neutron stars are compact stellar objects whose masses lie between $\sim 1-2$ solar masses with a radii of $\sim 10-15$ kilometers. Such a large mass enclosed in a small volume results in densities on the nuclear scales. Traditionally, non-rotating neutron stars are modeled with the assumption that they are perfect spheres whose said stellar properties are described by the well known Tolman-Oppenheimer-Volkoff (TOV) equation [1, 2],

$$\frac{dP}{dr} = -\frac{\varepsilon\left(1+\frac{P}{\varepsilon}\right) m \left(1+\frac{4\pi P r^3}{m}\right)}{r^2\left(1-\frac{2m}{r}\right)} , \tag{1}$$

where the normalized units of $G = c = 1$. In Eq. (1) P is the pressure, ε is the energy density, and m is the mass of a spherical shell of radius r described by

$$\frac{dm}{dr} = 4\pi r^2 \varepsilon . \tag{2}$$

Equations (1) and (2) can be easily solved numerically with a use of a given model for the equation of state (EoS), which describes the interior composition of a neutron star.

This assumption of perfect spherical symmetry may not always be correct. If high magnetic fields (up to approximately 10^{18-20} Gauss) are present such as for Magnetars [3–5], and/or the pressure of the matter in the cores of neutron stars is anisotropic, as predicted by some models of color superconducting quark matter [6–11], then deformation of neutron stars can occur [6, 12–17], thus making them oblong spheroids as shown in Fig. 1,

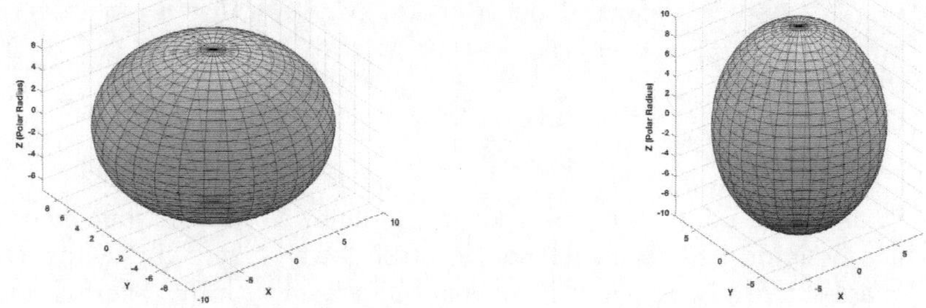

(a) (b)

Figure 1. (Color on-line) Graphical representation of an oblate (a) and prolate (b) spheroid. The polar direction (z) is orthogonal to the equatorial ($x-y$) plane.

where these oblate and prolate stars have distinct polar (z) and equatorial ($x-y$) directions. The geometry of these axial symmetric objects are described mathematically by the Weyl metric [18, 19]

$$ds^2 = \mathrm{e}^{2\lambda}dt^2 - \mathrm{e}^{-2\lambda}\left[\mathrm{e}^{2\nu}\left(dr^2 + dz^2\right) + r^2 d\phi^2\right] = g_{\mu\nu}dx^{\mu}dx^{\nu}\,, \tag{3}$$

where t is the time component and r, z, and ϕ are the spatial components in cylindrical coordinates. The terms λ and ν are the unknown metric functions that depend on both the radial r and polar z directions such that $\lambda = \lambda(r, z)$ and $\nu = \nu(r, z)$. Due to the fact that there are distinct polar and radial directions which are both coupled together, as described in Eq. (3), the hydrostatic equilibrium of deformed neutron stars must take these distinct directions into consideration.

Recent work on deformation models of compact stars suggest that the masses of these deformed objects can change appreciably from the spherical case [20–23]. In particular, oblate stars can have an increase in mass and prolate stars can have a decrease in mass [21–23]. In this work, we further examine the deformation of highly magnetized neutron stars by calculating the mass quadropole moment which is expected to be non-zero.

2. Non-Spherical Geometry for Oblong Spheroids

The stellar structure of oblong spheroids with distinct polar and radial directions can be described by applying a parametrization on the polar radius in terms of the equatorial radius along with deformation constant γ on the metric described by Eq. (3), the metric will then read as

$$ds^2 = -\mathrm{e}^{2\Phi(r)}dt^2 + \left(1 - \frac{2m(r)}{r}\right)^{-\gamma} dr^2 + r^2 d\theta^2 + r^2\sin^2(\theta)d\phi^2\,, \tag{4}$$

described in [21], which is analogous to the metric described in [19]. In Eq. (4), γ determines the degree of deformation either in the oblate ($\gamma < 1$) or prolate case ($\gamma > 1$) (see Fig. 1), where the parametrization is described by $z = \gamma r$. Using this parametrization, the mass quadropole moment is then

$$Q = \frac{\gamma}{3} M^3 \left(1 - \gamma^2\right) \,, \tag{5}$$

as generalized by [19] where M is the total gravitational mass of an oblate or prolate star. In order to calculate the gravitational mass quadropole moment, we first need to determine the total gravitational mass M. Second, we need values for the deformation constant γ. To determine the total gravitiational mass of a deformed neutron star M, we utilize a parameterized TOV equation as described in [21],

$$\frac{dP}{dr} = -\frac{(\epsilon + P)\left(\frac{1}{2}r + 4\pi r^3 P - \frac{1}{2}r\left(1 - \frac{2m}{r}\right)^{\gamma}\right)}{r^2\left(1 - \frac{2m}{r}\right)^{\gamma}} \,, \tag{6}$$

where in the case when $\gamma = 1$, Eq. (6) simplifies to the well-known Tolman-Oppenheimer-Volkoff equation [1, 2] which describes the stellar structure of perfectly spherically symmetric object. The gravitational mass of a deformed neutron star parameterized with the deformation constant γ is then described by

$$\frac{dm}{dr} = 4\pi r^2 \epsilon \gamma \,, \tag{7}$$

so that the total gravitational mass, M, of a deformed neutron star with an equatorial radius R follows as [19, 21] $M = \gamma\, m(R)$. For a given deformation, γ, this is a very straightforward and useful model for the description of deformed compact stars. Using any EoS model in the limiting case of isotropy, the parameterized structure equations can be easily solved numerically and produce stellar properties such as masses and radii.

Solving Eqs. (6) and (7), stellar properties such as masses and radii are produced as generalized by [20, 21] shown in Fig. 2.

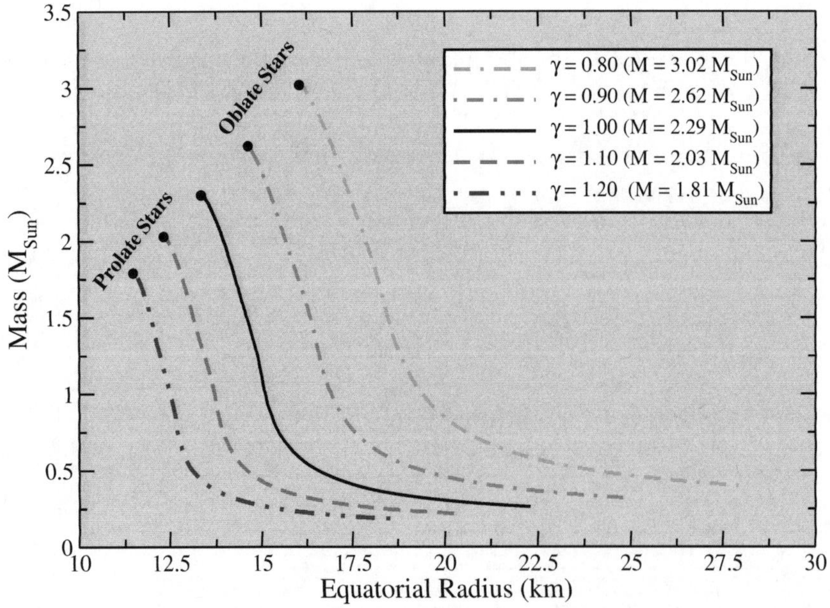

Figure 2. (Color-online) Mass-Radius relations for deformed neutron stars using a Quark-Hadron equation of state model [24, 25]. The dots on each curve represent the maximum mass for each stellar sequence.

IOP Conf. Series: Journal of Physics: Conf. Series **1239** (2019) 012007 doi:10.1088/1742-6596/1239/1/012007

The black solid line in Fig. 2, represents the spherical case which is in good agreement from results in the literature [25]. However, if we change the deformation constant γ by various amounts, our results show an appreciable change in the maximum mass from the spherical case. The mass increases as the oblateness increases and the mass decreases as the prolateness increases. The maximum mass can increase up to about approximately 13.5% for a 10% change in deformity in oblateness. For increasing prolateness, the mass decreases on the order of approximately 12% for the same change in deformity. These interesting results about the maximum masses leads us to investigate the deformity further by examining the gravitational quadrupole moment Q.

3. Deformity

The shapes of oblate and prolate neutron stars are distinct. One must be able to distinguish the differences from oblate and prolate stars. At a first glance at Fig. 1, it may appear that the oblate star and a prolate star is the same object just rotated 90 degrees. This is not the case in these models. The magnetic field is pointing from the polar (z) direction, therefore we must take that direction as our reference axis as shown in Figs. 3 and 4.

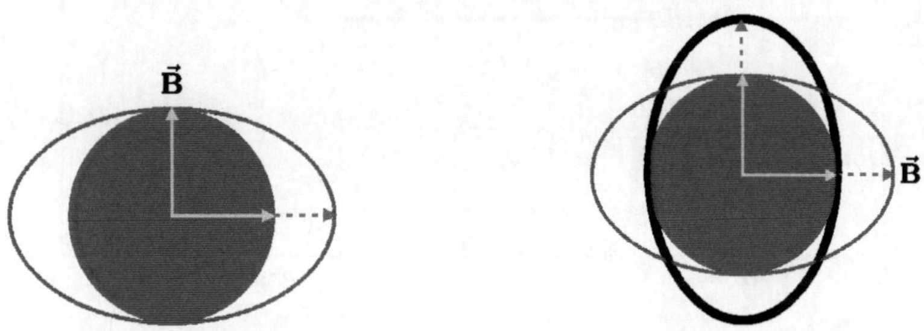

(a) (b)

Figure 3. (Color on-line) Illustration of oblateness for deformed neutron stars. The magnetic field $\vec{B}$ is oriented along the polar (z) – axis for both cases. If the star is being observed head-on (a), it is clear the star is oblate. If we view the star at 90 degrees (b), the star is still an oblate star since the magnetic field is still oriented along the polar axis.

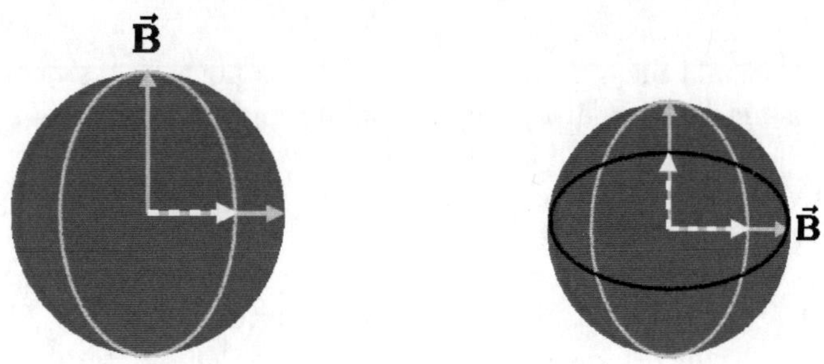

(a) (b)

Figure 4. (Color on-line) Same as Fig. 3 but for prolate stars. The magnetic field is oriented along the polor (z) – axis for both cases.

Both objects in Figs. 3 and 4 are oblate and prolate stars respectively, regardless of the angle from which they are viewed. The important distinction is the z-axis and where the magnetic field is oriented from.

IOP Conf. Series: Journal of Physics: Conf. Series **1239** (2019) 012007 doi:10.1088/1742-6596/1239/1/012007

4. Results and Conclusions

Using the same values of γ in addition to mass results from Fig. 2, we compute the gravitational mass quadropole moment as described by Eq. (5). We calculate this mass quadropole moment as we numerically construct the star, thus producing the graphically results shown in Fig. 5.

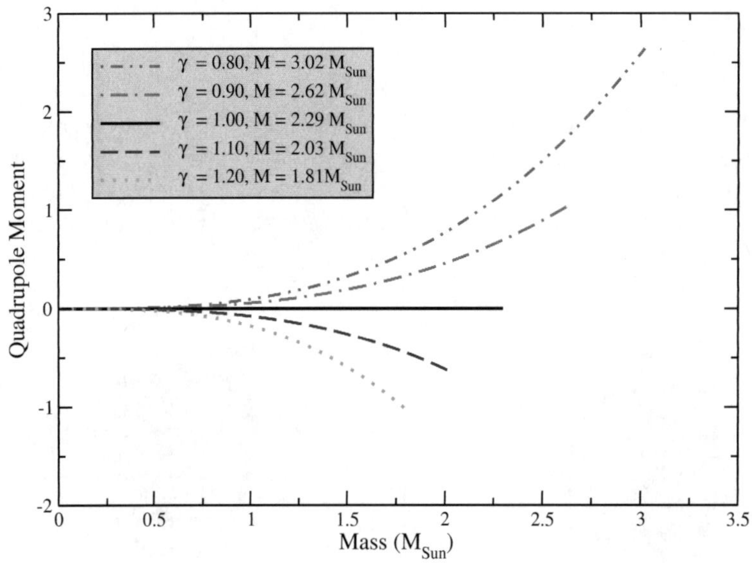

Figure 5. (Color-online) Quadropole moment as a function of mass for various values of γ using the Quark-Hadron EoS model. The black solid line represents the quadropole moment for a perfect spherically symmetric object. Oblate stars are for values of (γ<1) while prolate stars are for values of (γ>1). The results here show how asymmetric the mass distributions are for oblate versus prolate stars. The results shown here are in agreement with the change in masses from Fig. 2.

In Fig. 5, the gravitational mass quadropole moment is non-zero as expected for deformed neutron stars. There is also a clear distinction of oblate and prolate stars due to the fact that the quadropole moment is not symmetric indicating the mass distribution is different for oblate and prolate stars. If this was not the case then the all the curves corresponding to values of $\gamma \neq 1$ would be symmetric about the solid black line. This is obviously not the case. We also see that there is more uneven mass distribution as the star gets more oblate. For example, a 20% deformation in the oblate case results in a higher quadropole moment. This is in agreement with the results in Fig. 2 which indicate higher masses for oblate stars and lower masses for prolate stars. This can be more easily seen in the heatmaps (i.e. colormaps to show how the mass distribution is inhomogeneous throughout the star) in Figs. 6 and 7.

The heatmaps in Figs. 6 and 7 show how different the mass distribution is for oblate and prolate stars. Examining the right vertical axis of the mass quadrupole moment Q, it is apparent that the mass is inhomogeneously distributed throughout the star thus resulting in either an oblate or prolate star. We can also graphically illustrate this inhomogeneity another way by looking how the quadropole moment Q changes as function of polar and equatorial radii as shown in Fig. 8.

IOP Conf. Series: Journal of Physics: Conf. Series **1239** (2019) 012007 doi:10.1088/1742-6596/1239/1/012007

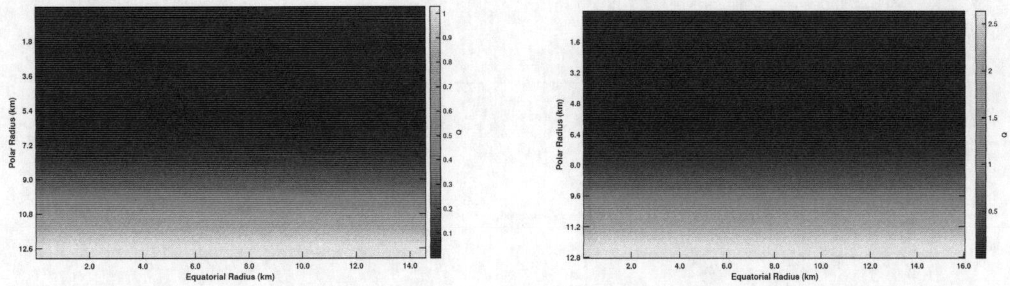

(a) (b)

Figure 6. (Color on-line) Heatmap for quadropole moment Q as a function of the polar z and equatorial r directions for $\gamma = 0.90$ (a) and $\gamma = 0.80$ (b) for oblate stars. In these figures, blue represents zero quadropole moments and thus even homogeneous mass distribution. The heatmap shows how the mass distribution is unevenly distributed along the equatorial and polar directions. The more we deviate from the spherical case, the more uneven the mass distribution gets for these oblate stars. However, we see the increase in deviation as we move vertically away from the equatorial plane.

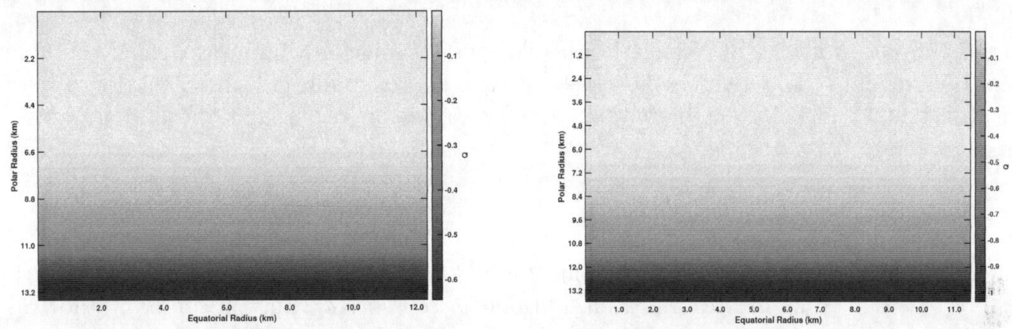

(a) (b)

Figure 7. (Color on-line) Heatmap for quadropole moment Q as a function of the polar z and equatorial r directions for $\gamma = 1.10$ (a) and $\gamma = 1.20$ (b) for prolate stars. In these figures, yellow represents zero quadropole moments and thus even homogeneous mass distribution. The heatmap shows how the mass distribution is unevenly distributed along the equatorial and polar directions. The more we deviate from the spherical case, the more uneven the mass distribution gets for these prolate stars. However, we see the increase in deviation as we move horizontally away from the polar direction direction.

In Fig. 8, the black solid line represents even mass distribution as expected for spherically symmetric stars. However, for oblate stars the mass distribution is more unevenly distributed in the polar direction than it is in the equatorial direction. That is, there is more deviation in the polar region than in the equatorial plane. For prolate stars, we see the opposite effect. The mass distribution is more unevenly distributed in the equatorial direction and we see more deviation in the equatorial plane than in the polar direction. From these results, we also see how asymmetric the deviations are allowing us to conclude that the mass distribution is not symmetric for oblate and prolate stars, thus these objects are distinct oblong spheroids.

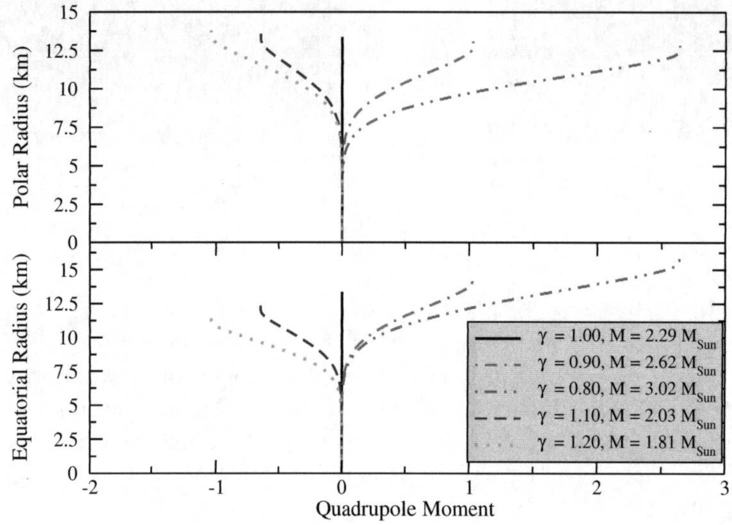

Figure 8. (Color-online) Quadropole-Radii relations for various values of γ using the Quark-Hadron EoS model. The black solid line represents the quadropole moment for a perfect spherically symmetric object. Oblate stars are for values of (γ<1) while prolate stars are for values of (γ>1).

4.1. Observational Implications

One should also investigate the observational implications of these deformed objects. The bulk stellar properties such as masses and radii obtained from solving Eqs. (6) and (7) allow us to compute the total gravitational redshift ($\mathcal{Z}$) which is described by

$$\mathcal{Z} = \frac{1}{\left(1 - \frac{2M}{R}\right)^{\gamma/2}} - 1 \, . \tag{8}$$

For the full derivation of Eq. (8), see [21]. In Eq. (8) M and R are the total gravitational mass and equatorial radii respectively which are obtained by solving Eqs. (6) and(7) for a given equation of state model. Using the same values of γ as in Fig. 2 , we compute the gravitational redshift numerically thus producing the results shown in Fig. 9. From these results it is clear that the gravitational redshift changes significantly from the spherical case due to these deformations. Since the redshift depends on the total mass, and that total mass is different for oblate and prolate stars, one can then see that these oblong spheriods are distinct objects.

5. Conclusions

Using the results from a 1-D parameterized model for deformed neutron stars, we were able to calculate the gravitational mass quadrupole moment of non-rotating neutron stars. From our results shown in Figs. 2 - 9, the stellar properties such as masses, radii, gravitational redshift, and quadropole moment are greatly affected by deformation. Thus deformation plays a pivotal role in the stellar stellar structure of these compact objects. Hence, the deformation does not need to be high to see significant changes in said stellar properties.

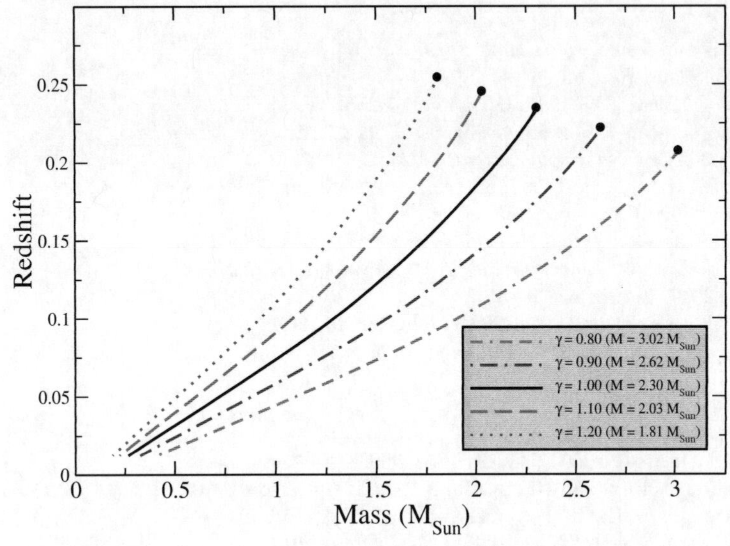

Figure 9. (Color-online) Gravitational Redshift vs. Mass for the Quark-Hadron equation of state model. The dots one each curve represent the maximum mass of each stellar sequence corresponding to the values in Fig. 2. The redshift values here are calculated from mass and radii values from Fig. 2 for the same values of γ. As it shows, the redshift values highly depend on the deformation of the object.

We investigated the inhomogeneity of the mass distribution in oblate and prolate stars and we conclude that the mass distribution is not symmetric among oblate and prolate stars, thus indicating that these oblong spheroids are distinct objects. Due to a non-zero quadropole moment, this work can lead to a more detailed study on these deformed objects and their possible connection to gravitational waves.

Deformation plays an important role on the stellar structure on highly magnetized compact stars. This work on deformed neutron stars can also be applied to highly magnetized White Dwarf stars. Recent publications on highly magnetized white dwarfs indicate that these objects can also be either oblate or prolate spheroids [11, 26–28]. The stellar properties such as masses, radii, and quadropole moment can be easily calculated by using the models presented here and in previous work performed on deformed neutron stars.

Acknowledgments
This work was conducted on the Department of Sciences high performance computer which was awarded to Dr. Omair Zubairi by Wentworth's EPIC mini grant program. Additional computing resources and software programs were provided by Wentworth Institute of Technology. Travel funding to the biannual IARD meeting was awarded to Professor Zubairi by the Office of the Provost at Wentworth Institute of Technology. Dr. Zubairi also would like acknowledge his former undergraduate students Alexis Romero from San Diego State University and Greg Shao from Wentworth Institute of Technology for their preliminary calculations on this project. Dr. F. Weber is supported by the National Science Foundation (USA) under Grant PHY-1714068.

The authors would like to acknowledge Dr. Vivian de la Incera and Dr. Efrain Ferrer for their initial motivation on this topic. The authors would also like to acknowledge Dr. Aurora Perez Martinez and Melinda Toth for their insightful discussions on this work. Lastly, the authors would like to thank Dr. Martin Land for the invitation to speak at the biannual 2018 IARD conference.

IOP Conf. Series: Journal of Physics: Conf. Series **1239** (2019) 012007 doi:10.1088/1742-6596/1239/1/012007

References

[1] Oppenheimer J R and Volkoff G M 1939 *Phys. Rev.* **55** 374.
[2] Tolman R C 1939 *Phys. Rev.* **55** 364.
[3] Thompson C and Duncan R C 1995 *MNRAS* **275** 255.
[4] Thompson C and Duncan R C 1996 *Astrophys. J.* **473** 322.
[5] Mereghetti S 2008 *Astron. Astrophys. Rev* **15** 225.
[6] Ferrer E, Incera V, Keith J, Portillo I, and Springsteen P 2010 *Phys. Rev. C* **82** 6.
[7] Chaichian M et al 2000 *PRL* **84** 5216.
[8] Felipe G R and Martínez A P and Rojas P H and Orsaria M 2008 *Phys. Rev. C.* **77** 015807.
[9] Paret D M and Horvath J E and Martínez A P 2015 *RAA* **15**, 7 975.
[10] Paret D M et al 2017 *Astron. Nachr.* **338** 1131.
[11] Terrero D A et al 2018 **arXiv:1807.09943v2 [astro-ph.HE]**.
[12] Chandrasekhar S and Fermi E 1953 *Astrophys. J.* **118** 116.
[13] Ferraro V C A 1954 *Astrophys. J.* **119** 407.
[14] Goossens M 1972 *Ap & SS* **16** 286.
[15] Katz J I 1989 *MNRAS* **239** 751.
[16] Payne D J B and Melatos A 2004 *MNRAS* **351** 569.
[17] Haskell B and Samuelsson S and Glampedakis K and Andersson N 2008 *MNRAS* **385** 531.
[18] Weyl H. 1918 *Ann. Phys.* (Leipzig), **54** 117.
[19] Herrera L Filipe M, and Santos N O, 1999 *J. Math Phys.* **40** 8.
[20] Weber F and Contrera G and Orsaria M and Spinella W and Zubairi O 2014 *Mod. Phys. Lett. A* **29** No. 23.
[21] Zubairi O and Romero A and Weber F 2015 *J. Phys. Conf.: Ser.* **615** 012003.
[22] Zubairi O et al 2015 **arXiv:1504.03006 [astro-ph.SR]**.
[23] Zubairi O and Wigley D and Weber F 2017 *Int. J. Mod. Phys.Conf.: Ser* **45** 1760029.
[24] Weber F 2015 *Prog. Part. Nucl. Phys.* **54** 193.
[25] Contrera G A and Spinella W and Orsaria M and Weber F 2014 **SLAC eConf/C1309292**.
[26] Paret D M and Horvath J E and Martínez A P 2015 *RAA* **15**, 10 975.
[27] Das U and Mukhopadhyay B 2012 *PRD* **86** 042001.
[28] Das U and Mukhopadhyay B 2013 *PRL* **110** 071102.

Relativistic Beaming as a Probe of Stellar and Planetary Masses

Ben Placek[1]

[1] Department of Sciences, Wentworth Institute of Technology, 550 Huntington Ave Boston, MA 02115, USA

E-mail: `placekb@wit.edu`[1]

Abstract. The primary method of extra-solar planet (exoplanet) detection and characterization is through planetary transits. These events occur when a planet is observed to pass in front of it's host star with respect to the observers line of sight, which causes a small dimming event. Transits alone yield information on the orbital properties such as period, inclination, semi-major axis as well as physical properties such as the planetary radius. With high-precision photometry, a new photometric effect has emerged as a probe of short-period exoplanet masses. This effect is known as relativistic Doppler beaming (or boosting), and has been used to estimate the masses and densities of numerous exoplanets and stars in binary systems. Here, this effect is discussed in detail along with the prospect of utilizing it with next generation space-based telescopes that will be devoted to the detection and characterization of exoplanets. Prospects for the characterization of binary systems will also be examined.

1. Introduction

The *Kepler* Space Telescope (*Kepler*) has revolutionized the study of extra-solar planets (exoplanets) — planets that orbit stars other than the Sun. *Kepler* was launched in 2009 with it's primary function of monitoring the brightness of approximately 250,000 stars continuously over the course of it's five year mission. The goal was to detect periodic dimming events caused by a planet transiting its host star and thus blocking a small portion of starlight. Perhaps the most revolutionary aspect of *Kepler* was the photometric precision that the telescope was able to achieve. With the ability to obtain photometric precisions down to $\sim$ 10 parts per million, *Kepler* was not only well capable of detecting the transits of exoplanets, but also much more subtle photometric effects associated with a subset of exoplanetary systems.

To date, *Kepler* has discovered 2342 extrasolar planets [1] along with 2878 eclipsing binary systems [14]. A subset of the *Kepler* planets are large, short-period exoplanets and display out of transit photometric effects in addition to the primary transit. First is the reflection of starlight and thermal emission from the dayside of the planet, which results in a quasi-sinusoidal signal as the planet is observed to go through phases over the course of it's orbital period. These mechanisms also give rise to what is known as the secondary eclipse, which occurs when the planet periodically disappears behind the star and any reflected/emitted light from the planet is blocked by the star. The secondary eclipse is often significantly ($> 10\times$) shallower than the primary transit. The second effect is known as ellipsoidal variations and is a result of the planet inducing tides on the surface of the host star. This also results in a sinusoidal photometric signal in the photometric at twice the orbital frequency of the planet. The third photometric effect

IOP Conf. Series: Journal of Physics: Conf. Series **1239** (2019) 012008 doi:10.1088/1742-6596/1239/1/012008

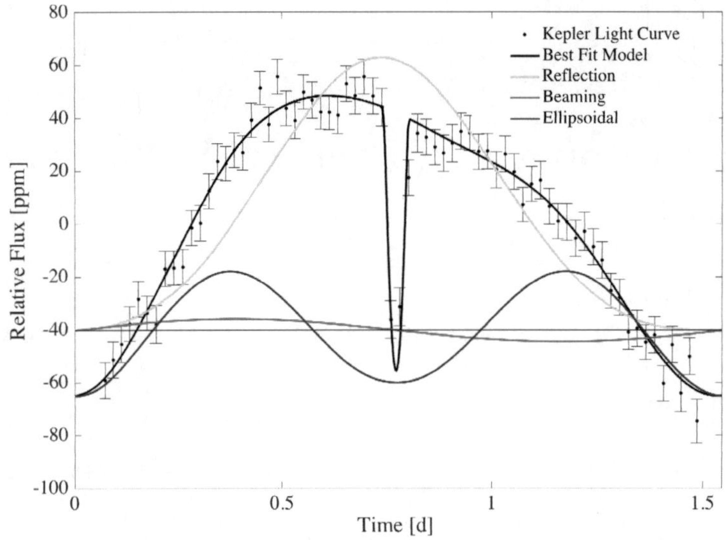

Figure 1. *Kepler* light curve of *Kepler*-76b. The relativistic beaming effect contributes 12ppm to the total observed variation.

is the relativistic Doppler beaming of light from the host star. Stars and planets collectively orbit about the systems barycenter. Therefore the host star will periodically advance toward and recede from an observer. Thus the brightness of the host star will vary sinusoidally at the orbital frequency of the planet. The resulting photometric variations due to these effects can be seen in Figure 1, which displays the *Kepler* light curve of the transiting hot Jupiter *Kepler*-76b. For this planet, the most dominant effect is reflection, which displays a peak to peak amplitude of $\sim$ 105 ppm, followed by ellipsoidal variations with a peak to peak amplitude of $\sim$ 44 ppm and finally Doppler boosting with an amplitude of $\sim$ 8 ppm.

Due to the fact that the beaming effect is caused by the motion about the center of mass of an exoplanet system, the amplitude of the effect can be used to estimate the mass of any companions to the host star provided the host star mass is known. This can be used to vet possible false-positive scenarios that mimic exoplanets, such as eclipsing binary systems for which the beaming amplitude has the potential to be much greater than in planet-hosting systems. It also allows one to constrain the density of exoplanets when paired with transit observations.

2. The Relativistic Beaming Effect

First observed by [17] in 2000, a theoretical treatment of the relativistic beaming effect was first described by [11] with the application of rotating white dwarfs, and later in the context of ultra-compact eclipsing binaries [27]. [29] and [12] also give a detailed theoretical treatment. [16] were the first to consider this effect in the context of exoplanet characterization with the *Kepler* Space telescope, followed by [33] with the aim of characterizing so-called *beaming binaries*. Additionally, the beaming effect has been used extensively to characterize both planetary [24, 25, 30, 6, 7, 2, 8, 18, 20, 21, 23] and binary star systems [31, 33, 3, 32, 10, 4] and is in reality a result of combination of three effects. First, the beaming of light in the direction of motion, second the increase/decrease in observed rate of photons emitted from the host star, and third the Doppler shift of the host stars spectrum causing it to move in an out of the observational bandpass.

For completeness the following section gives a derivation of the bolometric flux observed from a moving light source. This follows closely from the derivation given by [19].

2.0.1. The Bolometric Flux We can begin to quantify such effects in the bolometric case (across all wavelengths) by considering a spherical star that radiates isotropically in it's rest frame S'. It is assumed that the photons are emitted at a constant rate Γ' as measured in the rest frame of the star. Due to the aberration effect, the angles at which the photons will be emitted in S' will differ from those observed by someone moving relative to the star in S. This is given by

$$\cos\theta' = \frac{\cos\theta - \beta}{1 - \beta\cos\theta} \tag{1}$$

where β is the velocity of the star in units of c. Therefore, the solid angles into which the photons are emitted will also differ such that

$$d\Omega = \sin\theta d\theta \tag{2}$$

and

$$d\Omega' = \sin\theta' d\theta' \tag{3}$$

By differentiating (1), it can be shown that

$$\frac{d\Omega}{d\Omega'} = \gamma^2(1 - \beta\cos\theta)^2 \tag{4}$$

where γ is the Lorentz factor. Additionally, due to time dilation the ratio of the rates at which the photons are emitted as measured in the two frames is

$$\frac{d\Gamma'}{d\Gamma} = \gamma(1 - \beta\cos\theta). \tag{5}$$

The ratio of the observed flux between S and the rest frame of the star, S', is given by

$$\frac{dF}{dF'} = \frac{dE}{dE'}\frac{d\Gamma}{d\Gamma'}\frac{d\Omega'}{d\Omega} \tag{6}$$

where dE is the energy associated with the radiated photons. Since energy and momentum form a four-vector, energy transforms as $dE' = \gamma(1-\beta\cos\theta)dE$. This is simply a re-statement of the Doppler effect. Substituting (4) and (5) into (6) yields the boosted flux observed by S

$$F_b(t) = F_\star\left(\frac{1}{\gamma(1 - \beta\cos\theta(t))}\right)^4 \tag{7}$$

where $F_b(t)$ is the observed flux in S, $F_\star$ is the flux emitted by the star in its rest frame, and $\theta(t)$ is the angle between the stellar velocity and the line of sight. Since even in the most extreme cases, the velocities of planet-hosting stars about the center of mass doesn't exceed $10^2 - 10^3$ ms^{-1}, we take the non-relativistic limit to calculate the stellar-normalized boosted flux

$$\frac{F_b(t)}{F_\star} = 1 + 4\beta_r \tag{8}$$

where $\beta_r = \beta\cos\theta(t)$ is the component of the velocity of the host star along the observers line of sight (i.e. the radial velocity of the host star).

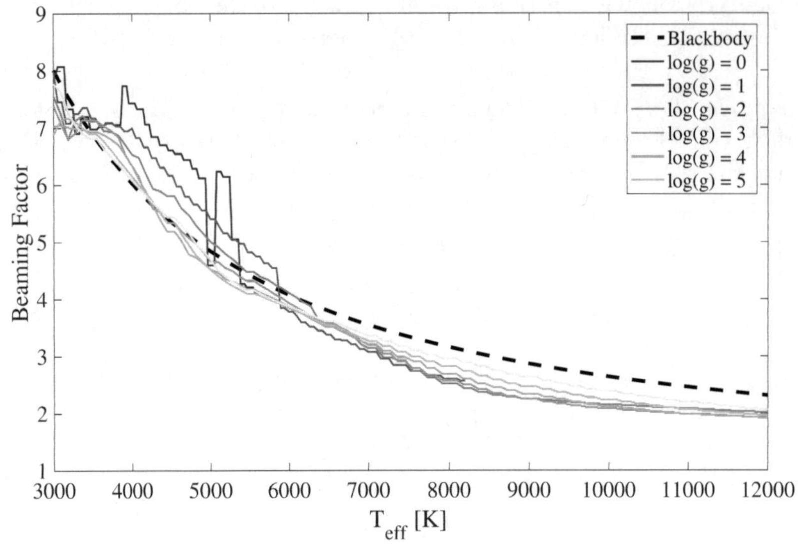

Figure 2. Beaming factors computed via Eq. (10) as a function of stellar effective temperature. The beaming factors were computed using synthetic spectra from the PHOENIX library.

2.0.2. The Bandpass-Integrated Flux Depending on the observational bandpass of the detector, only a portion of the bolometric flux will be observed. Therefore there is an additional bandpass-dependent component to the boosting effect, which is subject to the wavelength range over which one observes a beaming system. Due to the reflex motion of the host star, its spectrum will periodically Doppler shift causing portions of the spectrum to move in and out of the observational bandpass. If one assumed the flux to behave as a power law in frequency ($F_\nu \propto \nu^{-\alpha}$), the observed flux is given by

$$\frac{F_B(t)}{F_\star} = 1 + B\beta_r(t) \tag{9}$$

where B is the photon-weighted bandpass-integrated beaming factor given by

$$B = \frac{\int K(\lambda)\,(5+\alpha)\,\lambda F_{\lambda,\star} d\,\lambda}{\int K(\lambda)\lambda F_{\lambda,\star} d\,\lambda}. \tag{10}$$

Here, $K(\lambda)$ is the *Kepler* response function, λ is the wavelength of observed light, $F_{\lambda,\star}$ is the stellar spectrum, and α is the spectral index given by

$$\alpha = \frac{d\ln F_{\lambda,\star}}{d\ln\lambda}. \tag{11}$$

[16] modeled the stellar spectrum, $F_{\lambda,\star}$, as a blackbody, which neglects absorption features that could substantially affect the beaming factor. Figure 2 shows how the beaming factor changes with effective temperature for a variety of values of stellar log-surface gravity. The stellar metallicity was set to solar metallicity as the effect of changing metallicity was found to have little affect on the beaming factor. The spectra were obtained by interpolating the PHOENIX library of stellar spectra.

If one is able to estimate the amplitude of the beaming effect, the companion mass, M_c is immediately obtained due to the fact that the effect is proportional to the radial velocity, which

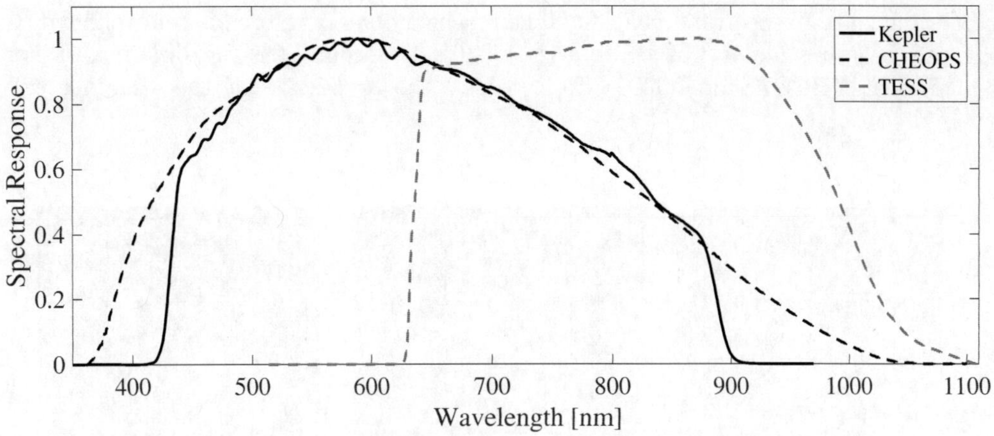

Figure 3. Spectral response functions for *Kepler*, CHEOPS, and TESS. The peak of each response function was normalized to unity. *Kepler* and CHEOPS are sensitive to similar wavebands, whereas TESS is sensitive to red/Near-IR wavelengths.

has a semi-amplitude

$$K = \left(\frac{2\pi G}{P}\right)^{1/3} \frac{M_c \sin i}{M_s^{2/3}} \frac{1}{\sqrt{1-e^2}} \tag{12}$$

where G is Newton's gravitational constant, P is the orbital period of the planet, i is the orbital inclination, and M_s and e are the host star mass and orbital eccentricity, respectively.

3. Doppler Beaming and Future Space-Based Platforms

The Transiting Exoplanet Survey Satellite (TESS) [26] was recently launched in April 2018, and the CHaracterizing ExOPlanets Satellite (CHEOPS) [9, 5] is currently set to launch in early 2019. The primary objective of the TESS mission is to observe the entire celestial sphere over two years searching for exoplanet transits in much the same way as *Kepler*. The sky will be binned into 26 sectors, each of which will be observed for approximately 27 days. CHEOPS on the other hand will target bright stars that are already known to harbor exoplanets. Since the amplitude of the beaming effect depends on the observational bandpass, one may ask whether or not it will be detectable with the next generation of telescopes given their expected photometric sensitivity. Figure 3 displays the observational bandpasses for *Kepler*, CHEOPS, and TESS. Both the *Kepler* and CHEOPS bandpasses peak in the visible at $\sim$ 600nm, while TESS is sensitive to redder wavelengths in general and peaks in the near-infrared at $\sim$ 900nm. The expected photometric precision of TESS is 200ppm in 1 hour of observations [15]. The expected photometric precision for CHEOPS is an order of magnitude better at 20ppm for a bright ($6 < V < 9$) star per six hours of integration time, and 85ppm for a dimmer (V = 12) star over just 3 hours of integration time.

3.0.1. Prospects for Exoplanet Characterization Figure 4 displays the expected beaming amplitudes for both CHEOPS (top) and TESS (bottom). Each pixel represents the expected beaming amplitude in parts-per-million, computed by varying the planetary mass $M_p = [0, 13]M_J$ and orbital period $P = [1, 10]$ days, and were produced for three different host star masses $M_s = [1, 1.5, 2]M_\odot$. Each of these assume circular, edge-on orbits with inclinations of $i = 90^o$. The effective temperatures were derived from the assumed host star masses by interpolating the table of [22], which lists measurements of masses and effective temperatures for 90 stars ranging from spectral type M to O. For each combination, the photon-weighted bandpass integrated beaming factor was computed (eq.10) to account for the differing observational bandpasses of CHEOPS and TESS from which the amplitude of the beaming effect was calculated.

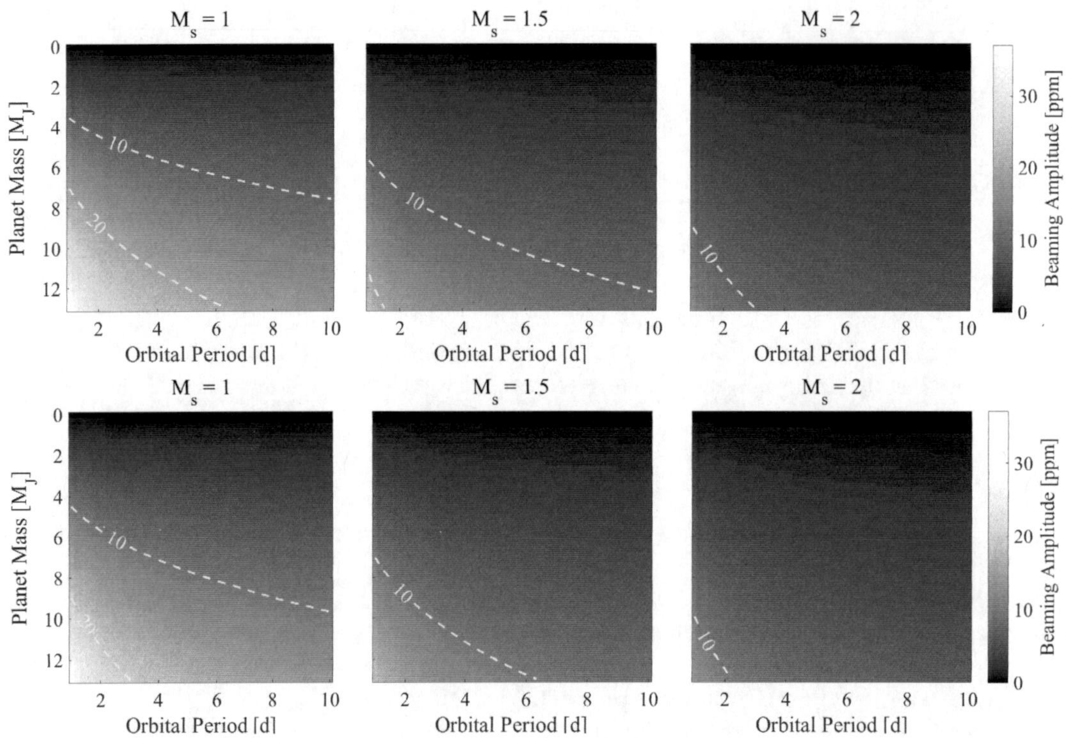

Figure 4. Predicted beaming amplitudes (in ppm) observed through the CHEOPS (top) and TESS (bottom) bandpasses. Amplitudes were computed in both bandpasses for combinations of host star mass, M_s, planetary mass M_p, and orbital period P. Contours of 10ppm and 20ppm are overlaid for perspective.

It seems that relativistic beaming will be difficult to detect in most situations with these two platforms. With CHEOPS, certain massive planets in short period configurations with a bright solar mass host star should be detectable with six hours integration time. If the photometric precision can be pushed to the 10ppm level, the beaming effect from planet-hosting stars with companions with masses as low as $\sim$ 4–7M_J should be detectable. Given TESS' expected photometric precision of 200ppm over one hour of integration, it seem highly unlikely that the beaming effect will be detectable for most if not all planetary systems. However, the short 2-minute cadence of the TESS observations allows one to bin the phase curve to increase the SNR making it possible to detect beaming amplitudes on the order of 10-100ppm. The extent to which one can "Bin down" the observations will depend on the number of orbits observed.

Coincidentally, the first phase curve observation from TESS has recently been published for WASP-18b. This is a massive 10.4M_J planet in an ultra-short $\sim$ 0.94 day orbit around a $1.46 \pm 0.29 M_\odot$ star. A beaming amplitude was detected with an amplitude of 18 ± 2 ppm [28], which is consistent with our theoretical estimate of 16.6ppm.

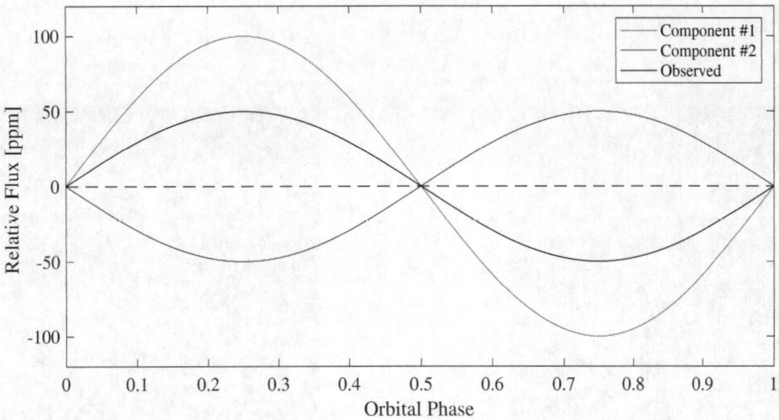

Figure 5. Plot of the beaming components from a hypothetical binary system where both components are displaying the beaming effect. The red and blue curves represent the beaming effect from the individual components of the system. The red and blue components have amplitudes of 100 and 50ppm, respectively. These two will partially cancel each other out yielding a smaller amplitude for the total observed flux (black curve).

3.0.2. Prospects for Eclipsing Binary Characterization While CHEOPS will be devoted solely to characterizing exoplanets, TESS will inevitably find a large number of eclipsing binaries as it monitors the entire celestial sphere. In such systems, the beaming effect could play a significant role depending on the nature of the components.

When it comes to determining the masses of the components in a binary system, the beaming effect has the potential to bring with it a significant degeneracy. Since both components are stars, they will both contribute their own beaming effects to the overall observed flux. Additionally, since the components co-orbit the barycenter in opposite directions, it is possible for the beaming from one component to cancel out all or a portion of the beaming from the other as shown in Figure 5. The extent to which this degeneracy will be relevant will depend on the spectral type of the stars in the system. It should be noted that this would only impact systems for which there are no available radial velocity observations and more specifically systems for which the spectral lines of the components cannot be disentangled. In such cases, the radial velocities of both stars can be measured and therefore the degeneracy in the beaming effect (if observed) will not be present.

Figure 6 shows the amplitude of the beaming effect as a function of effective temperature of the primary (vertical axis) and secondary (horizontal axis) as observed through the TESS bandpass. These were generated assuming circular, edge-on orbits with orbital periods of 5, 10, and 15 days. Greater orbital periods were not considered since TESS will only observe each sector of the sky in 30d intervals. First, the masses of the components were derived from the effective temperatures by again interpolating the table from [22]. Next, theoretical spectra were obtained by interpolating the PHOENIX spectral database and the bandpass-integrated beaming factors were computed for both. Finally, the observed beaming amplitude was computed as the difference between that of the primary and secondary components. Overlayed with the beaming amplitudes is the 200ppm contour, which highlights the approximate limit of the TESS photometer. The diagonal represents binary components of equal effective temperature, which will in turn yield beaming components that cancel out. Any points under the diagonal represents configurations where the host star is less massive than the secondary and were therefore removed. It should be noted that these calculations do not take into account post-main sequence stars.

IOP Conf. Series: Journal of Physics: Conf. Series **1239** (2019) 012008 doi:10.1088/1742-6596/1239/1/012008

As the orbital period of the companions increases, the top of the 200ppm contour shifts to the right by approximately 200K.

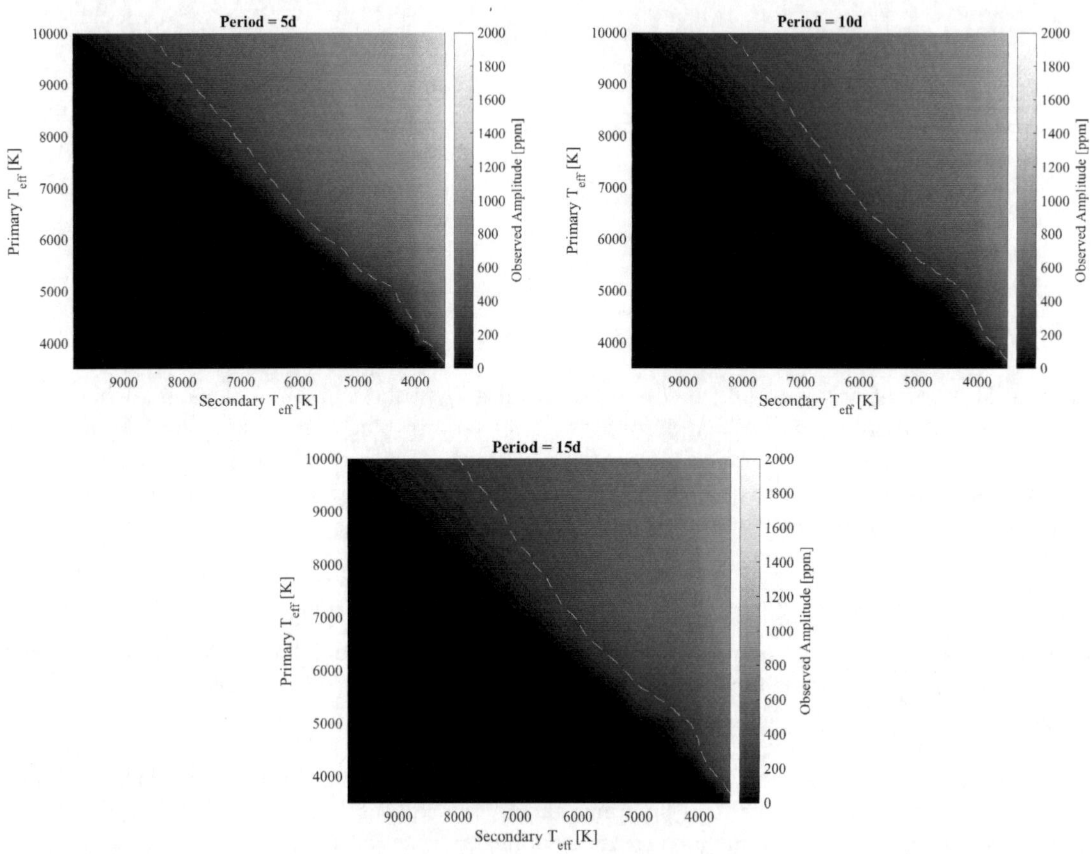

Figure 6. Expected observed beaming amplitudes as measured by the TESS photometer and as a function of effective temperature for the primary and secondary in a beaming binary.

4. Summary

The relativistic beaming effect causes a periodic quasi-sinusoidal variation in the flux from planet-hosting stars due to their reflex motion about the center of mass. Measuring the amplitude of this effect allows the observer to estimate the mass of the companion if physical properties of the host star are known along with certain orbital characteristics. As this is a bandpass-dependent effect, the amplitude will vary across observational platforms.

By analyzing the beaming effect for a wide range of system configurations, across two platforms (CHEOPS & TESS), it was shown that CHEOPS will be able to detect the beaming effect for certain massive, short period exoplanets. The ability to detect this effect will rely on how many orbits are observed. Based on theoretical amplitudes and the expected precision of the photometer, TESS will only be able to detect the beaming effect for certain extreme (high planet mass, short orbital period) planetary systems. However, the amplitude should be large enough to be detected by TESS for a wide variety of combinations of short-period eclipsing binaries.

With the sensitivity of the CHEOPS photometer, beaming should be detectable for massive short period planets in the $\sim 4\text{--}7M_J$ range. However, the primary science goal of CHEOPS is to constrain radii of planets in the range of $1 - 6R_E$ with known masses from radial velocity measurements. This class of planets will have beaming signals much smaller than 1ppm making it impossible to detect.

References

[1] Akeson R L et al. 2013. *PASP*, **125(930)**, 989.
[2] Angerhausen D, Delarme E, Morse, J A 2015. *PASP*, **127**, 1113-1130.
[3] Bloemen S et al. 2011. *MNRAS*, **410(3)**, 1787-1796.
[4] Bloemen S et al. 2012. *MNRAS*, **422(3)**, 2600-2608.
[5] Broeg C et al. 2013. *EPJ Web of Conferences*, **47**, 03005.
[6] Esteves L J, De Mooij E J, and Jayawardhana R 2013. *ApJ*, **772(1)**, 51.
[7] Esteves L J, De Mooij E J, and Jayawardhana R 2015. *ApJ*, **804(2)**, 150.
[8] Faigler S and Mazeh T 2011. *MNRAS*, **415(4)**, 3921-3928.
[9] Fortier A, Beck T, Benz W, Broeg C, Cessa V, Ehrenreich D, and Thomas N 2014. *SPIE*, 91432J.
[10] Herrero E, Lanza A F, Ribas I, Jordi C, Cameron A Collier, and Morales J C 2014. *A&A*, **563**,A104.
[11] Hills J G, and Dale T M 1974. *A&A*, **30**, 135-139.
[12] Hogg D W 1997. *Lecture Notes.*
[13] Husser T O, Wende-von Berg S, Dreizler S, Homeier D, Reiners A, Barman T, Hauschildt P H 2013. *A&A*, **553**, A6.
[14] Kirk B et al. 2016. *AJ*, **151(3)**, 68.
[15] Kraft Vanderspek R et al. 2010. *Bull. of the Amer. Astr. Soc.*, **42**, 459.
[16] Loeb A, and Gaudi S B 2003. *ApJL*, **588(2)**, L117.
[17] Maxted P F L, Marsh T R, and North R C 2000. *MNRAS*, **317(3)**, L41-L44.
[18] Mazeh T, Nachmani G, Sokol G, Faigler S, and Zucker S 2012. *A&A*, **541**, A56.
[19] Mccrea W H 1972. *MNRAS*, **157(4)**, 359-365.
[20] Mislis D and Hodgkin S 2012. *MNRAS*, **422(2)**, 1512-1517.
[21] Mislis D, Heller R, Schmitt J H M M, and Hodgkin S 2012. *A&A*, **538**, A4.
[22] Pecaut M J and Mamajek E E 2013. *ApJ Supplement Series*, **208(1)**, 9.
[23] Pfahl E, Arras P, Paxton B 2008. *ApJ*, **679(1)**, 783.
[24] Placek B, Knuth K H, and Angerhausen D 2014. *ApJ*, **795(2)**, 112.
[25] Placek B, Knuth K H, Angerhausen D, and Jenkins J M 2015. *ApJ*, **814(2)**, 147.
[26] Ricker G R et al. 2015. *Journal of Astr. Tel., Instr. & Sys.*, **1(1)**, 014003.
[27] Shakura N I and Postnov K A 1987. *A&A*, **183**, L21.
[28] Shporer A et al 2018. *ArXiv e-prints.* 1811.06020.
[29] Rybicki G B, and Lightman A P 1979. *John Wiley & Sons.*
[30] Shporer A, Jenkins J M, Rowe J F, Sanderfer D T, Seader S E, Smith J C, Still M D, Thompson S E, Twicken J D, and Welsh W F 2011. *AJ*, **142(6)**, 195.
[31] Shporer A, Kaplan D L, Steinfadt J D R, Bildsten L, Howell S B, and Mazeh T 2010. *ApJL*, **725(2)**, L200.
[32] Telting J H, Baran A S, Nemeth P, Østensen R H, Kupfer T, Macfarlane S, Heber U, Aerts C, and Geier S 2014. *A&A*, **570**, A129.
[33] Zucker S, Mazeh T, and Alexander T 2007. *ApJ*, **670(2)**, 1326.

IOP Conf. Series: Journal of Physics: Conf. Series **1239** (2019) 012009 doi:10.1088/1742-6596/1239/1/012009

Radial Acceleration and Tully-Fisher Relations in Conformal Gravity

James G. O'Brien[1], Thomas L. Chiarelli[2], Philip D. Mannheim[3], Mark A. Falcone[4], Muhannad H. AlQurashi[4], and Jordan Carter[4]

[1]Department of Mathematics, Physics and Computer Science, Springfield College, Springfield, MA 01109, USA
[2]Department of Electromechanical Engineering, Wentworth Institute of Technology, Boston, MA 02115, USA
[3]Department of Physics, University of Connecticut, Storrs, CT 06268, USA
[4]Department of Mechanical Engineering, Wentworth Institute of Technology, Boston, MA 02115, USA

E-mail:
jobrien7@springfieldcollege.edu,chiarellit@wit.edu,philip.mannheim@uconn.edu

Abstract. In 2016 McGaugh, Lelli and Schombert established a universal Radial Acceleration Relation for centripetal accelerations in spiral galaxies. Their work showed a strong correlation between observed centripetal accelerations and those predicted by luminous Newtonian matter alone. Through the use of the fitting function that they introduced, mass discrepancies in spiral galaxies can be constrained in a uniform manner that is completely determined by the baryons in the galaxies. Here we present a new empirical plot of the observed centripetal accelerations and the luminous Newtonian expectations, which more than doubles the number of observed data points considered by McGaugh et al. while retaining the Radial Acceleration Relation. If this relation is not to be due to dark matter, it would then have to be due to an alternate gravitational theory that departs from Newtonian gravity in some way. In this paper we show how the candidate alternate conformal gravity theory can provide a natural description of the Radial Acceleration Relation, without any need for dark matter or its free halo parameters. We discuss how the empirical Tully-Fisher relation follows as a consequence of conformal gravity.

1. Introduction

The rotation curves of spiral galaxies have been intensively studied for many years and serve as a key case study for the missing mass problem. Following the pioneering work of Freeman [1], of Roberts and Whitehurst [2], and of Rubin, Ford and Thonnard [3], there has been extensive study of the rotation curves of spiral galaxies, in particular in the region far beyond the optical disk. The cold dark matter (CDM) formalism, first developed for spiral galaxies by Navarro, Frenk and White (NFW) [4], explains mass discrepancies through the introduction of dark matter, to thus remain consistent with Newtonian gravity. Even with the ability of NFW to fit rotation curves, a lack of direct observational evidence for dark matter and the shrinking allowed parameter space still allowed for it invites alternate explanations of the missing mass problem. Some typical attempts are Modified Newtonian Dynamics (MOND) [5], Modified Gravity (MOG) [6], and Conformal Gravity (CG) [7]. Typical of these alternate theories is the ability not just to fit galactic rotation curves without dark matter, but to do so without the two free parameters per galactic halo that are needed in CDM-type theories. For the 236 galaxies considered here, dark matter theories currently require 472 free halo parameters, none of which are needed in the CG theory we study here.

Recently, McGaugh, Lelli and Schombert identified a universal relation between the observed centripetal accelerations in galaxies and the centripetal accelerations predicted by luminous Newtonian matter alone [8]. They reached this conclusion not by looking at individual rotation curves, but by plotting all of the centripetal accelerations on a single plot. In [8] it was found that the observed centripetal accelerations $g_{OBS} = v^2_{OBS}/r$ are correlated with the baryonic luminous Newtonian $g_{NEW} = v^2_{NEW}/r$ by a fitting function

$$g_{OBS} = \frac{g_{NEW}}{[1 - \exp(-(g_{NEW}/g^{\dagger})^{1/2})]}, \tag{1}$$

where $g^{\dagger}$ is a fitting parameter with a fitted value of $g^{\dagger} = 1.2 \times 10^{-10} ms^{-2}$. This correlation has been called the Radial Acceleration Relation (RAR).

By being able to include all data points in a single plot, the empirical study of McGaugh et al. is more extensive and comprehensive than previous empirical studies such as that of Persic, Salucci and Stel [9]. Implicit in the study of McGaugh et al. is the suggestion that there is a universal nature to the pattern of accelerations observed in spiral galaxies. As such, McGaugh et al. provide a universal formula which can be used to test and/or constrain any theory of galactic rotation curves. Since the RAR is an analysis of every data point in each rotation curve, it is more general than the well known Tully-Fisher (TF) relation, which is found to hold in many spiral galaxies [10] as

$$v^4_{OBS} \propto M, \tag{2}$$

where M is the total visible mass in each galaxy. As such, both the RAR and TF relation challenge any theory of galactic rotation curves. As noted by McGaugh et al. themselves, eq. (1) either constrains and challenges dark matter theory or implies alternate physics that goes beyond the standard Newton-Einstein theory. In this paper we discuss a particular alternate theory, namely conformal gravity, and show how it describes the RAR and the TF relation in a universal fashion.

2. Conformal Gravity Rotation Curve Fitting

By being a pure metric theory of gravity, conformal gravity, just like Einstein gravity, identifies gravity as spacetime curvature. However, unlike Einstein gravity, conformal gravity imposes an additional conformal invariance symmetry. In consequence, conformal gravity is derived from an action based on the square of the Weyl tensor as

$$I_{\rm W} = -\alpha_g \int d^4x(-g)^{1/2} C_{\lambda\mu\nu\kappa} C^{\lambda\mu\nu\kappa}, \tag{3}$$

where

$$C_{\lambda\mu\nu\kappa} = R_{\lambda\mu\nu\kappa} - \frac{1}{2}\left(g_{\lambda\nu}R_{\mu\kappa} - g_{\lambda\kappa}R_{\mu\nu} - g_{\mu\nu}R_{\lambda\kappa} + g_{\mu\kappa}R_{\lambda\nu}\right) + \frac{1}{6}R^{\alpha}{}_{\alpha}\left(g_{\lambda\nu}g_{\mu\kappa} - g_{\lambda\kappa}g_{\mu\nu}\right) \tag{4}$$

is the conformal Weyl tensor with dimensionless coupling constant α_g. The resulting field equations, $4\alpha_g\left(2\nabla_\kappa\nabla_\lambda C^{\mu\lambda\nu\kappa} - C^{\mu\lambda\nu\kappa}R_{\lambda\kappa}\right) = T^{\mu\nu}$ were solved by Mannheim and Kazanas in the region exterior to a static, spherically symmetric source [11], where it was shown that a point stellar mass produces a potential $V^*(r) = -\beta^* c^2/r + \gamma^* c^2 r/2$. To go from the single star to the prediction for rotational velocities of entire spiral galaxies, it was shown [12] that the resulting galactic velocity expectation is given by

$$v_{CG}(R) = \left[v^2_{NEW}(R) + \frac{M}{M_\odot}\frac{\gamma^* c^2 R^2}{2R_0} I_1\left(\frac{R}{2R_0}\right) K_1\left(\frac{R}{2R_0}\right) + \frac{\gamma_0 c^2 R}{2} - \kappa c^2 R^2\right]^{1/2}, \tag{5}$$

IOP Conf. Series: Journal of Physics: Conf. Series **1239** (2019) 012009 doi:10.1088/1742-6596/1239/1/012009

where M is the mass of the galaxy in solar mass units ($M_\odot$), R_0 is the galactic disk scale length, and $v_{NEW}(R)$ is the standard Freeman formula for a Newtonian disk:

$$v_{NEW}(R) = \left[\frac{M}{M_\odot}\frac{\beta^* c^2 R^2}{2R_0^3}\left[I_0\left(\frac{R}{2R_0}\right)K_0\left(\frac{R}{2R_0}\right) - I_1\left(\frac{R}{2R_0}\right)K_1\left(\frac{R}{2R_0}\right)\right]\right]^{1/2}. \quad (6)$$

Fits to 236 galaxies have been achieved through eq. (5) with the following fixed parameters: $\beta^* = 1.48 \times 10^5 cm$, $\gamma^* = 5.42 \times 10^{-41} cm^{-1}$, $\gamma_0 = 3.06 \times 10^{-30} cm^{-1}$ and $\kappa = 9.54 \times 10^{-54}\ cm^{-2}$ [12]. In actual applications of eq. (5), one typically includes gas contributions as well as bulge contributions (when applicable). For details on the specifics, see [12].

In recent years CG has enjoyed success in fitting rotational velocities of spiral galaxies without dark matter. New physics is encapsulated in the uniquely CG terms in eq. (5), terms that depend on just three new parameters γ^*, γ_0 and κ beyond the standard Newtonian β^*, with no fine tuning of input parameters being required. To date, all observed spiral galaxy rotation curves have been explained by conformal gravity in a universal manner despite varying morphologies, luminosities and masses of galaxies studied. Although McGaugh et al. provide a new analysis of rotation curves, it should be noted that the issue of universality in centripetal accelerations of spiral galaxies has been discussed previously in conformal gravity. Specifically, the values of $(v_{OBS}^2/R)_{last}$ for the last data points in each and every studied galaxy were found to be of the order of $3 \times 10^{-11} ms^{-2}$, with $(v_{OBS}^2/c^2R)_{last}$ being of order $3 \times 10^{-30} cm^{-1}$ (see for example, [12, 13] and the last column in Table 1). In the next section we will discuss the procedure for reconstructing the plots provided by McGaugh et al. while also extending their data sample substantially by using the 236 galaxies that CG has successfully fit.

3. Reproducing the Study of McGaugh, Lelli and Schombert

The analysis of McGaugh et al. uses the Spitzer Photometry and Accurate Rotation Curves (SPARC) database [14], which consists of 175 spiral galaxies, containing 3149 total rotation curve data points. The points (R, v_{OBS}) are the observed velocity v_{OBS} at a given distance from the galactic center R. Using these points one can calculate the observed centripetal accelerations as $g_{OBS} = v_{OBS}^2/R$. In order to calculate g_{NEW}, McGaugh et al. used fixed mass to light ratios of 0.5 for disks and 0.7 for bulges. After filtering the data, the net number of galaxies used by McGaugh et al. was reduced to 153, with a net total of 2693 points [15]. With g_{NEW} being calculated from luminous Newtonian matter alone, and g_{OBS} being based on the observed velocities, McGaugh et al. constructed a plot of g_{OBS} versus g_{NEW} for every data point in their sample. In such a plot, any deviation form the line of unity, namely the line of $g_{OBS} = g_{NEW}$, would reflect a mass discrepancy. The resulting 2693 point plot as given in [8] has exactly the same structure as the 2693 point Fig. 1(a) plot that we provide here, as well as having the same structure as the Fig. 1(b) plot that uses the entire 6377 data points considered here. In both Fig. 1(a) and Fig. 1(b) we have plotted eq. (1), to show that it provides a good characterization of the data.

IOP Conf. Series: Journal of Physics: Conf. Series **1239** (2019) 012009 doi:10.1088/1742-6596/1239/1/012009

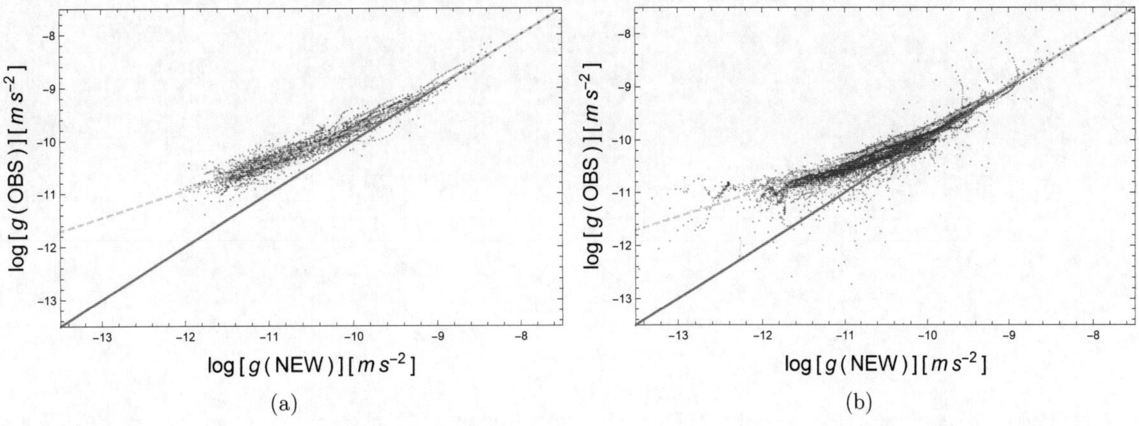

Figure 1. Fig. 1(a) shows the (g_{OBS}, g_{NEW}) plot using the 2693 points studied by McGaugh et al. Fig. 1(b) shows the (g_{OBS}, g_{NEW}) plot using the full 6377 points. The solid line is the line of unity $g_{OBS} = g_{NEW}$, and the dashed line in both Fig. 1(a) and Fig. 1(b) is eq. (1).

4. Our Data Sample

Since the work of McGaugh et al. provides a universal description of galactic rotation curves, it is not specific to the data set that they chose, and not only should it not be, as we see it is reflected in the larger data set we study. Our data sample includes all of the galaxies currently fit by CG in the past decade, as well as any galaxies that were included in [15] that were not yet fit by CG. Data sets previously fit by CG include surveys such as THINGS [16], LITTLE THINGS [17], the Ursa Major survey [18], and the WHISP Dwarfs survey [19]. It should be noted that many of these galaxies in the aforementioned samples are galaxies already included in the SPARC database. However, there are some galaxies that were included in the SPARC database that were not previously fit by CG. To this end, fits were performed on the remainder of the SPARC database, including the surveys of Spekkens et al. [20], Lelli et al. [21], Walsh [22], DeBlok et al. [23], and Noordermeer et al. [24], to thereby complete the CG fitting of the entire SPARC database. A typical selection of these new fits is shown in Fig. 2, and the entire set of these as yet unpublished conformal gravity rotation curve fits will be presented in a future work. The relevant parameters for producing the fits in Fig. 2 are included in Table 1.

Table 1. Properties of typical galaxies in the SPARC data set.

Galaxy	i	Dis.	Lum.	R_0	R_{last}	M_{HI}	M_d	M/L	$(v^2/c^2R)_{last}$
	(o)	(Mpc)	$(10^9 L_\odot)$	(kpc)	(kpc)	$(10^9 M_\odot)$	$(10^9 M_\odot)$	$(M_\odot L_\odot^{-1})$	$(10^{-30} cm^{-1})$
NGC0100	89	16.08	4.59	1.98	11.46	3.95	2.99	0.65	2.61
NGC1090	64	29.40	45.48	2.80	23.91	7.76	46.13	1.01	3.86
NGC2915	56	8.22	2.63	1.11	20.33	2.92	1.76	0.67	1.33
NGC7814	89	16.31	95.56	2.88	22.12	1.92	74.70	0.93	7.46
UGC05918	46	7.27	0.21	1.58	4.23	0.37	0.41	1.97	1.68
UGC09037	65	74.81	54.95	3.83	25.02	21.39	30.47	0.55	3.32
UGC12506	86	90.99	114.18	6.68	45.21	40.72	136.26	1.19	4.03
UGCA442	64	5.55	0.23	1.51	8.08	0.60	0.39	1.71	1.42

Table Columns: Galaxy name, reported inclination, NED distance, total B-Band luminosity, disk scale length, distance to last data point, HI gas mass, disk mass, mass to light ratio, centripetal acceleration of the last data point.

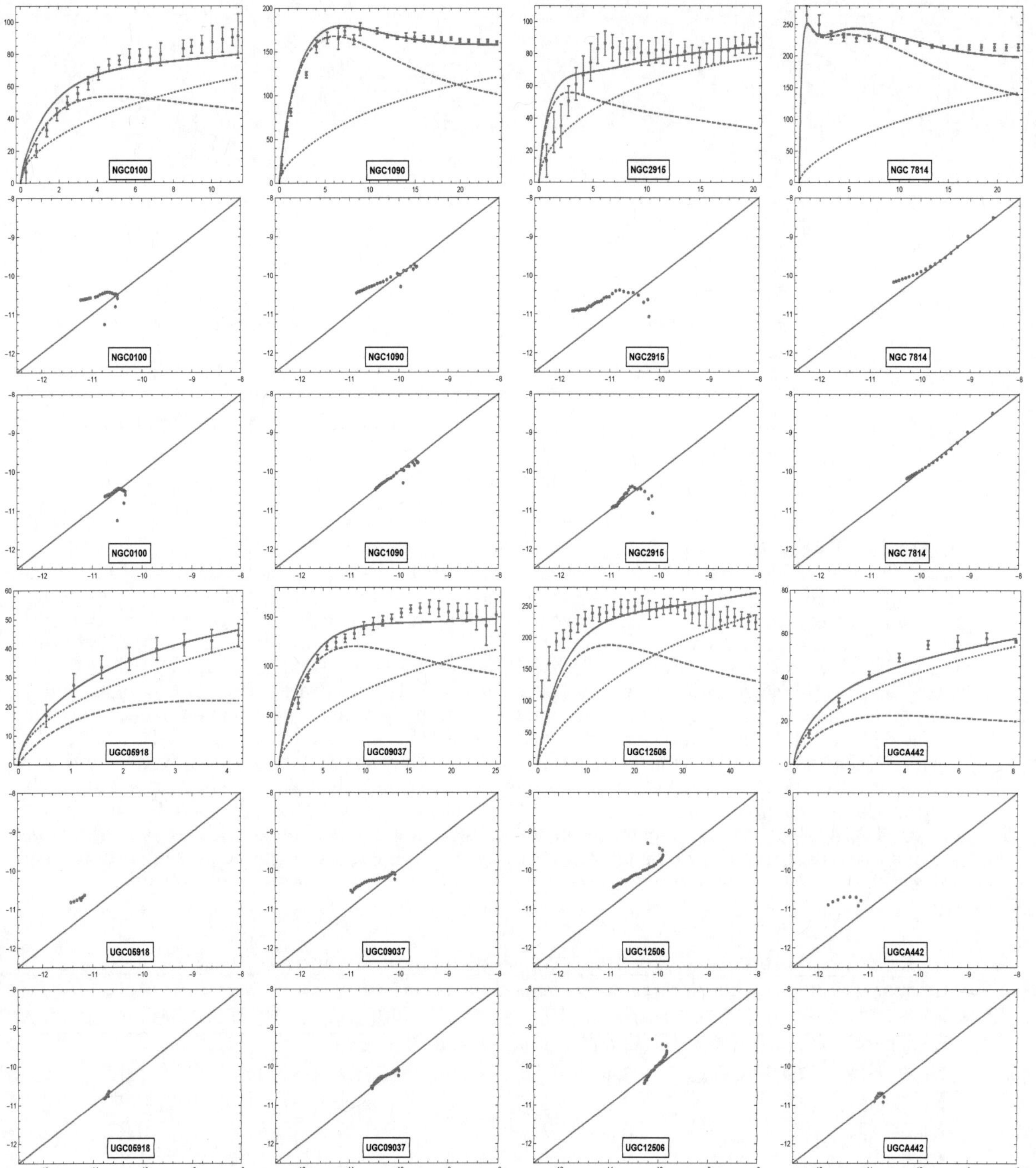

Figure 2. Fitting to the rotational velocities (in km sec^{-1}) of the selected eight galaxy sample with their reported errors as plotted as a function of radial distance (in kpc) is given in rows one and four. The dashed line shows the luminous Newtonian contribution, the dotted line is the contribution of the γ^*, γ^0 and κ terms in eq. (5), and the full curve is the conformal gravity fit. Rows two and five show the respective contributions of the eight galaxies to Fig. 1(b), and rows three and six show their respective contributions to Fig. 4(a).

IOP Conf. Series: Journal of Physics: Conf. Series **1239** (2019) 012009 doi:10.1088/1742-6596/1239/1/012009

To go from the total 236 rotation curves to the centripetal acceleration plots, we follow the same procedure outlined in [15], albeit with some important differences. The first major difference is that for the CG fitting we did not fix mass to light ratios to a particular value as McGaugh et al. had done. Instead, we used the fitting procedure outlined in [12] itself, and used the fitted masses that were obtained in those works. Table 1 shows that this procedure produces acceptable mass to light ratios of the order of the mass to light ratio in the local solar neighborhood. The fitted masses were then used to produce the values for g_{NEW} exhibited in Fig. 1. A second significant difference in constructing our plots versus those found in [15] is our use of the NASA Extragalactic Database (NED) average distance for each galaxy, as it provides a common baseline. For both individual rotation curve points and fitted mass to light ratios there is great sensitivity to the adopted distance as most parameters scale with distance. The last significant difference is our inclusion of the recent ultra high resolution rotation curves of the THINGS and LITTLE THINGS galaxies. Some of the galaxies in these two surveys were included in [15] but older data sets were used instead of the work of [16] and [17]. Since the THINGS and LITTLE THINGS surveys represent some of the highest quality rotation curve data, we choose to include them in our analysis. All other considerations in individual fits such as inclusion of gas and bulge when applicable remain consistent with our previous studies. In total our full data sample contains 236 galaxies with 6377 data points, thus significantly increasing the scope of the study of McGaugh et al.

5. The Conformal Gravity Fits

Fig. 1 illustrates the consistency of the original McGaugh et al. data sample with our data sample. Fig. 1(a) shows the shows the exact data set as taken from the SPARC database while Fig. 1(b) shows all the data in our sample. In Figs. 1(a) and 1(b) we display eq. (1), to show the correlation between the observed centripetal accelerations and the luminous Newtonian expectations. In the standard paradigm it is thought that dark matter is responsible for the fact that g_{OBS} is not equal to g_{NEW} for low g_{NEW}. In CG however, rotation curves are fit without dark matter, with the extra terms in eq. (5) capturing the needed alternate physics. Fig. 3 shows the CG prediction on a point by point basis overlaid on the data in Fig. 1(b). It can be seen from this plot that the CG fits fill out the entire width of the plot without the need for a fitting function such as eq. (1) that would approximate the data by a single curve. To show that CG does fit the data, in Fig. 4 we plot g_{OBS} against $g_{CG} = v_{CG}^2/R$. We see that when CG is used for the fitting we obtain a one to one correlation across the entire data set without any adjustment of free parameters.

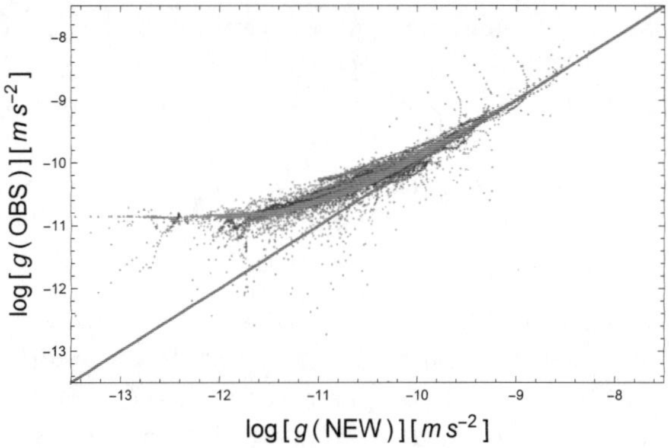

Figure 3. The (g_{OBS}, g_{NEW}) plot with the g_{CG} prediction overlaid. The solid line is the line of unity $g_{OBS} = g_{NEW}$.

Fig. 4(a) shows that there are some points which are off the line of unity. The vertical descending lines in Fig. 4(a) are due to some low rotational velocities at large distances in some dwarf galaxies (see for example [25]). There are also some vertical ascending lines in Fig. 4(a) that represent scatter in the data. To illustrate the negligible significance of the points that are off the line $g_{OBS} = g_{CG}$ in Fig. 4(b) we show a heat map of Fig. 4(a) with cooler colors representing low counts of points, and hotter colors representing high counts. This convenient visualization shows the dominance of the points on or near the line $g_{OBS} = g_{CG}$, and allows for binning of the data points. Out of the total 6377 points, 205 data points fall into the descending or ascending categories described above, which amounts to a total of 3.2% of the total sample. Hence, conformal gravity can predict the observed centripetal accelerations to a high degree of accuracy in 97% of the studied rotation curve data points. Fig. 4 uses no additional free parameters to create these plots, and uses only the physics embodied in eq. (5). The CG theory provides agreement with the Fig. 4 line of unity $g_{OBS} = g_{CG}$, with no dark matter being required at all.

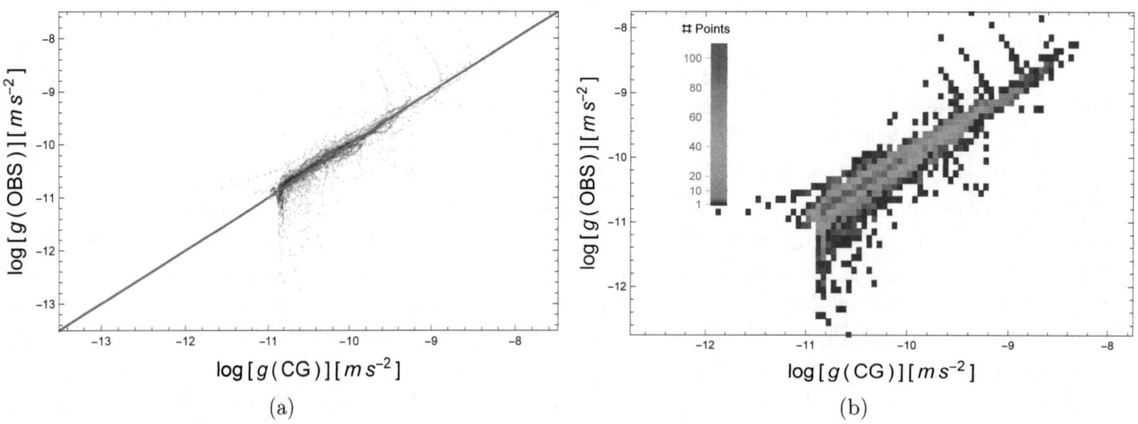

Figure 4. Fig. 4(a) shows the (g_{OBS}, g_{CG}) plot, with the solid line being $g_{OBS} = g_{CG}$. Fig. 4(b) shows a density binning of the points in Fig. 4(a).

6. Tully-Fisher Relation

The Tully-Fisher (TF) relation is an empirical relationship between the total luminous masses ($M = M_{disk} + M_{gas} + M_{bulge}$) and the observed rotation velocities as inferred from HI line widths. With the quadratic term in eq. (5) being negligible except at the largest radii, on evaluating the velocity at the point where the Newtonian Kepler fall off crosses the CG linear potential rise, we obtain a formula for the predicted v^4 at the crossover point of the form

$$v^4 = \frac{BM}{M_\odot}\left(1 + \frac{N^*}{D}\right), \tag{7}$$

where $B = 2c^2 M_\odot G\gamma_0 = 0.0074\ \mathrm{km^4 s^{-4}}$, $D = \gamma_0/\gamma^* = 5.65 \times 10^{10}$, and $N^* = M/M_\odot$ is the number of stars in the galaxy in solar mass units. As we see, the conformal gravity theory fixes the coefficient $B/M_\odot = 2c^2 G\gamma_0$ entirely in terms of fundamental quantities.

For small dwarf galaxies N^* is small compared to D, and for larger galaxies N^* is of the order of D, leading to $BM/M_\odot < v^4 < 2BM/M_\odot$. With there being little difference between the velocities at the crossover points and the velocities at the last data points, eq. (7) then shows that the TF relation is a natural consequence of conformal gravity. In Fig. 5 we construct the Tully-Fisher plot for the convenient last data point for all 236 galaxies in our data sample using the observed velocities and the fitted CG masses. In the figure we have plotted v_{OBS} against $M^{1/4}$ and incorporated the reported errors in the observed rotational velocities. We overlay both $v/M^{1/4}$ = constant and eq. (7). Fig. 5 shows that for most galaxies the CG eq. (7) is obeyed. We note a potentially testable departure from the pure $v/M^{1/4}$ = constant and the CG eq. (7) at the largest galactic masses. As we have noted above, accurate measurements of distance are imperative in rotation curve physics. If the distance is incorrect, we would have a scaling issue and points on Fig. 5 would shift. Moreover, in many galaxies, the TF relation is used to determine the distance in the first place. This leads to a sometimes circular argument where the distance derived using the TF relation is used to make mass models which are then shown to obey the TF relation. In future work, particular galaxies for which the TF relation is used to determine distance will be compared to other galaxies that use alternative distance measurements such as cepheids, in order to isolate the overall effects on modeling.

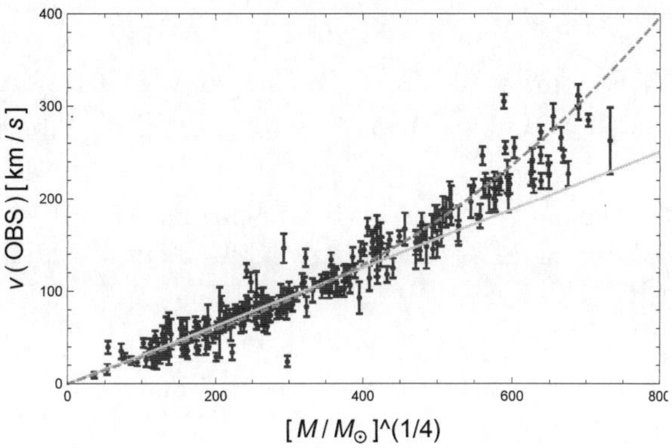

Figure 5. The Tully-Fisher plot for the 236 last data points of our sample as evaluated using the CG fitted masses M. Points are displayed with their reported velocity error. The solid line is the curve $v_{OBS}/M^{1/4}$ = constant, while the dotted line shows the CG prediction of eq. (7).

7. Conclusion

In McGaugh et al. [8] the RAR in galactic rotation curves was established via a set of 2693 total points. In this work we have shown that conformal gravity can universally fit the g_{OBS} versus g_{NEW} data in an even larger 6377 data point sample. Conformal gravity has successfully fitted over 97% of the 6377 data points across 236 galaxies without any filtering of points, fixing of mass to light ratios or modification of input parameters. Further, conformal gravity is shown to satisfy the $v^4 \propto M$ relation consistently found in rotation curve studies, while also providing a derivation and extension of the TF relation. We conclude by noting that there is a great deal of universality in rotation curve data. This universality does not obviously point in favor of dark matter, and is fully accounted for by the alternate conformal gravity theory.

Acknowledgments

J.G. O'Brien would like to thank the International Association of Relativistic Dynamics (IARD) for the opportunity to present this work at the IARD 2018 Conference in Merida, Mexico. J. G. O'Brien would like to thank J. Gagnon and Dr. P. Hafford for their help and support.

References

[1] Freeman K C 1970 *ApJ* **160** 811
[2] Roberts M S and Whitehurst R N 1975 *ApJ* **201** 327
[3] Rubin V C, Ford Jr W K and Thonnard N 1980 *ApJ* **238** 471

[4] Navarro J F, Frenk C S and White S D M 1996 *ApJ* **462** 563

[5] Milgrom M 1983 *ApJ* **270** 365

[6] Moffat J W 2006 *JCAP* **0603** 004

[7] Mannheim P D 2006 *Prog. Part. Nucl. Phys.* **56** 340

[8] McGaugh S S, Lelli F and Schombert J M 2016 *Phys. Rev. Lett.* **117** 201101

[9] Persic M, Salucci P and Stel F 1996 *MNRAS* **281** 27

[10] Tully R B, Verheijen M A W, Pierce M J, Huang J S and Wainscoat R J 1996 *AJ* **112** 2471

[11] Mannheim P D and Kazanas D 1989 *ApJ* **342** 635

[12] Mannheim P D and O'Brien J G 2012 *Phys. Rev. D* **85** 124020

[13] Mannheim P D and O'Brien J G 2011 *Phys. Rev. Lett.* **106** 121101

[14] Lelli F, McGaugh S S and Schombert J M 2016 *AJ* **152** 157

[15] Lelli F, McGaugh S S, Schombert J M and Pawlowski M S 2017 *ApJ* **836** 152

[16] Walter F, Brinks E, de Blok W J G, Bigiel F, Kennicutt R C, Thornley M D and Leroy A 2008 *AJ* **136** 2563

[17] Oh S H, Hunter D A, Brinks E, Elmegreen B G, Schruba A, Walter F, Rupen M P, Young L M, Simpson C E, Johnson M C, Herrmann K A, Ficut-Vicas D, Cigan P, Heesen V, Ashley T and Zhang H X 2015 *AJ* **149** 180

[18] Verheijen M and de Blok E 1999 *Ap&SS* **269** 673

[19] Swaters R A, van Albada T S, van der Hulst J M and Sancisi R 2002 *AA* **390** 829

[20] Spekkens K and Giovanelli R 2006 *AJ* **132** 1426

[21] Lelli F, Verheijen M and Fraternali F 2014 *AA* **566** A71

[22] Walsh W, Staveley-Smith L and Oosterloo T 1997 *AJ* **113** 1591

[23] de Blok W J G, McGaugh S S and van der Hulst J M 1996 *MNRAS* **283** 18

[24] Noordermeer E, van der Hulst J M, Sancisi R, Swaters R S and van Albada T S 2007 *MNRAS* **376** 1513

[25] O'Brien J G, Chiarelli T L, Dentico J, Stulge M, Stefanski B, Moss R and Chaykov S 2018 *ApJ* **852** 6

Gravitation in the surface tension model of spacetime

H A Perko[1]
[1]Office 14, 140 E. 4th Street, Loveland, CO, USA 80537

E-mail: hperko@kopparesearch.com

Abstract. A mechanical model of spacetime was introduced at a prior conference for describing perturbations of stress, strain, and displacement within a spacetime exhibiting surface tension. In the prior work, equations governing spacetime dynamics described by the model show some similarities to fundamental equations of quantum mechanics. Similarities were identified between the model and equations of Klein-Gordon, Schrödinger, Heisenberg, and Weyl. The introduction did not explain how gravitation arises within the model. In this talk, the model will be summarized, corrected, and extended for comparison with general relativity. An anisotropic elastic tensor is proposed as a constitutive relation between stress energy and curvature instead of the traditional Einstein constant. Such a relation permits spatial geometric terms in the mechanical model to resemble quantum mechanics while temporal terms and the overall structure of tensor equations remain consistent with general relativity. This work is in its infancy; next steps are to show how the anisotropic tensor affects cosmological predictions and to further explore if geometry and quantum mechanics can be related in more than just appearance.

1. Introduction

The focus of this research is to find a mechanism by which spacetime might curl, warp, or re-configure at small scales to provide a geometrical explanation for quantum mechanics while remaining consistent with gravity and general relativity. This research borrows techniques from physical chemistry of surfaces, mathematics of elasticity, and continuum mechanics. The research led to a proposed mechanical model of spacetime [1,2] with multi-dimensional surface tension (herein hypersurface tension). Equations governing energetic perturbations of stress, strain, and reconfiguration of spacetime with hypersurface tension were shown to resemble Schrödinger, Klein-Gordon, Heisenberg, and Weyl-like expressions. The previous work also provided several mathematical expressions for quantization and formation of particles from perturbations in space and time.

In this paper, the mechanical model discussed in [1] is extended to include further comparison with general relativity. It is shown that the model can provide geometrical equations resembling both quantum mechanics and general relativity by assuming a symmetric nondegenerate anisotropic constitutive relation between stress energy and curvature rather than the traditional scalar Einstein constant. The next step in development is to apply the model to describe physical systems and compare results with known experiments so that the model can be verified, falsified, or further refined.

IOP Conf. Series: Journal of Physics: Conf. Series **1239** (2019) 012010 doi:10.1088/1742-6596/1239/1/012010

2. Hypersurface Geometry

It is well known in physics that every three-dimensional volume is a three-surface in four-dimensional space [3]. To aid in visualization of this geometry, an arbitrary three volume is shown by the cube in the center of Figure 1. Such a volume may be filled with points of matter and energy moving along different trajectories and velocities. Each point contained within the three volume corresponds to a point in each of three projected surfaces plotted with respect to two coordinates of space and one coordinate of time. The projected surfaces are coincident with the three volume and are sometimes referred to as time-like hypersurfaces or slices.

In the previous work [1], the geometry in Figure 1 was described as being associated with the "synchronous reference system" defined by Landau and Lifshitz [4]. The author [1] only intended to highlight the diagonal metric but unintentionally implied stationary conditions. The reference frame described in Figure 1 is simply that of any arbitrary observer. Each arbitrary observer records the position of matter and energy in their coordinate of time.

A coordinate frame with a diagonal metric is called a "principle frame" in continuum mechanics. When solving problems in continuum mechanics, it is often simplest to start from the principle frame, write equations of motion, allow the system to evolve, and then transform as necessary to find the stress energy and manifold configuration in other frames [5]. Thus, the choice of principle frame should not be considered preferential for any reason other than mathematical convenience. To the engineer, the principle frame suppresses shear strains. To the physicist, the principle frame suppresses the gravitational equivalent of magnetic fields [6].

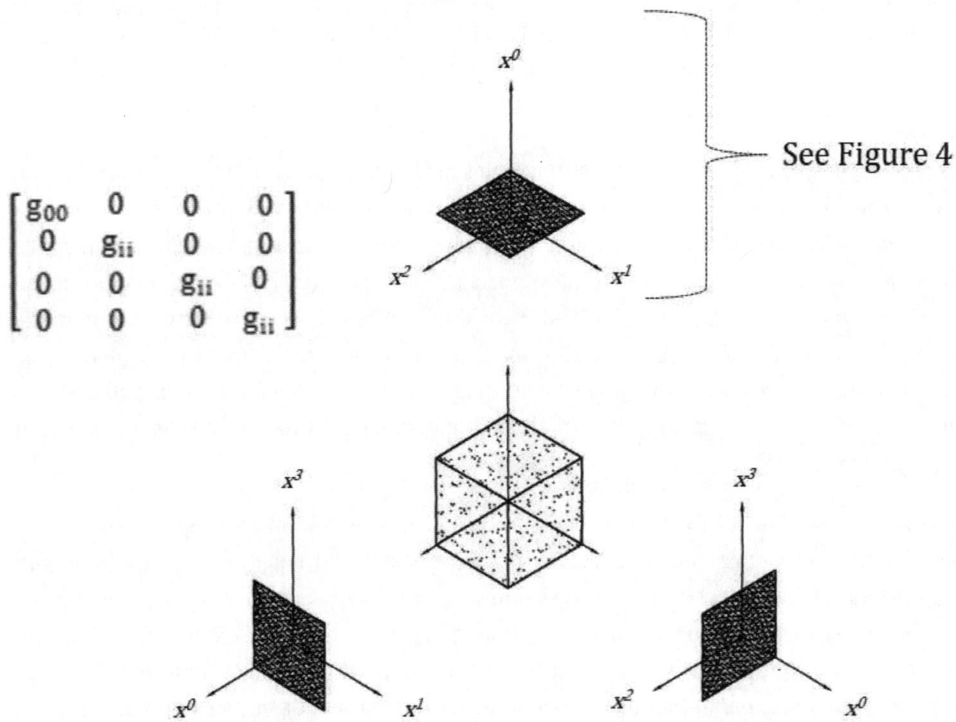

Figure 1. Hypersurface Projections in the Principle Frame

IOP Conf. Series: Journal of Physics: Conf. Series **1239** (2019) 012010 doi:10.1088/1742-6596/1239/1/012010

3. Physical Chemistry of Hypersurfaces

The diagram shown in Fig. 1 illustrates how interfaces with time or hypersurfaces arise in four-dimensions leading to a "fabric of spacetime". Fields and forces are truncated for the arbitrary observer due to limitations imposed by the speed of light. It is argued here that such truncation must result in surface tension.

Ordinary surface tension is a familiar concept for most people from everyday interactions with liquids. Surface tension causes membrane-like behavior at an interface. It is important to note that this behavior does not imply a real physical membrane nor is the root cause a chemical process. Rather, surface tension and membrane behavior arise from truncated fields and bound multiplicity of quantum states when matter and energy are arranged at a surface for an observer. Physical chemists using Boltzmann statistics have proven that surface tension exists at any stable interface, regardless of material composition and phase. The only prerequisite for surface tension is a non-dispersive interface [7].

To better understand surface tension, consider the example distribution of particles arranged in a continuum in Figure 2a. Each set of circles represents probability contours describing the likely location of a particle. Particle field interactions lead to a statistically uniform distribution as a consequence of entropy. Creation of an interface, as shown by the dark line in Figure 2b, leads to an increase in local probability at the boundary. Surface tension is the mathematical result of probability density field truncation that exists at any stable interface.

More explicitly, let us define a quantum state, s_1, of existence above the boundary in Figure 2 and another state, s_2, of existence below the boundary. In Figure 2a, the two states have equal probability. In Figure 2b, the state $s_1$is no longer accessible and the probability of s_2 increases. The increase in probability of state $s_2$is associated with a decrease in multiplicity of states when moving from the system in Figure 2a to Figure 2b. Decrease in multiplicity means a decrease in entropy and requires work. The work done on the system is the energy of formation of the interface or *surface energy*.

An underlying theorem of physical chemistry of surfaces is that surface energy is mathematically equivalent to surface tension [7]. Herein, surface energy and surface tension shall be denoted by the Greek letter Koppa, Ϙ. [1] To better understand the relationship between surface energy and surface tension, consider the simple mechanical model in Figure 3, which is adapted from [7]. If a window is opened in a boundary allowing access to quantum states above such that they can affect the probability of states below and the interface between states has surface energy, Ϙ, then the work done on the system is given by

$$Work = \text{Ϙ} dA$$

[1]In physical chemistry, the Greek letter gamma, γ, is typically used to represent surface tension. To avoid confusion with the Lorentz factor in physics, assignment of a new symbol was necessary. The archaic Greek letter Koppa, Ϙ, was chosen. Before being adopted and then omitted from the modern Greek alphabet, the letter Koppa, Ϙ, had its origins in ancient Hebrew. The ancestor Koppa symbol was drawn as the sun on the horizon and it signified light at the edge of time. Thus, Koppa seemed a fitting parameter to describe tension at the edge of time. In the Roman alphabet, Koppa corresponds to the letter *Q*.

IOP Conf. Series: Journal of Physics: Conf. Series **1239** (2019) 012010 doi:10.1088/1742-6596/1239/1/012010

which can be rewritten as a force pulling on length L moving a distance *dx*,

$$Work = ϘLdx$$

where $dA = Ldx$ represents the creation of additional surface area. In the second formulation, Ϙ appears as a tension per unit length or *surface tension*. Thus, surface tension and surface energy are mechanically equivalent concepts.

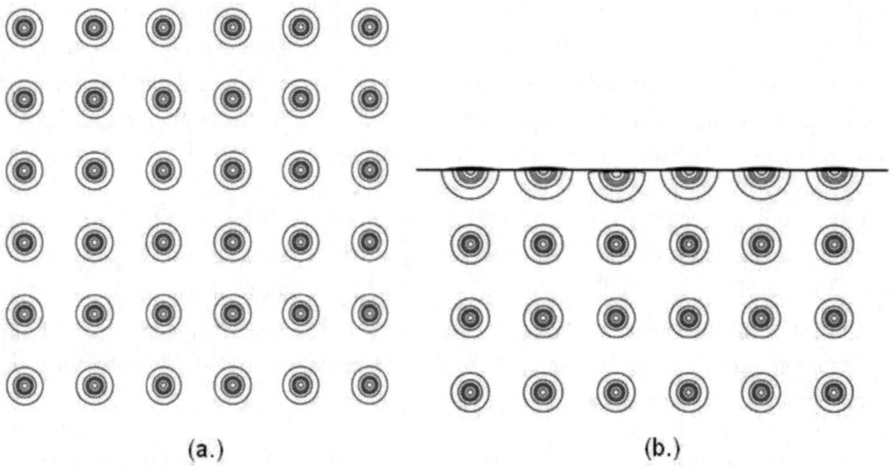

Figure 2. Probability Density of Interacting Particles (a. in continuum, b. with boundary)

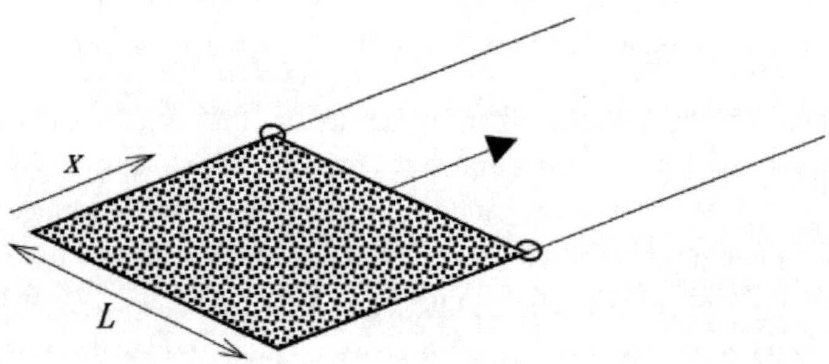

Figure 3. Hypersurface Stretched Across a Wire Frame Window with One Moveable Side

In physics, future states are inaccessible because nothing can travel faster than the speed of light. Therefore, a boundary in time exists for an observer and, according to the basic postulates of physical chemistry of surfaces, surface tension should and must exist regardless of dimensionality of a hypersurface. What is true for one observer must be true for all observers. Hypersurface tension should result in a uniform negative stress energy holding all matter and energy together within the construct of general relativity, as we shall see, enforcing causality, and avoiding complete temporal dispersion. In four-dimensions, the units of Koppa are J/m^3, which is equivalent to N/m^2[1].

IOP Conf. Series: Journal of Physics: Conf. Series **1239** (2019) 012010 doi:10.1088/1742-6596/1239/1/012010

4. Hypersurface Continuum Mechanics

It would be useful to apply traditional continuum mechanics to describe hypersurface geometry with surface tension. Continuum mechanics offers tools based in Riemannian geometry for relating stress energy to reconfiguration and surface evolution that satisfy conditions of covariance and uniqueness [8]. However, one cannot apply these tools without a change of coordinates. Spacetime geometry is pseudo-Riemannian and, in fact, hyperbolic. According to special relativity,

$$-c^2 d\tau^2 = -c^2 dt^2 + {dx^1}^2 + {dx^2}^2 + {dx^3}^2$$

where dx^j are spatial coordinates with index, j, running from 1 to 3, and t is the coordinate of time. This pseudo-Riemannian geometry can be transformed to a Riemannian geometry by rewriting the equation above in complex coordinates, as given by

$$d\tau'^2 = {dx^0}^2 + {dx^1}^2 + {dx^2}^2 + {dx^3}^2$$

where

$$x^0 = ict$$

$$\tau' = ic\tau$$

The parameter, τ, is proper time, a frame independent scalar. The parameter, τ', is referred to as proper distance, but will be called imaginary proper time herein for the following reason. Since all observers agree on the imaginary proper time, τ', one can write a set of equations for the evolution of the system according to τ'. Thus, imaginary proper time may serve as a type of frame independent "clock" in 4-dimensional continuum mechanics.

The substitution of complex coordinates allows traditional continuum mechanics concepts to be applied to spacetime. The remainder of this section is devoted to describing each of the tensor equations of continuum mechanics as they relate to spacetime with hypersurface geometry. Specifically, the Cauchy stress tensor, rate of deformation tensor, equations of motion, and constitutive relationships are presented. In the sections that follow, these geometrical equations are compared with quantum mechanics and general relativity.

Consider again the arbitrary distribution of matter and energy plotted with respect to space and time in Figure 1. One of the projections with time and two coordinates of space from Figure 1 is replotted in Figure 4. A similar plot can be made with time and each pair of spatial coordinates. Recall that dots in the surface represent a particle wave packet or some sort of matter or energy. The distribution of matter and energy in Figure 4 provides an energy density in the hypersurface. The question under consideration is not in regard to specific particles and waves comprising the surface, but how does spacetime evolve due to this bound *surface energy*.

In a rest frame or static spacetime, dx^i is assumed small, $d\tau' \approx dx^0$, and the hypersurface in Figure 4 would be flat. In this work, spacetime is assumed dynamic, and each infinitesimal element in spacetime is moving at some four-vector velocity, u^μ, according to the arbitrary observer. The metric is allowed to evolve beyond flat, deviations in imaginary proper time and coordinate time are permitted at a point, and hypersurfaces are irregular.

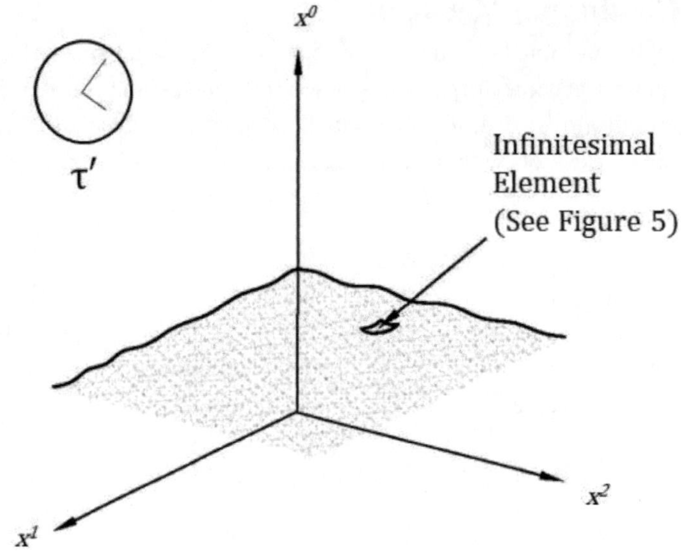

Figure 4. Hypersurface Showing Postulated Curvy Microstructure

An infinitesimal element of hypersurface from Figure 4 is depicted at larger scale in Figure 5. Surface tension is shown along the four spatial sides of the element. A pressure, *P*, is applied normal to the hypersurface and represents an external volumetric temporal pressure or encapsulated energy density. The infinitesimal element is depicted as curved and surface tension vectors are not co-linear. Shear stresses are omitted along the edges and upper and lower surfaces of the element by choice of principle frame.

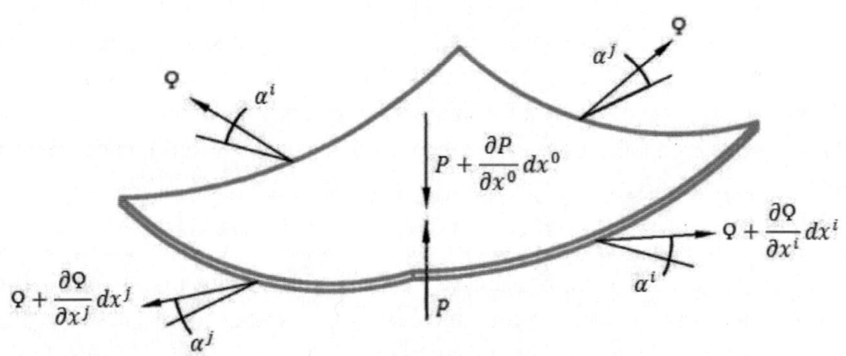

Figure 5. Hypersurface Tension on an Infinitesimal Element of Spacetime

The stresses about the infinitesimal element of spacetime in Figure 5 can be represented by a covariant hypersurface tension stress energy tensor, $T_{\mu\nu}$, given by,

$$T_{\mu\nu} = \begin{bmatrix} P & 0 & 0 & 0 \\ 0 & -♀ & 0 & 0 \\ 0 & 0 & -♀ & 0 \\ 0 & 0 & 0 & -♀ \end{bmatrix} \quad (1)$$

Note that the sign convention in (1) is opposite that in [1]. Surface tension is generally taken as positive in physical chemistry and membrane mechanics. A sign convention is used herein where tensile stress is negative and compression stress is positive. This convention is more consistent with usual treatments of general relativity.

Reconfiguration of a system in continuum mechanics is represented by a two-point rate of deformation tensor constructed from the gradient of spatial velocity and the conjugate transpose of the gradient. Velocity in continuum mechanics refers to the movement of "material" relative to some arbitrary inertial reference frame. Here, velocity refers to metric dynamics, or movement of spacetime, according to any arbitrary observer. Recall the four-vector velocity of the infinitesimal element of spacetime in Figure 5 is denoted by $u^\mu = \frac{dx^\mu}{d\tau'}$ and varies according to the observer. The two-point rate of deformation tensor, D^α_β, for a membrane-like infinitesimal element of spacetime for any observer is written as,

$$D^\alpha_\beta = \begin{bmatrix} \frac{\partial u^0}{\partial x^0} & \frac{1}{2}\left(\frac{\partial u^0}{\partial x^1}+\frac{\partial u^1}{\partial x^0}\right) & \frac{1}{2}\left(\frac{\partial u^0}{\partial x^2}+\frac{\partial u^2}{\partial x^0}\right) & \frac{1}{2}\left(\frac{\partial u^0}{\partial x^3}+\frac{\partial u^3}{\partial x^0}\right) \\ \frac{1}{2}\left(\frac{\partial u^1}{\partial x^0}+\frac{\partial u^0}{\partial x^1}\right) & \frac{\partial u^1}{\partial x^1}-\frac{1}{2}\left|\frac{\partial u^0}{\partial x^1}\right|^2 d\tau' & \frac{1}{2}\left(\frac{\partial u^1}{\partial x^2}+\frac{\partial u^2}{\partial x^1}\right)-\frac{1}{2}\left|\left(\frac{\partial u^0}{\partial x^1}\right)\left(\frac{\partial u^0}{\partial x^2}\right)\right| d\tau' & \frac{1}{2}\left(\frac{\partial u^1}{\partial x^3}+\frac{\partial u^3}{\partial x^1}\right)-\frac{1}{2}\left|\left(\frac{\partial u^0}{\partial x^1}\right)\left(\frac{\partial u^0}{\partial x^3}\right)\right| d\tau' \\ \frac{1}{2}\left(\frac{\partial u^2}{\partial x^0}+\frac{\partial u^0}{\partial x^2}\right) & \frac{1}{2}\left(\frac{\partial u^2}{\partial x^1}+\frac{\partial u^1}{\partial x^2}\right)-\frac{1}{2}\left|\left(\frac{\partial u^0}{\partial x^2}\right)\left(\frac{\partial u^0}{\partial x^1}\right)\right| d\tau' & \frac{\partial u^2}{\partial x^2}-\frac{1}{2}\left|\frac{\partial u^0}{\partial x^2}\right|^2 d\tau' & \frac{1}{2}\left(\frac{\partial u^2}{\partial x^3}+\frac{\partial u^3}{\partial x^2}\right)-\frac{1}{2}\left|\left(\frac{\partial u^0}{\partial x^2}\right)\left(\frac{\partial u^0}{\partial x^3}\right)\right| d\tau' \\ \frac{1}{2}\left(\frac{\partial u^3}{\partial x^0}+\frac{\partial u^0}{\partial x^3}\right) & \frac{1}{2}\left(\frac{\partial u^3}{\partial x^1}+\frac{\partial u^1}{\partial x^3}\right)-\frac{1}{2}\left|\left(\frac{\partial u^0}{\partial x^3}\right)\left(\frac{\partial u^0}{\partial x^1}\right)\right| d\tau' & \frac{1}{2}\left(\frac{\partial u^3}{\partial x^2}+\frac{\partial u^2}{\partial x^3}\right)-\frac{1}{2}\left|\left(\frac{\partial u^0}{\partial x^3}\right)\left(\frac{\partial u^0}{\partial x^2}\right)\right| d\tau' & \frac{\partial u^3}{\partial x^3}-\frac{1}{2}\left|\frac{\partial u^0}{\partial x^3}\right|^2 d\tau' \end{bmatrix} \quad (2)$$

The rate of deformation tensor shown here is similar to the ordinary three-dimensional dynamical strain tensor in continuum mechanics except it is expanded to four-dimensions and contains extra terms to account for temporal strain, $\frac{\partial x^0}{\partial x^\mu}$, and out-of-plane deformations, $\frac{1}{2}\left|\left(\frac{\partial u^0}{\partial x^i}\right)\left(\frac{\partial u^0}{\partial x^j}\right)\right| d\tau'$, in time. Out-of-plane deformations are movements in the time direction in Figure 5 and resulting curvature. Equations for out-of-plane deformations can be found in common texts on mechanics of thin shells without bending (membrane theory) [9,10,11] and are described in terms of Riemannian geometry in [8]. Spatial components of tensor (2) were derived in [1].

Rate of deformation of an element of spacetime over an infinitesimally small increment in imaginary proper time can be assumed to be very small compared to unity. The Riemannian metric is related to the rate of deformation tensor by the following approximation derived in [1] with opposite sign convention,

$$g^\alpha_\beta = \boldsymbol{I} - 2D^\alpha_\beta d\tau' \quad (3)$$

Written in this way, the reconfiguration of the system is a linearization of Lagrangian and Eulerian strains by a process known as infinitesimal strain theory. It has many similarities to linearized gravity. By the underlying assumption of small changes in the metric, one might conclude that infinitesimal strain theory cannot be used to describe dynamics of spacetime with a high degree of curvature. This conclusion is false as infinitesimal strain theory is used frequently in numerical modeling to describe highly curved systems provided numerical elements are small compared to the degree of curvature.

Equations of motion describing evolution of an infinitesimal element of spacetime with imaginary proper time were derived in [1]. The derivation followed from a simple application of Newton's laws in four-dimensions taking advantage of local Euclidian geometry in the tangent space at a point. These laws of motion can be written as,

$$\begin{bmatrix} \text{Ϙ}\left(\frac{\partial^2 x^0}{\partial x^{1^2}}+\frac{\partial^2 x^0}{\partial x^{2^2}}+\frac{\partial^2 x^0}{\partial x^{3^2}}\right)-dP & \frac{1}{2}\left(\frac{\partial \text{Ϙ}}{\partial x^1}-dP\right) & \frac{1}{2}\left(\frac{\partial \text{Ϙ}}{\partial x^2}-dP\right) & \frac{1}{2}\left(\frac{\partial \text{Ϙ}}{\partial x^3}-dP\right) \\ \frac{1}{2}\left(\frac{\partial \text{Ϙ}}{\partial x^1}-dP\right) & \frac{\partial \text{Ϙ}}{\partial x^1} & \frac{1}{2}\left(\frac{\partial \text{Ϙ}}{\partial x^1}+\frac{\partial \text{Ϙ}}{\partial x^2}\right) & \frac{1}{2}\left(\frac{\partial \text{Ϙ}}{\partial x^1}+\frac{\partial \text{Ϙ}}{\partial x^3}\right) \\ \frac{1}{2}\left(\frac{\partial \text{Ϙ}}{\partial x^2}-dP\right) & \frac{1}{2}\left(\frac{\partial \text{Ϙ}}{\partial x^2}+\frac{\partial \text{Ϙ}}{\partial x^1}\right) & \frac{\partial \text{Ϙ}}{\partial x^2} & \frac{1}{2}\left(\frac{\partial \text{Ϙ}}{\partial x^2}+\frac{\partial \text{Ϙ}}{\partial x^3}\right) \\ \frac{1}{2}\left(\frac{\partial \text{Ϙ}}{\partial x^3}-dP\right) & \frac{1}{2}\left(\frac{\partial \text{Ϙ}}{\partial x^3}+\frac{\partial \text{Ϙ}}{\partial x^1}\right) & \frac{1}{2}\left(\frac{\partial \text{Ϙ}}{\partial x^3}+\frac{\partial \text{Ϙ}}{\partial x^2}\right) & \frac{\partial \text{Ϙ}}{\partial x^3} \end{bmatrix}$$

$$= -\text{Ϙ}\begin{bmatrix} \frac{\partial u^0}{\partial \tau'} & \frac{1}{2}\left(\frac{\partial u^0}{\partial \tau'}+\frac{\partial u^1}{\partial \tau'}\right) & \frac{1}{2}\left(\frac{\partial u^0}{\partial \tau'}+\frac{\partial u^2}{\partial \tau'}\right) & \frac{1}{2}\left(\frac{\partial u^0}{\partial \tau'}+\frac{\partial u^3}{\partial \tau'}\right) \\ \frac{1}{2}\left(\frac{\partial u^1}{\partial \tau'}+\frac{\partial u^0}{\partial \tau'}\right) & \frac{\partial u^1}{\partial \tau'} & \frac{1}{2}\left(\frac{\partial u^1}{\partial \tau'}+\frac{\partial u^2}{\partial \tau'}\right) & \frac{1}{2}\left(\frac{\partial u^1}{\partial \tau'}+\frac{\partial u^3}{\partial \tau'}\right) \\ \frac{1}{2}\left(\frac{\partial u^2}{\partial \tau'}+\frac{\partial u^0}{\partial \tau'}\right) & \frac{1}{2}\left(\frac{\partial u^2}{\partial \tau'}+\frac{\partial u^1}{\partial \tau'}\right) & \frac{\partial u^2}{\partial \tau'} & \frac{1}{2}\left(\frac{\partial u^2}{\partial \tau'}+\frac{\partial u^3}{\partial \tau'}\right) \\ \frac{1}{2}\left(\frac{\partial u^3}{\partial \tau'}+\frac{\partial u^0}{\partial \tau'}\right) & \frac{1}{2}\left(\frac{\partial u^3}{\partial \tau'}+\frac{\partial u^1}{\partial \tau'}\right) & \frac{1}{2}\left(\frac{\partial u^3}{\partial \tau'}+\frac{\partial u^2}{\partial \tau'}\right) & \frac{\partial u^3}{\partial \tau'} \end{bmatrix}$$

$$- \begin{bmatrix} u^0\frac{\partial \text{Ϙ}}{\partial \tau'} & 0 & 0 & 0 \\ 0 & u^1\frac{\partial \text{Ϙ}}{\partial \tau'} & 0 & 0 \\ 0 & 0 & u^2\frac{\partial \text{Ϙ}}{\partial \tau'} & 0 \\ 0 & 0 & 0 & u^3\frac{\partial \text{Ϙ}}{\partial \tau'} \end{bmatrix} \quad (4)$$

Diagonal terms in (4) are those derived in [1] except that one-sided temporal pressure has been replaced by differential pressure across the hypersurface to be more rigorous. Off-diagonal terms describing rotation and shear have been added for completeness. The right side of (4) differs in sign from [1] due to convention (e.g. compression stress is positive).

In order to maintain a stable hypersurface for the arbitrary observer, it is proposed that spacetime behave in accordance with an elastic continuum. An elastic continuum exhibits a constitutive relationship between stress energy and rate of deformation tensors. The simplest example of a stable constitutive relationship is the isotropic and linear elastic tensor given by ***EI*** where ***I*** is the identity matrix and *E* is a constant known in continuum mechanics as Young's Modulus. This was the approach taken in [1] where, with some math errors corrected, the constitutive relationship was set as $T_{ii} = (c\hbar/2)D_i^{\,i}$ where $\hbar$ is the reduced Plank's constant. This isotropic relationship was chosen to enforce spatial symmetries and cause spacetime geometry to resemble the appearance of some equations in quantum mechanics.

Constitutive relationships between temporal components of stress energy and rate of deformation were omitted in the original model [1]. It is necessary to include temporal components for comparison with general relativity. However, if the scalar elastic moduli in [1] is applied to temporal terms, it would result in gravity that is about 57 orders of magnitude higher than reality. A scalar constitutive relation does not provide for a unified model for quantum mechanics and gravity.

In continuum mechanics, there are many stable constitutive relationships between stress energy and rate of deformation. To say a constitutive relationship is stable, is to say it is nondegenerate and invertible. A scalar elastic tensor is just one of many stable possibilities. Another possibility is a stable *anisotropic*

IOP Conf. Series: Journal of Physics: Conf. Series **1239** (2019) 012010 doi:10.1088/1742-6596/1239/1/012010

linear elastic tensor. Such a tensor would allow for different elastic moduli for spatial and temporal energy terms.

To maintain the dimensionality of spacetime, an anisotropic elastic tensor must have linearly independent eigenfunctions and, thus, be eigen-decomposable. A simple example of an eigen-decomposable tensor is diagonal. A diagonal elastic tensor has zero off-diagonal terms and is manifestly symmetric which ensures conservation of angular momentum (torsion free equations). The components of the diagonal should be spatially isotropic to ensure rotational symmetry in time and conservation of momentum and mass energy. To be invertible, the matrix must have a non-zero determinant. Any diagonalizable matrix with non-zero determinate can be written as the product of a scalar and a matrix with determinate equal to one. A possible elastic tensor meeting these criteria is shown in Voigt notation below

$$T_{\mu\nu} = C^{\beta}_{\mu\nu\alpha} \cdot D^{\alpha}_{\beta} \qquad (5)$$

$$\begin{bmatrix} T_{00} \\ T_{11} \\ T_{22} \\ T_{33} \\ \vdots \end{bmatrix} = A \begin{bmatrix} b^3 & 0 & 0 & 0 & \cdots \\ 0 & \frac{1}{b} & 0 & 0 & \cdots \\ 0 & 0 & \frac{1}{b} & 0 & \cdots \\ 0 & 0 & 0 & \frac{1}{b} & \cdots \\ \vdots & \vdots & \vdots & \vdots & \end{bmatrix} \begin{bmatrix} D^0_0 \\ D^1_1 \\ D^2_2 \\ D^3_3 \\ \vdots \end{bmatrix} \qquad (6)$$

where A and b are non-zero constants. Voigt notation is commonly used in continuum mechanics to represent a higher order elastic tensor in lower order. As can be seen, the determinant of the elastic tensor in (6) is equal to one for all values of b.

As in earlier work [1], a geometry suggestive of quantum mechanics can be derived if constants A and b are assigned values such that the spatial components of the constitutive relation equal $c\hbar/2$. In a later section, tangent space in continuum mechanics is pulled back to four-dimensional gravitational geometry. In order to match gravity at slow speed approximations, it is important that the time-time constitutive term in (6) have a value equal to the inverse of Einstein's constant since moving the elastic tensor to the stress energy side of (7) will invert the eigenvalues. Thus, an elastic tensor satisfying these conditions can be found by solving the following simultaneous equations,

$$Ab^3 = \frac{c^4}{8\pi G}$$

$$\frac{A}{b} = \frac{c\hbar}{2}$$

which yields, with some algebraic simplification,

$$C_{\mu\nu\alpha}^{\beta} = \frac{c\hbar}{2}\begin{bmatrix} \frac{1}{4\pi {l_p}^2} & 0 & 0 & 0 & \cdots \\ 0 & 1 & 0 & 0 & \cdots \\ 0 & 0 & 1 & 0 & \cdots \\ 0 & 0 & 0 & 1 & \cdots \\ \vdots & \vdots & \vdots & \vdots & \end{bmatrix} \tag{7}$$

where l_p is the Plank length. Note that the determinant of (7) is the Einstein constant. This is an important result that enables the elastic tensor to replace the Einstein constant in general relativity and maintain scale between energy and curvature for observers.

The elastic tensor for a four-dimensional continuum is actually a 4x4x4x4 matrix containing 256 terms. The problem was greatly simplified by imposing a maximum of internal symmetries. Also, shear moduli (diagonal terms to bottom right) are not shown in (7). Shear moduli may prove to be important in transformations and may need to be added at a later time. It is possible that nature may be more complex, so (7) should be viewed as a starting point for exploration of other possible spacetime anisotropies.

5. Quantum Mechanical Similarities

In this section, the mechanical model of spacetime with hypersurface tension is dissected and compared with mathematical expressions found in quantum mechanics. Similar comparisons were made in earlier work [1], but are shown here with more rigor. Concerns have arisen regarding the haphazard placement of imaginary number, *i*, in earlier work. The appearance of *i* is not haphazard, but arises from moving between the complex coordinates used here for Reimannian geometry of continuum mechanics and the normal "laboratory" coordinates of pseudo-Riemannian spacetime geometries. More clarity regarding the appearance of imaginary numbers is given below.

The first similarity between the mechanics of spacetime with hypersurface tension and quantum mechanics is found in the time-time component of (4). For problems in physical chemistry, the change in surface tension with time, $\frac{\partial \text{♀}}{\partial \tau}$, is generally ignored for short-duration, micro-scale phenomena as this term has to do with overall changes in temperature of the system. The temperature independent time-time component in (4) is given by,

$$\frac{\partial^2 x^0}{\partial {x^1}^2} + \frac{\partial^2 x^0}{\partial {x^2}^2} + \frac{\partial^2 x^0}{\partial {x^3}^2} - \frac{dP}{\text{♀}} = -\frac{\partial^2 x^0}{\partial \tau'^2}$$

When this equation is converted from complex coordinates back to natural units, we find,

$$ic\left(\frac{\partial^2 t}{\partial {x^1}^2} + \frac{\partial^2 t}{\partial {x^2}^2} + \frac{\partial^2 t}{\partial {x^3}^2}\right) - \frac{dP}{\text{♀}} = -ic\frac{\partial^2 t}{-c^2 \partial \tau^2}$$

which simplifies to,

$$\frac{\partial^2 t}{\partial {x^1}^2} + \frac{\partial^2 t}{\partial {x^2}^2} + \frac{\partial^2 t}{\partial {x^3}^2} - \frac{dP}{ic\text{♀}} = \frac{1}{c^2}\frac{\partial^2 t}{\partial \tau^2} \tag{8}$$

The wave function in (8) is the observer's natural coordinate of time, t, at different points in space. These waves are known as capillary waves in physical chemistry of surfaces. As can be seen in (8), the waves move at the speed of light independent of frequency and wavelength. Since capillary waves move along hypersurfaces, they exhibit polarization and follow the overall curvature of spacetime. Capillary waves do not carry mass. Electromagnetic radiation behaves in accord with this analog. Equation (8) is similar in mathematical structure to the Klein-Gordon equation in quantum mechanics.

A second similarity with quantum mechanics is found in the diagonal spatial terms of (1), (2), (4), (5) and (7). Begin by relating the diagonal spatial terms of (1) and (2) in accord with (5) and (7) to find,

$$-\text{Q} = \frac{c\hbar}{2}\left(\frac{\partial u^j}{\partial x^j} - \frac{1}{2}\left|\frac{\partial u^0}{\partial x^j}\right|^2 d\tau'\right)$$

where j is an index running from 1 to 3 to indicate spatial components only. Next, the partial derivative of the above expression for Q with respect to x^j is given by,

$$\frac{\partial \text{Q}}{\partial x^j} = -\frac{c\hbar}{2}\left(\frac{\partial^2 u^j}{\partial x^{j^2}} - \left|\frac{\partial u^0}{\partial x^j}\right|\frac{\partial^2 u^0}{\partial x^{j^2}} d\tau'\right)$$

The component $\frac{\partial^2 u^j}{\partial x^{j^2}}$ represents the change of in-plane[2] strain with spatial distance, which is a stretching of the hypersurface. The component $\left|\frac{\partial u^0}{\partial x^j}\right|\frac{\partial^2 u^0}{\partial x^{j^2}} d\tau'$ is the change in curvature with spatial distance. Substituting the simplified lead-order result for $\frac{\partial \text{Q}}{\partial x^j}$ into (4) and again ignoring $\frac{\partial \text{Q}}{\partial \tau}$ (temperature dependence) yields the energy dispersion equation,

$$\frac{c\hbar}{2}\left(\frac{\partial^2 u^j}{\partial x^{j^2}}\right) = \text{Q}\frac{\partial u^j}{\partial \tau'}$$

Converting this equation from complex coordinates to natural units, noting that $u^j = \frac{\partial x^j}{\partial \tau'} = \frac{1}{ic}\frac{\partial x^j}{\partial \tau}$, we find,

$$\frac{c\hbar}{2}\left(\frac{\partial^2 u^j}{\partial x^{j^2}}\right)\frac{1}{ic} = \text{Q}\frac{\partial u^j}{\partial \tau}\frac{1}{-c^2}$$

which simplifies to,

$$\frac{\hbar}{2}\frac{\partial^2 u^j}{\partial x^{j^2}} = -i\frac{\text{Q}}{c^2}\frac{\partial u^j}{\partial \tau} \qquad (9)$$

The wave function in (9) is the spatial velocity of a point in space relative to the arbitrary observer. Velocity is related to energy states in transformed reference frames. This equation governs how stress and, therefore, mass/energy moves in space. Equation (9) is similar in appearance to the one-dimensional time-dependent Schrödinger equation. Surface energy density divided by the speed of light

[2] The terms *in-plane* and *out-of-plane* refer to the plane of the hypersurface in spacetime located at the arbitrary observer's time coordinate as described in Figures 1, 4, and 5.

IOP Conf. Series: Journal of Physics: Conf. Series **1239** (2019) 012010 doi:10.1088/1742-6596/1239/1/012010

squared, $♀/c^2$, plays the role of mass density. In the geometry of continuum mechanics, mass/energy moves through space according to Schrödinger-like dispersion equations.

A third similarity with quantum mechanics is found in the constitutive relation for diagonal spatial terms in (5) and (7), which can be written as

$$\frac{T_{jj}}{D_j^j} = \frac{c\hbar}{2}$$

Recalling that hypersurface tension, T_{jj}, is the energy contained in a volume of spacetime and dynamical strain, D_j^j, has units of the inverse of imaginary proper time, this constitutive relation is simply a linear proportionality of energy and time. The expression holds for a change in time and a change in energy such that,

$$\Delta ♀ \, \Delta\tau' = \frac{c\hbar}{2}$$

which in natural units is given by,

$$\Delta ♀ \, \Delta\tau = -\frac{i\hbar}{2}$$

If both sides are squared, then we arrive at,

$$(\Delta ♀)^2 (\Delta\tau)^2 = \left(\frac{\hbar}{2}\right)^2 \qquad (10)$$

In continuum mechanics, equation (10) is essentially a version of Hooke's law. It governs how perturbations in spacetime oscillate and are quantized in accord with the elastic modulus. In this model, the elastic modulus for objects moving in space is the reduced Planck's constant. The form of (10) is similar to the generalized Heisenberg uncertainty principle with the exception of the equal sign.

The forth similarity with quantum mechanics is found by equating the off-diagonal spatial terms in (1) and (2) in accord with (5) and (7). One finds the following relation arising from the fluidity of spacetime,

$$\frac{1}{2}\left(\frac{\partial u_1}{\partial x_2} + \frac{\partial u_2}{\partial x_1}\right) - \frac{1}{2}\left|\left(\frac{\partial u_0}{\partial x_1}\right)\left(\frac{\partial u_0}{\partial x_2}\right)\right| = 0$$

In natural units, this expression reduces to,

$$\frac{\partial u_1}{\partial x_2} + \frac{\partial u_2}{\partial x_1} - ic\left|\left(\frac{\partial u_0}{\partial x_1}\right)\left(\frac{\partial u_0}{\partial x_2}\right)\right| = 0 \qquad (11)$$

Membrane-like behavior provides solutions in spacetime geometry where in-plane strains are counterbalanced by out-of-plane displacements leading to rotation and local curvature. In membrane mechanics, these rotating geometries are known as vortices. Vortices can have short duration rotational geometry or perpetual irrotational geometry. Vortices carry specific quantized amounts of angular momentum and energy that depend on the elastic properties of the hypersurface. The expression (11) is one of several possible perturbations in the rate of deformation tensor that resemble Weyl-type equations in quantum mechanics. The other possibilities are located in the off-diagonal temporal components and perturbations between P and $\wp$ along the diagonal.

One of the interesting mathematical solutions of (11) is a stable pair of geometrically entangled vortices. By geometrical entanglement, what is meant is a solution consisting of two oppositely rotating vortices whose flow lines are joined above or below the hypersurface. In this model, the flow lines would form a micro-wormhole that would appear as a string fixed on both ends to the hypersurface by Dirichlet boundary conditions.

Although continuum mechanics is founded on geometry with elasticity, and quantum mechanics is interpreted based on probability with uncertainty, the mathematical similarities between these two arenas are interesting and worth investigation. Many physicists are working to relate quantum mechanics and gravity through a probabilistic interpretation (i.e. quantum gravity); others study the stochastic mechanics of random fluctuations [12]. The direction taken here is to lay roots for a geometrical interpretation of quantum mechanics with gravitation. This work is in its infancy; a focus of future research will be to further investigate if continuum mechanics of spacetime in complex coordinates has additional similarities to quantum mechanics beyond just appearance.

6. Gravitational Similarities

The previous section showed that spacetime geometry with surface tension described in Equations (1) to (7) shares some similarities with equations in quantum mechanics. If the warping of spacetime geometry in the model resembles quantum mechanics, how does the model also describe warping of spacetime with gravitation? In this section, the model will be re-written in index notation and compared with general relativity.

The rate of deformation tensor (2) can be re-written as,

$$D^{\alpha}_{\beta} = \frac{1}{2}\left(\frac{\partial u_\alpha}{\partial x_\beta} + \frac{\partial u_\beta}{\partial x_\alpha}\right) - \frac{1}{2}\left|\left(\frac{\partial u_0}{\partial x_\alpha}\right)\left(\frac{\partial u_0}{\partial x_\beta}\right)\right| d\tau' \qquad (12)$$

In continuum mechanics, the right term on the right side of (12) represents membrane or thin shell curvature. In fact, it is the Gaussian Curvature in 3+1 space, which is half the Scalar Curvature, R. In continuum mechanics, the left term on the right side of (12) represents the change in volume or strain. The mathematical significance of this strain term was misquoted in [1]. According to [8], this traditional strain component is the push forward of the Lie derivative of the metric tensor. Thus, Equation (12) can be written in this form,

$$D^{\alpha}_{\beta} = \frac{1}{2} g^{\mu\nu} \mathcal{L}_{\mathrm{v}} \boldsymbol{g} - \frac{1}{2} R$$

In order to obtain the rate of deformation tensor with two covariant indices, it is necessary to lower one index by multiplication of both sides by the metric tensor with two covariant components,

$$D^{\alpha}_{\beta} g_{\mu\nu} = \frac{1}{2} g_{\mu\nu} g^{\mu\nu} \mathcal{L}_{\mathrm{v}} \boldsymbol{g} - \frac{1}{2} R g_{\mu\nu}$$

Note that the metric with two covariant components and the metric with two contravariant components cancel in the first term on the right side. According to [13], the Lie derivative of the metric, $\frac{1}{2}\mathcal{L}_v\boldsymbol{g}$, is the extrinsic curvature in 3+1 space, which in its natural form is the Ricci Curvature Tensor, $R_{\mu\nu}$, in four-dimensions [14]. Since α and β are dummy indices, one can change indices such that $D_\beta^\alpha g_{\mu\nu} = D_\mu^\alpha g_{\alpha\nu} = D_{\mu\nu}$. Hence,

$$D_{\mu\nu} = R_{\mu\nu} - \frac{1}{2}Rg_{\mu\nu} \qquad (13)$$

The rate of deformation tensor with two covariant components is the Einstein Tensor given in [14].

Returning attention to the constitutive relationship (5), it can be seen that,

$$T_{\mu\nu} = C_{\mu\nu\alpha}^{\beta} \cdot D_\beta^\alpha = C_{\mu\nu\alpha}^{\beta} \cdot \frac{D_{\mu\nu}}{g_{\mu\nu}}$$

Contracting the elasticity tensor, moving it to the stress energy side, and inserting (13) yields an equation closely analogous to general relativity,

$$T_{\mu\nu} \cdot C_\beta^\alpha = R_{\mu\nu} - \frac{1}{2}Rg_{\mu\nu} \qquad (14)$$

except the Einstein constant from [15] is replaced by a symmetric nondegenerate anisotropic elasticity tensor,

$$C_\beta^\alpha = g_{\mu\nu}C_\beta^{\mu\nu\alpha} = \frac{2}{c\hbar}\begin{bmatrix} 4\pi l_p{}^2 & 0 & 0 & 0 \\ 0 & 1 & 0 & 0 \\ 0 & 0 & 1 & 0 \\ 0 & 0 & 0 & 1 \end{bmatrix} \qquad (15)$$

The proposed anisotropic constitutive relationship provides an energetic spatial geometry resembling quantum mechanics and a lower energy temporal geometry consistent with general relativity. The fact that (15) has independent eigenvalues implies that whatever energy is inserted as rest mass, *P*, in the time-time term of (1) will continue to be multiplied by Einstein's constant through Lorentz transformations. Whereas the negative stress energy inserted for surface tension, *Q*, will continue to be multiplied by Planck's constant through Lorentz transformations.

7. Discussion

The concept that spacetime itself might behave as a multi-dimensional membrane long has been suggested [16]. Much of the interest in relativistic membranes in recent decades stems from the realization that action in membrane dynamics is equivalent to Nambu-Goto action in bosonic string theory. Earlier works require membranes to be embedded in higher dimensional spaces. The work [16] and the present work suggest spacetime itself may exhibit membrane-like properties.

IOP Conf. Series: Journal of Physics: Conf. Series **1239** (2019) 012010 doi:10.1088/1742-6596/1239/1/012010

In the Stueckelberg covariant quantum theory [17, 18], time and space are dynamical observables that evolve according to an invariant parameter, τ, related to proper time on the mass shell. The mass shell is described as the tangent space of freely falling clocks of general relativity. Proper time and the parameter, τ, are allowed to deviate on the mass shell and mass becomes dynamic. Of interest to the present work, the Stueckelberg theory alludes to a tension-like mechanism that holds a particle's mass to its original mass-shell value and drives parameter, τ, back to the proper time outside the influence of an electric field. The earlier works of [17, 18] provide more sophisticated treatments of electromagnetic interactions and could serve as a roadmap for further development of the current model.

Others have compared continuum mechanics with general relativity as recently as [19, 20, 21]. There is a general consensus on the definition of the linearized spacetime metric with infinitesimal strain. The main difference between past endeavors [19, 20] and the present is the work here is based on membrane behavior without plate bending stresses. Plate bending requires simultaneity of more than one hypersurface. In [19], a Poisson's ratio of 1 was required to explain gravitational waves. The present work used a Poisson's ratio of 0 and has yet to consider gravitational waves.

Few examples of spacetime anisotropic elastic behavior are found in literature. A form of spacetime elastic anisotropy was considered in a recent work [22]. To find a material explanation for dark energy, [22] decomposed the stress energy tensor into potential energy and viscosity components and suggested separate elastic moduli for each. The two elastic tensors in this prior work were isotropic, but acted on different components of the stress energy tensor. The twin elastic moduli suggested in [22] are mathematically similar to the anisotropic constitutive relation proposed here in combined tensor form.

Despite differences in approach, several authors of past works [19, 22] make excellent arguments about the importance in the study of continuum mechanics in general relativity. The authors [19] state that, "Over the past century, Solid Mechanics and General Relativity have advanced independently". The authors [19] also state that models of spacetime based on solid mechanics, "allows General Relativity problems to be formulated as Solid Mechanics problems" and vice-versa, "Thus, ideas, methodologies and tools from each field become available to the other field." The author [22] describes their work as a "gate-way" for importing the physics of solid mechanics into cosmology. The mathematical similarities between continuum mechanics and general relativity are intriguing and should inspire others to join in the study of this approach.

8. Conclusions

Spacetime is a three surface in the observer's coordinate of time. Simple statistical thermodynamics requires that surfaces of any dimension must exhibit surface tension. Ordinary continuum mechanics is applied in four-dimensions to write tensors governing the evolution of spacetime with surface tension. Tensors for stress energy, rate of deformation, and equations of motion for an infinitesimal element of spacetime are presented. In the model, the stress energy tensor, $T_{\mu\nu}$, with surface tension contains positive mass and negative stress components and has signature +,-,-,-.

A stable anisotropic elastic tensor is proposed for relating stress energy, $T_{\mu\nu}$, to curvature. Such a constitutive relation provides gravitation in the manifold of the observer while also providing wave equations of motion resembling quantum mechanics in the tangent space. The fact that equations of motion of spacetime resemble quantum mechanics does not alone give access to the quantum theory.

More research is needed to evaluate if quantum mechanics can be fully embodied in complex geometry. The proposed negative stress terms in the stress energy tensor and stable anisotropic elastic tensor represent a significant modification to equations of general relativity that require additional careful evaluation.

9. References

[1] Perko, H.A. (2017) "Introducing Surface Tension to Spacetime", 10th Biennial Conference on Classical and Quantum Relativistic Dynamics of Particles and Fields, Ljubljana, Slovenia, *Journal of Physics: Conference Series*, Volume 845, 012003, IOP Science.

[2] Perko, H.A. (2017) "Surface Tension of Spacetime", Session H3:New Directions in Gravity, Vol. 62, No. 1, APS April Meeting, Washington, D.C., American Physical Society.

[3] Hartle, J.B. (2003) Gravity: An Introduction to Einstein's General Relativity, Addison Wesley, San Francisco

[4] Landau, L. D. and Lifshitz, E. M. (1988) Classical Theory of Fields, 7th Edition, Vol. 2, Nauka, Moscow

[5] Boresi A P and Schmidt R J (2003) Advanced Mechanics of Materials, 6th Edition (New York: Wiley)

[6] Hamilton, A. J. S. (2017) "A Covariant Hamiltonian Tetrad Approach to Numerical Relativity", *Physical Review D*, 96, 124027

[7] Adamson, A. W. (1990) Physical Chemistry of Surfaces, 5th Edition, Wiley Interscience, New York

[8] Marsden, J.E. and Hughes, T.J.R. (1983) Mathematical Foundations of Elasticity, Dover, New York

[9] Ugural, A. C. (1998) Stresses in Plates and Shells, 2nd Edition, McGraw-Hill, New York

[10] Timoshenko, S. and Woinowsky-Krieger, S. (1987) Theory of Plates and Shells, 2nd Edition McGraw-Hill, New York

[11] Ventsel, E. and Krauthammer, T. (2001) Thin Plates and Shells: Theory, Analysis, and Applications, Marcel Dekker, Inc., New York

[12] Nelson, E. (2012) "Review of Stochastic Mechanics", Emergent Quantum Mechanics 2011, *Journal of Physics: Conference Series*, 361, 012011, IOP Publishing

[13] Misner, C.W., Thorne, K.S., and Wheeler, J.A. (2017) Gravitation, Princeton University Press, Princeton, NJ

[14] Isreal, W. (1966) "Singular Hypersurfaces and Thin Shells in General Relativity", *Nuovo Cimento* 44 B, 1-14, and correction in 48 B, 463.

[15] Einstein, A. (1916) "The Foundation of the General Theory of Relativity", *The Collected Papers of Albert Einstein*, Volume 6, A.J. Kox, M. J. Klein, and R. Schulmann Eds., Princeton University Press, New Jersey, 1997

[16] Pavsic, M. (1997) "Spacetime as an Unconstrained Membrane", *Gravitation and Cosmology*, Vol. 3, pp. 305-311.

[17] Stueckelberg, E.C.G. (1941) *Helvetica Physica Acta*, Vol. 14, pp. 51-80

[18] Horwitz, L.P. and Oron, O. (2004) "Classical Gravity as an Eikonal Approximation to a Manifestly Lorentz Covariant Quantum Theory with Brownian Interpretation", Relativity, Gravitation, Cosmology, V.V. Dvoeglazov and A.A. Espinoza Garrido, Eds., Nova Science

[19] Tenev, T.G. and Horstemeyer, M.F. (2017) "The Mechanics of Spacetime - A Solid Mechanics Perspective on the Theory of General Relativity", *International Journal of Modern Physics D*, World Scientific Publishing Company, December.

[20] Kokarev, S.S. (2000) "Space-Time as Multidimensional Elastic Plate", *Nuovo Cimento B*, Vol. 113, Springer, pp. 1339-1350

IOP Conf. Series: Journal of Physics: Conf. Series **1239** (2019) 012010 doi:10.1088/1742-6596/1239/1/012010

[21] Visser, M., Barcelo, C., and Liberati, S. (2002) "Analogue Models of and for Gravity", General Relativity and Gravitation, Vol. 34, Issue 10, pp 1719-1734, Springer
[22] Pearson, J.A. (2014) "Material Models of Dark Energy", *Annalen der Physik*, Vol. 526, Issue 7-8, Wiley, pp. 318-339, DOI 10.1002

IOP Conf. Series: Journal of Physics: Conf. Series **1239** (2019) 012011 doi:10.1088/1742-6596/1239/1/012011

Geometrization of Newtonian Dynamics

Yaakov Friedman and Tzvi Scarr
Jerusalem College of Technology
Departments of Mathematics and Physics
P.O.B. 16031 Jerusalem 91160, Israel

E-mail: friedman@jct.ac.il, scarr@g.jct.ac.il

Abstract. Riemann's principle "force equals geometry" provided the basis for Einstein's General Relativity - the geometric theory of gravitation. In this paper, we follow this principle to derive the dynamics for any conservative force. We introduce the *relativity of spacetime*: an object lives in its own spacetime, whose geometry is determined by all of the forces affecting it. We also introduce the *Generalized Principle of Inertia* which unifies Newton's first and second laws and states that: *An inanimate object moves freely, that is, with zero acceleration, in* its own *spacetime.* We derive the metric of an object's spacetime in two ways. The first way uses conservation of energy to derive a Newtonian metric. We reveal a physical deficiency of this metric (responsible for the inability of Newtonian dynamics to account for relativistic behavior), and remove it. The dynamics defined by the corrected Newtonian metric leads to a new *Relativistic Newtonian Dynamics (RND)* for both massive objects and massless particles moving in any static, conservative force field, not necessarily gravitational. In the case of the gravitational field of a static, spherically symmetric mass distribution, this metric turns out to be the Schwarzschild metric. This dynamics reduces in the weak field, low velocity limit to classical Newtonian dynamics and also exactly reproduces the classical tests of General Relativity. In the second way, we obtain the *RND* metric directly, without first obtaining a Newtonian metric. Instead of conservation of energy, we use conservation of angular momentum, a carefully defined Newtonian limit and Tangherlini's condition. The non-static case is handled by applying Lorentz covariance to the static case.

1. Introduction

Bernhard Riemann, although best known as a mathematician, became interested in physics in his early twenties. His lifelong dream was to develop the mathematics to unify the laws of electricity, magnetism, light and gravitation. At an 1894 conference in Vienna, the mathematician Felix Klein said:

> "I must mention, first of all, that Riemann devoted much time and thought to physical considerations. Grown up under the tradition which is represented by the combinations of the names of Gauss and Wilhelm Weber, influenced on the other hand by Herbart's philosophy, he endeavored again and again to find a general mathematical formulation for the laws underlying all natural phenomena .... The point to which I wish to call your attention is that these physical views are the mainspring of Riemann's purely mathematical investigations [1]."

Riemann's approach to physics was geometric. As pointed out in [2], "one of the main features of the local geometry conceived by Riemann is that it is well suited to the study of gravity and

IOP Conf. Series: Journal of Physics: Conf. Series **1239** (2019) 012011 doi:10.1088/1742-6596/1239/1/012011

more general fields in physics." He believed that the forces at play in a system determine the geometry of the system. For Riemann, *force equals geometry.*

The application of Riemann's mathematics to physics would have to wait for two more essential ideas. While Riemann considered how forces affect *space*, physics must be carried out in *spacetime.* One must consider trajectories in spacetime, not in space. For example, in flat spacetime, an object moves with constant velocity if and only if his trajectory in *spacetime* is a straight line. On the other hand, knowing that an object moves along a straight line in *space* tells one nothing about whether the object is accelerating. As Minkowski said, "Henceforth, space by itself, and time by itself, are doomed to fade away into mere shadows, and only a kind of union of the two will preserve an independent reality [3]." This led to the second idea. Riemann worked only with positive definite metrics, whereas Minkowski's metric on spacetime is not positive definite. The relaxing of the requirement of positive-definiteness to non-degeneracy led to the development of *pseudo-Riemannian geometry.*

Fifty years after Riemann's death, Einstein used pseudo-Riemannian geometry as the cornerstone of General Relativity (*GR*). Acknowledging his reliance on Riemann, Einstein said:

> "But the physicists were still far removed from such a way of thinking; space was still, for them, a rigid, homogeneous something, incapable of changing or assuming various states. Only the genius of Riemann, solitary and uncomprehended, had already won its way to a new conception of space, in which space was deprived of its rigidity, and the possibility of its partaking in physical events was recognized. This intellectual achievement commands our admiration all the more for having preceded Faraday's and Maxwell's field theory of electricity [4]."

GR is a direct application of "force equals geometry." In *GR*, the gravitational force curves spacetime. Since, by the Equivalence Principle, the acceleration of an object in a gravitational field is independent of its mass, curved spacetime can be considered a *stage* on which objects move. In other words, the geometry is the same for all objects. However, the Equivalence Principle holds only for gravitation. In this way, *GR* singles out the gravitational force from other forces which are not treated geometrically. For example, the potential of an electric force depends on the charge of the particle, and the particle's acceleration depends on its charge-to-mass ratio. Thus, the electric field does not create a common stage on which all particles move. Indeed, a neutral particle does not feel any electric force at all. The way spacetime curves due to an electric potential depends on *both* the potential *and* intrinsic properties of the object. This was also recognized in the geometric approach of [5]. How, then, are we to apply Riemann's principle of "force equals geometry" to other forces?

We answer this question here for any conservative force. One of the main new ideas is the *relativity of spacetime.* This means that spacetime is an object-dependent notion. An object lives in its own spacetime, its own geometric world, which is defined by the forces which affect it. For example, in the vicinity of an electric field, a charged particle and a neutral particle exist in different worlds, in different spacetimes. In fact, for the neutral particle, the electric field does not exist. Likewise, in the vicinity of a magnet, a piece of iron and a piece of plastic live in two different worlds.

The second new idea is the *Generalized Principle of Inertia.* It is based on the fact that an inanimate object has no internal mechanism with which to change its velocity. Hence, it has constant velocity, or zero acceleration, in *its own* world (spacetime). The *Generalized Principle of Inertia* unifies Newton's first and second laws and states that: **An inanimate object moves freely, that is, with zero acceleration, in *its own* spacetime, whose geometry is determined by all of the forces affecting it**. This is a generalization, or more accurately, a relativization of Einstein's idea.

In *GR*, a freely-falling object in a gravitational field moves along a geodesic determined by the metric of the spacetime. With respect to this metric, the object's acceleration is zero. The

Generalized Principle of Inertia extends this idea so that *every* object moves along a geodesic in *its* spacetime. This geodesic is with respect to an appropriate metric, which we call the *metric of the object's spacetime.* The metric in GR is determined solely by the gravitational sources, while in our model, the metric of the object's spacetime depends on all of the forces affecting the object. In the case of static, conservative forces, the metric will depend only on the potentials of these forces, defined in some preferred inertial frame, which may be the rest frame of the Universe.

Since the motion is by geodesics, we use a variational principle and derive Euler-Lagrange type equations and the ensuing conservation laws.

> "Many results in both classical and quantum physics can be expressed as variational principles, and it is often when expressed in this form that their physical meaning is most clearly understood. Moreover, once a physical phenomenon has been written as a variational principle, ... it is usually possible to identify conserved quantities, or symmetries of the system of interest, that otherwise might be found only with considerable effort [6]."

In the classification [7] of alternative gravitation theories, our theory is a preferred frame, Lagrangian-based metric theory.

We call our theory *Relativistic Newtonian Dynamics (RND).* RND has the following features:

(i) It is based on the classical, unmodified Newtonian potential.
(ii) It avoids the complicated field equations of GR.
(iii) It reveals the physical mechanism responsible for relativistic phenomena.
(iv) For the gravitational field of a non-rotating, spherically symmetric mass distribution, the RND metric is the Schwarzschild metric
(v) RND exactly reproduces the classical tests of GR.
(vi) It is also valid for non-spherically symmetric fields.
(vii) It does not rely on the Equivalence Principle, and so is applicable to *any combination* of conservative force fields whose potentials vanish at infinity.

We note that the RND metric was also derived in [8] using the escape trajectory and the influence of the potential on spacetime.

In the literature, there are other alternative approaches to reproducing the relativistic gravitational features of GR. One approach uses modified Newtonian-like potentials. This so-called "pseudo-Newtonian" approach, introduced in [9], is much simpler mathematically than GR, with no need for covariant differentiation and complicated tensorial equations. Numerous authors ([10, 11, 12, 13, 14, 15, 16]) have proposed various modified Newtonian-like potentials. However, none of these potentials are able to reproduce the tests of GR, even in the weak field regime. Moreover, as stated in [17], most of these modified potentials "are arbitrarily proposed in an ad hoc way" and, more fundamentally, are "not a physical analogue of local gravity and are not based on any robust physical theory and do not satisfy Poisson's equation."

More recently, the above shortcomings were addressed in [18]. Using a metric approach and hypothesizing a generic relativistic gravitational action and a corresponding Lagrangian, the authors derive a velocity-dependent relativistic potential which generalizes the classical Newtonian potential. For a static, spherically symmetric geometry, this potential exactly reproduces relativistic gravitational features, including the tests of GR. Even more recently, one finds a fundamental grounding to these velocity-dependent pseudo-Newtonian potentials in [19]. The authors generalize the pseudo-Newtonian approach to any stationary spacetime. They also include additional forces, such as the electromagnetic force.

IOP Conf. Series: Journal of Physics: Conf. Series **1239** (2019) 012011 doi:10.1088/1742-6596/1239/1/012011

2. A geometric approach to dynamics

We present the preferred frame, Lagrangian approach to geometrize the motion of an object of mass m in a force field. The results here are quite general and may be applied to any force field. Later, we will consider the particular case of a static, spherically symmetric, attractive force field.

Let

$$ds^2 = g_{\alpha\beta}(q)dq^\alpha dq^\beta \tag{1}$$

be the metric of the object's spacetime, where $q^\alpha, \alpha = 0, 1, 2, 3$ are the coordinates of an inertial observer far removed from the sources of the field. This is similar to Earth-based observations of motion within a distant galaxy and to observations within the solar system measured with respect to the far-removed stars.

In order to define the length of the trajectory, we introduce the following function L of the eight independent variables $q^\alpha, \acute{q}^\alpha$:

$$L(q, \acute{q}) = mc\sqrt{g_{\alpha\beta}(q)\acute{q}^\alpha \acute{q}^\beta}, \tag{2}$$

where the constant mc has been chosen to be compatible with the classical notion of momentum.

Let $q : \sigma \to x, a \leq \sigma \leq b$ be a trajectory of an object of mass m, where σ is an arbitrary parameter. The length $l(q)$ of the trajectory q is given by

$$l(q) = mc\int_a^b \frac{ds}{d\sigma}d\sigma = \int_a^b L\left(q, \frac{dq}{d\sigma}\right)d\sigma. \tag{3}$$

It is well known that the length of the trajectory does not depend on the parametrization.

From the Generalized Principle of Inertia, the length of the trajectory $q(\sigma)$ is extremal. By a standard argument, it follows that $q(\sigma)$ satisfies the Euler-Lagrange equations

$$\frac{\partial L}{\partial q^\mu} - \frac{d}{d\sigma}\frac{\partial L}{\partial \acute{q}^\mu} = 0, \tag{4}$$

where L is defined by (2) and $\acute{q} = \frac{dq}{d\sigma}$. In this case, the conjugate momentum p_μ is

$$p_\mu = \left.\frac{\partial L}{\partial \acute{q}^\mu}\right|_{\acute{q}=\frac{dq}{d\sigma}} = \frac{mcg_{\mu\beta}\frac{dq^\beta}{d\sigma}}{ds/d\sigma} = mcg_{\mu\beta}\frac{dq^\beta}{ds}. \tag{5}$$

Note that the second term in equation (4) contains differentiation by two parameters on the curve. The first differentiation is by s, as seen in equation (5). The second differentiation is by σ. In order to obtain a differential equation with a single parameter, we will choose σ to be proportional to s. More precisely, we choose σ to be the parameter

$$\tau = c^{-1}s, \tag{6}$$

called *proper time*, which is proportional to s and reduces to the coordinate time t in the classical limit.

Using τ will turn equation (4) into a second-order differential equation. We denote differentiation of q with respect to τ by $\dot{q}$. From (4) and (5), the Euler-Lagrange equations become

$$\frac{\partial L}{\partial q^\mu} = \frac{dp_\mu}{d\tau}, \tag{7}$$

where

$$p_\mu = mg_{\mu\beta}\dot{q}^\beta. \tag{8}$$

Note that momentum is a covector, while velocity is a vector. Therefore, the metric is necessary in (8) in order to lower the index. It can be shown that (7) is equivalent to the geodesic equation.

The following proposition follows immediately from equation (7).

Proposition 1 If the metric coefficients $g_{\alpha\beta}$ do not depend on the coordinate q^μ, then the μ component $p_\mu = mg_{\mu\beta}\dot{q}^\beta$ of the conjugate momentum is conserved on the trajectory.

3. Method I: Relativistic Metric from Newtonian metric

3.1. Geometric Formulation of Newtonian Dynamics

We begin our derivation by applying the geometric approach to classical Newtonian dynamics. Based on the discussion in the previous section, we replace the coordinate time t (the classical evolution parameter) by the proper time τ. In this modification, Newton's second law for a force with potential U becomes

$$m\frac{d^2\mathbf{x}}{d\tau^2} = -\nabla U. \tag{9}$$

Taking the Euclidean dot product of both sides of (9) with $\dot{\mathbf{x}}$ gives

$$m\ddot{\mathbf{x}}\cdot\dot{\mathbf{x}} = -\nabla U\cdot\dot{\mathbf{x}},$$

which, upon integration, yields

$$\frac{m\dot{\mathbf{x}}^2}{2} + U(\mathbf{x}) = E, \tag{10}$$

where the integral of motion E is the total energy on the object's worldline. The only difference between equation (10) and the classical energy conservation equation is the kinetic energy term, in which $d\mathbf{x}/dt$ has been replaced by $\dot{\mathbf{x}}$. In the classical limit, we have $dt = d\tau$, and then equation (10) reduces to the classical energy conservation equation.

Assume now that the potential $U(\mathbf{x}) \le 0$ and vanishes at infinity. Introduce the *dimensionless potential*

$$u(\mathbf{x}) = \frac{-2U(\mathbf{x})}{mc^2}, \tag{11}$$

where c denotes the speed of light. With this definition, equation (10) yields the dimensionless energy conservation equation

$$\frac{\dot{\mathbf{x}}^2}{c^2} - u = \mathcal{E}, \tag{12}$$

where $\mathcal{E}$ denotes the dimensionless total energy on the worldline. The total energy is a sum of kinetic energy, depending on the magnitude of the velocity, and potential energy, depending on position.

We turn now to the construction of the metric of the object's spacetime, for motion satisfying (9), where U is the potential of a static force. In our inertial lab frame, the metric is of the form

$$ds^2 = f(\mathbf{x})c^2dt^2 - g(\mathbf{x})d\mathbf{x}^2, \tag{13}$$

where $f(\mathbf{x})$ and $g(\mathbf{x})$ depend solely on $\mathbf{x}$. Note that there are no anisotropic terms in the metric because in Newtonian dynamics, space is isotropic. Moreover, assuming Einstein synchrony, a straightforward argument shows that there are no time-space cross terms (see [20], page 187).

For $u(\mathbf{x}) \ll 1$, the worldlines are approximately straight, implying that this metric is asymptotically Minkowski. Hence, $f(\mathbf{x}) \to 1$ and $g(\mathbf{x}) \to 1$ as $u(\mathbf{x}) \to 0$.

Since the metric is static, Proposition 1 implies that the zero component of the conjugate momentum is conserved. Thus,

$$f(\mathbf{x})\dot{t} = k \quad , \quad \dot{t} = \frac{k}{f(\mathbf{x})}, \tag{14}$$

IOP Conf. Series: Journal of Physics: Conf. Series **1239** (2019) 012011 doi:10.1088/1742-6596/1239/1/012011

for some constant k related to the total energy on the worldline. The square of the norm with respect to (13) of the four-velocity $(\dot{t}, \dot{\mathbf{x}})$ is

$$f(\mathbf{x})c^2\dot{t}^2 - g(\mathbf{x})\dot{\mathbf{x}}^2 = c^2,$$

which, by the use of (14), leads to

$$\frac{k^2}{f(\mathbf{x})} - g(\mathbf{x})\frac{\dot{\mathbf{x}}^2}{c^2} = 1. \tag{15}$$

This can be considered a conservation equation on the worldline.

We can now determine the metric coefficients $f(\mathbf{x})$ and $g(\mathbf{x})$ by comparing this conservation to the energy conservation. Adding 1 to both sides of equation (12) and dividing by $-(1-u)$, we obtain

$$\frac{1}{1-u}\left(\mathcal{E} + 1 - \frac{\dot{\mathbf{x}}^2}{c^2}\right) = 1. \tag{16}$$

Comparing (15) and (16), and using $f(\mathbf{x}) \to 1$ as $u \to 0$, one obtains

$$g(\mathbf{x}) = \frac{1}{1-u} \quad , \quad f(\mathbf{x}) = 1 - u \quad , \quad k = \sqrt{\mathcal{E}+1}. \tag{17}$$

From (13) and (17), we obtain the *Newtonian metric*

$$ds^2 = (1 - u(\mathbf{x}))c^2dt^2 - \frac{1}{1-u(\mathbf{x})}d\mathbf{x}^2. \tag{18}$$

Reversing our argument shows that a trajectory which is minimal with respect to this metric satisfies Newton's second law (9).

In order to complete the spacetime description of the worldline, from (14) and (17) we obtain

$$\dot{t} = \frac{k}{1-u(\mathbf{x})} = \frac{\sqrt{\mathcal{E}+1}}{1-u(\mathbf{x})}. \tag{19}$$

3.2. Preferability of our Newtonian metric

In the *GR* literature, we find several metrics, like (18), whose first-order approximation leads to Newtonian dynamics. For example, in terms of u, the metric in [20] is

$$ds^2 = (1 - u(\mathbf{x}))c^2dt^2 - d\mathbf{x}^2, \tag{20}$$

and in [21] and [22], the metric is

$$ds^2 = (1 - u(\mathbf{x}))c^2dt^2 - (1 + u(\mathbf{x}))d\mathbf{x}^2. \tag{21}$$

Lagrangians L_1 and L_2 are said to be *equivalent* ($L_1 \sim L_2$) if they lead to the same equation of motion. It is obvious that if $L_2 = a + bL_1$, for constants a and b, then $L_1 \sim L_2$. For the three metrics (18), (20) and (21), we have

$$L = \frac{ds}{dt} = c\sqrt{1 - u - f(u)\beta^2},$$

where $\beta^2 = v^2/c^2$ and $f(u) = \frac{1}{1-u}, 1$ and $1+u$, respectively. In the low velocity ($\beta^2 \ll 1$) and weak field ($u \ll 1$) limit, the first-order approximation in u and β^2 is

$$L = c\left(1 - \frac{1}{2}u - \frac{1}{2}\beta^2\right).$$

Dividing by c, subtracting 1, and then multiplying by $-mc^2$, yields the classical Lagrangian

$$L = \frac{1}{2}mv^2 - U.$$

Hence, in the first-order approximation, all three metrics lead to classical Newtonian dynamics. From this point of view, there is no reason to prefer any one of the metrics over the other two.

We thus compare the equations of motion for these metrics. The equation of motion for our metric (18) is equation (9), while for the metrics (20) and (21), the equations of motion are

$$m\ddot{\mathbf{x}} = -\nabla U \frac{1 + \frac{\dot{\mathbf{x}}^2}{c^2}}{1 + \frac{2U}{mc^2}},$$

and

$$m\ddot{\mathbf{x}} = -\nabla U \frac{1 - \frac{4U\dot{\mathbf{x}}^2}{mc^4}}{1 - \frac{4U^2}{m^2c^4}},$$

respectively. The relative simplicity of equation (9) makes our metric (18) preferable over the others.

3.3. The deficiency of the Newtonian metric and the corrected metric

The huge success of Newtonian dynamics implies that the Newtonian metric (18) is close to the one that governs the laws of Nature. Nevertheless, the observed astrophysical deviations from the predictions of this dynamics indicate that this metric has a deficiency and needs to be corrected.

It is natural to assume that time intervals are influenced by the potential and should be altered by a factor defined by this potential when translated to the inertial frame. This influence is handled by the coefficient $1-u$ of c^2dt^2 in (18) and accurately predicts the known gravitational time dilation in a spherically symmetric gravitational field. It is also natural to assume that the space increments in the direction of the gradient ∇U are influenced by the potential and should be altered by a factor defined by this potential when translated to the inertial frame. This influence is also present in (18).

However, the metric (18) is deficient in that it is isotropic - it alters the spatial increments equally in *all* spatial directions. The potential, on the other hand, influences only the direction of the gradient ∇U and has no influence on the spatial increments in the directions transverse to the gradient. To remove this problem, we alter the metric so that the potential affects *only* the direction of the force, in the same way as is in (18), and leaves the transverse directions unaffected.

More precisely, introduce at each $\mathbf{x}$ where $\nabla U(\mathbf{x}) \neq \mathbf{0}$ a normalized vector

$$\mathbf{n}(\mathbf{x}) = \frac{\nabla U(\mathbf{x})}{|\nabla U(\mathbf{x})|} \tag{22}$$

in the direction of the gradient of $U(\mathbf{x})$, or the negative of the direction of the force. Let $d\mathbf{x}_n = (d\mathbf{x} \cdot \mathbf{n})\mathbf{n}$ and $d\mathbf{x}_{tr} = d\mathbf{x} - (d\mathbf{x} \cdot \mathbf{n})\mathbf{n}$, respectively, denote the projections of the spatial increment $d\mathbf{x}$ in the parallel and transverse directions to $\mathbf{n}(\mathbf{x})$. With this notation, the *Relativistic Newtonian Dynamics (RND) metric* is

$$ds^2 = (1 - u(\mathbf{x}))c^2dt^2 - \frac{1}{1 - u(\mathbf{x})}d\mathbf{x}_n^2 - d\mathbf{x}_{tr}^2. \tag{23}$$

At the points $\mathbf{x}_0 = x_0^j$ where $\nabla U(\mathbf{x}_0) = \mathbf{0}$, the normalized vector $\mathbf{n}$ is not defined. We claim that on any smooth trajectory $q(\sigma)$ in spacetime with $q(\sigma_0) = (x_0^0, \mathbf{x}_0) = x_0$, the metric (23) can be extended continuously to the point x_0. The Taylor expansion of the potential $U(\mathbf{x})$ at $\mathbf{x}_0$ to second order is $U(\mathbf{x}) \approx U(\mathbf{x}_0)+\frac{1}{2}U_{,jk}(\mathbf{x}_0)(x^j-x_0^j)(x^k-x_0^k)$. The limit of $\mathbf{n}(\sigma)$ as $\Delta\sigma = \sigma-\sigma_0 \to 0$ on the trajectory $q(\sigma)$ can be calculated by the limit along its tangent line $x_0^\alpha + a^\alpha\Delta\sigma$, where a is the tangent vector to the trajectory, and use of the second-order approximation of $U(\mathbf{x})$ to calculate $\nabla U(\mathbf{x})$. This yields $\nabla U(\mathbf{x}(\Delta\sigma))_k = \frac{1}{2}U_{,jk}(\mathbf{x}_0)(a^j\Delta\sigma) = \mathbf{b}\Delta\sigma$. Hence,

$$\lim_{\Delta\sigma\to 0+} \mathbf{n}(\Delta\sigma) = -\lim_{\Delta\sigma\to 0-} \mathbf{n}(\Delta\sigma).$$

Since the metric is not affected by the sign of $\mathbf{n}$, this proves our claim. Moreover, since, in general, the measure of such points $\mathbf{x}_0$ is zero, the length of the trajectory is not affected by the metric at these points.

In the case of the gravitational field of a non-rotating, spherically symmetric body of mass M, in spherical coordinates with origin at its center, the potential is $U(r) = -GmM/r$, and the dimensionless potential (11) is $u(r) = 2GM/c^2r = r_s/r$, where $r_s = 2GM/c^2$ is the Schwarzschild radius. In this case, the metric (23) is

$$ds^2 = \left(1 - \frac{r_s}{r}\right)c^2dt^2 - \frac{1}{1-r_s/r}dr^2 - r^2d\theta^2 - r^2\sin^2\theta d\varphi^2, \tag{24}$$

which is the well-known Schwarzschild metric ([23]). It is shown in section 5 that RND passes the classical tests of GR.

3.4. The Equations of Relativistic Newtonian Dynamics

We now obtain the dimensionless and dimensional *energy conservation equations* and *equations of motion* of RND. The derivation is similar to the reversal of the derivation in Section 3.1 for the metric (18). Since the metric (23) is static, Proposition 1 implies here, as in (14), that

$$\dot{t} = \frac{k}{1-u(\mathbf{x})}. \tag{25}$$

The square of the norm with respect to (23) of the four-velocity $\dot{x}$ is

$$c^2\frac{k^2}{1-u} - \frac{1}{1-u}\dot{\mathbf{x}}_n^2 - \dot{\mathbf{x}}_{tr}^2 = c^2. \tag{26}$$

Multiplying by $\frac{1-u}{c^2}$, using $\dot{\mathbf{x}}^2 = \dot{\mathbf{x}}_n^2 + \dot{\mathbf{x}}_{tr}^2$ and rearranging terms, we obtain the *dimensionless energy conservation equation*

$$\frac{\dot{\mathbf{x}}^2}{c^2} - u - u\frac{\dot{\mathbf{x}}_{tr}^2}{c^2} = k^2 - 1. \tag{27}$$

The corresponding *dimensional energy conservation equation* is

$$\frac{m\dot{\mathbf{x}}^2}{2} + U(\mathbf{x}) + U(\mathbf{x})\frac{\dot{\mathbf{x}}_{tr}^2}{c^2} = E, \tag{28}$$

where the integral of motion E is the total energy on the worldline. As in the energy conservation equation (10) of modified Newtonian dynamics, equation (28) has a kinetic energy term and a potential energy term. But in (28), there is also a *mixed term* which depends on both the velocity of the object and the potential. This means that in order to reproduce relativistic effects, one can no longer distinguish between potential and kinetic energy, as in Newtonian dynamics. This also explains the need to include the velocity in the modified Newtonian potentials proposed in

IOP Conf. Series: Journal of Physics: Conf. Series **1239** (2019) 012011 doi:10.1088/1742-6596/1239/1/012011

[17, 18, 24, 25, 26]. The mixed term in (28) is approximately $\beta^2 U(\mathbf{x})$ and is therefore only seen for high velocities or in high-precision experiments. Note that we can also write (28) as

$$\frac{m}{2}\left(\dot{\mathbf{x}}_n^2 + (1-u)\dot{\mathbf{x}}_{tr}^2\right) + U(\mathbf{x}) = E. \tag{29}$$

This is the usual "kinetic plus potential" energy from of energy conservation, only now the square of the velocity is computed with respect to the metric (23). Notice also that the force in RND is no longer conservative.

Let $\phi = U/m$ denote the potential per unit mass. Differentiating equation (28) with respect to τ, one obtains the *equation of motion* of RND

$$\ddot{\mathbf{x}} = -\nabla\phi - \nabla\phi\frac{\dot{\mathbf{x}}_{tr}^2}{c^2} + 2\frac{\phi(\mathbf{x})}{c^2}(\dot{\mathbf{x}}\cdot\dot{\mathbf{n}})\mathbf{n}, \tag{30}$$

which has now two additional terms not appearing in the corresponding classical equation (9). In the classical regime, both of these terms are small and have therefore gone unrecognized.

For potentials such as the gravitational potential, for which ϕ is independent of m, equations (27) and (30) can be extended to massless particles as well by using the symbol ε, which equals 1 for *objects* with non-zero mass and 0 for massless *particles*. However, for massless particles, the proper time τ is not defined. Instead, we will use an affine parameter (see, for example, [23]). There is no need here to specify this parameter, because to test our theory, we obtain parameter-free equations.

For massless particles, the norm of the four-velocity $\dot{x} = \left(\dot{t}, \dot{\mathbf{x}}\right)$ is 0. Replacing c^2 by 0 on the right-hand side of equation (26), we obtain

$$\frac{\dot{\mathbf{x}}_n^2}{c^2} + (1-u)\left(\frac{\dot{\mathbf{x}}_{tr}^2}{c^2} + \varepsilon\right) = k^2. \tag{31}$$

Differentiating with respect to τ yields, in turn,

$$\ddot{\mathbf{x}} = -\varepsilon\nabla\phi - \nabla\phi\frac{\dot{\mathbf{x}}_{tr}^2}{c^2} + 2\frac{\phi(\mathbf{x})}{c^2}(\dot{\mathbf{x}}\cdot\dot{\mathbf{n}})\mathbf{n}, \tag{32}$$

which is valid everywhere except on the Schwarzschild horizon. Equations (31) and (32) are, respectively, the *dimensionless energy conservation equation* and the *dimensionless equation of motion for objects/particles* in RND. Equations (25) and (32) provide a complete description of a worldline in RND.

Note that even though the classical force is zero for a massless particle, the second and third terms in the equation of motion (32) nevertheless remain and account for phenomena such as gravitational lensing and the Shapiro time delay. In the classical regime, both of these terms are small and have therefore gone unrecognized. Details of the derivation of (30) can be found in [27] and [28].

It is clear from these equations that RND reduces in the low velocity, weak field limit to classical Newtonian dynamics.

4. Method II: Relativistic metric from Tangherlini's condition

In this section, we derive the RND metric directly, that is, without first obtaining a Newtonian metric. The derivation here is based on a carefully defined Newtonian limit and Tangherlini's condition, which states that the metric coefficients must satisfy $g_{tt}g_{rr} = -1$ [29].

4.1. The metric of an object moving in a static, spherically symmetric, attractive central force field

Here, we derive the metric of the spacetime of an object of mass m moving in a static, spherically symmetric, attractive central force field. Let K_∞ be an inertial frame whose origin is at infinity. Let K be the same frame but with the origin shifted to the center of symmetry of the field. We use standard spherical coordinates ct, r, θ, φ in K. Let $U(r)$ denote the potential of this field. We assume that $U \leq 0$ and vanishes at infinity.

4.2. The implications of spherical symmetry

We turn now to the construction of the metric of the object's spacetime. From spherical symmetry, the metric in K is of the form

$$ds^2 = f(r)c^2dt^2 - g(r)dr^2 - h(r)r^2(d\theta^2 + \sin^2\theta d\varphi^2). \tag{33}$$

Since the force is static, the metric coefficients do not depend on t. Moreover, there are no time-space cross terms (see [20], page 187). By spherical symmetry, the functions f, g and h cannot depend on φ. For $u(r) \ll 1$, the worldlines are approximately straight, implying that this metric is asymptotically the Minkowski metric

$$\eta_{\alpha\beta} = \mathrm{diag}(1, -1, -1, -1). \tag{34}$$

Hence,

$$\lim_{r\to\infty} f(r) = \lim_{r\to\infty} g(r) = \lim_{r\to\infty} h(r) = 1. \tag{35}$$

Without loss of generality, we may assume that the motion is in the plane $\theta = \pi/2$. Since the metric coefficients do not depend on φ, Proposition 1 implies that the φ component of the conjugate momentum is conserved on the trajectory of the object. Thus,

$$h(r)r^2\dot{\varphi} = J, \tag{36}$$

for some constant J. Since the force is central, the acceleration $\ddot{\mathbf{r}}$ is parallel to $\mathbf{r}$ implying that the angular momentum per unit mass $\mathbf{r} \times \dot{\mathbf{r}} = r^2\dot{\varphi}$ is also conserved on the trajectory. Thus $h(r)$ remain constant for any values r on the trajectory of the object. Since around any value $r > 0$ there can be a trajectory of the object, this imply that $h(r)$ is constant for all r and from (35)

$$h(r) = 1. \tag{37}$$

4.3. Newtonian approximation for radial motion

Since angular momentum conservation does not provide any information about radial motion ($\varphi =$ constant), we now consider radial motion. The function $L(x, \dot{x})$ defined by (2) in this case becomes

$$L(x, \dot{x}) = mc\sqrt{f(r)c^2\dot{t}^2 - g(r)\dot{r}^2}. \tag{38}$$

From (8), the r-momentum is $p_r = -mg(r)\dot{r}$, and the r component of the Euler-Lagrange equation (7) is

$$\frac{m}{2}\left(f'(r)c^2\dot{t}^2 - g'(r)\dot{r}^2\right) + mg'(r)\dot{r}^2 + mg\ddot{r} = 0 \tag{39}$$

Dividing (33) by $d\tau^2$, we obtain the norm equation

$$f(r)c^2\dot{t}^2 - g(r)\dot{r}^2 = c^2, \tag{40}$$

implying that $c^2\dot{t}^2 = \frac{c^2}{f(r)} + \frac{g(r)}{f(r)}\dot{r}^2$. Substituting this into (39), we obtain

$$\ddot{r} = -\frac{1}{2}\left(\left(\frac{f'(r)}{f(r)} + \frac{g'(r)}{g(r)}\right)\dot{r}^2 + \frac{c^2 f'(r)}{f(r)g(r)}\right). \tag{41}$$

In Newtonian dynamics, the acceleration of an object depends only on the forces acting on it and on the object's mass, but is independent of its velocity. As explained in the previous section, in the geometric approach to dynamics, the acceleration should be $\ddot{r} = \frac{d^2r}{d\tau^2}$. Can we define the metric so that this acceleration for radial motion will be independent on the velocity of the object? From equation (41), it follows that this can be achieved if we set

$$\frac{f'(r)}{f(r)} + \frac{g'(r)}{g(r)} = 0, \tag{42}$$

which implies that $\ln(f(r)g(r)) = const$ and $f(r)g(r) = const$. Using (35), this implies that

$$f(r)g(r) = 1. \tag{43}$$

The condition (43) was also used in [29]. For some interesting conditions equivalent to (43), see [30] and [31]. Under this condition, (41) becomes

$$\ddot{r} = -\frac{c^2 f'(r)}{2}. \tag{44}$$

The function $f(r)$ will be defined from the classical limit. Let r_0 be an arbitrary value of r. Consider the radial motion of an object whose velocity at r_0 is $v(r_0) = 0$. We now connect this object's acceleration $\ddot{r}$ with the classical acceleration $\frac{d^2r}{dt^2}$ at r_0. Since near the point r_0, we have $d\tau = \sqrt{f(r)}dt$, the acceleration $\ddot{r}$ at r_0 is

$$\ddot{r} = \frac{d^2r}{d\tau^2} = \frac{d}{d\tau}\left(\frac{1}{\sqrt{f}}\frac{dr}{dt}\right) = \left(-\frac{1}{2}f^{-3/2}f'\left(\frac{dr}{dt}\right)^2 + \frac{1}{\sqrt{f}}\frac{d^2r}{dt^2}\right)f^{-1/2}.$$

Since $v(r_0) = 0$, we have

$$\ddot{r}(r_0) = \frac{1}{f(r_0)}\frac{d^2r}{dt^2}\bigg|_{r_0}. \tag{45}$$

The Newtonian radial acceleration in tensorial form is $\frac{d^2r}{dt^2} = m^{-1}\eta^{1\beta}U_{,\beta}$ (see [32]). Since the object's spacetime is not flat, this formula should be replaced by $\frac{d^2r}{dt^2} = m^{-1}g^{1\beta}U_{,\beta}$. For radial motion with metric (33) satisfying (43), this acceleration is $\frac{d^2r}{dt^2} = -\frac{1}{mg(r)}\frac{dU}{dr} = -\frac{f(r)}{m}\frac{dU}{dr}$. Hence, the acceleration $\ddot{r}$ at r_0 is

$$\ddot{r}(r_0) = \frac{1}{f(r_0)}\frac{d^2r}{dt^2}\bigg|_{r_0} = -\frac{1}{f(r_0)}\frac{f(r_0)}{m}\frac{dU}{dr}\bigg|_{r_0} = -\frac{1}{m}\frac{dU}{dr}\bigg|_{r_0}. \tag{46}$$

This is the analog of Newton's second law which, for geometric dynamics, holds only in the case in which the velocity is in the direction of the acceleration. Equation (46) holds for any radial motion, not just motion with has zero velocity at some point, since the acceleration $\ddot{r}$ which satisfies (44) is independent of the velocity of the object. Moreover, since r_0 was arbitrary, the radial acceleration is

$$\ddot{r} = -\frac{1}{m}\frac{dU}{dr}. \tag{47}$$

For radial motion, the geometrization of Newton's second law involves only replacing the time parameter with the proper time parameter. Our derivation assumes only that $\ddot{r}$ is independent of the velocity, that Newton's second law holds for motion with initial zero velocity and that the raising the index of the covector $\frac{dU}{dr}$ to the acceleration vector is done via the metric of the object's spacetime.

From (44), we obtain $\frac{c^2 f'(r)}{2} = \frac{\frac{dU}{dr}}{m}$. Thus, $f(r) = \frac{2U}{mc^2} + const$, and from (35), $f(r) = 1 + \frac{2U}{mc^2}$. Introducing the *dimensionless potential*

$$u(r) = \frac{-2U(r)}{mc^2}, \tag{48}$$

we have

$$f(r) = 1 - u(r), \tag{49}$$

and from (43),

$$g(r) = \frac{1}{1 - u(r)}. \tag{50}$$

Thus, using $u(r)$ defined by (48), the metric (33) is

$$ds^2 = (1 - u(r))c^2 dt^2 - \frac{1}{1 - u(r)} dr^2 - r^2 (d\theta^2 + \sin^2\theta d\varphi^2), \tag{51}$$

which in the case of the gravitational field of a spherically symmetric mass distribution is the Schwarzschild metric. This metric was also derived without using Einstein's equation in [33], from assumptions different from those used here.

The metric (51) can be generalized to motion under any conservative force with potential $U(\mathbf{x})$, as follows. As in the previous section, introduce at each $\mathbf{x}$ where $\nabla U(\mathbf{x}) \neq \mathbf{0}$ a normalized vector

$$\mathbf{n}(\mathbf{x}) = \frac{\nabla U(\mathbf{x})}{|\nabla U(\mathbf{x})|} \tag{52}$$

in the direction of the gradient of $U(\mathbf{x})$, or the negative of the direction of the force. Let $d\mathbf{x}_n = (d\mathbf{x} \cdot \mathbf{n})\mathbf{n}$ and $d\mathbf{x}_{tr} = d\mathbf{x} - (d\mathbf{x} \cdot \mathbf{n})\mathbf{n}$, respectively, denote the projections of the spatial increment $d\mathbf{x}$ in the parallel and transverse directions to $\mathbf{n}(\mathbf{x})$. With this notation, our metric (51) becomes

$$ds^2 = (1 - u(\mathbf{x}))c^2 dt^2 - \frac{1}{1 - u(\mathbf{x})} d\mathbf{x}_n^2 - d\mathbf{x}_{tr}^2, \tag{53}$$

which is the RND metric.

5. The Classical Tests of GR

In this section, we show that RND passes the following classical tests of GR:

(1) the anomalous precession of Mercury

(2) the periastron advance of a binary star

(3) gravitational lensing

(4) the Shapiro time delay

For the gravitational field of a non-rotating, spherically symmetric body of mass M, the unit vector $\mathbf{n}$, defined by (52), is the radial direction, so $\dot{\mathbf{x}}_n = \dot{r}$. If the initial position and velocity of the object/particle are in the plane $\theta = \pi/2$, then they will remain in this plane throughout the motion ([34]). Thus, one may chose the coordinate system so that $\theta = \pi/2$.

Moreover, the metric coefficients (24) are independent of φ. Hence, by Proposition 1, we have

$$r^2\dot{\varphi} = J, \tag{54}$$

where, for objects with non-zero mass, the constant J has the meaning of angular momentum per unit mass. This implies that $\dot{\mathbf{x}}_{tr}^2 = r^2\dot{\varphi}^2 = \frac{J^2}{r^2}$, and one can rewrite equation (31) as

$$\frac{\dot{r}^2}{c^2} + (1-u(r))\left(\frac{J^2}{c^2r^2} + \varepsilon\right) = k^2. \tag{55}$$

This, together with the definition of J, leads to the *path equation for a central force*

$$\left(\frac{J}{cr^2}\frac{dr}{d\varphi}\right)^2 + (1-u(r))\left(\frac{J^2}{c^2r^2} + \varepsilon\right) = k^2, \tag{56}$$

which depends on the two integrals of motion k and J and coincides with the geodesic equation of the Schwarzschild metric ([23]).

Furthermore, from (25) and (54), the *time dependence equation for a central force* on this path is

$$\frac{dt}{d\varphi} = \frac{kr^2}{J(1-u(r))}. \tag{57}$$

For a non-rotating, spherically symmetric object of mass M, the dimensionless gravitational potential defined by (11) is $u(r) = \frac{r_s}{r}$, where $r_s = \frac{2GM}{c^2}$ is its Schwarzschild radius.

1. To describe the trajectory of Mercury ($\varepsilon = 1$) in the gravitational field of the Sun, we rewrite (56) in terms of the dimensionless potential energy u by substituting $r = r_s/u$. Defining the orbit parameter $\mu = \frac{1}{2}\left(\frac{cr_s}{J}\right)^2$, we obtain the *RND equation for the planetary orbit*

$$\left(\frac{du}{d\varphi}\right)^2 = u^3 - u^2 + 2\mu u + 2\mu(k^2-1), \tag{58}$$

which is similar to the corresponding equation in GR (see, for example, [35]).

The corresponding classical Newtonian equation for this orbit is

$$\left(\frac{du}{d\varphi}\right)^2 = -u^2 + 2\mu u + 2\mu(k^2-1). \tag{59}$$

For a bounded orbit, the maximum and minimum values of u are the roots u_p, u_a of the quadratic on the right-hand side of this equation, corresponding to the perigee and apogee, respectively. This equation has the obvious classical solution $u_{cl}(\varphi) = \mu(1 + e\cos(\varphi - \varphi_0))$, where φ_0 is the polar angle of the perigee and e is the eccentricity of the orbit. Here, $u_p = \mu(1+e)$ and $u_a = \mu(1-e)$. Then $\mu = (u_p+u_a)/2$ is the average energy on the trajectory. In polar coordinates, we have

$$r_{cl}(\varphi) = \frac{r_s/\mu}{1 + e\cos(\varphi - \varphi_0)}, \tag{60}$$

which is a non-precessing ellipse. The reason there is no precession is that the radial and angular periodicities are both equal to 2π. This is no longer the case in RND dynamics, due to the anisotropy of the metric (24).

The RND solution (the solution of (58)) is of the form $u(\varphi) = \mu(1 + e\cos\alpha(\varphi))$, where the angle α satisfies $r_{cl}(\alpha) = r(\varphi)$. As in the classical case, u_p, u_a are again roots of the cubic on the right-hand side of equation (58), but this cubic has an additional root $u_3 = 1-(u_p+u_a) = 1-2\mu$. Substituting this into (58), one obtains $d\alpha/d\varphi = \sqrt{1-3\mu-\mu e\cos\alpha}$ and the explicit dependence

$$\varphi(\alpha) = \varphi_0 + \int_0^\alpha (1-3\mu-\mu e\cos\tilde{\alpha})^{-1/2} d\tilde{\alpha}, \tag{61}$$

which eventually yields the known perihelion precession formula ([23])

$$\varphi(2\pi) - \varphi(0) - 2\pi \approx 3\pi\mu \frac{rad}{rev}. \tag{62}$$

Substituting the value of μ for Mercury, we obtain its observed anomalous precession.

2. Since, in RND, we use the unmodified Newtonian potential, the potential of a binary star is the same as the potential of a classical two-body problem. Therefore, we can reduce the problem to a one-body problem in the gravitational field of an object with mass M, the combined mass of the binary, located at the center of mass of the binary. Hence, the RND treatment of the binary is the same as for Mercury and will once again produce precessing elliptic orbits for each component of the binary, with precession given by (62). As shown in [36], this leads to a periastron advance

$$\dot{\omega} = 3\frac{(GM)^{2/3}}{c^2(1-e^2)}\left(\frac{P_b}{2\pi}\right)^{-5/3}, \tag{63}$$

where P_b is the orbital period of the binary and ω is the angular position of the periastron. This formula is identical to the post-Keplerian formula for the relativistic advance of the periastron found in [37].

3. Gravitational lensing and the Shapiro time delay (or gravitational time delay) describe the deflection of a light ray and the slowing of a light pulse ($\varepsilon = 0$) as it moves from a point A to a point B in the gravitational potential of a spherically symmetric massive object of mass M. Denote by r_0 the distance from the point P on the trajectory closest to the massive object. Since $\frac{dr}{d\varphi} = 0$ at the point P, it follows from (56) that

$$\frac{J^2}{c^2k^2} = -\frac{r_0^2}{1-r_s/r_0}. \tag{64}$$

To obtain the formula for gravitational lensing, substitute (64) into (56), which yields

$$\left(\frac{r_0}{r^2}\frac{dr}{d\varphi}\right)^2 + \left(1-\frac{r_s}{r}\right)\frac{r_0^2}{r^2} = 1 - \frac{r_s}{r_0}. \tag{65}$$

For any angle φ on the trajectory, one may associate an angle $\alpha(\varphi)$ for which $r(\varphi) = \bar{r}(\alpha)$, where $\bar{r}(\alpha) = \frac{r_0}{\sin\alpha}$ is the straight-line approximation of the trajectory at the point P, chosen to be the x direction. This suggests the substitution $r = \frac{r_0}{\sin\alpha}$, which implies $\frac{dr}{d\varphi} = -\cos\alpha\frac{r^2}{r_0}\frac{d\alpha}{d\varphi}$ and

$$\frac{d\varphi}{d\alpha} = \left(1 - \frac{r_s}{r_0}\left(\sin\alpha + \frac{1}{1+\sin\alpha}\right)\right)^{-1/2}. \tag{66}$$

Hence, the deflection angle of a light ray moving from A to B is

$$\delta\phi = \int_{\alpha_A}^{\alpha_B} \left(1 - \frac{r_s}{r_0}\left(\sin\alpha + \frac{1}{1+\sin\alpha}\right)\right)^{-1/2} d\alpha - \pi, \tag{67}$$

where α_A, α_B are the α values of the points A and B, respectively. Assuming that these points are very remote from the massive body ($\alpha_A \approx \pi, \alpha_B \approx 0$) and that $r_s/r_0 \ll 1$, the weak deflection angle becomes

$$\delta\phi \approx \frac{2r_s}{r_0} = \frac{4GM}{c^2 r_0}, \tag{68}$$

which is identical to the angle given by Einstein's formula for weak gravitational lensing using GR ([23, 38]).

4. To obtain the formula for the Shapiro time delay, one substitutes the value of J/ck from (64) into the RND time dependence equation for a central force (57). Hence, the time of passage of light from the point P to B is given by

$$c(T_B - T_P) = \int_{\varphi_B}^{\pi/2} \frac{r^2\sqrt{1 - r_s/r_0}}{(1 - r_s/r)r_0} d\varphi\,. \tag{69}$$

For the common case $r_s/r_0 \ll 1$, we work in first order in r_s/r_0. Then, using (66) and the same substitution $r = \frac{r_0}{\sin\alpha}$ as above, the time propagation between P and B is

$$c(T_B - T_P) \approx x_B + r_s \ln\frac{r_B + x_B}{r_0}, \tag{70}$$

where x_B denotes the x coordinate of B. Using this approximation, the Shapiro time delay for a signal traveling from A to B and back is

$$r_s \ln\frac{4x_B|x_A|}{r_0^2}, \tag{71}$$

which is the known formula for the Shapiro time delay ([23, 38]), confirmed by several experiments.

6. The RND metric of several objects moving in an inertial frame

If the inertial frame K is the rest frame of the Universe, and the rest frame K' of the Sun moves uniformly with respect to K, then as pointed out in [39], the theory will pass the tests of GR only if it is Lorentz covariant. Therefore, we now extend the previous results by imposing Lorentz covariance. In electromagnetism, the potential of a moving charge can be defined from the potential of a stationary charge by applying Lorentz covariance. This leads to the Liénard-Wiechert retarded four-potential. Our approach here is similar.

Consider a spherically symmetric object of rest mass M moving uniformly in an inertial frame K. Denote by $P = x^\mu$ the spacetime point in K at which we want to define the four-potential. Let $L : (x')^\mu(\tau)$ denote the worldline of the object generating the gravitational field. Let the point $Q = x'(\tau(x))$ be the unique point of intersection of the past light cone at P with the worldline $x'(\tau)$, see Figure 1. The time on the worldline of the object corresponding to this intersection is uniquely determined by the point P. It is called the *retarded time* and will be denoted by $\tau(x)$. Note that radiation emitted at $Q = x'(\tau(x))$ at the retarded time will reach P at time $t = (x^0 - (x')^0(\tau(x)))/c$.

The four-potential per unit mass at P, in an inertial frame, is given by

$$A_\alpha(P) = \frac{-GMw_\alpha}{(x'(\tau(x)) - x) \cdot w}, \tag{72}$$

where $x'(\tau(x)) - x$ is the relative position of P and the position of the object at the retarded time, and w is its four-velocity at that time. Formula (72) certainly holds in an inertial frame

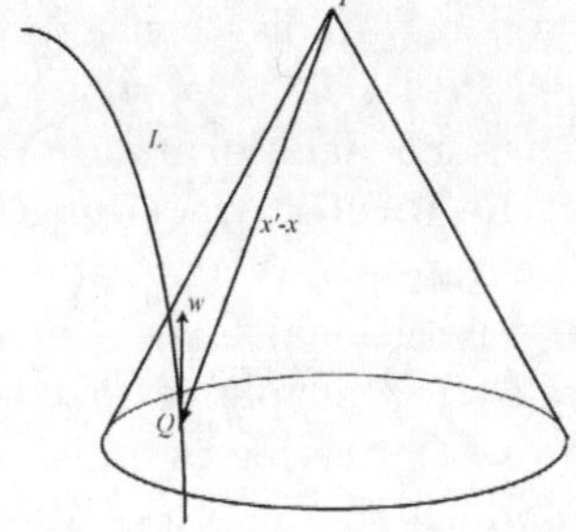

Figure 1. The four-vectors $x' - x, w$ associated with an observer and a moving object.

in which M is at rest. By Lorentz invariance, this formula holds in any inertial frame. The derivation parallels that of the Liénard-Wiechert potentials of the field of an arbitrarily moving charge (see, for example, [40], pp. 174-6).

Now consider a gravitational field generated by several objects of masses M_j with worldlines $x'_j(\tau)$. The total four-potential of the sources is the sum of the individual four-potentials (see [42]). To calculate the total four-potential, we have to define first their retarded times $\tau_j(x)$, their relative positions $x'_j(\tau_j(x)) - x$, and their four-velocities w_j at the retarded time. Then, the total gravitational four-potential at P is

$$A_\alpha(P) = \Sigma_j \frac{-GM_j(w_j)_\alpha}{(x'_j(\tau_j(x)) - x) \cdot w_j}. \tag{73}$$

Note that this four-potential tends to 0 when P is far removed from the sources.

Denote the time and spatial components of the four potential by

$$A_\alpha = (\Phi, -\mathbf{A}), \tag{74}$$

where the spatial part has a minus sign because it is a co-vector. Since each individual four-potential is a timelike vector, the total four-potential is also timelike. Therefore, we can apply a Lorentz transformation Λ from K to an inertial frame K' such that $\Lambda(A) = (U, 0, 0, 0)$. In K', the four-potential behaves like a "static" potential since it has zero spatial component. Therefore, in K', the metric is defined by formula (23), with u defined by (11) and $\mathbf{n}$ defined by (52). To obtain the metric in K, apply the inverse Lorentz transformation Λ^{-1} to the metric in K'.

7. Discussion

Riemann's approach to unify the laws of electricity, magnetism, light and gravitation was geometric. He believed that the forces at play in a system determine the geometry of the system. Put simply, for Riemann, *force equals geometry.* His quest failed, unfortunately, because he considered how forces affect *space*, not *spacetime*. Nevertheless, his geometric approach led to the development of *pseudo-Riemannian geometry* which fifty years later provided the cornerstone for Einstein's *GR*. However, *GR* singles out the gravitational force from other forces which are not treated geometrically.

In this paper we introduced the relativity of spacetime in order to apply Riemann's principle of "force equals geometry" to the dynamics under any conservative force. We accomplished

IOP Conf. Series: Journal of Physics: Conf. Series **1239** (2019) 012011 doi:10.1088/1742-6596/1239/1/012011

this by describing the geometry of the spacetime of a moving object via a metric derived from the potential of the force field acting on the object. Since an inanimate object has no internal mechanism with which to change its velocity, it has constant velocity in *its* own world. This led us to formulate our new *Generalized Principle of Inertia*, which states that: **An inanimate object moves freely, that is, with zero acceleration, in *its own* spacetime, whose geometry is determined by all of the forces affecting it.**

This is a generalization, or more accurately, a relativization of what Einstein accomplished. In GR, an object freely falling in a gravitational field is in free motion. Along a geodesic, the acceleration is zero. The Generalized Principle of Inertia states that *every* object is in free motion in *its* own world, determined by the forces which affect it. Thus, we assumed the existence of a metric with respect to which the length of the object's trajectory is extremal, enabling us to use a variational principle and conserved quantities to calculate trajectories.

Specifically, we began by treating classical Newtonian dynamics within this framework. Using second Newton's law and energy conservation, we derived a *Newtonian metric*. Nevertheless, this Newtonian metric is still deficient, since it fails to reproduce the tests of GR. The deficiency lies in the fact that this metric is isotropic, while the potential influences only the direction of the force and has no influence on directions transverse to the force. We removed this deficiency and obtained a corrected Newtonian metric (23). The dynamics built on the corrected metric is called *Relativistic Newtonian Dynamics (RND)*. We derived the *dimensionless energy conservation equation* (31) and the *dimensionless equation of motion* (32) of RND, for both massless particles and objects with non-zero mass. For a gravitational field of a spherically symmetric massive body, the RND metric is the Schwarzschild metric.

It is clear from these equations that this dynamics reduces in the low velocity, weak field limit to classical Newtonian dynamics. Moreover, as a partial validation of our approach, we have shown in section 5 that for a gravitational potential, RND exactly reproduces the classical tests of GR. The derivation of our metric is much simpler than in GR and uses potentials defined by the sources via Poisson's equation for the static case. We expect RND to be useful for studying relativistic gravitational astrophysical (or other) phenomena.

The first approach reveals why Newtonian dynamics cannot produce relativistic behavior and what modification is needed to produce such behavior. However, this approach is based on energy conservation which relies on a flat metric and not on the Generalized Principle of Inertia. This led us to the second approach which uses only the metric-free parts of Newtonian dynamics.

In our second approach, we took advantage of the symmetry of the problem and angular momentum conservation, which helped to derive the form of the metric in the direction transverse to the force. The remaining part of the metric was obtained from considering radial motion. We defined the metric so that for radial motion, the proper acceleration is independent of the velocity, as in the classical case. This property, which is equivalent to Tangherlini's condition, cannot be kept for arbitrary motion, only for motion in the direction of the force. We used the classical limit to complete the derivation of the metric.

In classical physics, the total energy has two mutually exclusive contributions: the kinetic energy depending only on the magnitude of the velocity of the object, and the potential energy depending on the field and the object's position. This leads to the isotropy of the spatial part of the Newtonian metric (18). In RND, on the other hand, the metric is not spatially isotropic because space is not isotropic. The direction of the force is a preferred direction for the given object. This is reflected in the additional term

$$U(\mathbf{x})\frac{\dot{\mathbf{x}}_{tr}^2}{c^2}$$

of the energy conservation equation (28), which depends on both the position and the velocity of the moving object. This implies that in order to reproduce relativistic effects, one can no longer separate these contributions. Indeed, some authors [17, 18, 24, 25, 26] have defined velocity-dependent "potentials" in order to reproduce relativistic effects. However, these potentials are not potentials in the true sense. They are "Newtonian analogous potentials" (see [18]).

The non-static case is handled as follows. First, we consider a single source which is moving in our inertial frame. Using Lorentz invariance, the potential of this source at a given spacetime point becomes a four-potential, which depends on the position of the source at the retarded time with respect to the given spacetime point. For several sources, the metric of the object's spacetime depends on the sum of the four-potential of the sources which affect our object. The formula for the four-potentials is similar to the that of the Liénard-Wiechert potentials. Since the total four-potential is timelike, there is an inertial frame in which it behaves like a scalar potential. In this frame, the metric is defined as in the static case. The metric in the original frame is then computed via a Lorentz transformation. A similar approach can be found in [42].

At this stage, the RND model does not describe modern tests of GR beyond the classical tests. In particular, since the potential of a collapsing binary is not static, the model in its current form does not provide a mechanism for the recently discovered gravitational waves by the LIGO team. We hope to extend the model to handle these modern tests as well.

Our methodology here differs from the standard approach in the way we define the radial coordinate r. The standard approach (see [20, 44], for example) is to *define* r so that a sphere of radius r has surface area $4\pi r^2$. Our approach, on the other hand, is to measure r in the inertial frame at infinity. This is a more natural way to define the radial coordinate, since, in practice, this is what is measured.

References

[1] Klein, F.: Riemann und seine Bedeutung für die Entwicklung der modernen Mathematik. In: Klein, F. Gesammelte mathematische Abhandlungen. Springer-Verlag, Berlin (1923) (reprinted 1973) 3: 482497. English translation: Riemann and his significance for the development of modern mathematics. Bull. Amer. Math. Soc. 1 No. 5, 165-180 (1895)

[2] Papadopoulos, A.: Physics in Riemann's Mathematical Papers. In: Ji, L., Papadopoulos A., Yamada S. (eds.) From Riemann to Differential Geometry and Relativity, pp. 151-207. Springer, Cham, Switzerland (2017)

[3] Minkowski, H.: Raum und Zeit (Space and Time). Phys Z 10, 7588 (19081909)

[4] Einstein, A.: Ideas and opinions, based on C. Seeling (ed.) Mein Weltbild. New translations and revisions by S. Bargmann. Crown Publishers, Inc., New York (1954)

[5] Duarte, C.: The classical geometrization of the electromagnetism. International Journal of Geometric Methods in Modern Physics 12 (2015)

[6] Riley, K.F., Hobson, M.P., Bence, S.J.: Mathematical Methods for Physics and Engineering, Third Edition. Cambridge University Press, New York (2006) p. 787

[7] Will, C.M., Nordtvedt, K., Jr.: Astrophys. J. 177, 757-774 (1972)

[8] Friedman, Y., Steiner, J. M.: Predicting Mercury's precession using simple relativstic Newtonian dynamics. Europhys Lett. 113, 39001 (2016)

[9] Paczyński, B., Wiita, P.J.: Thick accretion disks and supercritical luminosities. Astron Astrophys. 88, 23-31 (1980)

[10] Abramowicz, M.A., Czerny, B., Lasota, J.P., Szuszkiewicz, E.: Slim accretion disks. Astrophys J. 332, 646-658 (1988)

[11] Chakrabarti, S., Titarchuk, L.G.: Spectral Properties of Accretion Disks around Galactic and Extragalactic Black Holes. Astrophys J. 455, 623 (1995)

[12] Hawley, J.F., Balbus, S.A.: The Dynamical Structure of Nonradiative Black Hole Accretion Flows. Astrophys J. 573, 738-748 (2002)

[13] Lee, W.H., Ramírez-Ruiz, E.: Accretion Modes in Collapsars: Prospects for Gamma-Ray Burst Production. Astrophys J. 641, 961-971 (2006)

[14] MacFadyen, A.I., Woosley, S.E.: Collapsars: Gamma-Ray Bursts and Explosions in "Failed Supernovae". Astrophys J. 524, 262-289 (1999)

[15] Matsumoto, R., Kato, S., Fukue, J., Okazaki, A.T.: Viscous transonic flow around the inner edge of geometrically thin accretion disks. Astron Soc Japan 36, 71-85 (1984)

[16] Rosswog, S., Ramírez-Ruiz, E., Hix, W.R.: Tidal Disruption and Ignition of White Dwarfs by Moderately Massive Black Holes. Astrophys J. 695, 404-419 (2009)

[17] Ghosh, S., Sarkar, T., Bhadra, A.: Newtonian analogue of corresponding spacetime dynamics of rotating black holes: implication for black hole accretion. Mon Not R Astron Soc. 445, 4463-4479 (2014)

[18] Ghosh, S., Sarkar, T., Bhadra, A.: Exact Relativistic Newtonian Representation of Gravitational Static Spacetime Geometries. Astrophys J. 828, 6-9 (2016)

[19] Witzany, V., Lämmerzahl, C.: Pseudo-Newtonian Equations for Evolution of Particles and Fluids in Stationary Space-times. Astrophys J. 841, 105-118 (2017)

[20] Rindler, W.: Relativity, Special, General and Cosmological. Oxford, New York (2001)

[21] Bartelmann, M.: General Relativity. Heidelberg: Institut fr Theoretische Astrophysik Universitt Heidelberg, www.ita.uni-heidelberg.de/research/bartelmann/files/relativity.pdf, 1970

[22] Carroll, S.: Spacetime and Geometry: An Introduction to General Relativity. Pearson, London (2003)

[23] Misner, C., Thorne, K., Wheeler, J.: Gravitation. Freeman, San Francisco (1973)

[24] Tejeda, E., Rosswog, S.: An accurate Newtonian description of particle motion around a Schwarzschild black hole. Mon Not R Astron Soc. 433, 1930-1940 (2013)

[25] Sarkar, T., Ghosh, S., Bhadra, A.: Newtonian analogue of Schwarzschild–de Sitter spacetime: Influence on the local kinematics in galaxies. Phys Rev D 90, 063008-063027 (2014)

[26] Ghosh, S., Sarkar, T., Bhadra, A.: Newtonian analogue of static general relativistic spacetimes: An extension to naked singularities. Phys Rev D 92, 083010-083032 (2015)

[27] Friedman, Y.: Relativistic Newtonian Dynamics for objects and particles. Europhys. Lett. 117, 49003 (2017)

[28] Friedman, Y., Steiner, J. Relativistic Newtonian Dynamics. J. Phys. Conf. Ser. 845 012028 (2017)

[29] Tangherlini, F.R.: Postulational Approach to Schwarzschild's Exterior Solution with Application to a Class of Interior Solutions. Il Nuovo Cimento 25 (5), 10811105 (1962)

[30] Guendelman, E., Rabinowitz, A.: Linearity, non self-interacting spherically symmetric gravitational fields, the "sphereland equivalence principle" and Hamiltonian bubbles. Gen. Rel. Grav. 28(2), 117 (1996)

[31] Jacobson, T.: When is $g_{tt}g_{rr} = -1$? Class. Quant. Grav. 24, 5717 (2007)

[32] Itin, Y.: Premetric representation of mechanics, electromagnetism and gravity. International Journal of Geometric Methods in Modern Physics 15 (2018) 1840002

[33] Guendelman, E., Rabinowitz, A., Bank, A.: Finding Schwarzschild Metric Component $g_r r$ and FLRW's k without Solving the Einstein Equation, Rather by a Synergistic Matching between Geometric Results Enfranhised by Newtonian Gravity. Bulg. J. Phys. 45, 180 (2018)

[34] Grøn, Ø.: Lecture Notes on the General Theory of Relativity. Springer, New York (2009)

[35] Brown, K.: Reflections on Relativity. lulu.com (2015)

[36] Friedman, Y., Livshitz, S., Steiner, J.M.: Predicting the relativistic periastron advance of a binary without curving spacetime. Europhys Lett. 116, 59001-59006 (2016)

[37] Damour, T., Deruelle, N.: General Relativistic Celestial Mechanics of binary systems I. The post-Newtonian motion. Ann Inst Henri Poincare A 43, 107-132 (1985)

[38] Will, C.M.: The Confrontation between General Relativity and Experiment. Liv Rev Rel 17, 4 (2014)

[39] Nordtvedt, K., Jr., Will, C.M.: Astrophys. J. 177, 775-792 (1972)

[40] Landau, L., Lifshitz, E.: The Classical Theory of Fields. Addison-Wesley (1951)

[41] Friedman, Y., with the assistance of Scarr, T.: Physical Applications of Homogenous Balls. Birkhauser(2005)

[42] Mashhoon, B.: Gravitoelectromagnetism: A Brief Review in Pascual-Sanchez, J.-F., Floria, L., San Miguel, A., Vicente, F. eds. Reference Frames and Gravitomagnetism. World Scientific (2001) pp. 121-32

[43] Friedman, Y., Scarr, T., Steiner, J.M.: Int. J. Geom. Meth. Mod. Phys to appear

[44] Rindler, W.: Counterexample to the Lenz-Schiff Argument. Am J Phys 36 (6), 540-544 (1968)

IOP Conf. Series: Journal of Physics: Conf. Series **1239** (2019) 012012 doi:10.1088/1742-6596/1239/1/012012

How Quasars with Thick Accretion Disk Eject Relativistic Jets

N Ben-Amots
P.O.Box 3193, Haifa 3103201, Israel
E-mail: benamots@alumni.technion.ac.il

Abstract. One of the unsolved problems in physics is how thick accretion disks produce and eject relativistic jets. "Jet-producing disks are thick" [1]. Another unsolved problem is how quasars and active galactic nuclei persist at least millions of years, while their masses are up to milliards of solar masses. Here we suggest that these problems are connected and solved simultaneously by a series of processes not recognized until now, some of them involving relativistic effects. The thick toroidal accretion disk around a massive central body is derived analytically by Newtonian conservation of angular momentum. Radiation emitted from the accretion disk enhances the particles of the jets to relativistic velocity. The particles are expelled as jets from the poles of the rotating central body, whose mass is millions to trillions of sun masses. The central body possesses radius, as predicted by exponential gravitation [2], [3], [4], [5], [6], [7], [8] and [9] and proved by the discovery of gravitational waves from colliding massive bodies. The source of energy at the center of the massive central body is explained, as well as the formation of the inner jets and how they expel from the poles. Various implications are discussed.

1. Introduction

Djorgovski et al [10] reviewed and summarized the observations and the accepted theories on quasars. See references therein. The question of why do the thick accretion disks surrounding certain astronomical objects, such as quasars and the nuclei of active galaxies, emit relativistic jets along their polar axes remains unsolved [11]. Tchekhovskoy, Narayan et al [1] asked: "Jet-producing disks are thick... Why do only *thick* accretion disks produce jets?" This paper explains both the quasars' structure and their jets and thick accretion disks beyond the currently accepted theories.

There are a few relevant processes that occur simultaneously in a few components: a massive central body, a thick accretion disk and two opposite jets expelled from the central body. They are combined together to produce two other huge very long relativistic jets both associated with the accretion disk.

IOP Conf. Series: Journal of Physics: Conf. Series **1239** (2019) 012012 doi:10.1088/1742-6596/1239/1/012012

Each of the relevant processes is not easy to understand and is explained separately. Some of the processes involve rotation in unexpected ways which are not easy to visualize. Nature combined all the simultaneous processes and stages to produce the final products: two opposite up to two millions of light years long relativistic jets.

Section 2 describes the structure of the *thick* accretion disk and explains its focusing effect that enhances the jet at the axis of symmetry of the accretion disk. Section 3 derives analytically the toroidal shape of the thick accretion disk and the inner conical surfaces. Section 4 describes the inner structure of the massive central body, and how the inner jets are formed and expelled from the poles. Section 5 roughly estimates the velocity of the jet when expelled from the poles. Section 6 discusses the implication on the abundance of helium in the universe. Section 7 describes how short-lived jets are formed in the massive central body. Section 8 discusses celestial bodies with a large mass. Section 9 mentions the relativistic effects involved. Section 10 suggests rough estimations of how the collision of photons radiated from the accretion disk with the particles of the jet amplifies the motion of these particles up to relativistic velocity. Section 11 remarks about the advantage of rough estimations in general. Section 12 concludes the paper.

2. Processes in and near a quasar

We claim that the processes that happen inside and near the quasar are (to be explained in detail later in the paper):

(i) A source of energy situated at the center of the quasar that supplies energy which
a. stabilizes the quasar thermally, and
b. decomposes the iron nuclei into helium nuclei (alpha particles) and hydrogen nuclei (protons). Massive stars have layers of nuclei of several sorts of elements. The accepted model is Hoyle's "onion" model [12], in which the heaviest is the iron layer in the center of the star. Decomposition of iron to helium at a very large pressure and a very high temperature in the center of massive stars was explained 1955 [12].

(ii) We deduce that the helium and hydrogen nuclei which are produced from the decomposed iron nuclei accumulate and ascend along the axis of rotation of the quasar, and then are expelled from the poles as jets of helium and hydrogen nuclei. In many cases these jets were observed expelled from the poles of quasars or active galactic nuclei.

(iii) Many quasars have a thick toroidal accretion disk around them. A section of this accretion disk is shown in Figure 1, representing our calculation detailed in the next section 3.

(iv) This thick toroidal accretion disk is shaped inside in the form of two conical surfaces symmetric with respect to the torus plane (Figure 1). These conical surfaces emit energetic electromagnetic radiation *perpendicular* to the surface inward towards the torus axis.

(v) This energetic radiation is produced by friction between rotating layers of the accretion disk.

(vi) This radiation is focused and collimated along the axis of symmetry.

(vii) This energetic radiation meets and further collimates the jets of helium and hydrogen nuclei expelled at the poles of the quasar, and accelerates them outwards along the axis of symmetry farther and further, up to relativistic velocity.

(viii) The symmetrically emitted radiation that is coming *perpendicularly* to the conical surface meets the nuclei emitted axially from the poles of the central body as jets, which were observed in many cases. This radiation accelerates and collimates the nuclei on the axis direction to velocities that approach the speed of light c (see Figure 1). These nuclei are accelerated upwards to form two fast opposite relativistic jets in the direction of the rotational axis of the torus.

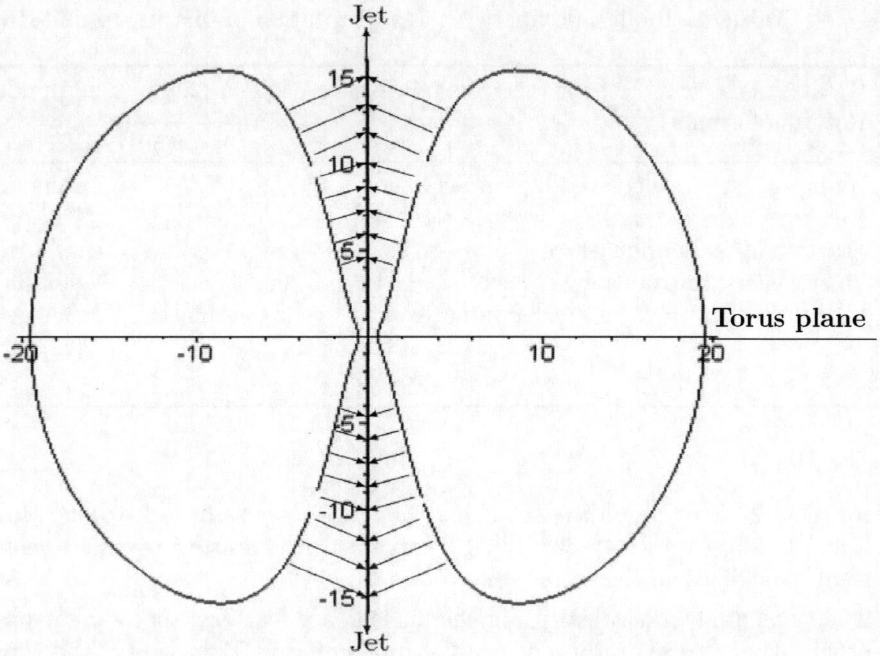

Figure 1. Section of the accretion disk. Radiation from the conical inner surface collimates and accelerates the jets. The axes units depends on the actual size of the thick accretion disk. In large accretion disks each unit represents hundreds of light years! The shape is the grapho-analytical representation of the analytical solution obtained in the next section 3 for the equation derived there

(ix) This process exists in quasars and active galactic nuclei as well as in microquasars. For example, see Harpaz's [13] explanation of how radiation accelerates the hydrogen atom component of the jets of the microquasar SS433. For the original explanation see [14]. These explanations [13] [14] use the lowest excitation level 10.2 eV of a hydrogen atom that enables it to absorb radiation energy in order to increase its velocity without considering the cones. Similarly to these explanations [13] [14], we deduce that every atom or nucleus has excitation level(s) that may enable it to absorb radiation energy in order to increase its velocity.

(x) The decomposition of iron nuclei to helium nuclei [12] and hydrogen (see also items (i)b and (ii) above) also explains why the spectrum of the jets of microquasars shows helium and hydrogen but no other elements (Margon [15]).

(xi) In this way quasars and active galactic nuclei produce and disperse a significant portion of the helium in the universe. This partly explains the abundance of helium in the universe.

(xii) The composition of cosmic rays includes 92% hydrogen nuclei and 6% helium nuclei (MacKeown and Weeks [16]). This similar composition shows that the origin of most cosmic rays may be jets from quasars, galactic nuclei and microquasars.

(xiii) The source of energy at the center of the quasar or the active galactic nucleus involves quarks. The high pressure in the core of the massive body may cause close contact between quarks. This happens because quarks have a radius. (Present accepted theories assume pointlike quarks). MacGregor, using $v = \omega r$, [17] [18] [19] and Ben-Amots, using Franklin's

IOP Conf. Series: Journal of Physics: Conf. Series **1239** (2019) 012012 doi:10.1088/1742-6596/1239/1/012012

Table 1. Radii and kinetic energy of rotation of quarks **up** and **down**.

Rotation type	Not relativistic	Rosen [20]	Franklin [21]
Rotation formula	$v = \omega r$	$v = \frac{\omega r}{[1+(\omega r/c)^2]^{\frac{1}{2}}}$	$v = c \tanh(\omega r/c)$
Radius of quark calculated by	MacGregor		Ben-Amots
Refs for radius of quark	[17] [18] [19]		[22] [2]
Fraction of mass of quark as kinetic energy of rotation	1/3	96.5%	99.3%
Calculated by	MacGregor [17] [18] [19] Leonard [23]	Ben-Amots [2]	Ben-Amots [22] [2]

rotation [21] $v = c\tanh(\omega r/c)$, [22] [2] calculated the radius of **up** and **down** quarks. See Table 1. Macgregor [24] [25] [26] [19] explained why scattering experiments found a much smaller radius than the correct one.

(xiv) The spinning quarks constituting the nucleon are fermions that cannot pass through each other. They possess radii, and at high pressure and high density the rotation of each of the three quarks within the nucleons is quenched [22], [2], [5]. We presume that then the circumferential layers of the rotating quarks interpenetrate (Ben-Amots [22], [2], [5]).

(xv) The interpenetration occurs between the external surfaces of the quarks. The external surfaces of the spinning quarks have relativistic velocities, as calculated in Chapter 4 of Ben-Amots [2].

(xvi) Burbidge and Hoyle [27] have already suggested that when quarks are converted into neutrinos and photons immense amounts of kinetic energy are released and given to these neutrinos and photons. They did not consider the relativistic rotation of the quarks.

(xvii) The penetration causes intensive Cherenkov radiation [28] in this case, which becomes significant.

(xviii) The energy needed for Cherenkov radiation within the star is taken from the kinetic energy of the rotation of the quarks. This rotation energy of the spinning quark constitutes more than 99% of the quark mass (calculated by Ben-Amots [22], [5], [2]). See Table 1.

(xix) High energy Cherenkov photons are created that supply energy to decompose the iron nuclei into helium and hydrogen nuclei at the center of the quasar.

(xx) The collimation effect of the cone of the inner surface of the toroidal accretion disk is similar to the collimation of the shaped charge or hollow charge explosive that generates a fast plasma jet [29] [30]. The hollow space in the explosive charge is also usually cone-shaped. The missile includes an inner cone surface that collimates the penetrating jet along the axis of symmetry in a forward direction. The shock of the explosion is directed *perpendicular* to the surface of the inner cone, and is focused alomg the axis of symmetry of the cone. The concentrated shock has a component of momentum in a forward direction along the axis of the cone. The resultant momentum and the resultant velocity are in a forward direction along the axis of the cone. All the shock waves concentrate and collimate at the axis of symmetry, with component forward, building plasma "super shock" jet in a forward direction. As the jet moves forward in the direction of the axis of the cone it continues to absorb more and more energy and momentum from other parts of the cone. The plasma jet penetrates the armor of the tank.

The hollow charge explosive was invented in 1883 [31] [29] [30]. In the twentieth century a bazooka hollow charge was widely used as an anti-tank weapon because when hitting a tank it can penetrate the tank's armor. Other hollow charge explosive weapons are widely used at present. Hollow charge explosives also have a few civilian uses, for example [32].

(xxi) We deduce that like in a hollow charge, the cone shape of the inner surface of the torus of the accretion disk produces radiation of photons perpendicular to the cone surface with a resultant in the direction of the axis of symmetry. This accelerates the hydrogen and helium atoms by Compton effect. Therefore, similar to hollow explosive charges but by a continuous process of emission, the two jets emerge in two polar opposite directions from the hollow cones in accretion disks around quasars and active galactic nuclei. The emission is not a single event as in hollow charge weapon, but a continuous phenomenon as long as there is a supply of particles from the quasar or the galactic nucleus and radiation energy from the accretion disk [2].

(xxii) The somewhat "convex cone" of the accretion disk and the continuous emission give a further continuous symmetric boost to the particles in the jet during their path along the jet, up to very far distances, increasing their speed to relativistic velocity and collimating their direction.

(xxiii) The processes as described above are based on a theory that quasars and active galactic nuclei are *not* point-like "black holes," but bodies with some dimensions, in spite of their enormous mass of up to milliards of sun masses. See Ben-Amots [22], [5], [2], [3], [33], [4]. Collisions of "black holes" orbiting each other were already observed by LIGO [34] (Laser Interferometer Gravitational waves Observatory, USA). Such a collision is not possible between two point-like masses orbiting each other without contradicting the law of conservation of angular momentum, thus proving that "black holes" posses radii.

3. A proof of how the cone is created

We consider a cloud of small masses m orbiting around the axis z, and gravitationally attracted to a heavy central point mass M (Figure 2).

We use *spherical* coordinates r, θ, ϕ (Figure 3) where each small mass has constant angular momentum

$$J = mvr\cos\theta \tag{1}$$

From (1):

$$v = \frac{J}{mr\cos\theta} \tag{2}$$

The potential of an orbiting point body in spherical coordinates r θ ϕ is the kinetic energy divided by m

$$\frac{v^2}{2} = \frac{J^2}{2m^2r^2\cos^2\theta} \tag{3}$$

where we ignored the kinetic energy of the smaller motion in the z direction. The gravitational potential (= Energy/Mass) is

$$-\frac{MG}{r} \tag{4}$$

Together, the potential is

$$-\frac{MG}{r} + \frac{J^2}{2m^2r^2\cos^2\theta} \tag{5}$$

Where this potential is constant, we get a constant potential surface:

$$-\frac{MG}{r} + \frac{J^2}{2m^2r^2cos^2\theta} = k \tag{6}$$

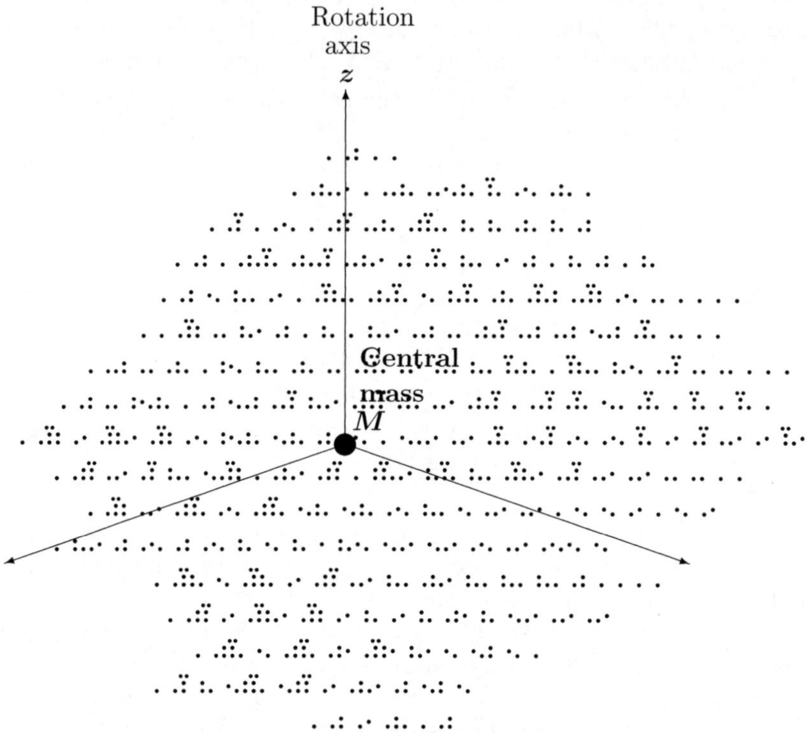

Figure 2. A cloud of homogeneous particles (schematic) rotating around axis z. In a *spherical* coordinate system the distance of a small mass m from the central mass M is r

where k=constant is associated with energy.

Solving for $r(\theta)$

$$r = \frac{GM \pm \sqrt{G^2M^2 - 2\,kJ^2/(m^2\cos^2\theta)}}{2k} \tag{7}$$

For M=0 there is no stable solution for r, meaning that a torus can exist only around a central mass M.

Normalizing Equation (7) by choosing $GM = 1$ and $J/m = 1$ and K instead $2k$ we obtain:

$$r = \frac{1}{K}\left(1 \pm \sqrt{1 - \frac{K}{cos^2\theta}}\right) \tag{8}$$

Normalization in such cases is a mathematical tool used mainly to obtain values appropriate for numerical calculation. The values obtained by normalization should be convenient to calculate, in the order of 1. The scale in the following figures is normalized. The actual scale depends on the size of the accretion disk. In large accretion disks each unit represents hundreds of light years.

The solution is torus-like. Slow rotation velocity means low K. High K means fast rotation velocity.

Looking at Equation (8) one immediately deduces that K should be limited in order to avoid a square root of a negative value, meaning that the rotation velocity is limited in order to sustain

IOP Conf. Series: Journal of Physics: Conf. Series **1239** (2019) 012012 doi:10.1088/1742-6596/1239/1/012012

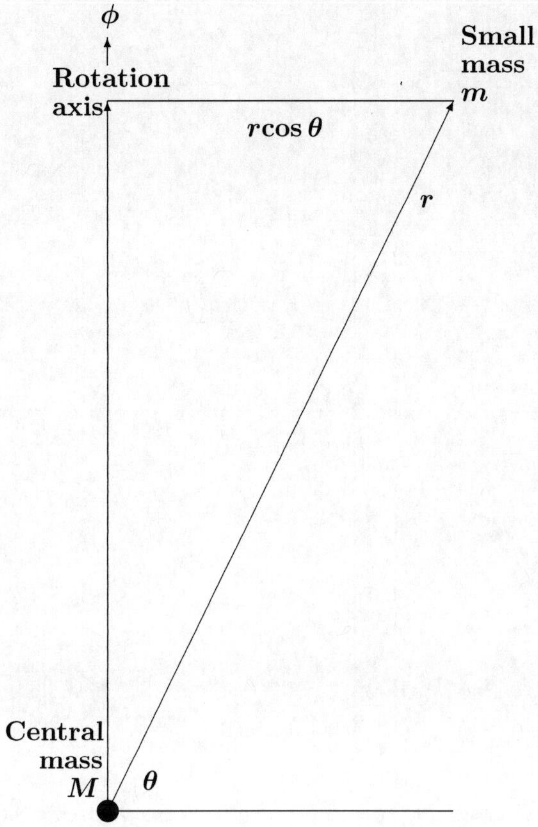

Figure 3. Rotation in *spherical* coordinate system. The distance of an element of mass m from the *central mass* M is r. The distance of an element of mass m from the axis of rotation is $r\cos\theta$

a stable solution. A careful interpretation of the similar (but not normalized) equation (7) leads to the same conclusion.

For K=0.01 (=slow rotation) the shape approaches a spheroid (Figure 4).

IOP Conf. Series: Journal of Physics: Conf. Series **1239** (2019) 012012 doi:10.1088/1742-6596/1239/1/012012

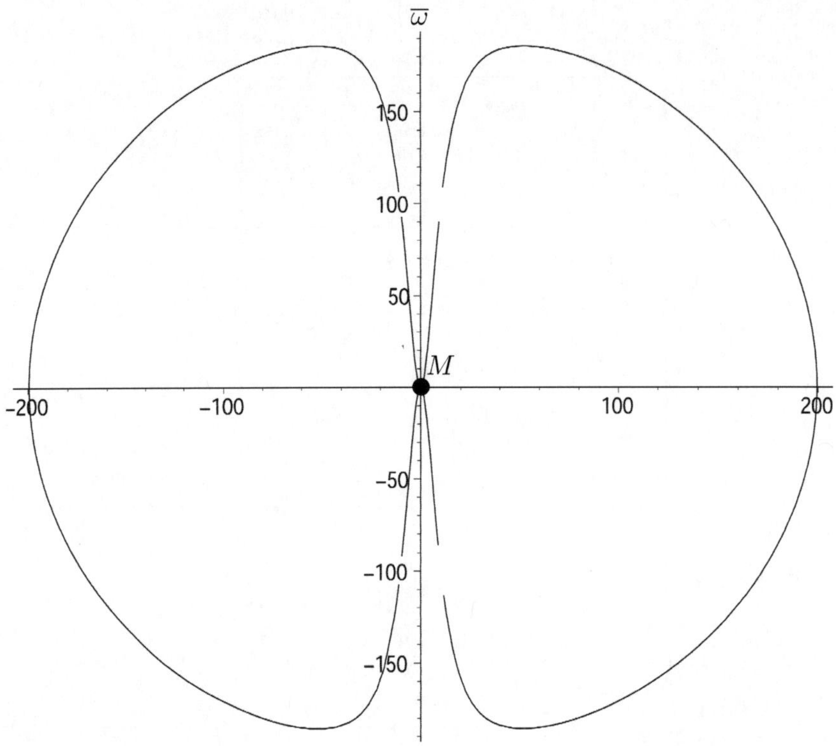

Figure 4. Section through a slowly rotating toroidal accretion disk with K=0.01 gravitationally attracted to a central massive body M. This shape is a grapho-analytical representation of the numerical calculation of Equation (8) for k=0.01 (=slow rotation)

Plotting for K =0.1 (Figure 5). The shape in Figure 5 is a grapho-analytical representation of the numerical calculation of Equation (8).

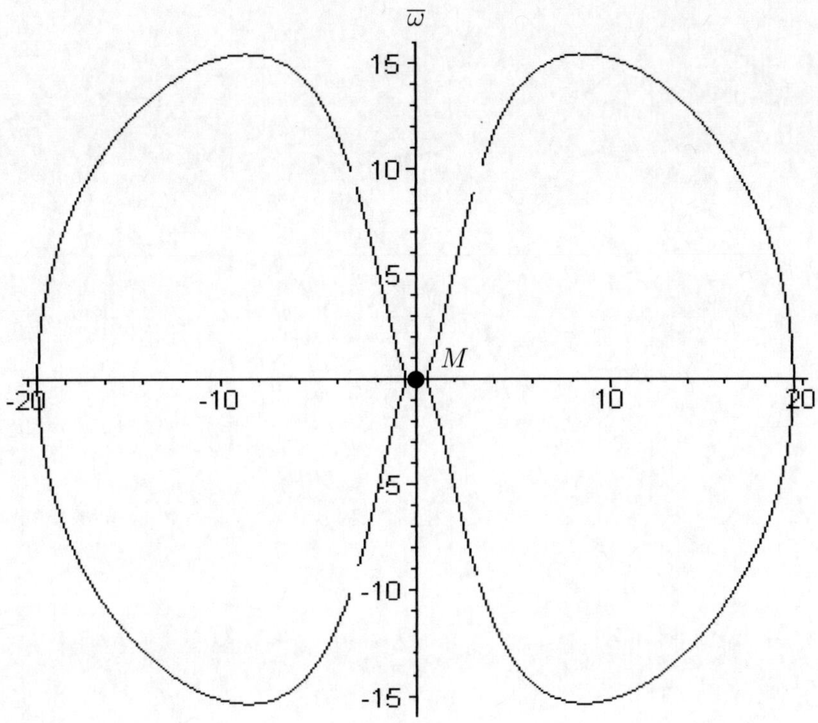

Figure 5. Section through a rotating toroidal accretion disk with K=0.1 gravitationally attracted to a central massive body M. The two inner cone surfaces are the result of the numerical calculation of Equation (8)

See quasi-tori with K=0.3-0.9, 0.99 (Figure 6). These shapes are the grapho-analytical representation of Equation (8) for K=0.3-0.9, 0.99

The torus structure for some astronomical objects was first suggested by Lynden-Bell [35] in 1962. Tori were observed in NGC1068 in 1985 [36], and in NGC4258 in 1995 [37]. Observations of tori were reported later in many papers. In most cases no attention was given to the relevance of the *torus cone* for the collimation of astronomical *jets.*

Other tori were also obtained by other researchers using different methods, but the few who paid attention to the cones explained that each cone is produced because the jet cuts its way through the inner side of the torus.

We claim that on the contrary the two cones *produce* the two jets.

According to our analytical solution (8):

One condition for obtaining torus-like shapes is that the angular momentum per unit of mass (J/m) should be large, but not too large so as to cause the disintegration of the torus. Therefore, the term under the square root sign in the relevant equation above must be positive:

$$G^2M^2 - 2K(J/m)^2/\cos^2\theta > 0 \tag{10}$$

This limits the specific angular momentum J/m relative to the mass of the central body represented by GM.

Another major condition is a massive central mass M. No torus is possible without a massive central mass.

Without a massive central mass the solution is a spheroid without the hollow pipe [2].

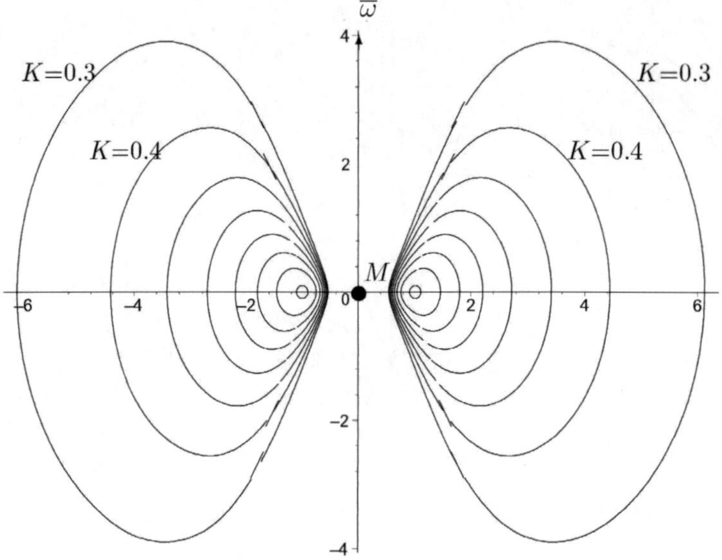

Figure 6. Section through a torus-like rotating accretion disk showing sections of surfaces with K=0.3-0.99. The inner cones with a similar tilt represent the results of the numerical calculations of Equation (8) for various values of K. The thick toroidal accretion disk is also the result of the numerical calculation. The thickness and the cones were obtained for all the values of M, m and J that we tried for $K < 1$ (This is not shown here. See in the text). One conclusion is that when matter in the accretion disk loses energy and angular momentum (by friction between the layers) it "falls" to layers with a smaller K. Only matter in a layer with a very small K may meet the central body M

The central mass may be a quasar or a galactic nucleus, whether active or not.

The collimation is as in Figure 1:

Two cones in a torus collimate radiation to two opposite jets:

The radiation (arrows) is emitted *perpendicular to the conical surfaces* of the torus and is collimated to two opposite jets (K=0.1 in our example).

Usually photons pass easily through other photons, because they may share one place. Particles are necessary for the collimation effect of a cone. Our discussion above does not take into account photon-photon scattering, in which two interacting photons need at least 0.511 MeV for each photon to produce a pair of electron and positron. See Section 10.

Hollow-charge missiles cannot be stabilized in the presence of rotation, because then the collimation of the jet is lost [30]. Therefore apparently the particles necessary for collimating a jet do *not* originate in the rotating torus [2].

We deduce that the rotating torus emits *radiation*, while matter particles have to be already in the axis of rotation, ready to be kicked or boosted by the radiation collimated by the torus.

Thick accretion disks possess cones, but flat accretion disks lack cones. Therefore, flat accretion disks cannot produce or enhance relativistic jets, but "Thick accretion disks produce jets." [1] .

IOP Conf. Series: Journal of Physics: Conf. Series **1239** (2019) 012012 doi:10.1088/1742-6596/1239/1/012012

Choosing $J/m=1$ for normalization fits a certain angular velocity that makes the sections of the tori in figures 1, 4, 5 and 6 look like a circle. For faster angular velocities the sections of the tori would look somewhat different but their inner surfaces would be similar to cones. These cones collimate the radiation that enhances the jets as well. The faster angular velocity is limited in order to avoid a square root of a negative number. A square root of a negative number means in this case disintegration of the torus due to too strong centrifugal forces. For example, for $GM=1$ and a potential $K=0.1$ the ratio J/m must not exceed $\sqrt{5}$. For this potential, near $J/m=2$ the section of the torus is very small, meaning that there is no effective collimation.

Instead of using the Newtonian equation (1), one may use the appropriate relativistic equation, which is not shown here, and obtain another potential equation instead of equation (6). It is solvable, however the solution is too cumbersome but similar.

Fragile [38] [39] and Hawley [40] conducted numerical simulations including magnetic fields and obtained sections through tori similar to our figure 6 (Font [41] for review).

Hawley's 3-D simulations [40] found that "things in the disk seem pretty independent of [magnetic] field topology." This fits our conclusion that gravity alone determines the shape of the thick accretion disk.

In 1982 Paczyński analytically obtained a *similar* torus by similarly using Newtonian potential. See Abramowicz's review [42] and Frank et al's [43] textbook. Paczyński and also Rees et al [44] suggested that the cones they obtained in the inner surfaces of the torus collimate relativistic jets. See Abramowicz [42] for review.

Observations identified the cases in which the directions of rotation were different for the central bodies and their accretion disks (Fragile [38]). The observations revealed that in these cases the jets start in the direction of the axis of rotation of the central body, and farther adjust to the axis of rotation of the accretion disk. Our theory explains that the particles of the two jets are first produced in the non-black-hole supermassive central body and then are expelled in its poles. Later they are accelerated and collimated by the radiation from the inner cones of the torus of the thick accretion disk. Thus there are earlier stages of the jets, before they are further accelerated outside the central mass as described above.

4. Processes inside quasars and galactic nuclei

Formerly we dealt with the amplification and acceleration of the jet *outside* of the central mass (quasar or active galactic nucleus). The outside jet is the *fourth stage*. The *first stage* is production of the jets *inside the central mass near its center.* The *second stage* is the moving of the *inner* jets outward along the axis of rotation. These two early stages are hidden inside the central mass. The *third stage* is expelling the opposite jets at the poles, as already observed in a few cases. These three stages are explained in the following text.

The particles of the jet originate near the center of the central massive body, although and because its mass may be millions or milliards of sun masses. These massive bodies possess radii. The LIGO observation of gravitational waves from *colliding massive bodies* [34] also proved that they possessed radii. We mentioned that if they were pointlike they could not collide without disobeying the law of conservation of angular momentum.

A modification of Hoyle's onion model [12] (1955) takes place. Hoyle [12] showed that the fusion of elements supplies energy and builds "onion" layers of heavier elements within the star. Iron is the inner and the heaviest element close to the center of the star. Hoyle [12] mentioned that at higher temperatures at the center of the massive star iron nuclei decompose to helium nuclei. This decomposition *demands energy.* Hoyle's onion model [12] of the structure of massive stars is accepted, including the decomposition of iron nuclei to helium nuclei in extreme conditions at the center of massive stars.

We claim that in order to sustain stability of a supermassive body an energy source at the center of the massive star supplies energy also to decompose iron nuclei to helium nuclei (alpha particles) and hydrogen nuclei (protons).

(i) Helium and hydrogen nuclei are produced around the center of the supermassive body.

(ii) By buoyancy, the hot helium and hydrogen nuclei flow outward continuously and/or accumulate in huge bubbles that occasionally burst outward between layers 1-8. These bubbles were observed [45].

(iii) See our modified model of layers (Figure 7) for a celestial body of thousands up to milliards of solar masses (a quasar or an active galactic nucleus).

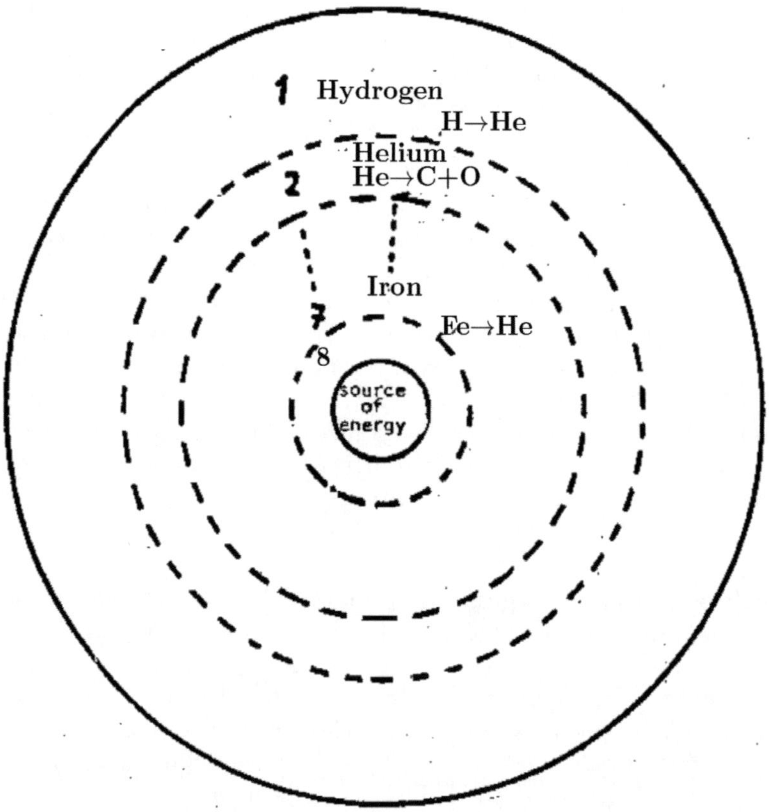

Figure 7. "Onion" layer model of a celestial body of millions to milliards of solar masses (not to scale). Strong convection currents of huge bubble bursts of helium exist between layers 1-8.

The hot helium and hydrogen nuclei are much lighter than the upper layers of the nuclei of metals. The hot helium and hydrogen nuclei ascend to the surface of the massive celestial body as a continuous flow or as accumulated huge bubbles [45].

Huge bubbles may explain the variable luminosity of quasars, because a huge hot bubble that arrives to the surface increases the radiation emitted by the ejected jet.

In which outward direction do they go? We will explain this in the following text.

Centrifugal force would push *heavier* matter to the equatorial bulge, but the super hot helium and hydrogen are *lighter* than the upper layers. The heavier upper layers squeeze the lighter helium and hydrogen nuclei toward to the axis of rotation and upwards *along the axis of rotation* until they are expelled. There the helium and hydrogen nuclei are ready to be the *particles* in the axis of rotation that are pushed farther and faster by the *collimating radiation* from the cones of the torus.

The gravitational attraction force of the central mass M acts *spherically* inward on a mass m and is equal to mMG/r^2

The centrifugal force acts *cylindrically outward* on the mass m and is equal to $m\omega^2 r\cos\theta$ and is obviously zero at the axis of rotation z where $\cos\theta$=0.

Therefore, at the axis of rotation the gravitational attraction force inward is maximal and is equal to $\Delta mMG/z^2$ for a mass Δm *denser* than the surrounding environment.

In addition, at the axis of rotation the total *buoyancy* force *outward* is *maximal* for a mass *lighter* than the surrounding environment. A *lighter* mass will *ascend* along the axis of rotation outward toward the two poles and be expelled there as *two opposite jets* of hydrogen and helium.

Then the expelled *helium and hydrogen* will be collimated as two opposite jets on the axis directions by a hollow charge mechanism and accelerated to relativistic velocities by the absorption of energy supplied by the *radiation from the cones.*

See more details in Ben-Amots [2].

Porco [46] explained clearly the much weaker Enceladus's polar jet: "a warm region, with lower than average density, will naturally drift toward the axis of rotation. Furthermore, a warmer zone under the south pole would rise in convective motion under the uppermost brittle layer of the ice shell...". The flow of the lighter fraction *inward* to the axis of rotation as in Enceladus and *ascending upwards* along the axis of rotation towards the poles as in Enceladus exists in all rotating stars, quasars and active galactic nuclei. Enceladis is a moon of Saturn. Its diameter is ≈500 km. The much larger dimensions than those of Enceladus mean that there is a similar but much stronger and faster flow *upward* in *inner jets* along the axis of rotation of supermassive bodies, even without any calculation. See rough estimation in the next section 5.

Interpreting the hydrodynamics in a supermassive rotating body one may compare it with the analytical solution of the flow in a star rotating with uniform rotation (Section 9.4 in [13]). Figure 9.2 in [13] presents the characteristic flow. The radius in which the flow changes direction depends on the density. For a supermassive rotating body the radius is relatively close to the surface of the rotating supermassive body. The flow near the axis of rotation ascends along most of the axis of rotation almost up to the external surface.

The buoyancy of the light fraction of hydrogen and helium produces an upward flow along the axis of rotation similarly to Porco's [46] explanation. This upward buoyancy flow is added to the upward flow of the analytical solution in Section 9.4 of [13].

To apply this analysis on a supermasive rotating body one should accept the fact that a supermassive body is *not* a pointlike "black hole." See exponential gravitation in [2], [3], [4]. [5], [6], [7], [8] and [9].

5. Velocity estimation

For a very *rough* estimation we assume a constant acceleration a equal to the acceleration g on Earth 980 [cm/sec], along the radius r of a quasar, presumed to be 5×10^{13} cm (half a milliard km. See [48] p 50). Using the simple Newtonian formula

$$v = \sqrt{2ar} \tag{11}$$

the estimated velocity is approximately 3×10^8 cm/sec or 3000 km/sec or $0.01c$, a fast jet but not a relativistic jet. These particles on the rotation axis are further accelerated to relativistic velocities by the radiation collimated by the cone of the torus of the accretion disk as described above in sections 2 and 3.

The acceleration is not constant, yet in small radii the gravitational buoyancy outward is decreased because the active attracting mass is smaller, but increased because of the large difference between the atomic weight of the helium and hydrogen nuclei and between the atomic weight of the iron layer. In large radii the attractive mass is larger and increases the buoyancy, but the environment consists of light atoms with a smaller difference between the atomic weights that decreases the buoyancy.

6. Abundance of helium

Transforming hydrogen to helium is a process that occurs in most of the stars by fusion. It is a slow process. It is accepted [13] that only stars more massive than 8 M_{sun} endure supernova explosions, which disperse the helium produced in the star. These supernovae cannot produce the 25% abundance of helium observed in the universe [47] during the accepted age of the universe, which is 14 milliard years. The accepted theory assumes that most of the helium was produced in a short time during the so-called "Big Bang."

Margon [15] reported that jets ejected from the microquasar SS433 expel only hydrogen and helium. We showed above that quasars and active galaxies nuclei produce and eject huge jets and bubbles of helium nuclei.

Ben-Amots (Section 8.10.7 in [2]) showed that the Cosmic Microwave Background was ended approximately 90 milliard years ago. During 90 milliard years the production and dispersion of helium by the stars and all the quasars, active galactic nuclei, microquasars and other jet emitting celestial bodies could reach the observed 25% of the matter in the universe. Therefore, "big bang" production of helium is redundant.

7. Short lived jets during supernovae

A supernova type II starts with a fast collapse of the core of a large star. Then a bounceback upwards occurs (Ben-Amots [5], [2]) as a shockwave. The fast collapse causes the upper layers of the star to lose their support and collapse as well. We deduce that due to the angular momentum of these layers, a torus with two inner cone surfaces is formed *within the star*, in a similar process as was described in sections 2 and 3. (see Figure 8 and Ben-Amots [2]). The heat of the friction between the layers in the torus is radiated perpendicular to the surfaces of the cones, and *collimates* the matter in the axis of rotation. Thus two opposite ultra relativistic jets *within the star* are formed. These jets break out and are expelled from the poles of the star as two opposite ultra relativistic jets.

A short time later the shock wave outwards reaches the newly born torus *in the star* and disrupts it. This *stops the collimation* and the production of the jets, making them the observed short-lived extreme jets (see more details in Ben-Amots [2]).

8. Celestial bodies with a larger mass

Massive stars of 100-200 sun masses were already observed. For example the estimated mass of Eta Carinae is 100 sun masses [49]. Tracing progenitors of exceptionally luminous supernovae found stars with up to 200 sun masses (Gal-Yam [50]). Larson and Bromm [51] assumed that it is possible that there are ancient stars of 1000 sun masses.

On the other hand, the masses of active galactic nuclei are millions of sun masses and more, up to milliards of sun masses. Quasars possess milliards up to trillions of sun masses. There also exist celestial bodies whose mass is between about 1000 and a million of sun masses. Greene [52] calls celestial bodies in the range of 1000-2000000 M_{sun} "middleweight black holes." Lemonick [53] mentions one million M_{sun}.

Gal-Yam [50] explains that very massive stars endure exceptionally strong supernovae explosions early in their life. This is caused by the electron-positron pair production that occurs and initiates the fusion of oxygen nuclei. The pair production instability as a cause of

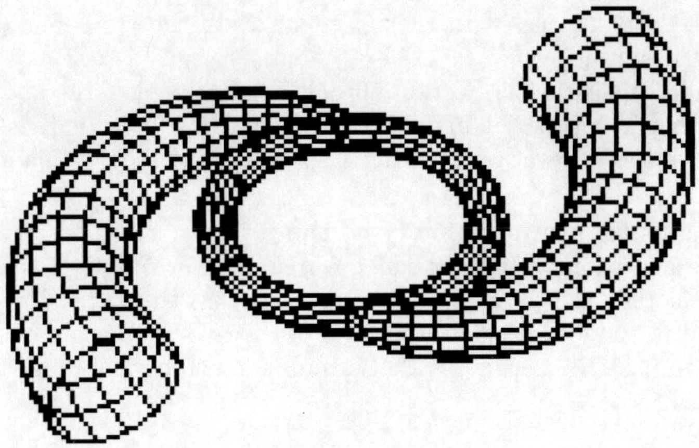

Figure 8. Two stream tubes converge into a torus during collapse of a supernova (schematic). Actually there are many such simultaneous stream tubes converging into the same torus [2]

supernova explosions of very massive stars was predicted and calculated by Barkat et al [54], Rakavy and Shaviv [55], [56] and Shaviv and Kovetz [57] and was confirmed by observations about fifty years later (Gal-Yam [50], [58], Gal-Yam and Leonard [59] and Gal-Yam et al [60]).

This early explosion apparently prevents existence of stars more massive than about a thousand sun masses. That is why there are no stars more massive than a limit of hundreds or a thousand sun masses.

However, the much more massive galactic nuclei and quasars, whose mass is millions of sun masses or much more, avoided and still avoid both the electron-positron pair production instability and the iron decomposing instability. The question of how the pair instability and the iron decomposing instability are bypassed during the formation of active galactic nuclei and quasars exists for any theory of quasar formation, or galactic nucleus formation, regardless of how they are stable later.

9. Relativistic effects

(i) The accepted beliefs that quasars and galactic nuclei are "black holes," and that nothing can be expelled out of a "black hole" led to the belief that the accretion disks supply matter to the jets, bypassing the presumed "black holes."
In this paper it is shown that the accretion disks supply *radiation* to the jets and matter to the quasars and the active galactic nuclei. The matter (hydrogen and helium) is supplied to the jets by expelling from the quasars and the active galactic nuclei. For these explanations the belief in "black holes" should be ruled out. For relativistic rotation and appropriate relativistic theories without "black holes" the interested reader can consult recent papers by Ben-Amots [22], [5], [2], [3], [33], [4], Majerník [6], [7], Walker [8], [9] and earlier references therein.

(ii) Many works were done on various aspects and phenomena involved in the astronomical jets in order to explain the observations of these jets. Among them are various suggestions of processes in order to explain how the velocities of these jets are enhanced even further, how the jets keep moving collimated along thousands to millions of light years and how they disperse at last to giant lobes. Collisions between the nuclei in the jet that increase the

velocity of a small portion of the nuclei in the jet to relativistic velocities were suggested [48]. This work does not deal with these late stages of the astronomical jets (fifth stage according to our counting).
However, a rough calculation in Section 10 below suggests that the above fourth stage of enhancing the velocity of the jet may be sufficient to accelerate it to relativistic velocity, making the fifth stage relevant only during dispersing of the jets to giant lobes.

10. Examples of calculating the velocity of the jet

We mentioned the calculation for the jets of the microquasar SS433 [13] [14]. Radiation of hydrogen atoms from the surface of the microquasar along the axis of rotation at 13.6 eV (912 Å) is redshifted to the lowest excitation level of the hydrogen atom of 10.2 eV (1200 Å), accelerating the jet to $0.26c$. Accordingly, the formula for microquasar that takes into account the relativistic redshift is

$$v = c\,\frac{(13.6/10.2) - 1}{(13.6/10.2) + 1} \tag{12}$$

which gives $v = 0.26c$ [13] [14], as measured in the microquasar SS433 [15]. 13.6 eV is in the ultra-violet radiation range.

If faster jets have a similar process for enhancing their velocity, then by substituting radiation stronger than 13.6 eV from an accretion disk around a quasar or an active galactic nucleus we may use Equation (12) to get approximations for their faster jets.

Looking at figures 1 and 5, for K=0.1 the radiation perpendicular to the surface of the cone has a *component* of approximately 1/4 in the direction of the axis of symmetry of the cone. We assume that this component may accelerate an atom in the jet in the direction of the jet, but in the usual case the components *perpendicular to the jet* do not accelerate the atoms in the jet. We calculate for the hydrogen atom, whose lowest excitation level is 10.2 eV.

Examples of calculating approximations:

(i) Substituting of 850 eV (14.6 Å) in Equation (12) for an example of K=0.1 and radiation perpendicular to the surface of the cone, while considering only 204 eV of it as the resultant contributing to the acceleration of a hydrogen atom in the direction of the jet, a velocity of about $0.995c$ is obtained. 850 eV is in the X rays radiation range.

(ii) Substituting of 527 KeV (0.0235 Å) in Equation (12) for an example of K=0.1 and radiation perpendicular to the surface of the cone, while considering only 126.5 KeV of it as the resultant contributing to the acceleration of a hydrogen atom in the direction of the jet, a velocity of about $0.99999999c$ is obtained. 527 KeV is in the γ rays radiation range.
In this case the component perpendicular to the direction of the jet is 511.1 KeV. The head on collision of two opposite photons of 511 KeV or more is capable of producing a pair of electron and positron. A nucleus of hydrogen in the jet may absorb such an electron and may become an atom, that may be accelerated by absorbing 10.2 eV, which is its lowest excitation level.
Another assumption may be avoiding the necessity of mass expelled from the central body, but producing of jets of electrons and positrons from γ rays focused by the cones of the accretion disks.
The scenarios assumed in item (ii) may be realistic only if the accretion disk emits γ rays stronger than 527 KeV.

(iii) The calculation in this section may be valid for *hydrogen atoms*, or for helium *atoms* when the appropriate excitation levels of helium *atoms* are considered instead those of hydrogen. The lowest energy level of He_4 nucleus is higher than 20 MeV, but apparently weaker photons in the ultraviolet radiation range or X-ray range are capable of directing jets to

the axis of rotation of the thick accretion disk. Therefore, for hydrogen and helium *nuclei* a different mechanism should be considered:
Each photon possesses momentum in direction perpendicular to the surface of the cone. When many photons collide with a nucleus near the axis of rotation, each photon delivers its momentum to the nucleus. During many collisions the nucleus receives momenta from many photons. The resultant momentum of all the photons that collide with a nucleus is in the direction of the axis of rotation of the thick accretion disk, which is also the axis of symmetry of the cone. This way collisions with many photons increase the velocity of every nucleus in the direction of the jet.
The resultant momentum is small for a too slow rotation of the accretion disk (Figure 4). The section of the torus is small for a too fast rotation of the accretion disk (Figure 6). The conclusion is that the rotation of the thick accretion disk should be effective in producing a fast jet for $0.8 > K > 0.05$.
The photons radiated from the rotating accretion disk are scattered by collisions with the nuclei in arbitrary directions and frequencies. This arbitrary scattering enables us to observe the jets. Therefore, we deduce that apparently these collisions continue to occur along the jets up to the lobes (fourth stage according to our counting).

11. Remark
This paper includes some approximations. In this context we cite
"Thumb's first postulate
It is better to solve a problem with a crude approximation and know the truth, $\pm 10\%$, than to demand an exact solution and not know the truth at all." [61]

12. Conclusions
This paper details the processes in the center of massive celestial bodies like quasars and galactic nuclei. These processes stabilize them and prevent implosions. Production of hydrogen and helium near the center of these bodies is a by-product of these processes. These helium and hydrogen nuclei (or atoms in microquasars) are lighter than their surroundings. A region with lower than the average density will naturally drift toward the axis of rotation. There they rise upward in a very fast convective motion and by buoyancy along the axis of rotation and are expelled at the poles as jets.

This paper also proves that a *thick* accretion disk around such celestial body takes the shape of a torus that includes inner cone surfaces. The torus emits radiation produced by friction between the rotating layers of the accretion disk. The *radiation* that is emitted through the surfaces of the inner cones collimates the helium and hydrogen nuclei (or atoms in microquasars) that are moving along the axis of rotation, and accelerates them by absorption of the radiation to two opposite relativistic jets.

In this way the paper explains the stability of quasars and galactic nuclei as well as the production of relativistic jets, and also adds them as producers of helium that explain a significant part of the abundance of helium in the universe.

Acknowledgments
Thanks to the anonymous referee for his wise comments.

References
[1] Tchekhovskoy A, Narayan R and McKinney J C 2013 Relativistic jets formation by spinning black holes *Jets2013 Conf.* Granada, Spain
http://jets2013.iaa.es/sites/jets2013.iaa.es/files/imagecashe/tchekhovskoy_jets2013.pdf

[2] Ben-Amots N 2017 *Relativity, Gravitation, and Relativistic Rotation: Clarifying some Paradoxes of Relativity at the Extreme* (Bergenfield, NJ: Technology Dynamics) chapters 10, 4, 5
http://www.netsivi.org/book.htm
https://www.technologydynamicsinc.com/book-relativity-2017/

[3] Ben-Amots N 2007 Relativistic exponential gravitation and exponential potential of electric charge *Found. Phys.* **37** 773-87

[4] Ben-Amots N 2011 Some features and implications of exponential gravitation *J. Phys.: Conf. Series* **330** 012017

[5] Ben-Amots N 2015 Energy accumulation in relativistic sub-Bohr orbitals, Franklins relativistic rotation of quarks and gravitational field bounceback as processes relevant to explosion of supernovae *J. Phys.: Conf. Series* **615** 012012

[6] Majerník V 2006 Energy conservation at the gravitational collapse *Preprint* astro-ph/0609313

[7] Majerník V 2009 An alternative to classical naked singularity and black hole *Phys. Essays* **22** 413-9

[8] Walker C 2010 Calculation of gravitational potential by the method of multiple redshifts
http://sites.google.com/site/revisingnewton

[9] Walker C 2012 Newtonian potential and the Einstein equivalence principle *Foundational Questions Contest*
http://www.fqxi.org/data/essay-contest-files/Walker_shells.pdf

[10] Djorgovski S G, Volonteri M, Springel V, Bromm V and Meylan G 2008 The origins and the early evolution of quasars and supermassive black holes, in: ed Klement *et al Proc. XI Marcel Grossmann Meeting on General Relativity, Berlin 2006* (Singapore: World Scientific) (*Preprint* astro-ph/0803.2862)

[11] Wikipedia 2018 List of unsolved problems in physics (Retrieved 2018)
https://en.wikipedia.org/wiki/List_of_unsolved_problems_in_physics

[12] Hoyle F 1955 *Frontiers of Astronomy* (London: Heinman) p 211

[13] Harpaz A 1994 *Stellar Evolution*, (Wellesly, Massachusetts: A.K. Peters) sections 12.5 and 9.4, and fig 9.2

[14] Shapiro P R, Milgrom M and Rees M J 1986 The radiative acceleration of astrophysical jets - line locking in SS433 *Astroph. J. Supp.* **60** 393-431

[15] Margon B 1980 The bizarre spectrum of SS433 *Sci. Am.* **243** (4) 44-55

[16] MacKeown P K and Weeks T C 1985 Cosmic rays from Cygnus X-3 *Sci. Am.* **253** (5) 40-9

[17] MacGregor M H 1972 A spectroscopy for hadrons, *Nuo. Cim.* A **8** 235-97. See pp 258-9, 294

[18] MacGregor M H 1974 Light-quark hadron spectroscopy: Experimental systematics and angular momentum systematics, *Phys. Rev.* D **9** 1259-329. See p 1316

[19] MacGregor M H 1990 An elementary particle constituent-quark model, *Nuo. Cim.* A **103** 983-1052. See p 1047

[20] Rosen N 1946 Note on the problem of uniform rotation, *Phys. Rev.*, **70** 93-4

[21] Franklin Philip 1922 The meaning of rotation in the special theory of relativity, *Proc. Nat. Acad. Sci. USA* **8** 265-8

[22] Ben-Amots N 2003 Basic aspect of relativistic rotation: Franklin rotation of a sphere *Found. Phys.* **33** 1369-72

[23] Leonard C M L 1984 The relativistic classical increase of energy and angular momentum with rotation, *Am. J. Phys.* **52** 527-31

[24] MacGregor M H 1987 (June 17), UCRL 94777

[25] MacGregor M H 1987 *Bull. Am. Phys. Soc.* **32** 31

[26] MacGregor M H 1989 On the interpretation of the electron anomalous magnetic moment, *Found. Phys. Lett.* **2** 577-89. See pp 586-7

[27] Burbidge G and Hoyle F 1966 The problem of quasi-stellar objects *Sci. Am.* **215** (6) 40-52. See p 52

[28] Cherenkov P A 1934 Visible emission of clean liquids by action of γ radiation, *Doklady Akademi Nauk SSSR* **2** 451-4. Reprint: *Selected Papers of Soviet Physicists* 1967 *Usp. Fiz. Nauk* **93** 385-8 (Both in Russian)

[29] Wikipedia 2018 Shaped charge (Retrieved 2018)
https://en.wikipedia.org/wiki/Shaped_charge

[30] Wikipedia 2018 High-explosive anti-tank warhead (Retrieved 2018)
https://en.wikipedia.org/wiki/High-explosive_anti-tank_warhead

[31] Kennedy D R 1983 *History of the Shaped Charge Effect, The First 100 Years - USA - 1983*, Defense Technology Support Services Publications

[32] Powell K M 2004 Hollow charge explosive device particularly for avalanche control, *USA Patent* 6786157

[33] Ben-Amots N 2008 A new line element derived from the variable rest mass in gravitational field *Preprint* gen-ph/0808.2609

[34] Abbott B P et al LIGO Scientific Collaboration and Virgo Collaboration 2016 Observation of gravitational waves from a binary black hole merger *Phys. Rev. Lett.* **116**(6), 061102

[35] Lynden-Bell D 1962 Stellar dynamics: Exact solution of the self-gravitation equation *Mon. Not. R. Astron. Soc.* **123** 447-58

[36] Antonucci R R J and Miller J S 1985 Spectropolarimetry and the nature of NGC1068 *Astroph. J.* **297** 621-32. See Figure 5 in p 628
[37] Greenhill L J, Jiang D R, Moran J M, Reid, M J, Lo K Y and Claussen M J 1995 Detection of a subparsec diameter disk in the nucleus of NGC4258 *Astroph. J.* **440** 619-27
[38] Fragile C 2005 Tilted rotating disks around rapidly rotating black holes *Jets and Disks KITP Conf.* Santa Barbara, California. See figure in transparent 7 http://online.itp.ucsb.edu/jet+disk_c05/fragile/pdf/Fragile_KITP.pdf
[39] Fragile C 2008 Jet formation in MHD simulations *Preprint* 0810.0526 figure 8
[40] Hawley J F 2008 (June 18) Magnetic fields and jet formation, Workshop on MRI turbulence, University of Virginia. https://slideplayer.com/slide/7755862
[41] Font J A 2008 Numerical hydrodynamics and magnetohydrodynamics in general relativity *Living Rev. Relativity* **11** lrr-2008-7 Figure in p 73
[42] Abramowicz M A 2008 Five ideas on black hole accretion disks *Preprint* 0812.3924
[43] Frank J, King A and Raine D 2002 *Accretion Power in Astrophysics* (Cambridge University Press) chapter 10: "Thick disks" and figures 10.4 and 10.5
[44] Rees M J, Begelman R D, Blandford R D and Phinney E S 1982 Ion-supported tori and the origin of radio jets *Nature* **295** 17-21. See figures in p 19
[45] Mirabel I F and Rodriguez L F 1994 A superluminal source in the galaxy, *Nature* **371** 46-8. See figure on the front page of the issue.
[46] Porco C 2008 The restless world of Enceladus *Sci. Am.* **299** (6) 52-63. See p 62
[47] Hoyle F and Tayler R J 1964 The mystery of the cosmic helium abundance *Nature* **203** 1108-10
[48] Levinson A 2012 *Fireworks in dark universe* (Tel-Aviv: Matar) 225-6 (In Hebrew)
[49] Wikipedia, 2018 Eta Carinae (Retrieved 2018) https://en.wikipedia.org/wiki/Eta_Carinae#mass
[50] Gal-Yam A 2012 Super supernovae *Sci. Am.* **306** (6) 44-9
[51] Larson R B and Bromm V 2001 The first stars in the universe *Sci. Am.* **285** (6) 64-71
[52] Greene J E 2012 Goldilock black holes *Sci. Am.* **306** (1) 40-7
[53] Lemonick M D 2014 The first starlight *Sci. Am.* **310** (4) 38-45
[54] Barkat Z, Rakavy G and Sack N 1967 Dynamics of supernova explosion resulting from pair formation *Phys. Rev. Lett.* **18** 379-81
[55] Rakavy G and Shaviv G 1967 Instabilities in highly evolved stellar models *Astrophys. J.* **148** 803-16
[56] Rakavy G and Shaviv G 1968 Isentropic models for final stages of stellar evolution *Astrophy. Space Sci.* **1** 429-41
[57] Shaviv G and Kovetz A 1970 Relativistic isentropic stellar methods *Astrophy. Space Sci.* **7** 416-23
[58] Gal-Yam A 2012 Pair instability explosions: observational evidence *Preprint* astro-ph[CO]/1206.2157
[59] Gal-Yam A and Leonard D C 2009 A massive hypergiant star as the progenitor of the supernova SN2005gi *Nature* **458** 865-7
[60] Gal-Yam A, Mazzali P, Ofek E O et al 2009 Supernova 2007bi pair instability explosion *Nature* **462** 624-7
[61] Nuclear News, August 1971 Thumb's first and second postulates *Nuclear News*

IOP Conf. Series: Journal of Physics: Conf. Series **1239** (2019) 012013 doi:10.1088/1742-6596/1239/1/012013

Relative and Absolute Time

Tepper L. Gill[1]

[1]Department of EECS, Mathematics and Computational Physics Laboratory, Howard University, Washington DC 20059 USA, tgill@howard.edu; tgill@access4less.net

Abstract. In this paper, we derive a new identity that relates the proper time and the observer time as a contact transformation on configuration space. This leads to a consistent relativistic extension of the special theory to include Newtonian mechanics. The basic conclusions of this paper are:

(i) There always exist two unique inertial frames for any given system of particles: one with a unique speed of light and time relative and one with unique simultaneity and the speed of light relative.

(ii) If the ratio of the total energy of the universe to the total mass energy of the universe is constant, then a unique definition of Newtonian time exists for the universe with zero set at the big bang.

(iii) There are two mathematically consistent representations for the same physical system.

(iv) Data collected from distant objects in the universe does not have a unique physical interpretation.

We also use the isotopic dual of $\mathbb{R}$ to provide a basic improvement in the big bang model. This leads to a model with conservation of energy, momentum, angular momentum, charge, a natural arrow for time and a natural explanation for the lack of antimatter in our universe. We predict that protons and antiprotons (respectively electrons and positrons) are gravitationally repulsive.

1. Background and History

In the beginning of the 1900's, a major problem in physics was to reconcile the group transformation problem between the theories of Newton and Maxwell. At that time, it was unthinkable that Maxwell's theory had any serious flaws. Lorentz [1, 2] had recently shown that all of the macroscopic phenomena of electrodynamics and optics could be accounted for based on an analysis of the microscopic behavior of electrons and ions.

Independent of Lorentz and Poincaré, Einstein took a deeper look at the problem. He derived the Lorentz transformations from kinematical arguments, as opposed to the symmetry properties of Maxwell's equations. As Brown points out [3], Einstein took this approach because he was not sure that Maxwell's theory would survive the existence of photons. Einstein believed that the Lorentz transformations were fundamental and would survive any failures in Maxwell's theory.

(At that time, Maxwell's theory was still not accepted by all, see [4] and references therein.)

Einstein [5] observed that the constant c appears in Maxwell's equations for all inertial observers. At that time experimental information about the speed of light was meager, being restricted to macroscopic and astronomical studies. Einstein was the first to realized that a formal postulate on the velocity of light was necessary. His proposal was that all physical theories should satisfy the (now well-known) postulates of special relativity:

(1) The physical laws of nature and the results of all experiments are independent of the particular inertial frame of the observer (in which the experiment is performed).

(2) The speed of light in empty space is constant and is independent of the motion of the source or receiver.

Like those before him, Einstein realized that the concept of absolute space was neither knowable nor necessary. The first postulate abandons this idea. Einstein's second postulate forces the abandonment of the idea of absolute time.

Poincaré was singular among first class physicists, who were equally well trained and accomplished in engineering, mathematics, physics and the philosophy of science. (E. T. Bell called Poincaré "The Last Universalist", because he excelled in all these fields as they existed in his lifetime [6].) Poincaré discovered an error in Lorentz's analysis, realized that after correction the transformations form a group, which he named for Lorentz [7]. As early as 1906, Poincaré recognized that the Lorentz group could be treated as a rotation in four-dimensional space if time is treated as an imaginary coordinate and introduced the metric (proper-time) now attributed to Minkowski (see [8]). However, his insight and understanding of the difference between mathematics and physics helped him to resist the temptation to use a "physically unjustified" mathematical observation as a (necessary) tool for the representation of physical reality. This leap was later made (without physical justification and much philosophical fanfare) by Minkowski, a well-known number theorist with few accomplishments in physics and a strong belief in Hilbert's program to geometrize physics [9]. A complete discussion of Minkowski's motivation, his prior knowledge of Poincaré's work and his background in physics can be found in Walters [10]. *Thus, we make explicit Minkowski's unacknowledged postulate for the special theory of relativity*:

(3) The correct implementation of the first two postulates requires that time be treated as a fourth coordinate, and the relationship between components so constrained as to satisfy the natural invariance induced by the Lorentz group of electrodynamics, (Minkowski space).

1.1. Newtonian Mechanics

The four-geometry postulate became very popular, but Einstein, Lorentz, Poincaré, Ritz other important thinkers maintained that space and time have distinct physical properties. Einstein was the first to oppose Minkowski's postulate openly. In 1908, Einstein and Laub published two papers which offered a different approach, that was simpler and did not depend on the spacetime formalism (see [11], [12]). They argued that the spacetime formalism was too complicated and did not add any new physics.

Sommerfeld later simplified Minkowski's complicated formulation, making it easy for physicists to understand. The theory also became attractive to the mathematics community, which made it even more popular. In the excitement, no one seem to notice that the theory did not work for more then one particle and thus was far from a complete extension of Newton's theory. By the time problems in attempts to merge the special theory of relativity with quantum theory forced researchers to take a deeper look at the foundations of classical electrodynamics, the Minkowski postulate had already become sacred. By the time Einstein considered the extension of the special to the general theory, the only approach of interest was one that extended the Minkowski postulate (see Pais [13] and [14]).

Once it was accepted that the proper Newtonian theory should be invariant under the Lorentz group, work on this problem was actually ignored until after World War Two when it was realized that quantum theory did not solve the problems left open by the classical theory. A canonical center of mass is fundamental for the extension of the special theory to include a many-particle Newtonian or a many-particle quantum theory.

In classical electrodynamics, Dirac partially by-passed many of the problems by replacing particles with fields (see [15]). However, this approach led to the first example of a divergent theory (infinite self-energy). This divergency was the motivation for the Wheeler-Feynman approach to classical electrodynamics (see [16]). Their theory solved the divergency problem, but could not be used as the foundations for quantum theory. The first study of the center of mass problem was conducted by Pryce [17]. This led to the implication that the canonical center-of-mass cannot be the three-vector part of a four-vector. (This will be made explicit later.) After the investigations of Pryce, Bakamjian and Thomas [18] showed that they could construct a global interacting quantizable many-particle theory that satisfied the first two postulates of Einstein (but not Minkowski's postulate). They conjectured that Minkowski's postulate would only be compatible with free particles.

1.1.1. No-Interaction There are a two major no-interaction theorems: the first was due to Haag [19] and applies to the foundations of quantum field theory. It is often confused with the one of interest here, first proved by Currie et al [20] and shows that the Bakamjian-Thomas conjecture was correct. The theorem has since been extended to an arbitrary number of particles by Leutwyler [21]. We present the general form.

Theorem 1.1. (No-Interaction Theorem) *Let n particles have a representation $(\mathbf{x}_i, \mathbf{p}_i)$ on $6n$-dimensional phase space. Supposed that the following is satisfied:*

(i) The system has a Hamiltonian representation.

(ii) The system has a canonical representation of the Poincare group.

(iii) The each coordinate $\mathbf{x}_i$ is the vector part of a four-vector.

Then these assumptions are only compatible with free particles.

IOP Conf. Series: Journal of Physics: Conf. Series **1239** (2019) 012013 doi:10.1088/1742-6596/1239/1/012013

All attempts to keep Minkowski's postulate, avoid The No-Interaction Theorem and merge the special theory with quantum mechanics have failed.

1.2. The 2.7 °K MBR and Mach's Principle

Penzias and Wilson discovered, the 2.7 °K microwave background radiation 1965. We know that this radiation represents a unique preferred frame of rest, which exists throughout the universe and is available to all observers (see [22]). This radiation is highly isotropic with anisotropy limits set at 0.001%. Furthermore, direct measurements have been made of the velocity of our Solar System and Galaxy (370 and 600 km/sec respectively, see Peebles [23]). One can only speculate as to what impact this information would have had on the thinking of the founding fathers, Einstein, Lorentz, Poincaré, Ritz and the many other serious investigators of the early 1900's who were concerned with truly understanding the foundations of electrodynamics and mechanics.

Peebles has conjectured that the special theory is valid with or without a preferred frame of reference, so that the MBR does not violate the special theory (see [23]). This statement is not obvious, in addition general relativity predicts that at each point we can adjust our acceleration locally to find a freely falling frame where the special theory holds. In this frame, all observers with constant velocity are equivalent. Thus, according to the general theory we have an infinite family of freely falling frames. Within this context, the Penzias and Wilson findings show that, we can actually set the acceleration equal to zero.

Mach's principle was originally conceived to show that there is no difference between the rotation of the earth with respect to the fixed stars or the fixed stars with respect to the earth. The MBR shows that the fixed stars are not needed and Newton's bucket experiment does gives the correct view.

2. Global World View

For the global world view, we begin with an interacting system of n-particles, observed in an inertial frame by O. We assume that O is able to identify the ith particle with coordinates $(\mathbf{x}_i, t)$. We assume that the natural position variable representing the system as a whole is the center of mass position vector. Pryce, found that there are three possible definitions for this position vector. However, only one of them is canonical and independent of the frame in which it is defined. This is the natural and necessary choice if we want a theory that provides the same physics to all observers and is compatible with quantum mechanics. Before discussing this vector, we want to construct a unique clock for each particle and one for the system as a whole.

2.0.1. One-Particle Clock If $\mathbf{w}_i$ is the velocity of particle i, let $\gamma^{-1}(\mathbf{w}_i) = \sqrt{1 - \mathbf{w}_i^2/c^2}$. The proper time of the ith particle is defined by:

$$d\tau_i = \gamma^{-1}(\mathbf{w}_i)dt, \quad \mathbf{w}_i = \frac{d\mathbf{x}_i}{dt}, \quad d\tau_i^2 = dt^2 - \tfrac{1}{c^2}d\mathbf{x}_i^2. \tag{2.1}$$

We can rewrite the last term to get:

$$dt^2 = d\tau_i^2 + \tfrac{1}{c^2}d\mathbf{x}_i^2, \Rightarrow cdt = \left(\sqrt{\mathbf{u}_i^2 + c^2}\right) d\tau_i, \quad \mathbf{u}_i = \frac{d\mathbf{x}_i}{d\tau_i} = \gamma(\mathbf{w}_i)\mathbf{w}_i. \tag{2.2}$$

If we let $b_i = \sqrt{\mathbf{u}_i^2 + c^2}$, the second term in equation (2.2) becomes $cdt = b_i d\tau_i$. This leads to our first identity:

$$\frac{1}{c}\frac{d}{dt} \equiv \frac{1}{b_i}\frac{d}{d\tau_i} \tag{2.3}$$

This identity provides the correct way to define the relationship between the proper time and the observer time. This represents a contact transformation on configuration space, also known as a tangency transformation ([24]). We show in [4], that it is a canonical transformation which leaves phase space invariant. We note that, while the left hand-side of (2.3) is only defined in an inertial frame, the right hand side is well-defined when $\dot{u}_i \neq 0$. If we apply the identity to $\mathbf{x}_i$, we obtain our second identity:

$$\frac{\mathbf{w}_i}{c} = \frac{1}{c}\frac{d\mathbf{x}_i}{dt} \equiv \frac{1}{b_i}\frac{d\mathbf{x}_i}{d\tau_i} = \frac{\mathbf{u}_i}{b_i}.$$

The particle coordinates transform from $(\mathbf{x}_i, t) \to (\mathbf{x}_i, \tau_i)$, while configuration space transforms from $(\mathbf{x}_i, \mathbf{w}_i) \to (\mathbf{x}_i, \mathbf{u}_i)$. In the new representation, $\mathbf{x}_i$ is uniquely defined relative to O, while τ_i is uniquely defined relative to the ith particle. Using $\gamma(\mathbf{w}_i) = H_i/m_i c^2$, we can also write $d\tau_i = \left(m_i c^2/H_i\right) dt$. The ith particle momentum remains invariant, since $\mathbf{p}_i = m_i\gamma(\mathbf{w}_i)\mathbf{w}_i = m_i\mathbf{u}_i$, where m_i is the rest mass. (Thus, phase space is left invariant.)

2.0.2. Many-Particle Clock To construct the many-particle clock, we assume the system is closed, $n > 1$ with particle proper clocks τ_i and with obvious notation, Hamiltonians $H_i = \sqrt{c^2\left(\mathbf{p} - \frac{e}{c}\mathbf{A}\right)^2 + m_i^2c^4} + V_i(\mathbf{x}_i)$. We defined the total Hamiltonian by $H = \sum_{i=1}^n H_i$, define the effective mass M and total momentum $\mathbf{P}$ by

$$Mc^2 = \sqrt{H^2 - c^2\mathbf{P}^2}, \quad \mathbf{P} = \sum_{i=1}^{n} \mathbf{p}_i.$$

We can then represent H as $H = \sqrt{c^2\mathbf{P}^2 + M^2c^4}$.

The canonical position of the center of mass $\mathbf{X}$, for a system of particles in the O frame is defined by (see [25]):

$$\mathbf{X} = \frac{1}{H}\sum_{i=1}^{n} H_i\mathbf{x}_i + \frac{c^2\left(\mathbf{S}\times\mathbf{P}\right)}{H\left(Mc^2 + H\right)}, \tag{2.4}$$

where $\mathbf{S}$ is the global spin of the system of particles relative to O. (It is clear that (2.4) cannot represent the vector part of a four-vector.) If there is no interaction, S, H and M are constant, with no implicit dependence on the $\{\mathbf{x}_i\}$ variables, so that:

$$\{X_i, X_j\} = \sum_{k=1}^{n} \frac{\partial X_i}{\partial \mathbf{p}_k}\cdot\frac{\partial X_j}{\partial \mathbf{x}_k} - \frac{\partial X_j}{\partial \mathbf{p}_k}\cdot\frac{\partial X_i}{\partial \mathbf{x}_k} \equiv 0.$$

However, when interaction is present, S, H and M may all depend implicitly on the $\{\mathbf{x}_i\}$ variables, so that in general $\{X_i, X_j\} \neq 0$. Since $\mathbf{X}$ is the canonical conjugate of $\mathbf{P}$, it precisely what we need for a consistent merge with quantum mechanics.

Let $\mathbf{V}$ be the velocity of $\mathbf{X}$ with respect to O. It follows that H also has the representation $H = Mc^2\gamma(\mathbf{V})$, so that $\gamma(\mathbf{V})^{-1} = (Mc^2/H)$. In this representation, we see that $d\tau = \gamma(\mathbf{V})^{-1}dt = (Mc^2/H)dt$ does not depend on the number of particles in the system. It follows that, as long as Mc^2/H is fixed, τ is invariant, so that the number of particles n, can increase or decrease without changing τ. (This means that number n is not conserved and, in some cases of physical interest, may even be a integer-valued random variable).

From $dt^2 = d\tau^2 + d\mathbf{X}^2/c^2$, we see that $(\mathbf{U} = d\mathbf{X}/d\tau)$

$$c^2dt^2 = \left(c^2 + \mathbf{U}^2\right) d\tau^2 \quad \Rightarrow \quad cdt = \left(\sqrt{c^2 + \mathbf{U}^2}\right) d\tau.$$

It is easy to see that $\mathbf{U} = \gamma(\mathbf{V})\mathbf{V}$, so that $\mathbf{U}$ is constant. If we define $b = \sqrt{\mathbf{U}^2 + c^2}$, we can write $cdt = bd\tau$. Since b is constant we have: $ct = b\tau$.

For any other observer O' the same system has center of mass velocity $\mathbf{V}'$ and, by the same calculations, we obtain $ct' = b'\tau$, where $b' = \sqrt{\mathbf{U}'^2 + c^2}$.

Theorem 2.1. *Given any closed system of interacting particles, there always exists two distinct sets of inertial frame coordinates for each observer, to describe the particles in the system and the system as a whole. The following is true:*

(i) In one frame, the speed of light is an invariant upper bound and time is relative, while in the other, time is invariant and the speed of light b, is relative with no upper bound.

(ii) For the whole system and for each particle, the equations of motion are mathematically equivalent.

Proof. Observer O has $(\{\mathbf{x}_i\}_{i=1}^n, t)$ (respectively $(\mathbf{X}, t)$) and, since b is constant, he also has $(\{\mathbf{x}_i\}_{i=1}^n, \tau)$ (respectively $(\mathbf{X}, \tau)$) to study the same system of particles. The same is also true for observer O', since b' is constant. This proves the first part. To prove (1), we need only note that, in general, $b \neq b'$ and each is $\geq c$. A proof of (2) can be found in [4]. □

This theorem shows that a unique (operational) measure of time is available to all observers. From the definition of τ, we obtain our third identity:

$$\frac{1}{c}\frac{d}{dt} \equiv \frac{1}{b}\frac{d}{d\tau} \equiv \frac{1}{b_i}\frac{d}{d\tau_i} \tag{2.5}$$

Let $\mathbf{W}$ be the relative velocity between observer O and O'. Since τ is the same for both we only need the relationship between the two scale factors b and b' to satisfy the first postulate. Starting with $t' = \frac{b'}{c}\tau = \gamma(\mathbf{W})\left(\frac{b}{c}\tau - \mathbf{X}\cdot\mathbf{V}/c^2\right)$, we see that, since $\mathbf{U} = (\mathbf{X}/\tau)$, we get:

$$b' = \gamma(\mathbf{W})\left[b - (\mathbf{X}/\tau)\cdot(\mathbf{W}/c)\right] = \gamma(\mathbf{W})\left(b - \mathbf{U}\cdot\mathbf{W}/c\right).$$

A similar calculation shows that $b = \gamma(\mathbf{W})\left(b' + \mathbf{U}'\cdot\mathbf{W}/c\right)$. This shows that each observer can have direct access to all information available to any other observer once they know their relative velocities. (Thus the first postulate of Einstein is satisfied.)

Definition 2.2. *A theory is said to be Einsteinian if at least one representation exists, which satisfies the two postulates of the special theory.*

This definition allows each observer an additional degree of freedom, while the difference between either choice is a constant change of scale.

Remark 2.3. *For many experiments (e. g., high energy particle studies) the center of mass is the natural frame of choice. In this case, $t = \tau$ and one has a constant speed of light for all events associated with the system of (interacting) particles.*

This distinction may also prove important in the future, because there continues to appear research in cosmology, applied physics and engineering, suggesting that the constant c is not an upper bound in all cases (see for example [26, 27, 28, 29]).

Theorem 2.4. *Any closed system of particles is Einsteinian and independent of the Minkowski postulate.*

Since we do not require the particle coordinates to transform as four-vectors (manifest Lorentz invariance), the no-interaction theorem does not apply.

2.0.3. Cannonical Hamiltonian If we let $\mathbf{L}$ be the boost (generator of pure Lorentz transformations) and define the total angular momentum $\mathbf{J}$ by

$$\mathbf{J} = \sum_{i=1}^{n} \mathbf{x}_i \times \mathbf{p}_i,$$

we then have the following Poisson Bracket relations characteristic of the Lie algebra for the Poincaré group (when we use the observer clock):

$$\frac{d\mathbf{P}}{dt} = \{H, \mathbf{P}\} = \mathbf{0} \qquad \frac{d\mathbf{J}}{dt} = \{H, \mathbf{J}\} = \mathbf{0} \qquad \{P_i, P_j\} = 0$$

$$\{J_i, P_j\} = \varepsilon_{ijk} P_k \qquad \{J_i, J_j\} = \varepsilon_{ijk} J_k \qquad \{J_i, L_j\} = \varepsilon_{ijk} L_k$$

$$\frac{d\mathbf{L}}{dt} = \{H, \mathbf{L}\} = -\mathbf{P} \qquad \{P_i, L_j\} = -\delta_{ij} H/c^2, \qquad \{L_i, L_j\} = -\varepsilon_{ijk} J_k/c^2.$$

It is easy to see that M commutes with H, $\mathbf{P}$, and $\mathbf{J}$, and to show that M commutes with $\mathbf{L}$.

Since we desire complete compatibility with quantum theory, it is natural to require that any change from the observer clock to the local clock of the observed system be a canonical change of variables. The key concept to our approach may be seen by examining the time evolution of a dynamical parameter $W(\mathbf{X}, \mathbf{P})$, via the standard formulation of classical mechanics, described in terms of the Poisson brackets:

$$\frac{dW}{dt} = \{H, W\}. \tag{2.6}$$

We can also represent the dynamics using the proper (or local) time of the system by using the representation $d\tau = (1/\gamma)dt = (Mc^2/H)dt$, so that:

$$\frac{dW}{d\tau} = \frac{dt}{d\tau}\frac{dW}{dt} = \frac{H}{Mc^2}\{H, W\}.$$

IOP Conf. Series: Journal of Physics: Conf. Series **1239** (2019) 012013 doi:10.1088/1742-6596/1239/1/012013

Since Mc^2 is a well-defined (invariant) rest energy for the system, we determine the canonical proper-time Hamiltonian K such that:

$$\{K, W\} = \frac{H}{Mc^2}\{H, W\}, \quad K|_{\mathbf{p}=0} = H|_{\mathbf{P}=0} = Mc^2.$$

Using

$$\begin{aligned}\{K, W\} &= \left[\frac{H}{Mc^2}\frac{\partial H}{\partial \mathbf{P}}\right]\frac{\partial W}{\partial \mathbf{X}} - \left[\frac{H}{Mc^2}\frac{\partial H}{\partial \mathbf{X}}\right]\frac{\partial W}{\partial \mathbf{P}} \\ &= \frac{\partial}{\partial \mathbf{p}}\left[\frac{H^2}{2Mc^2} + a\right]\frac{\partial W}{\partial \mathbf{X}} - \frac{\partial}{\partial \mathbf{X}}\left[\frac{H^2}{2Mc^2} + a'\right]\frac{\partial W}{\partial \mathbf{P}},\end{aligned}$$

we get that $a = a' = \frac{1}{2}Mc^2$, so that (assuming no explicit time dependence)

$$K = \frac{H^2}{2Mc^2} + \frac{Mc^2}{2}, \quad \text{and} \quad \frac{dW}{d\tau} = \{K, W\}.$$

Thus, we can use the same definitions for $\mathbf{P}$, $\mathbf{J}$, and $\mathbf{L}$ to obtain our new commutation relations:

$$\frac{d\mathbf{P}}{d\tau} = \{K, \mathbf{P}\} = \mathbf{0}, \quad \frac{d\mathbf{J}}{d\tau} = \{K, \mathbf{J}\} = \mathbf{0}, \quad \{P_i, P_j\} = 0,$$

$$\{J_i, P_j\} = \varepsilon_{ijk}P_k, \quad \{J_i, J_j\} = \varepsilon_{ijk}J_k, \quad \{J_i, L_j\} = \varepsilon_{ijk}L_k,$$

$$\frac{d\mathbf{L}}{d\tau} = \{K, \mathbf{L}\} = \frac{-H}{Mc^2}\mathbf{P}, \quad \{P_i, L_j\} = -\delta_{ij}H/c^2, \quad \{L_i, L_j\} = -\varepsilon_{ijk}J_k/c^2.$$

It follows that, except for a constant scale change, the inhomogeneous proper-time group is generated by the same Lie algebra as the Poincaré group. This result is not surprising given the close relation between the two groups. Thus, the form of K is fully relativistic.

3. Applications

3.1. Newtonian Time

In his work, Newton assumed there exists a unique observer-independent measure of time for the universe.

Theorem 3.1. *Suppose that the observable universe is representable in the sense that the observed ratio of mass energy to total energy is constant and independent of our observed portion of the universe. Then the universe has a unique clock that is available to all observers.*

Proof. Let O be any observer anywhere in the universe. From his vantage point, H/Mc^2 is constant, so that he can define τ via $ct = b\tau$. Since all other observers can also define τ, the proof follows. □

Remark 3.2. *The assumption of Theorem 3.1 is equivalent to the standard assumption that the universe is homogenous and isotropic.*

Since all observers can transform to a frame at rest with respect to the 2.7° K mbr, if they did, all clocks would run at the same rate. In this case, any two observers view of the universe would differ by (at most) a translation and rotation. If in addition, the big bang is true, it defines a zero point for time, making the clock attached to the 2.7° K mbr unique.

Theorem 3.3. *(Peebles) Suppose all observers choose a frame that is at rest with respect to the 2.7 °K microwave background radiation, then all the laws of physics will be invariant and not just covariant with respect to Lorentz transformations.*

In the study of physical systems one is sometimes not interested in the behavior of the global system, but only in some subsystem. The cluster decomposition property is a requirement of any theory purporting to be a possible representation of the real world. Basically this is the property that, if any two or more subsystems become widely separated, then they may be treated as independent systems (clusters).

Theorem 3.4. *Suppose the system of particles can be decomposed into two or more clusters. Then there exists a unique (local) clock and corresponding canonical Hamiltonian for each cluster.*

Proof. We assume that the subsystems are sufficiently separated that all observers can agree that they are distinct. In this case, each observer can identify effective masses M_1, M_2 and Hamiltonians H_1, H_2. It follows that $d\tau_1 = [(M_1c^2)/H_1]dt$ and $d\tau_2 = [(M_2c^2)/H_2]dt$, so that each observer can construct a local-time theory for each cluster. □

Actually, the theorem is true without the assumption that the systems are weakly interacting. This makes the theorem less difficult to apply than the various phenomenological approaches, which require both model justification and consistency analysis prior to use. This theorem also allows us to prove a weaker version of Theorem (3.1), in the sense that we replace the assumption of homogeneity and isotropy of the energy and mass density for a possible infinite universe by finite energy and mass density for a possibly inhomogeneous universe.

Theorem 3.5. *Suppose the universe has finite mass and energy density and that each observer can choose a local inertial frame from which his/her region of the universe is at rest relative to the observed system. Then there exists a unique proper clock for the universe.*

Proof. Applying the cluster decomposition theorem, our observer can identify masses M_1 for his/her region of the universe and M_2 for the complement region, along with Hamiltonians H_1 and H_2. It follows that $H = H_1 + H_2$, $M = M_1 + M_2$ and $d\tau = [(Mc^2)/H]dt$ define the total mass, Hamiltonian and proper clock for the universe. We can now construct our canonical proper-time Hamiltonian K. Since M and H are fixed, and invariant for all observers, we see that both K and τ are unique and invariant for all observers. □

Remark 3.6. *It should be remarked that M_i and H_i, $i = 1, 2$, will vary with observers, reflecting the non-uniqueness of inertial frames.*

3.2. The Big Bang

The currently accepted cosmological model for the universe assumes that our universe began around 13.8 billion years ago with a big bang. From a minimalist point of view, this model should be consistent with our current experimentally obtained information about the universe:

(i) Whenever an antiparticle is observed in experiment, we always find that it is also accompanied by a particle.

IOP Conf. Series: Journal of Physics: Conf. Series **1239** (2019) 012013 doi:10.1088/1742-6596/1239/1/012013

(ii) Whenever an interaction is observed, a complete analysis always shows that momentum, angular momentum, charge and energy are conserved.

In this section, we suggest a slight alteration of the beginning, which makes the big bang event consistent our experimental understanding of the world.

We first revisit our conceptual view of the real numbers and their representation. Recall that a field is a set A that has two binary operations $\oplus$ and $\odot$ that satisfies all our common experience with real numbers. Formally:

Definition 3.7. *The real numbers is a triplet* $(\mathbb{R}, +, \cdot)$, *which is a field, with* 0 *as the additive identity (i.e.,* $a + 0 = a$ *for all* $a \in \mathbb{R}$*) and* 1 *as the multiplicative identity (i.e.,* $a \cdot 1 = a$ *for all* $a \in \mathbb{R}$*).*

This structure was developed by mathematicians without regard to its possible use in physics. As a consequence the asymmetry in this structure went unnoticed and physicists used it without investigation until Santilli [30] defined the isodual number field. His definition is more general. For our purpose, we only need the following.

Definition 3.8. *The isodual real numbers* $(\hat{\mathbb{R}}, +, *)$ *is a field, with* $0 = \hat{0}$ *as the additive identity (i.e.,* $\hat{a} + \hat{0} = \hat{a}$ *for all* $-a = \hat{a} \in \hat{\mathbb{R}}$*) and* $-1 = \hat{1}$ *as the multiplicative identity (i.e.,* $\hat{a} * \hat{1} = \hat{a}$ *for all* $\hat{a} \in \hat{\mathbb{R}}$*).*

Example 3.9. *A simple example from quantum theory is the following: the evolution of particle is defined on a Hilbert space* $\mathcal{H}$ *over the complex numbers* $\mathbb{C} = \mathbb{R} + i\mathbb{R}$, *with Hamiltonian* H *by the equation:*

$$i\hbar \frac{\partial \psi}{\partial t} = H\psi.$$

The conjugate equation is:

$$-i\hbar \frac{\partial \psi^*}{\partial t} = H\psi^*.$$

If we use $\hat{\mathbb{C}}$ *as our number field, we can write the above equation as:*

$$i * \hat{\hbar} * \frac{\partial \psi^*}{\partial \hat{t}} = \hat{H} * \psi^*$$

Thus, this approach allows us to naturally view anti-particles as time reversed particles, with their evolution define on $\mathcal{H}^*$ *over* $\hat{\mathbb{C}}$.

We note that we can obtain the isodual of any physical quantity $\hat{A}$ from the equation $A + \hat{A} = 0$.

Remark 3.10. *Santilli [30] has shown that charge conjugation and isoduallity are equivalent for the particle-antiparticle symmetry operation. However, use of isoduallity allows us to view the existence of antimatter and charge conservation as fundamental aspects of the universe, while also explaining why large amounts of antimatter is not found in this universe.*

If our view of the beginning is correct, we would predict that the proton and antiproton (respectively electron and positron) should be gravitationally repulsive.

In the diagram below, we provide a new picture of the big bang beginning. In this case, two universes are created, one going forward in τ_N and one going backward in τ_N (Newtonian time), relative to our reality.

$$\xleftarrow{\hat{\tau}_N} \begin{matrix} \nwarrow & \tau_N=0 & \nearrow \\ \diagdown & | & \diagup \\ \swarrow & & \searrow \end{matrix} \xrightarrow{\tau_N} .$$

This view has the following advantages: we obtain

(i) a natural arrow for time, with a zero initial point.

(ii) a natural explanation for the lack of antimatter in this universe.

(iii) antiparticles as particles moving backward in (Newtonian) proper time.

(iv) conservation of charge, energy, linear and angular momentum.

It is important to be clear that our assumption does not imply that there are any other symmetries or necessary similarities between the two universes.

Conclusion

We have shown that the correct relationship between the proper time and observer time is a contact transformation on configuration space. This approach is consistent with the first two postulates of Einstein, Peebles conjecture and Newton's universal time. We have also shown that, by introducing a symmetric view of the number line, we are able to introduce an improved version of the big bang, which explains the lack antimatter in the universe, introduces an the arrow for time and, provides conservation of energy, linear and angular momentum.

The most important conclusion from this investigation is embodied in Theorem 2.1: "given any closed system of interacting particles, there always exists two distinct sets of coordinates for each observer to describe the particles in the system and the system as a whole". Thus, the physical interpretation of experimental data is not unique. In order to make this above statement explicit in a very powerful manner, recall that many measurements are based on the dimensionless ratio $\beta = \mathbf{w}/c$. However, $\mathbf{w}/c \equiv \mathbf{u}/b$, so we see that measurements of velocity and the speed of light for distant objects are totally ambiguous. For example, ([31], pp. 556-561), the red shift factor z, used to determined distances in astronomy, is defined by:

$$z = \sqrt{\frac{1+\frac{\mathbf{w}}{c}}{1-\frac{\mathbf{w}}{c}}} - 1 \equiv \sqrt{\frac{1+\frac{\mathbf{u}}{b}}{1-\frac{\mathbf{u}}{b}}} - 1.$$

We thus conclude that distant objects may have much higher velocities and light may have velocities higher then c, from the same data, without any contradiction.

A known problem with the 2.7 °K MBR is that there is no known way for the early universe to instantly reach equilibrium following the big bang. One solution is to introduce the inflationary hypothesis. It has many critics and other approaches have been suggested. The varying speed of light assumption introduced by Moffet [32] is consistent with the use of $b = \sqrt{U^2 + c^2}$ in the beginning (see also Magueijo [33] and references there in). As time progresses and after equilibrium, U can slow down to zero, while b reduces to c.

Acknowledgements

The author would like to thank Professors Elliott Lieb and Larry Horwitz for a number of helpful suggestions. In addition many thanks to the referees for idenifying errors of fact, omission and oversight.

This paper is an extended version of a talk given at the The 11th Biennial Conference on Classical and Quantum Relativistic Dynamics of Particles and Fields held in Merida, Mexico, June 2018.

References

[1] H. A. Lorentz, *Archives Neerlandaises des Sciences Exactes et Naturelles*, **25**, 353 (1892).
[2] H. A. Lorentz, *The Theory of Electrons* B. G. Teubner, Leipzig, 1906; (reprinted by Dover, New York, 1952).
[3] H.R Brown, Eur. J. Phys. **26**, S85 (2005).
[4] T. L. Gill and W. W. Zachary, *Two Mathematically Equivalent Versions of Maxwell's Equations*, Foundations of Physics, DOI 10.1007/s10701-009-9331-8, (2011).
[5] A. Einstein, Ann. d. Phys. **17**, 891 (1905).
[6] E. T. Bell, *Men of Mathematics* , Simon & Schuster, New York, (1937).
[7] H. Poincaré, C.R. Acad. Sci. Paris **140**, 1504 (1905).
[8] H. Poincaré, *Sur la dynamique de l'électron* , Rendiconti del Circolo matematico Rendiconti del Circolo di Palermo, **21**, 129-176 (1906).
[9] H. Minkowski, Physikalische Zeitschrift **10**,104 (1909).
[10] S. Walters, In: Gonner, H., Renn, J., Ritter, J. (eds.) *The Expanding Worlds of General Relativity.* Einstein Studies, vol. 7, pp. 45-86. Birkhäuser, Boston (1999).
[11] A. Einstein and J Laub, *Über die elektromagnetischen Grundgleichugen für bewegte Köper*, Ann. d. Phys. **33**(8), 532-540 (1908).
[12] A. Einstein and J Laub, *Über die im elektromagnetischen Felde auf ruhende Köper ausgeübten ponderomotorischen Kräfte*, Ann. d. Phys. **331**(8), 551-550 (1908).
[13] A. Pais, *Subtle is the lord: the science and life of Albert Einstein*, Oxford University Press (1982).
[14] A. Einstein and M. Grossmann, *Entwurf einer verallgemeinerten Relativitätstheorie und einer Theorie der Gravitation*, Leipzig, Berlin: Teubner, (1913).
[15] P. A. M. Dirac, *Classical theory of radiating electrons*, Proceedings of the Royal Soc. of London A **167** (1938) 148-168.
[16] J.A. Wheeler and R.P. Feynman, *Interaction with the absorber as the mechanism of radiation* Rev. Mod. Phys. **17** (1949) 157-181.
[17] M.H.L. Pryce, Proc. Roy. Soc. London A **195**, 400 (1948).
[18] B. Bakamjian and L. H. Thomas, Phys. Rev. **92**, 1300 (1953).
[19] R. Haag, K. Dan. Vidensk. Silsk. Mat. Fys. Medd. **29** 12 (1955).
[20] D. G. Currie, T. F. Jordan, and E. C. G. Sudarshan, Rev. Mod. Phys. **35**, 350 (1963).
[21] H. Leutwyler. *A no-interaction theorem in classical relativistic hamiltonian particle mechanics*, Nuovo Cim., **37**, 556 (1965).
[22] A. A. Penzias and R. W. Wilson, Ap. J. **142**, 419 (1965).
[23] P.J.E. Peebles, *Principles of Physical Cosmology*, Princeton University Press, London (1993).
[24] E. T. Whittaker, *A Treatise on the Analytical Dynamics of Particles and rigid bodies*, Cambridge, University Press, Cambridge (1917).
[25] G. Longhi, L. Lusanna, and J. M. Pons, J. Math. Phys. 30, 1893 (1989).
[26] B. M. Bolotovskii and V. L. Ginzburg, *The Vavilov-Cerenkov effect and the Doppler effect in the motion of sources with superluminal velocity in vacuum*, SOVPU **15**, 184-192 (1972).

[27] J. Singleton, A. Ardavan, H. Ardavan, J. Fopma, D. Halliday and W. Hayes *Experimental demonstration of emission from a superluminal polarization current, a new class of solid-state source for MHz-THz and beyond*, IEEE Digest **04EX857**, 591-592 (2004).

[28] B. M. Bolotovskii and A. V. Serov, *Radiation of superluminal sources in empty space*, Phys. Usp. **48** 903-915 (2005).

[29] A. V. Bessarab, A. A. Gorbunov, S. P. Martynenko and N. A. Prudkoy, *Faster-than-light EMP source initiated by short X-ray pulse of laser plasma*, IEEE Trans. Plasma Sci. **32**, 1400-1403 (2004).

[30] R. M. Santilli, *Isonumbers and genonumbers of dimension 1, 2, 4, 8, their isoduals, and pseudoduals, and "hidden numbers" of dimension 3, 5, 6, 7*, Algebras, Groups and Geometries, Vol. **10**, 273-322 (1993).

[31] C. Payne-Gaposchkin and K. Haramundanis, *Introduction to Astronomy*, Prentice-Hall Inc., Englewood Cliffs, NJ, second Edition, (1970).

[32] J. Moffat, *Superluminary Universe: A Possible Solution to the Initial Value Problem in Cosmology*, J. Mod. Phys. D. **2**(3) 351-366 (1988).

[33] J. Magueijo, *Covariant and locally Lorentz-invariant varying speed of light theories*, Phys. Rev. **D62**(10) 10352 (2000).

An Elementary Canonical Classical and Quantum Dynamics for General Relativity

L. P. Horwitz
School of Physics, Tel Aviv University, Ramat Aviv 69978, Israel
Department of Physics, Bar Ilan University, Ramat Gan 52900, Israel
Department of Physics, Ariel University, Ariel 40700, Israel
E-mail: larry@tauex.tau.ac.il

Abstract. A consistent canonical classical and quantum dynamics in the framework of special relativity was formulated by Stueckelberg in 1941, and generalized to many body theory by Horwitz and Piron in 1973 (SHP). In this paper, using local coordinate transformations, following the original procedure of Einstein, this theory is embedded into the framework of general relativity (GR) both for potential models (where the potential appears as a spacetime mass distribution with dimension of mass) and for electromagnetism (emerging as a gauge field on the quantum mechanical Hilbert space). The canonical Poisson brackets of the SHP theory remain valid (invariant under local coordinate transformations) on the manifold of GR, and provide the basis, following Dirac's quantization procedure, for formulating a quantum theory. The theory is developed both for one and many particles.

Keywords: Relativistic dynamics, General relativity, Quantum theory on curved space, $U(1)$ gauge, many body theory in general relativity.
Classification PACS: 03.30.+p, 03.65.-w, 04.20 Cr, 04.60 Ds, 04.90.+e

1. Introduction

Einstein's procedure to obtain general relativity was to assume that a *free* particle (event) with coordinates $\{\xi^\mu\}$ in Minkowski spacetime travels on a worldline described by (timelike, metric $(-,+,+,+)$)

$$ds^2 = -\eta_{\mu\nu} d\xi^\mu d\xi^\nu, \tag{1}$$

where ds is the interval of proper time, so that the "proper velocity" satisfies

$$\dot\xi^\mu \dot\xi^\nu \eta_{\mu\nu} = -1 \tag{2}$$

and that for no forces acting

$$\ddot\xi^\mu = 0. \tag{3}$$

He then assumed that this motion is in a locally flat tangent space of a manifold with nontrivial metric and connection form with coordinates $\{x^\mu\}$; it then follows from (2) and (3), by transformation of coordinates, that the physical motion follows geodesic curves, providing the theory of general relativity as we know it. The structure of the coordinates, to be determined by solutions of the Einstein equations, was obtained by the construction of a second rank geometrical tensor set equal to the energy momentum tensor.

IOP Conf. Series: Journal of Physics: Conf. Series **1239** (2019) 012014 doi:10.1088/1742-6596/1239/1/012014

We study here this structure in case the motion in the locally flat coordinate system follows the relativistic canonical Hamiltonian dynamics of Stueckelberg, Horwitz and Piron (SHP) [1] with world scalar potential. The theory was originally formulated for a single particle by Stueckelberg in 1941 [?]. Stueckelberg envisaged the motion of a particle along a world line in spacetime that can curve and turn to flow backward in time, resulting in the phenomenon of pair annihilation in classical dynamics. The world line was then described by an invariant monotonic parameter τ. The theory was generalized by Horwitz and Piron in 1973 [3] (see [1])to be applicable to many body systems by assuming that the parameter τ is universal (as for Newtonian time [4]), enabling them to solve the two body problem classically, and later, a quantum solution was found by Arshanksy and Horwitz [5], both for bound state and scattering theory [6].

Performing a coordinate transformation to general coordinates, along with the corresponding transformation of the momenta (the cotangent space of the original Minkowski manifold), we obtain the SHP theory in a curved space of general coordinates and momenta with a canonical Hamilton-Lagrange (symplectic) structure; the method is also generalized here to a $U(1)$ Abelian gauge theory. This procedure provides a fundamental derivation of the framework assumed by Horwitz, Gershon and Schiffer [7] in their discussion[1] of the Bekenstein-Sanders fields [8] introduced into the TeVeS theory of Bekenstein and Milgrom [9], a geometrical way of obtaining the MOND theory introduced by Milgom [10] to explain the rotation curves of galaxies.

Birrell and Davies [11] have discussed fields on curved spacetime, and considerable progress has been made, as discussed by Poisson [12] in the formulation of Hamiltonian dynamics of such dynamical fields using Lagrangian functionals associated with the curvature of spacetime. The approach used in this paper is fundamentally different in that it studies a canonical dynamics (both Hamiltonian and Lagrangian) of *particles* on a curved spacetime.

This method is applied also to the many body case, for which the SHP Hamiltonian is a sum of terms quadratic in four momentum with a many body potential term. Each particle is assumed to move locally in a flat Minkowski space, the tangent space of the general manifold of motions at that point; these local motions can then be mapped at each point x^μ by coordinate transformation into the curvilinear coordinates reflecting the curvature induced by the Einstein equations.

Throughout most of our development, we assume a τ independent background gravitational field; the local coordinate transformations from the flat Minkowski space to the curved space are taken to be independent of τ, consistently with an energy momentum tensor that is τ independent. In a more dynamical setting, when the energy momentum tensor depends on τ, the spacetime evolves nontrivially in τ; the transformations from the local Minkowski coordinates to the curved space coordinates then depend on τ. We discuss this situation in an Appendix; many of the results for the τ independent case remain (such as the Poisson bracket relations), but the geodesic type equations, for example, are modified.

We also discuss the quantum theory that emerges from a quantization of this structure, and the electromagnetic theory for both single particle and many body systems.

2. Single particle in external potential

We write the SHP Hamiltonian as

$$K = \frac{1}{2M}\eta^{\mu\nu}\pi_\mu\pi_\nu + V(\xi) \tag{4}$$

where $\eta^{\mu\nu}$ is the flat Minkowski metric $(- + ++)$ and π_μ, ξ^μ are the spacetime canonical momenta and coordinates in the local tangent space, following Einstein's use of the equivalence

[1] A non-Abelian gauge was discussed there, and then an Abelian limit was taken, leaving a term that could cancel caustic singularities.

IOP Conf. Series: Journal of Physics: Conf. Series **1239** (2019) 012014 doi:10.1088/1742-6596/1239/1/012014

principle. The existence of a potential term (which may be Lorentz invariant), representing non-gravitational forces, implies that the "free fall" condition is replaced by a local dynamics carried along by the free falling system (an additional force acting on the particle within the "elevator" according to the coordinates in the tangent space)).

The canonical equations are

$$\dot{\xi}^\mu = \frac{\partial K}{\partial \pi_\mu} \qquad \dot{\pi}_\mu = -\frac{\partial K}{\partial \xi^\mu} = -\frac{\partial V}{\partial \xi^\mu}, \tag{5}$$

where the dot here indicates $\frac{d}{d\tau}$, with τ the invariant universal "world time". Since then

$$\dot{\xi}^\mu = \frac{1}{M}\eta^{\mu\nu}\pi_\nu, \tag{6}$$

$$\text{or} \quad \pi_\nu = \eta_{\nu\mu}M\dot{\xi}^\mu, \tag{7}$$

the Hamiltonian can then be written as

$$K = \frac{M}{2}\eta_{\mu\nu}\dot{\xi}^\mu\dot{\xi}^\nu + V(\xi). \tag{8}$$

We now transform the local coordinates (contravariantly) according to the diffeomorphism

$$d\xi^\mu = \frac{\partial \xi^\mu}{\partial x^\lambda}dx^\lambda \tag{9}$$

to attach small changes in ξ to corresponding small changes in the coordinates x on the curved space, so that

$$\dot{\xi}^\mu = \frac{\partial \xi^\mu}{\partial x^\lambda}\dot{x}^\lambda. \tag{10}$$

The Hamiltonian then becomes

$$K = \frac{M}{2}g_{\mu\nu}\dot{x}^\mu\dot{x}^\nu + V(x), \tag{11}$$

where $V(x)$ is the potential at the point ξ corresponding to the point x (actually the function $V(\xi)$ could be labeled $V_x(\xi)$, a function of ξ in a small neighborhood of the point x), and

$$g_{\mu\nu} = \eta_{\lambda\sigma}\frac{\partial \xi^\lambda}{\partial x^\mu}\frac{\partial \xi^\sigma}{\partial x^\nu} \tag{12}$$

Since V has significance as the source of a force in the local frame only through its derivatives, we can make this pointwise correspondence with a globally defined function $V(x)$.[2] We shall assume in most of the work of this paper that the geometric structure does not depend on τ, and is concerned with the study of the covariant dynamical evolution of a system in a background gravitational field. We study the case of a τ dependent metric in the Appendix.

The corresponding Lagrangian is then

$$L = \frac{M}{2}g_{\mu\nu}\dot{x}^\mu\dot{x}^\nu - V(x), \tag{13}$$

[2] Since $V(x)$ has dimension of mass, one can think of this function as a scalar mass field, reflecting forces acting in the local tangent space at each point. It may play the role of "dark energy" [7]. If $V = 0$, our discussion reduces to that of the usual general relativity, but with a well-defined canonical momentum variable.

IOP Conf. Series: Journal of Physics: Conf. Series **1239** (2019) 012014 doi:10.1088/1742-6596/1239/1/012014

In the locally flat coordinates in the neighborhood of x^μ, the symplectic structure of Hamiltonian mechanics (*e.g.* da Silva [13]) implies that the momentum π_μ, lying in the cotangent space of the manifold $\{\xi^\mu\}$, transforms covariantly under the local transformation (9), *i.e.*, as does $\frac{\partial}{\partial \xi^\mu}$, so that we may define

$$p_\mu = \frac{\partial \xi^\lambda}{\partial x^\mu} \pi_\lambda. \tag{14}$$

This definition is consistent with the transformation properties of the momentum defined by the Lagrangian (13):

$$p_\mu = \frac{\partial L(x, \dot{x})}{\partial \dot{x}^\mu}, \tag{15}$$

yielding

$$p_\mu = M g_{\mu\nu} \dot{x}^\nu. \tag{16}$$

The second factor in the definition (12) of $g_{\mu\nu}$ acts on $\dot{x}^\nu$; with (10) we then have (as in (14))

$$\begin{aligned} p_\mu &= M \eta_{\lambda\sigma} \frac{\partial \xi^\lambda}{\partial x^\mu} \dot{\xi}^\sigma \\ &= \frac{\partial \xi^\lambda}{\partial x^\mu} \pi_\lambda. \end{aligned} \tag{17}$$

We now discuss the geodesic equation obtained by studying the condition

$$\ddot{\xi}^\mu = -\frac{1}{M} \dot{\pi}_\mu = -\frac{1}{M} \eta^{\mu\nu} \frac{\partial V(\xi)}{\partial \xi^\nu}. \tag{18}$$

On the other hand,[3]

$$\begin{aligned} \ddot{\xi}^\mu &= \frac{d}{d\tau} \left(\frac{\partial \xi^\mu}{\partial x^\lambda} \dot{x}^\lambda \right) \\ &= \frac{\partial^2 \xi^\mu}{\partial x^\lambda \partial x^\gamma} \dot{x}^\gamma \dot{x}^\lambda \\ &+ \frac{\partial \xi^\mu}{\partial x^\lambda} \ddot{x}^\lambda \\ &= -\frac{1}{M} \eta^{\mu\nu} \frac{\partial x^\lambda}{\partial \xi^\nu} \frac{\partial V(x)}{\partial x^\lambda}, \end{aligned} \tag{19}$$

so that, after multiplying by $\frac{\partial x^\sigma}{\partial \xi^\mu}$ and summing over μ, we obtain

$$\begin{aligned} \ddot{x}^\sigma &= -\frac{\partial x^\sigma}{\partial \xi^\mu} \frac{\partial^2 \xi^\mu}{\partial x^\lambda \partial x^\gamma} \dot{x}^\gamma \dot{x}^\lambda \\ &- \frac{1}{M} \eta^{\mu\nu} \frac{\partial x^\lambda}{\partial \xi^\nu} \frac{\partial x^\sigma}{\partial \xi^\mu} \frac{\partial V(x)}{\partial x^\lambda} \end{aligned}. \tag{20}$$

Finally, with (12) and the usual definition of the connection

$$\Gamma^\sigma{}_{\lambda\gamma} = \frac{\partial x^\sigma}{\partial \xi^\mu} \frac{\partial^2 \xi^\mu}{\partial x^\lambda \partial x^\gamma} \tag{21}$$

[3] Here we assume ξ^μ locally a function of $x(\tau)$ only. If the spacetime evolves (τ dependent energy momentum tensor), then it is an explicit function of τ as well,*i.e.*,$\xi(x(\tau), \tau)$. We show in the Appendix how $\frac{\partial \xi(x(\tau),\tau)}{\partial \tau}$ is related to $\frac{\partial g_{\mu\nu}(x(\tau),\tau)}{\partial \tau}$.

IOP Conf. Series: Journal of Physics: Conf. Series **1239** (2019) 012014 doi:10.1088/1742-6596/1239/1/012014

we obtain the modified geodesic type equation

$$\ddot{x}^\sigma = -\Gamma^\sigma{}_{\lambda\gamma}\dot{x}^\gamma\dot{x}^\lambda - \frac{1}{M}g^{\sigma\lambda}\frac{\partial V(x)}{\partial x^\lambda}, \tag{22}$$

from which we see that the derivative of the potential $V(\xi)$ is mapped, under this coordinate transformation into a "force" resulting in a modification of the acceleration along the geodesic-like curves.

The procedure that we have carried out here provides a canonical dynamical structure for the motions in the curvilinear coordinates. We first remark that the Poisson bracket remains valid for the coordinates $\{x, p\}$. In the local coordinates $\{\xi, \pi\}$, the τ derivative of a function $F(\xi, \pi)$ is

$$\begin{aligned} \frac{dF(\xi,\pi)}{d\tau} &= \frac{\partial F(\xi,\pi)}{\partial \xi^\mu}\dot{\xi}^\mu + \frac{\partial F(\xi,\pi)}{\partial \pi_\nu}\dot{\pi}_\mu \\ &= \frac{\partial F(\xi,\pi)}{\partial \xi^\mu}\frac{\partial K}{\partial \pi_\mu} - \frac{\partial F(\xi,\pi)}{\partial \pi_\mu}\frac{\partial K}{\partial \xi^\nu} \\ &\equiv [F, K]_{PB}(\xi, \pi). \end{aligned} \tag{23}$$

If we replace in this formula

$$\begin{aligned} \frac{\partial}{\partial \xi^\mu} &= \frac{\partial x^\lambda}{\partial \xi^\mu}\frac{\partial}{\partial x^\lambda} \\ \frac{\partial}{\partial \pi_\mu} &= \frac{\partial \xi^\mu}{\partial x^\lambda}\frac{\partial}{\partial p_\lambda}, \end{aligned} \tag{24}$$

we immediately (as assured by the invariance of the Poisson bracket under local coordinate transformations) obtain

$$\frac{dF(\xi,\pi)}{d\tau} = \frac{\partial F}{\partial x^\mu}\frac{\partial K}{\partial p_\mu} - \frac{\partial F}{\partial p_\mu}\frac{\partial K}{\partial x^\nu} \equiv [F, K]_{PB}(x, p) \tag{25}$$

In this definition of Poisson bracket we have, as for the ξ^μ, π_ν relation,

$$[x^\mu, p_\nu]_{PB}(x, p) = \delta^\mu{}_\nu. \tag{26}$$

Furthermore (we drop the (x, p) label henceforth),

$$[p_\mu, F(x)]_{PB} = -\frac{\partial F}{\partial x^\mu}, \tag{27}$$

so that p_μ acts infinitesimally as the generator of *translation* along the coordinate curves and

$$[x^\mu, F(p)]_{PB} = \frac{\partial F(p)}{\partial p_\mu}, \tag{28}$$

so that x^μ is the generator of translations in p_μ.

This structure clearly provides a phase space which could serve as the basis for statistical mechanics, and lends itself to the construction of a canonical quantum theory on the curved spacetime, as we discuss below.

We now turn to a discussion of the dynamics introduced into the curved space by the procedure outlined above. We start by developing the relation between $\dot{p}_\mu$ and the geodesic equations for x^μ, and show that the result agrees with the direct Hamiltonian calculation.

Recall from (5) that

$$p_\mu = M g_{\mu\lambda}\dot{x}^\lambda, \tag{29}$$

IOP Conf. Series: Journal of Physics: Conf. Series **1239** (2019) 012014 doi:10.1088/1742-6596/1239/1/012014

so that

$$\dot{p}_\mu = M(\frac{\partial g_{\mu\lambda}}{\partial x^\gamma}\dot{x}^\gamma\dot{x}^\lambda + g_{\mu\sigma}\ddot{x}^\sigma). \tag{30}$$

Since, by Eq.(22),

$$\ddot{x}^\sigma = -\Gamma^\sigma{}_{\lambda\gamma}\dot{x}^\gamma\dot{x}^\lambda - \frac{1}{M}g^{\sigma\lambda}\frac{\partial V(x)}{\partial x^\lambda}, \tag{31}$$

Eq. (1.25)) becomes

$$\dot{p}_\mu = -\frac{\partial V(x)}{\partial x^\mu} + M(\frac{\partial g_{\mu\lambda}}{\partial x^\gamma}\dot{x}^\gamma\dot{x}^\lambda - g_{\mu\sigma}\Gamma^\sigma{}_{\lambda\gamma}\dot{x}^\gamma\dot{x}^\lambda). \tag{32}$$

We now use the relation

$$\Gamma^\sigma{}_{\lambda\gamma} = \frac{1}{2}g^{\sigma\eta}(\frac{\partial g_{\eta\lambda}}{\partial x^\gamma} + \frac{\partial g_{\eta\gamma}}{\partial x^\lambda} - \frac{\partial g_{\lambda\gamma}}{\partial x^\eta}) \tag{33}$$

to obtain

$$\dot{p}_\mu = -\frac{\partial V(x)}{\partial x^\mu} + M(\frac{\partial g_{\mu\lambda}}{\partial x^\gamma}\dot{x}^\gamma\dot{x}^\lambda - \frac{1}{2}(\frac{\partial g_{\mu\lambda}}{\partial x^\gamma} + \frac{\partial g_{\mu\gamma}}{\partial x^\lambda} - \frac{\partial g_{\lambda\gamma}}{\partial x^\mu})\dot{x}^\gamma\dot{x}^\lambda). \tag{34}$$

The first term in the brackets with coefficient M, symmetrized under multiplication by $\dot{x}^\gamma\dot{x}^\lambda$, cancels the first two terms of the contribution from the connection form leaving the rather remarkable result

$$\dot{p}_\mu = -\frac{\partial V(x)}{\partial x^\mu} + \frac{M}{2}\frac{\partial g_{\lambda\gamma}}{\partial x^\mu}\dot{x}^\gamma\dot{x}^\lambda. \tag{35}$$

The 3D form of the second term plays an important role in the study of (3D) Hamiltonian stability analysis in the work of Horwitz, Ben Zion, Lewkowicz and Levitan [14], where it is called the "reduced connection" form.

Eq. (35) displays the "force" induced from the local potential function along with the acceleration induced by the geometry. Since the derivative of the metric diverges in the neighborhood of the black hole solution, (35) provides a direct interpretation of the geometrical configuration as resulting in an "actual", from the point of view of the SHP theory, very large *force* on the particle at the black hole horizon as would be seen in coordinates of this type. We do not study here Kruskal type coordinates [15] which do not display a singularity at the horizon.

We may also write (35) in terms of the full connection form by noting that with (12),

$$\frac{\partial g_{\lambda\gamma}}{\partial x^\mu} = \eta_{\alpha\beta}(\frac{\partial^2\xi^\alpha}{\partial x^\lambda\partial x^\mu}\frac{\partial\xi^\beta}{\partial x^\gamma} + \frac{\partial\xi^\alpha}{\partial x^\lambda}\frac{\partial^2\xi^\beta}{\partial x^\gamma\partial x^\mu}). \tag{36}$$

Multiplying by $\dot{x}^\gamma\dot{x}^\lambda$, the two terms combine to give a factor of two. We then return to the original definition of Γ in (21)in the form

$$\frac{\partial^2\xi^\alpha}{\partial x^\lambda\partial x^\mu} = \frac{\partial\xi^\alpha}{\partial x^\sigma}\Gamma^\sigma{}_{\lambda\mu}, \tag{37}$$

so we can write

$$\begin{aligned}\frac{\partial g_{\lambda\gamma}}{\partial x^\mu}\dot{x}^\gamma\dot{x}^\lambda &= 2\eta_{\alpha\beta}\frac{\partial\xi^\alpha}{\partial x^\sigma}\frac{\partial\xi^\beta}{\partial x^\gamma}\Gamma^\sigma{}_{\lambda\mu}\dot{x}^\gamma\dot{x}^\lambda \\ &= 2g_{\sigma\gamma}\Gamma^\sigma{}_{\lambda\mu}\dot{x}^\gamma\dot{x}^\lambda.\end{aligned} \tag{38}$$

We therefore have

$$\dot{p}_\mu = -\frac{\partial V(x)}{\partial x^\mu} + Mg_{\sigma\gamma}\Gamma^\sigma{}_{\lambda\mu}\dot{x}^\gamma\dot{x}^\lambda. \tag{39}$$

We now return to the Hamiltonian (11) and carry out the calculation directly. Since $\dot{x}^\mu$ is, in general, a function of x^μ, we write the Hamiltonian (using (16)) in terms of the momenta, assured by the canonical structure to be independent variables,

$$K = \frac{1}{2M} g^{\alpha\beta} p_\alpha p_\beta + V(x). \tag{40}$$

Then,

$$\dot{p}_\mu = -\frac{1}{2M} \frac{\partial g^{\alpha\beta}}{\partial x^\mu} p_\alpha p_\beta - \frac{\partial V(x)}{\partial x^\mu}. \tag{41}$$

Returning to the form with $\dot{x}$ again, we have, with $p_\alpha = M g_{\alpha\lambda}\dot{x}^\lambda$,

$$\dot{p}_\mu = \frac{M}{2} \frac{\partial g^{\alpha\beta}}{\partial x^\mu} g_{\alpha\lambda} g_{\beta\gamma} \dot{x}^\gamma \dot{x}^\lambda - \frac{\partial V(x)}{\partial x^\mu}. \tag{42}$$

Since, however,

$$\frac{\partial g^{\alpha\beta}}{\partial x^\mu} g_{\alpha\lambda} g_{\beta\gamma} = -\frac{\partial g_{\lambda\gamma}}{\partial x^\mu}, \tag{43}$$

we recover the form obtained in (35).

We now turn to discussion of the many body problem.

3. The many body system with interaction potential

The many body Hamiltonian of the SHP theory is

$$K = \Sigma_{i=1}^N \frac{1}{2M_i} \eta^{\mu\nu} \pi_{\mu i} \pi_{\nu i} + V(\xi_1, \xi_2, \ldots \xi_N), \tag{44}$$

where the potential $V(\xi_1, \xi_2, \ldots \xi_N)$ is a function of the locally flat coordinates in the neighborhood of each of the particles at $\{x_i\}$. Although this function is Lorentz scalar, Poincaré invariance is, in general, inapplicable (even in the two-body case), unless all of the particles are in a sufficiently small neighborhood to be able to neglect the effects of curvature.

The Hamilton equations are (in the tangent space in the neighborhood of each particle at the point x_i)

$$\dot{\xi}_i^\mu = \frac{\partial K}{\partial \pi_{\mu i}} \qquad \dot{\pi}_{\mu i} = -\frac{\partial K}{\partial \xi^\mu{}_i} = -\frac{\partial V}{\partial \xi^\mu{}_i}, \tag{45}$$

We then have

$$\begin{aligned} \dot{\xi}_i^\mu &= \frac{1}{M_i} \eta^{\mu\nu} \pi_{\nu i}, \\ \text{or} \quad \pi_{\nu i} &= \eta_{\nu\mu} M_i \dot{\xi}_i^\mu, \end{aligned} \tag{46}$$

Following the procedure we used for the one-body case above, we may substitute this expression into the Hamiltonian to obtain

$$K = \Sigma_{i=1}^N \frac{M_i}{2} \eta_{\mu\nu} \dot{\xi}_i^\mu \dot{\xi}_i^\nu + V(\xi_1, \xi_2, \ldots \xi_N), \tag{47}$$

At the location $x^\mu{}_i$ of the ith particle, since in this neighborhood, $\xi^{\mu i}$ is a function locally of $x^\mu{}_i$, we can then make a local coordinate transformation

$$d\xi^\sigma{}_i = \frac{\partial \xi^\sigma{}_i}{\partial x^\mu{}_i} dx^\mu{}_i. \tag{48}$$

IOP Conf. Series: Journal of Physics: Conf. Series **1239** (2019) 012014 doi:10.1088/1742-6596/1239/1/012014

Defining

$$g_{\mu\nu}(x_i) = \eta_{\sigma\lambda}\frac{\partial\xi^\sigma{}_i}{\partial x^\mu{}_i}\frac{\partial\xi^\lambda{}_i}{\partial x^\nu{}_i}, \tag{49}$$

one obtains the Hamiltonian in terms of the four-velocities; changing notation for the arguments of the potential,[4]

$$K = \Sigma_{i=1}^N \frac{M_i}{2} g_{\mu\nu}(x_i)\dot{x}_i^\mu \dot{x}_i^\nu + V(x_1, x_2, \ldots x_N), \tag{50}$$

with corresponding Lagrangian

$$L = \Sigma_{i=1}^N \frac{M_i}{2} g_{\mu\nu}(x_i)\dot{x}_i^\mu \dot{x}_i^\nu - V(x_1, x_2, \ldots x_N). \tag{51}$$

As for the one body case, we can find the equations for the geodesic motion of the particles. Since (10) is valid for each of the particle coordinates,

$$\dot{\xi}_i^\mu = \frac{\partial\xi^\mu{}_i}{\partial x^\lambda{}_i}\dot{x}_i^\lambda, \tag{52}$$

from which we similarly obtain

$$\begin{aligned}\ddot{\xi}_i^\mu &= \frac{d}{d\tau}\Big(\frac{\partial\xi^\mu{}_i}{\partial x^\lambda{}_i}\dot{x}_i^\lambda\Big)\\ &= \frac{\partial^2\xi^\mu{}_i}{\partial x^\lambda{}_i\partial x^\gamma{}_i}\dot{x}_i^\gamma\dot{x}_i^\lambda\\ &+ \frac{\partial\xi^\mu{}_i}{\partial x^\lambda{}_i}\ddot{x}_i^\lambda\\ &= -\frac{1}{M_i}\eta^{\mu\nu}\frac{\partial x^\lambda{}_i}{\partial\xi^\nu{}_i}\frac{\partial V(x_1, x_2, \ldots x_N)}{\partial x^\lambda{}_i}.\end{aligned} \tag{53}$$

The many body geodesic curves are therefore described by

$$\begin{aligned}\ddot{x}_i^\sigma &= -\frac{\partial x^\sigma{}_i}{\partial\xi^\mu{}_i}\frac{\partial^2\xi^\mu{}_i}{\partial x^\lambda{}_i\partial x^\gamma{}_i}\dot{x}_i^\gamma\dot{x}_i^\lambda\\ &- \frac{1}{M_i}\eta^{\mu\nu}\frac{\partial x^\lambda{}_i}{\partial\xi^\nu{}_i}\frac{\partial x^\sigma{}_i}{\partial\xi^\mu{}_i}\frac{\partial V(x_1, x_2, \ldots x_N)}{\partial x^\lambda{}_i}.\end{aligned} \tag{54}$$

We can consider the Jacobian for the local mapping (48) as a *field*, a mapping defined over all $\{x^\mu\}$, in (48) evaluated at the point $x^\mu{}_i$ where the ith particle is found.

We then define a local connection form *at the point* x_i as

$$\Gamma^\sigma{}_{\lambda\gamma}(x_i) = \frac{\partial x^\sigma{}_i}{\partial\xi^\mu{}_i}\frac{\partial^2\xi^\mu{}_i}{\partial x^\lambda{}_i\partial x^\gamma{}_i} \tag{55}$$

also, since it is a property of the manifold, as a *field* evaluated at the point $x^\mu{}_i$, so that the geodesic equations can be written as

$$\ddot{x}_i^\sigma = -\Gamma^\sigma{}_{\lambda\gamma}(x_i)\dot{x}_i^\gamma\dot{x}_i^\lambda - \frac{1}{M_i}g^{\sigma\lambda}(x_i)\frac{\partial V(x_1, x_2, \ldots x_N)}{\partial x^\lambda{}_i}. \tag{56}$$

[4] We assume that (general covariance) $V(x_1, x_2, \ldots x_N)$ is a scalar function under local diffeomorphisms of any of the variables.

IOP Conf. Series: Journal of Physics: Conf. Series **1239** (2019) 012014 doi:10.1088/1742-6596/1239/1/012014

The connection form in this case then also satisfies (33) at each point x_i. Since this connection form coincides with Einstein's,the same method can be used to construct a Ricci tensor; the resulting Einstein equations will therefore have the same form, although there will necessarily be differences in the structure of the energy momentum tensor.[5] The empty space solution [16] will be applicable in this framework as well, providing an interesting example for application [17], and the homogeneous case of Robertson Friedman and Walker [18] would have a similar form to the well-known solution. Applications of this type will be investigated in succeeding papers.

Following the same procedure as for (25), with general functions $F(x_1, x_2, \dots x_N, p_1, p_2, \dots p_N)$ with variables $\xi_1, \xi_2, \dots \xi_N, \pi_1, \pi_2, \dots \pi_N$ in the tangent bundles assigned to the points $x_1, x_2, \dots x_N, p_1, p_2, \dots p_N$ in the general phase space for the N-body system, the Poisson bracket is defined by

$$\begin{aligned}\frac{dF(\xi_1, \xi_2, \dots \xi_N, \pi_1, \pi_2, \dots \pi_N)}{d\tau} &= \Sigma_i\Big(\frac{\partial F(\{\xi, \pi\})}{\partial \xi^\mu{}_i}\dot{\xi}_i^\mu + \frac{\partial F(\{\xi, \pi\})}{\partial \pi_{\nu i}}\dot{\pi}_{\mu i} \\ &= \frac{\partial F(\{\xi, \pi\})}{\partial \xi^\mu{}_i}\frac{\partial K}{\partial \pi_{\mu i}} - \frac{\partial F(\{\xi, \pi\})}{\partial \pi_{\mu i}}\frac{\partial K}{\partial \xi^\mu{}_i}\Big) \\ &\equiv [F, K]_{PB}(\{\xi, \pi\}).\end{aligned} \tag{57}$$

The local transformations on the differentials cancel as for the one particle case (at each point x_i), so the Poisson bracket remains in the same form on the $\{x_i, p_i\}$ phase space. We therefore have

$$\begin{aligned}\frac{dF(x_1, x_2, \dots x_N, p_1, p_2, \dots p_N)}{d\tau} &= \Sigma_i \frac{\partial F(\{x, p\})}{\partial x^\mu{}_i}\frac{\partial K}{\partial p_{\mu i}} - \frac{\partial F(\{x, p\})}{\partial p_{\mu i}}\frac{\partial K}{\partial x^\mu{}_i} \\ &\equiv [F, K]_{PB}(\{x, p\}).\end{aligned} \tag{58}$$

In general, for two functions $A(\{x, p\})$ and $B(\{x, p\})$, the many body Poisson bracket is then

$$[A, B]_{PB} = \Sigma_i\Big(\frac{\partial A(\{x, p\})}{\partial x^\mu{}_i}\frac{\partial B(\{x, p\})}{\partial p_{\mu i}} - \frac{\partial A(\{x, p\})}{\partial p_{\mu i}}\frac{\partial B(\{x, p\})}{\partial x^\mu{}_i}\Big). \tag{59}$$

Since the variables $x_1, x_2, \dots x_N, p_1, p_2, \dots p_N$ are to be considered as kinematically independent, we obtain the canonical bracket

$$[x_i{}^\mu, p_{j\nu}]_{PB} = \delta_{ij}\delta^\mu{}_\nu \tag{60}$$

We now turn to the equations of motion for $p_{\mu i}$. At the point $x_i{}^\mu$, as we have argued above,

$$p_{\mu i} = \frac{\partial \xi^\lambda{}_i}{\partial x_i{}^\mu}(x_i)\pi_{\lambda i}. \tag{61}$$

It therefore follows, in the same way that we obtained (39), that

$$\begin{aligned}\dot{p}_{\mu i} &= -\frac{\partial V(x_1, x_2, \dots x_N, p_1)}{\partial x^\mu{}_i} + M_i g_{\sigma\gamma}(x_i)\Gamma^\sigma{}_{\lambda\mu}(x_i)\dot{x}_i^\gamma \dot{x}_i^\lambda \\ &= -\frac{\partial V(x_1, x_2, \dots x_N, p_1)}{\partial x^\mu{}_i} + \frac{M_i}{2}(g_{\sigma\gamma}(x_i)\Gamma^\sigma{}_{\lambda\mu}(x_i) + g_{\sigma\lambda}(x_i)\Gamma^\sigma{}_{\gamma\mu}(x_i))\dot{x}_i^\gamma \dot{x}_i^\lambda.\end{aligned} \tag{62}$$

As for the one body case, this result also follows from the Lagrangian (51).

The time rate of change of the momentum is coupled, as for the geodesic motions of the x_i, to the other $N-1$ particles through the potential function.

[5] Note that even in the absence of a potential function, the solutions for $g_{\mu\nu}(x)$ would reflect the many body structure of the energy momentum tensor through the Einstein equations.

4. Quantum Theory on the Curved Space

The Poisson bracket formulas (27) and (28) can be considered as a basis for defining a quantum theory with canonical commutation relations

$$[x^\mu, p_\nu] = i\hbar\delta^\mu{}_\nu, \tag{63}$$

so that

$$[p_\mu, F(x)] = -i\hbar\frac{\partial F}{\partial x^\mu}, \tag{64}$$

and

$$[x^\mu, F(p)] = i\hbar\frac{\partial F(p)}{\partial p_\mu}. \tag{65}$$

The transcription of the SHP Schrödinger equation for a wave function $\psi_\tau(x)$ can be taken to be (see also Schwinger and deWitt [19])

$$i\frac{\partial}{\partial\tau}\psi_\tau(x) = K\psi_\tau(x), \tag{66}$$

where the operator valued Hamiltonian can be taken to be the Hermitian form, on a Hilbert space defined with scalar product (with invariant measure; we write $g = -det\{g^{\mu\nu}\}$),

$$(\psi, \chi) = \int d^4x\sqrt{g}\psi^*{}_\tau(x)\chi_\tau(x), \tag{67}$$

and

$$K = \frac{1}{2M\sqrt{g}}p_\mu g^{\mu\nu}p_\nu + V(x). \tag{68}$$

The normalization condition over the manifold $\{x\}$ is not a trivial transcription of the Euclidean condition on the SHP quantum theory [1]. If we think of the integral (67) as constructed from summing over coordinate components, a large excursion along a coordinate of the curved space may bring one, perhaps many times, to a nearby neighborhood of some point. The integration (67) must be considered as a total volume sum on the invariant measure on the whole space, consistent with the idea that the norm is the sum of probability measures on every subset contained. The procedure for carrying out such integrals would, of course, depend on the geometrical structure of the manifold.

This construction can be carried over to the many body case directly, *i.e*, with the operator properties of the coordinates and momenta

$$[x^\mu{}_i, p_{\nu i}] = i\hbar\delta^i{}_j\delta^\mu{}_\nu, \tag{69}$$

and therefore

$$[p_{\mu i}, F(\{x)\}] = -i\hbar\frac{\partial F(\{x\})}{\partial x^\mu{}_i}, \tag{70}$$

and

$$[x^\mu{}_i, F(\{p)\}] = i\hbar\frac{\partial F(\{p\})}{\partial p_{\mu i}}. \tag{71}$$

The scalar product is then (the flat space Lorentz invariant d^4x goes over to the local diffeomorphism invariant $d^4x\sqrt{g}$)

$$(\psi, \chi) = \int \Pi_i\{d^4(x_i)\sqrt{g(x_i)}\}\psi^*{}_\tau(x_1, x_2, \ldots x_N)\chi_\tau(x_1, x_2, \ldots x_N). \tag{72}$$

IOP Conf. Series: Journal of Physics: Conf. Series **1239** (2019) 012014 doi:10.1088/1742-6596/1239/1/012014

In this scalar product, the Hamiltonian

$$K = \Sigma_i \frac{1}{2M_i\sqrt{g(x_i)}} p_{\mu i} g^{\mu\nu}(x_i) p_{\nu i} + V(x_1, x_2, \dots x_N) \tag{73}$$

is formally Hermitian. The procedure of integration by parts in the scalar product involves a term by term cancellation of the factors $\sqrt{g(x_i)}$, thus permitting, in each term, integration by parts for the p_i.

We now turn to discuss the introduction of electromagnetism, for a single particle and for many particle systems.

5. Electromagnetism

As C.N. Yang [20] wrote, electromagnetism can be thought of as a $U(1)$ fiber bundle. The electromagnetic potential vector field emerges as a section on the fiber bundle in the gauge transformations of the quantum theory. To illustrate this idea, consider what happens to Eq. (66) if we consider, instead of the function $\psi_\tau(x)$, the function $\psi'_\tau(x)$ resulting from a unitary transformation $e^{i\Lambda(x,\tau)}$ defined locally on the Hilbert space at each value of τ. Since p_μ acts like a derivative on x^μ, it differentiates $\Lambda(x,\tau)$, just as for the corresponding computation in the flat Minkowski space. As for the flat space case, we must add a gauge compensation term so that

$$(p_\mu - a'_\mu(x,\tau))e^{i\Lambda(x,\tau)}\psi_\tau(x) = e^{i\Lambda(x,\tau)}(p_\mu - a_\mu(x,\tau))\psi_\tau(x), \tag{74}$$

i.e., assuring that $(p_\mu - a_\mu(x,\tau))\psi_\tau(x)$ is an element of the Hilbert space and therefore undergoes the same unitary transformation as $\psi_\tau(x)$. Carrying out the derivative implied by the action of p_μ (with $\hbar = 1$ here), we find the condition that

$$a'_\mu(x,\tau) = a_\mu(x,\tau) + \frac{\partial\Lambda(x,\tau)}{\partial x^\mu}, \tag{75}$$

the usual form of a gauge transformation. From the scalar nature of the wave function, we have implicitly assumed Einstein's property of general covariance for the fields $a_\mu(x,\tau)$.

Unless we restrict ourselves to the so-called "Hamilton gauge" (with Λ independent of τ), the form of (66) implies the existence of a fifth field [1][21] $a_5(x,\tau)$, for which we must have

$$\{i\frac{\partial}{\partial\tau} + a'_5(x,\tau)\}\psi'_\tau(x) = e^{i\Lambda(x,\tau)}\{i\frac{\partial}{\partial\tau} + a_5(x,\tau)\}\psi_\tau(x) \tag{76}$$

By the same argument, we then have

$$a'_5(x,\tau) = a_5(x,\tau) + \frac{\partial}{\partial\tau}\Lambda(x,\tau). \tag{77}$$

The Stueckelberg-Schrödinger equation then becomes

$$i\frac{\partial}{\partial\tau}\psi_\tau(x) = \{\frac{1}{2M\sqrt{g}}(p_\mu - a_\mu(x,\tau))g^{\mu\nu}(p_\nu - a_\nu(x,\tau)) - a_5(x,\tau)(x)\}\psi_\tau(x), \tag{78}$$

where the scalar field of the potential model is now replaced by the generally τ dependent $a_5(x,\tau)$.

In the usual way, we can define in the flat tangent space, a gauge invariant field strength[1][21]

$$\tilde{f}_{\alpha\beta}(\xi,\tau) = \partial_\alpha a_\beta(\xi,\tau) - \partial_\beta a_\alpha(\xi,\tau), \tag{79}$$

where $\alpha, \beta = (0,1,2,3,5)$, which satisfies the equation

$$\partial^\alpha \tilde{f}_{\alpha\beta}(\xi,\tau) = j_\beta(\xi,\tau). \tag{80}$$

The first four components of the current have the form of a Jackson type construction [22] before integration over τ, and the fifth component is the density $\rho(\xi,\tau) \propto \psi^*{}_\tau(\xi)\psi_\tau(\xi)$ in the SHP theory.

It is easy to see that a coordinate transformation leads to the rule of replacement of derivatives by covariant derivatives so that in the curved space

$$\begin{aligned} f_{\mu\nu}(x,\tau) &= a_{\mu;\nu} - a_{\nu;\mu} = \frac{\partial\xi^\sigma}{\partial x^\nu}\frac{\partial\xi^\lambda}{\partial x^\mu}\tilde{f}_{\lambda\sigma} \\ f_{\mu 5} &= \partial_\mu a_5 - \partial_5 a_\mu, \end{aligned} \tag{81}$$

since a_5 is a Lorentz scalar.

For the four- and fifth -components, we have

$$\begin{aligned} f^{;\mu}_{\mu\nu}(\xi,\tau) + \partial^5 f_{5\nu} &= j_\nu(\xi,\tau); \\ f_{\mu 5}{}^{;\mu}(x,\tau) &= j_5(x,\tau) = \rho(x,\tau), \end{aligned} \tag{82}$$

where the last is analogous to the non-relativistic $\nabla\cdot\mathbf{E} = \rho$. Clearly, the covariant divergence of $j_\nu(\xi,\tau)$ vanishes.

We now study the structure of the corresponding current. To do this, we write an action for which the variation with respect to $\psi^*{}_\tau(x)$ yields the Stueckelberg-Schrödinger equation (78),

$$\begin{aligned} S = \int d\tau d^4x\sqrt{g}\{ &i\psi^*{}_\tau(x)\frac{\partial}{\partial\tau}\psi_\tau(x) - i\psi_\tau(x)\frac{\partial}{\partial\tau}\psi^*{}_\tau(x) + a_5(x,\tau)\psi^*{}_\tau(x)\psi_\tau(x) \\ &- \{\psi^*{}_\tau(x)\{\frac{1}{2M\sqrt{g}}(p_\mu - a_\mu(x,\tau))g^{\mu\nu}(p_\nu - a_\nu(x,\tau)) - a_5(x,\tau)\}\psi_\tau(x) \end{aligned} \tag{83}$$

where $f^{\mu\nu}(x,\tau) = g^{\mu\lambda}g^{\nu\sigma}f_{\lambda\sigma}(x,\tau)$. We add to the action a purely electromagnetic part

$$S_{em} = +\frac{1}{4\sqrt{g}}(f^{\mu\nu}f_{\mu\nu} + f^{\mu 5}f_{\mu 5}), \tag{84}$$

where (since a_5 is scalar its covariant derivative is an ordinary derivative)

$$f_{\mu 5} = \partial_\mu a_5 - \partial_\tau a_\mu \tag{85}$$

and $a^5 = \pm a_5$ depending on the metric for $O(4,1)$ or $O(3,2)$ chosen for the $5D$ manifold. As for the nonrelativistic theory on $3D$, where the gauge fields make accessible the $(3,1)$ manifold of Minkowski space, the gauge fields of the $(3+1)D$ theory make accessible a $(4,1)$ or $(3,2)$ manifold. As we shall see in our discussion of the many body problem, the assumption of universality in τ does not admit such a higher symmetry.

In 1995, Land, Shnerb and Horwitz [23] studied the consequences of assuming covariant commutation relations between x^μ and $\dot{x}^\nu$ on a manifold using a theorem of Hojman and Sheply [24] extending and generalizing the work of Tanimura [25]. Their results, including the development of the 5D theory, agree in the one particle sector with what we have presented here.

To fully treat such a development with the methods we have used here, one would have to start with a one degree higher dimensional Stueckelberg equation; its gauge fields would open the possibility of a $6D$ manifold as a result of gauge invariance. We shall truncate this sequence here at the level of $4D$, retaining τ as the universal invariant parameter of evolution.

We now obtain the current by variation of a_μ in the action. Integrating by parts in the kinetic term (for $p_\mu = -i\frac{\partial}{\partial x^\mu}$), we have

$$\begin{aligned} S_{kin} &= -\frac{1}{2M}\int d\tau d^4x((p_\mu - a_\mu)\psi)^* g^{\mu\nu}(p_\nu - a_\nu)\psi \\ &= +\frac{1}{2M}\int d\tau d^4x((p_\mu + a_\mu)\psi^*)g^{\mu\nu}(p_\nu - a_\nu)\psi, \end{aligned} \tag{86}$$

so that

$$\frac{\delta S_{kin}}{\delta a_\mu} = \frac{1}{2Mi}(\psi^* g^{\mu\nu}(\partial_\nu - ia_\nu)\psi - ((\partial_\nu + ia_\nu)\psi^*)g^{\mu\nu}\psi). \tag{87}$$

However, $\sqrt{g}\psi^*\psi(x,\tau)$ is the probability to find the particle (event) in the invariant volume element $\sqrt{g}d^4x$, so that $\psi^*\psi(x,\tau)$ must go over to $(\sqrt{g})^{-1}\delta^4(x-x')$ in the classical limit (see Weinberg [26]). Therefore, we must define the current as[6]

$$j_\mu(x,\tau) = \frac{1}{2Mi\sqrt{g}}(\psi^* g^{\mu\nu}(\partial_\nu - ia_\nu)\psi - ((\partial_\nu + a_\nu)\psi^*)g^{\mu\nu}\psi), \tag{89}$$

in agreement in form with the corresponding known nonrelativistic formula ($g \to 1$ in the nonrelativistic limit). Note further that the integral of the current over a hypersurface with the invariant measure $d^4x\sqrt{g}$ has well-defined physical meaning.

We now study the variation of the action with respect to a_5. The variation of the full action (both S_m and S_{em}) with respect to a_5 then yields the field equation

$$f_{\mu 5}{}^{;\mu}(x,\tau) \equiv (\sqrt{g})^{-1}\rho(x,\tau) = \psi_\tau{}^*\psi_\tau(x,\tau), \tag{90}$$

in analogy to the standard Maxwell equation $\nabla\cdot\mathbf{E} = \rho$.

Furthermore, the variation with respect to a_μ, with the definition of the current (89), yields the covariant field equations (as one would conclude from the application of general covariance [26]) The variation of a_μ in (85) leads to the second field equation of (81).

$$f_{\mu\nu}{}^{;\mu}(x,\tau) = j_\nu(x,\tau), \tag{4.17)(91}$$

or, equivalently,

$$\partial_\mu(\sqrt{g}f^{\mu\nu}) = \sqrt{g}j^\nu \tag{92}$$

The Lorentz force (see also [23]) follows by directly transcribing the flat space formula (for charge unity),

$$F^\mu = f^\mu{}_\nu\frac{dx^\nu}{d\tau} \tag{93}$$

[6] Note that if we follow the method of Jackson [22], defining the macroscopic current

$$J^\mu(x) = (\sqrt{g})^{-1}\int d\tau\dot{x}^\mu\delta^4(x - x(\tau)), \tag{88}$$

then $\partial_\mu(\sqrt{g}J^\mu) = 0$.

IOP Conf. Series: Journal of Physics: Conf. Series **1239** (2019) 012014 doi:10.1088/1742-6596/1239/1/012014

6. The many body problem for electromagnetism

In the following, we generalize this structure to the many body problem.

The many-body wave function can be written as the span of the direct product of wave functions associated with isomorphic one particle Hilbert spaces (which also may be used in the construction of the Fock space on the manifold). The norm and orthogonality follow from the properties of the one particle spaces as above (with the rule that corresponding elements of the sequences are contracted by scalar product). We may therefore write

$$\psi_\tau(x_1, x_2, \dots x_N) = \Sigma a_{\alpha_1,\alpha_2\dots\alpha_N}\phi_{\alpha_1,\tau}(x_1)\phi_{\alpha_2,\tau}(x_2)\cdots\phi_{\alpha_N,\tau}(x_N). \tag{94}$$

We now argue that a local unitary transformation of the form $e^{i\Lambda(x,\tau)}$ should act, with the *same* function $\Lambda(x,\tau)$ in each of the factor spaces.[7] This construction provides a convenient mechanism for the gauge transformations of the Bose-Einstein or Fermi-Dirac Fock spaces (and, in general, for linear combinations). Furthermore, as we shall see below, it enables us to define a *field* $a_\mu(x,\tau)$. We therefore define the gauge transformation $\psi \to \psi'$ as

$$\begin{aligned}\psi'_\tau(x_1, x_2, \dots x_N) &= \Sigma a_{\alpha_1,\alpha_2\dots\alpha_N} e^{i(\Lambda(x_1,\tau)+\Lambda(x_2,\tau)+\cdots\Lambda(x_N,\tau))} \\ &\quad\times \phi_{\alpha_1,\tau}(x_1)\phi_{\alpha_2,\tau}(x_2)\cdots,\phi_{\alpha_N,\tau}(x_N) \\ &= e^{i(\Lambda(x_1,\tau)+\Lambda(x_2,\tau)+\cdots\Lambda(x_N,\tau))}\psi_\tau(x_1, x_2,\dots,x_N)\end{aligned} \tag{95}$$

The remaining argument is the same as for the one body case. At each point x_i of the wave function associated with the ith factor, we have

$$\begin{aligned}(-i\frac{\partial}{\partial x_i{}^\mu} - a'_\mu(x_i,\tau))\psi'_\tau(x_1, x_2, \dots x_N) &= e^{i(\Lambda(x_1,\tau)+\Lambda(x_2,\tau)+\cdots\Lambda(x_N,\tau))} \\ &\quad (-i\frac{\partial}{\partial x_i{}^\mu} - a_\mu(x_i,\tau))\psi_\tau(x_1, x_2, \dots x_N),\end{aligned} \tag{96}$$

so that

$$a'_\mu(x_i,\tau) = a_\mu(x_i,\tau) + \frac{\partial\Lambda(x_i,\tau)}{\partial x_i{}^\mu}, \tag{97}$$

It therefore follows that this procedure leads to a local covariant *field* for the electromagnetic potential vector.

If we call the compensation function for the τ evolution $a_5(x_1, x_2, \dots x_N, \tau)$ and

$$\Lambda(x_1, x_2, \dots x_N, \tau) = \Lambda(x_1,\tau) + \Lambda(x_2,\tau) + \cdots \Lambda(x_N,\tau) \tag{98}$$

then it follows that

$$a'_5(x_1, x_2, \dots x_N, \tau) = a_5(x_1, x_2, \dots x_N, \tau) + \frac{\partial}{\partial\tau}\Lambda(x_1, x_2, \dots x_N, \tau). \tag{99}$$

The fifth gauge function is clearly not a property of the individual particles, and in this sense it appears not to correspond to a local field on the individual particles (it is a local field, however, on the configuration space $(x_1, x_2, \dots x_N)$).

The field strengths associated with the fifth field form a set

$$f^i{}_{\mu 5}(x_1, x_2, \dots x_N) = \partial_{\mu i} a_5(x_1, x_2, \dots x_N) - \partial_\tau a_\mu(x_i). \tag{100}$$

[7] One can think of this procedure as the action of an *operator* $(e^{i\Lambda})^N = (e^{i\Lambda N})$ acting on the N particle state.

Under gauge transformation,

$$\begin{aligned}\partial_{\mu i}(a_5(x_1,x_2,\dots x_N,\tau)+\frac{\partial}{\partial\tau}\Lambda(x_1,x_2,\dots x_N,\tau))-\partial_\tau(a_\mu(x_i,\tau)+\frac{\partial}{\partial x_i{}^\mu}\Lambda(x_i\tau))\\ =\partial_{\mu i}a_5(x_1,x_2,\dots x_N,\tau)-\partial_\tau a_\mu(x_i,\tau),\end{aligned} \tag{101}$$

since the derivative $\frac{\partial}{\partial x_i{}^\mu}$ selects the term in the sum that cancels $\frac{\partial}{\partial x_i{}^\mu}\Lambda(x_i\tau)$.

To be able to write the elements of this set in a uniform way in the arguments, we define a field on $x_1,x_2,\dots x_N$ for each τ such that the projection

$$a^i{}_\mu(x_1,x_2,\dots x_N,\tau)=a_\mu(x_i,\tau). \tag{102}$$

We can then write

$$f^i{}_{\mu 5}(x_1,x_2,\dots x_N)=\partial_{\mu i}a_5(x_1,x_2,\dots x_N)-\partial_\tau a^i{}_\mu(x_1,x_2,\dots x_N). \tag{103}$$

7. Electromagnetic Currents and Field Equations for the Many Body System.

In the following we discuss the electromagnetic current and field equations for the N body case.

We write the action for an N particle system in the presence of electromagnetism as

$$\begin{aligned}S=\int d\tau\Pi_i d^4x_i\sqrt{g(x_i)}\{i\psi^*{}_\tau(x_1,x_2,\dots x_N)\\ \frac{\partial}{\partial\tau}\psi_\tau(x_1,x_2,\dots x_N)-i\psi_\tau(x_1,x_2,\dots x_N)\frac{\partial}{\partial\tau}\psi^*{}_\tau(x_1,x_2,\dots x_N)\\ +a_5(x_1,x_2,\dots x_N,\tau)\psi^*{}_\tau(x_1,x_2,\dots x_N)\psi_\tau(x_1,x_2,\dots x_N)\\ -\{\psi^*{}_\tau(x_1,x_2,\dots x_N)\Sigma_i\{\frac{1}{2M_i\sqrt{g(x_i)}}(p_\mu-a_\mu(x_i,\tau))g^{\mu\nu}(x_i)(p_\nu-a_\nu(x_i,\tau))\\ -a_5(x_1,x_2,\dots x_N,\tau)\}\psi_\tau(x_1,x_2,\dots x_N)\end{aligned} \tag{104}$$

As for the one-particle case, we add an electromagnetic part

$$\begin{aligned}S_{em}=+\int d\tau\Sigma_i\int d^4x_i\frac{1}{4\sqrt{g}}(f^{\mu\nu}(x_i,\tau)f_{\mu\nu}(x_i,\tau)\}\\ +f^i{}_{\mu 5}(x_1,x_2,\dots x_N)f^{i\mu 5}(x_1,x_2,\dots x_N)).\end{aligned} \tag{105}$$

The variation with respect to a_5 (taking into account the factor $g^{\mu\nu}$ in raising the index) yields the equation of motion

$$\begin{aligned}f_{\mu 5}{}^{;\mu}(x_1,x_2,\dots x_N,\tau)=\psi_\tau{}^*(x_1,x_2,\dots x_N)\psi_\tau(x_1,x_2,\dots x_N)\\ =\Pi_i g(x_i)^{-\frac{1}{2}}\rho(x_1,x_2,\dots x_N,\tau)\end{aligned}, \tag{106}$$

the density on the full space $(x_1,x_2,\dots x_N)$ at each τ.

Finally, the variation with respect to $a_\mu(x_i,\tau)$ of the interaction term, since $f_{\mu\nu}(x,\tau)$ is a one particle quantity, the field equations

$$f_{\mu\nu}{}^{;\nu}(x_i,\tau)=j_\mu(x_i,\tau), \tag{107}$$

where (the variation in $a_\mu(x_i,\tau)$ fixes τ but not the coordinates except for x_j)

$$\begin{aligned}j_\mu(x_i,\tau)=\Pi_{j\neq i}d^4x_j\sqrt{g(x_j)}\\ \frac{1}{2Mi\sqrt{g(x_i)}}(\psi_\tau{}^*(x_1,x_2,\dots x_N)g^{\mu\nu}(x_i)(\partial_{\nu i}-ia_\nu(x_i))\psi_\tau(x_1,x_2,\dots x_N)\\ -((\partial_{\nu i}+a_\nu(x_i))\psi_\tau{}^*(x_1,x_2,\dots x_N))g^{\mu\nu}(x_i)\psi_\tau(x_1,x_2,\dots x_N)).\end{aligned} \tag{108}$$

The local one-particle field equations are then

$$f_{\mu\nu}{}^{;\mu}(x_i, \tau) = j_\nu(x_i, \tau) \tag{109}$$

or, as in the one-particle case,

$$\partial_{\mu i}(\sqrt{g(x_i)} f^{\mu\nu}(x_i, \tau)) = \sqrt{g(x_i)} j^\nu(x_i, \tau). \tag{110}$$

The Lorentz force acting on a particle in this many body framework is then

$$F^\mu(x_i, \tau) = f^\mu{}_\nu(x_i, \tau) \frac{dx_i{}^\nu}{d\tau}, \tag{111}$$

providing a basis, for example, for writing Vlasov equations in general relativistic statistical mechanics.

8. Summary and Outlook

We have shown that the SHP theory can be embedded by local coordinate transformations into the framework of general relativity. The Minkowski spacetime coordinates of the SHP theory are considered to lie in the tangent space of a manifold with metric and connection form derived from the coordinate transformations on the equations of motion for particles moving on the locally flat Minkowski spacetime, parametrized by a universal monotonic world time τ. The four momentum is well-defined on the manifold, and a formula for its τ derivative, which may be understood as a "force", is obtained, displaying the effect of the potential as well as the curvature (through the connection form).

For the many body system, each particle, at the points $\{x_i{}^\mu\}$ is assumed to move locally on a flat Minkowski space, which is then transformed by local coordinate transformation to the manifold of GR with coordinates x^μ. Since particles with (flat Minkowski) coordinates $\xi_1, \xi_2 \ldots \xi_N$ lie in different local tangent spaces at the points $x_1, x_2 \ldots x_N$ of the curvilinear coordinatization of GR, Poincaré invariance of the potential function is not applicable.

Since the Poisson bracket of the SHP theory is unchanged in form under local diffeomorphisms, it forms the basis of a quantum theory in which the momentum operator generates infinitesimal translations along the local coordinates, and p_μ is therefore represented by $-i\frac{\partial}{\partial x^\mu}$ on a Hilbert space of functions $\psi_\tau(x)$, square integrable over the invariant measure $d^4x\sqrt{g(x)}$.

This Hilbert space provides a basis for a local $U(1)$ gauge, for which the compensation fields (sections on the bundle) correspond to the classical $5D$ electromagnetic fields [21]. We obtain field equations for the electromagnetic fields and associated currents from an action (τ integrated Lagrangian).

The many body quantum theory is treated by constructing a tensor product space and the associated electromagnetic theory is developed assuming that each factor in the tensor product carries the same gauge transformations. This enables us to define a gauge compensation field $a_\mu(x)$ for the four components which can be evaluated on each particle, but due to the universality of τ, the fifth component must depend on the coordinates of all of the particles as a locally defined function on the full configuration space, similar to the function $V(x_1, x_2, \ldots x_N)$ of the potential model.

The work of this paper is primarily restricted to describing a relativistic dynamics in a τ independent gravitation field, *i.e.*, the metric is assumed independent of τ. Since the connection form has the same structure as the usual GR, one can write Einstein's equations in the same way [26]. Therefore, the energy momentum tensor, determining $g^{\mu\nu}(x)$, should be, in this case, independent of τ. To achieve this, one may use partially integrated currents, taking into account correlations [27], or the zero modes extracted from full integration yielding $4D$ conserved currents [1][22]. In the more general case, where the structure of spacetime is dynamical (for example, collision between black holes or for unstable stars such as supernova) the energy momentum tensor would depend on τ. We show in the Appendix how, in such cases, the corresponding explicit dependence of the local transformations from the Minkowski space coordinates to the curved coordinates on τ can be expressed in terms of the now τ dependent metric tensor.

The classical results of this paper provide an eight dimensional phase space for general relativity, just as the SHP theory provides for special relativity,and therefore a general relativistic statistical mechanics can be formulated. The assumptions necessary to construct Gibbs ensembles [28] in this context must be carefully examined.

The existence of a two body quantum theory makes accessible the possibility of constructing a quantum scattering theory, for which the formulation of asymptotic conditions, as for the construction of the Gibbs ensemble, poses an interesting challenge. The many body problem, generalizing the work of Horwitz and Arshansky [29], can be formulated in this framework. The many body Hilbert space can be used to construct a Fock space as the basis for a quantum field theory.

We finally remark that the results of the work of this paper provide a basis for the approach of Bekenstein and Sanders [8] and the associated discussion of nonabelian gauge fields given by Horwitz, Gershon and Schiffer [7]; it will therefore be of interest to follow the development of the $U(1)$ gauge theory given here with a study of non-Abelian gauge theories (see [23]).

Acknowledgments
I wish to thank Asher Yahalom, Yossi Strauss, Igal Aharonovich, and Martin Land for helpful discussions.

Appendix

We study here the effect of a τ evolving spacetime, a situation which would occur if the energy momentum tensor depends on τ. For example, using τ dependent momenta.

Generally, at a point x^λ, the velocity of a particle is $\dot{x}^\lambda$, just a motion on the coordinates $\{x\}$. If the spacetime is changing, we think of the tangent space as reflecting this change; the particle moves on $x^\lambda(\tau)$, even as the meaning of x^λ changes, but at every point there is a change in the local flat tangent space coordinates. Therefore, ξ changes as the particle moves *and* as the world coordinates evolve. At each τ, it is still true that

$$d\xi^\mu = \frac{\partial \xi^\mu}{\partial x^\lambda} dx^\lambda, \tag{A.1}$$

but $\frac{\partial \xi^\mu}{\partial x^\lambda}$ changes as the particle moves and as the spacetime evolves. We can write

$$\frac{\partial \xi^\mu}{\partial x^\lambda} = \frac{\partial \xi^\mu}{\partial x^\lambda}(x(\tau), \tau) \tag{A.2}$$

so that

$$\frac{d}{d\tau}\frac{\partial \xi^\mu}{\partial x^\lambda} = \frac{\partial^2 \xi^\mu}{\partial x^\lambda \partial x^\sigma}((x(\tau), \tau)\dot{x}^\sigma + \frac{\partial}{\partial \tau}\frac{\partial \xi^\mu}{\partial x^\lambda}(x(\tau), \tau), \tag{A.3}$$

where the second term is due to the change in orientation of the ξ^μ coordinates in τ.

The canonical structure postulated in (4) and (5), and the definition of $g_{\mu\nu}$ remains the same but, as a function of $\frac{\partial \xi^\mu}{\partial x^\lambda}(x(\tau, \tau)$, it now becomes an function of τ. We therefore have

$$K = \frac{M}{2} g_{\mu\nu}(x(\tau), \tau)\dot{x}^\mu \dot{x}^\nu + V(x). \tag{A.4}$$

IOP Conf. Series: Journal of Physics: Conf. Series **1239** (2019) 012014 doi:10.1088/1742-6596/1239/1/012014

We now calculate from $(A.1)$ the total τ derivative

$$\begin{aligned}\ddot{\xi}^\mu &= \frac{d}{d\tau}\Big(\frac{\partial \xi^\mu}{\partial x^\lambda}\dot{x}^\lambda\Big)\\ &= \frac{\partial^2 \xi^\mu}{\partial x^\lambda \partial x^\gamma}\dot{x}^\gamma \dot{x}^\lambda\\ &+ \frac{\partial}{\partial \tau}\frac{\partial \xi^\mu}{\partial x^\lambda}\dot{x}^\lambda + \frac{\partial \xi^\mu}{\partial x^\lambda}\ddot{x}^\lambda\\ &= -\frac{1}{M}\eta^{\mu\nu}\frac{\partial x^\lambda}{\partial \xi^\nu}\frac{\partial V(x)}{\partial x^\lambda},\end{aligned} \tag{A.5}$$

where the last equality, as before, follows from the canonical structure of the tangent space. Multiplying by $\frac{\partial x^\sigma}{\partial \xi^\mu}$, and solving for $\ddot{x}^\sigma$, we find a geodesic type equation as before but with an additional (apparently dissipative type) term

$$\ddot{x}^\sigma = -\Gamma^\sigma{}_{\lambda\gamma}\dot{x}^\gamma \dot{x}^\lambda - \frac{1}{M}g^{\sigma\lambda}\frac{\partial V(x)}{\partial x^\lambda} - \frac{\partial x^\sigma}{\partial \xi^\mu}\frac{\partial}{\partial \tau}\Big(\frac{\partial \xi^\mu}{\partial x^\lambda}\Big)\dot{x}^\lambda. \tag{A.6}$$

We now show that $\frac{\partial}{\partial \tau}\big(\frac{\partial \xi^\mu}{\partial x^\lambda}\big)$ can be expressed in terms of $\frac{\partial}{\partial \tau}g_{\mu\nu}(x(\tau),\tau)$, which then carries the information, from the Einstein equations, about the evolution of the spacetime.

From the definition (12) we compute

$$\begin{aligned}\frac{\partial g_{\mu\nu}}{\partial \tau}(x(\tau),\tau) = \eta_{\sigma\gamma}\Big[&\frac{\partial}{\partial \tau}\Big(\frac{\partial \xi^\sigma}{\partial x^\mu}\Big)\frac{\partial \xi^\gamma}{\partial x^\nu}\\ &+ \frac{\partial \xi^\gamma}{\partial x^\mu}\frac{\partial}{\partial \tau}\Big(\frac{\partial \xi^\sigma}{\partial x^\nu}\Big)\Big],\end{aligned} \tag{A.7}$$

where we have used the σ, γ symmetry of $\eta_{\sigma\gamma}$ in the second term.

Now, define

$$\frac{\partial}{\partial \tau}\Big(\frac{\partial \xi^\mu}{\partial x^\lambda}\Big) \equiv t^\mu{}_\lambda. \tag{A.8}$$

Then we can write $(A.7)$ as

$$\begin{aligned}\frac{\partial g_{\mu\nu}}{\partial \tau}(x(\tau),\tau) &= \eta_{\sigma\gamma}\Big[t^\sigma{}_\mu \frac{\partial \xi^\gamma}{\partial x^\nu} + t^\sigma{}_\nu \frac{\partial \xi^\gamma}{\partial x^\mu}\Big]\\ &= \eta_{\sigma\gamma}\Big(\delta_\mu{}^\lambda \frac{\partial \xi^\gamma}{\partial x^\nu} + \delta_\nu{}^\lambda \frac{\partial \xi^\gamma}{\partial x^\mu}\Big)t^\sigma{}_\lambda\\ &\equiv M_{\mu\nu}{}^\lambda{}_\sigma t^\sigma{}_\lambda.\end{aligned} \tag{A.9}$$

This equation can be inverted with the matrix $N_\lambda{}^{\sigma\mu\nu}$ satisfying

$$N_{\lambda'}{}^{\sigma'\mu\nu}M_{\mu\nu}{}^\lambda{}_\sigma = \delta'_\lambda{}^\lambda \delta_\sigma{}^{\sigma'}. \tag{A.10}$$

It is a possible exceptional case that $N_\lambda{}^{\sigma\mu\nu}$ could be singular; this would correspond to a singular development of the transformation function $\frac{\partial \xi^\sigma}{\partial x^\nu}$ in τ, which we do not treat here.

Multiplying $(A.9)$ by $N_{\lambda'}{}^{\sigma'\mu\nu}$, we obtain

$$N_{\lambda'}{}^{\sigma'\mu\nu}\frac{\partial g_{\mu\nu}}{\partial \tau}(x(\tau),\tau) = t^{\sigma'}{}_{\lambda'} = \frac{\partial}{\partial \tau}\Big(\frac{\partial \xi^{\sigma'}}{\partial x^{\lambda'}}\Big). \tag{A.11}$$

The solution that we have found, as well as many of the relations we have used in our study, involve the transformation functions $\frac{\partial \xi^\gamma}{\partial x^\mu}$; as pointed out by Weinberg [24], these functions enter quadratically in $g_{\mu\nu}$, and are therefore determined up to Lorentz transformation, at any given τ, by $g_{\mu\nu}$.

IOP Conf. Series: Journal of Physics: Conf. Series **1239** (2019) 012014 doi:10.1088/1742-6596/1239/1/012014

References

[1] Lawrence Horwitz, *Relativistic Quantum Mechanics*, Fundamental Theories of Physics 180, Springer, Dordrecht (2015).

[2] E.C.G. Stueckelberg, Helv. Phys. Acta **14**, 372, 585.
E.C.G. Stueckelberg, Helv. Phys. Acta **15**, 23 (1942).

[3] L.P. Horwitz and C. Piron, Helv. Phys. Acta **66**, 316 (1973).
R.E. Collins and J.R. Fanchi, Nuovo Cim. **48A**, 314 (1978),
J.R. Fanchi, *Parametrized Relativistic Quantum Theory*, Kluwer, Dordrecht (1993).

[4] Isaac Newton, *Philosophia Naturalis Principia Mathematica*, London 1687
I.B. Cohen and A. Whitman *The Principia: Mathematical Principles of Natural Philosophy: A New Translation*, University of California Press, Berkeley (1999).

[5] R.I. Arshansky and L.P. Horwitz, Jour. Math. Phys. **30**, 66 (1989).

[6] R.I. Arshansky and L.P. Horwitz, Jour. Math. Phys. **30** 380 (1989).
R.I. Arshansky and L.P.Horwitz, Jour. Math. Phys. **30**, 213 (1989).

[7] L.P. Horwitz, A. Gershon and M. Schiffer, Found. Phys. **41**, 141 (2010).
A. Gershon and L.P. Horwitz, Jour. Math. Phys. **50**, 102704 (2009).

[8] J.D. Bekenstein and R.H. Sanders, Astrophys. Jour. **429**, 480 (1994).

[9] J.D. Bekenstein and M. Milgrom, Astrophys. Jour.**286**, 7 (1984)
J.D. Bekenstein, Phys. Rev. **D70**, 083509 (2004)
J.D. Bekenstein, Contemporary Physics **47** 387 (2006) for a review, references, and further development.

[10] M. Milgrom, Asrophys. Jour. **270**,365,371,384 (1983).

[11] N.B. Birrell and P.C.W. Davies, *Quantum Fields in Curved Space*, Cambrdge Monographs on Mathematical Physics, Cambridge (1982).

[12] E. Poisson, *A Relativist's Toolkit*, Cambridge (2004).

[13] Ana Cannas de Silva, *Lectures on Symplectic Geometry*, Lecture Notes in Mathematics 1764, Springer (2006).

[14] L.P. Horwitz, Y. Ben Zion, M. Lewkowicz, M. Schiffer and J.Levitan, Phys. ev. Lett. **98** 234301 (2007).

[15] M.D. Kruskal, Phys. Rev. **119** 1743.

[16] K. Schwarzschild, Sitzungsberichte der Königlich Preussischen Akademie der Einsteinschen Theorie **1**:424 (1916)

[17] D. Ludwin and L.P. Horwitz, Jour. Math. Phys.**52** 012303 (2011).

[18] A. Friedman. Zeits. fur Physik **A 21**, 326 (1924)
G. Lemaitre, Monthly Notices of the Roy. Astro. Soc. **91**, 483 (1931)[trans. from Ann. de la Societe Scientifique de Bruxelles **A 47** 49 (1927)]
G. Lemaitre, Ann. de la Societe Scientifique de Bruxelles **A 53**, 51 (1933)
H.P. Robertson, Astrophys. Jour. **82**, 284 (1935), **83**, 187 (1936), **83**, 257 (1936).
A.G. Walker, Proc. London Math. Soc. **42** 90 (1936).

[19] J. Schwinger, Phys. Rev. **82**, 664 (1951)
B.S. DeWitt, Physics Reports **19**, 295 (1975)
B.S. DeWitt, *The Global Approach to Quantum Field Theory*, Oxford University Press, Oxford (2002).

[20] C.N. Yang, Annals of the New York Academy of Science **294** 86 (1977).

[21] D. Saad, L.P. Horwitz and R.I. Arshansky, Foundations of Physics, **19** 1125 (1989).

[22] J.D. Jackson, *Classical Electrodynamics*, 2nd edition, John Wiley and Sons, New York (1974).

[23] M. C. Land, N. Shnerb and L.P. Horwitz, Jour. Math. Phys. **36**, 3263 (1995).

[24] S.A. Hojman and L.C. Sheply, Jour. Math. Phys. **32**, 142 (1991).

[25] S. Tanimura, Ann. Phys. **220**229 (1992).

[26] S. Weinberg, *Gravitation and Cosmology*, John Wiley and Sons, New York (1972).

[27] M. Land, IOP Conference Series:Jour. of Physics Conf. Series **845**, 012024 (2017).

[28] L.P. Horwitz, W.C. Schieve and C. Piron, Ann. of Phys. **137**, 306 (1981).
L.P. Horwitz, S. Shashoua and W.C. Schieve, Physica **A 161**, 300 (1989).

[29] Lawrence P. Horwitz and Rafael I. Arshansky, *Relativistic Many-Body Theory and Statistical Mechanics*, IOP Concise Physics, A. Morgan and Claypool, Bristol (2018).

IOP Conf. Series: Journal of Physics: Conf. Series **1239** (2019) 012015 doi:10.1088/1742-6596/1239/1/012015

Time dispersion in quantum mechanics

John Ashmead
Visiting Scholar, University of Pennsylvania, USA
E-mail: jashmead@seas.upenn.edu

Abstract. In quantum mechanics the time dimension is treated as a parameter, while the three space dimensions are treated as observables. This assumption is both untested and inconsistent with relativity. From dimensional analysis, we expect quantum effects along the time axis to be of order an attosecond. Such effects are not ruled out by current experiments. But they are large enough to be detected with current technology, if sufficiently specific predictions can be made. To supply such we use path integrals. The only change required is to generalize the usual three dimensional paths to four. We predict a large variety of testable effects. The principal effects are additional dispersion in time and full equivalence of the time/energy uncertainty principle to the space/momentum one. Additional effects include interference, diffraction, and entanglement in time. The usual ultraviolet divergences do not appear: they are suppressed by a combination of dispersion in time and entanglement in time. The approach here has no free parameters; it is therefore falsifiable. As it treats time and space with complete symmetry and does not suffer from the ultraviolet divergences, it may provide a useful starting point for attacks on quantum gravity.

"Wheeler's often unconventional vision of nature was grounded in reality through the principle of radical conservatism, which he acquired from Niels Bohr: Be conservative by sticking to well-established physical principles, but probe them by exposing their most radical conclusions." – Kip Thorne [1].

"You can have as much junk in the guess as you like, provided that the consequences can be compared with experiment." – Richard P. Feynman [2]

1. Introduction

1.1. Should the wave function extend in time as it does in space?

In relativity, time and space enter on a basis of formal equivalence. In special relativity, the time and space coordinates rotate into each other under Lorentz transformations. In general relativity, the time and the radial coordinate change places at the Schwarzschild radius (for instance in Adler [3]). In wormholes and other exotic solutions to general relativity, time can even curve back on itself as in Gödel or Thorne [4, 5].

But in quantum mechanics "time is a parameter not an operator" (Hilgevoord [6, 7]). This is clear in the Schrödinger equation:

$$\imath \frac{d}{d\tau} \psi_\tau (\vec{x}) = \hat{H} \psi_\tau (\vec{x}) \tag{1}$$

Here the wave function is indexed by time: if we know the wave function at time τ we can use this equation to compute the wave function at time $\tau + \epsilon$. The wave function has in general

IOP Conf. Series: Journal of Physics: Conf. Series **1239** (2019) 012015 doi:10.1088/1742-6596/1239/1/012015

non-zero dispersion in space, but is always treated as having zero dispersion in time. This would appear to be inconsistent with special relativity.

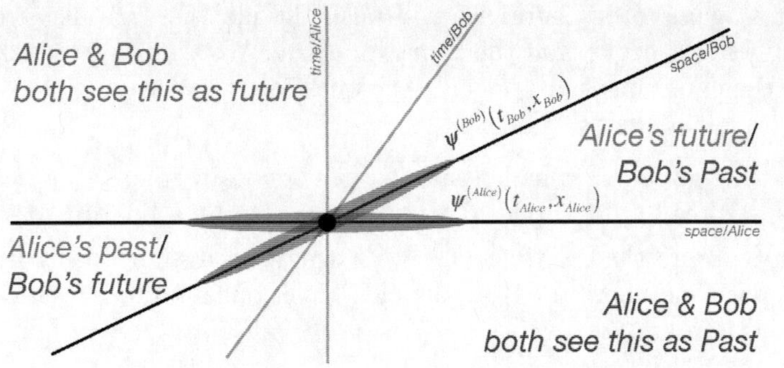

Figure 1. Wave functions for Alice and Bob

Consider Alice in her laboratory, her co-worker Bob jetting around like a fusion powered mosquito. Both are studying the same system but with respective wave functions:

$$\begin{gathered}\psi^{(Alice)}\left(t_{Alice}, x_{Alice}\right) \\ \psi^{(Bob)}\left(t_{Bob}, x_{Bob}\right)\end{gathered} \tag{2}$$

Alice and Bob center their respective wave functions on the particle:

$$\begin{gathered}\langle t_{Alice}\rangle = \langle t_{Bob}\rangle = 0 \\ \langle x_{Alice}\rangle = \langle x_{Bob}\rangle = 0\end{gathered} \tag{3}$$

These are distinct wave functions but give the same predictions for all observables. And do so to an extremely high degree of reliability.

But at Alice's time zero, Bob's wave function extends into her past and future. And at Bob's time zero her wave function extends into his past and future.

There are at least two problems here.

One is that in quantum mechanics there is a strict "plane of the present". The quantum mechanical wave function is non-localized in space but is strictly localized in time. What if Alice decided to work with Bob's wave function, rather than her own? She will get by hypothesis all the same predictions, but will be using a wave function that from her point of view slops into past and future.

The other is that from the point of view of special relativity, there should not be a strict "plane of the present" in the first place. We should be able to rotate between the four dimensional references frames of Alice and Bob as easily as we rotate between references frames for the three space dimensions.

What happens if we shift to four dimensional wave functions?

$$\psi\left(\vec{x}\right) \rightarrow \psi\left(t, \vec{x}\right) \tag{4}$$

Assume the coordinate systems for Alice and Bob are related by a Lorentz transformation Λ:

$$x^{(Bob)} = \Lambda x^{(Alice)} \tag{5}$$

Then their wave functions can be related by a Lorentz transformation of their coordinates:

$$\psi^{(Bob)}\left(x^{(Bob)}\right) = \psi^{(Alice)}\left(\Lambda x^{(Alice)}\right) \tag{6}$$

and matters are much more straightforward.

We make this our basic hypothesis: *the quantum mechanical wave function should be extended in the time direction on the same basis as it is extended along the three space dimensions.*

We are playing a "game of if" here: we will push the idea as hard as we can and see what breaks. We are not going to argue that this is or is not true. We are going to look for experimental tests and then let the experimentalists decide the question.

There are two principal effects:

(i) Dispersion in time appears on same basis as dispersion in space. Physical wave functions are always a bit spread out in space; they will now also be a bit spread out in time.

(ii) The uncertainty principle for time/energy is treated on same basis as the uncertainty principle for space/momentum. If a particle's position in time is well-defined, its energy will highly uncertain and vice versa.

1.1.1. Dispersion in time If the wave functions normally have an extension in time then every time-specific measurement should show additional dispersion in time.

Suppose we are measuring the time-of-arrival of a particle at a detector. Define the average time-of-arrival as:

$$\left\langle \tau^{(TOA)} \right\rangle \equiv \int_{-\infty}^{\infty} d\tau \tau p^{(TOA)}(\tau) \tag{7}$$

with an associated uncertainty:

$$\left\langle \Delta\tau^{(TOA)} \right\rangle \equiv \sqrt{\int_{-\infty}^{\infty} d\tau \tau^2 p^{(TOA)}(\tau) - \left\langle \tau^{(TOA)} \right\rangle^2} \tag{8}$$

The probability distribution for the particle will normally be spread out in space, so its arrival times will also be spread out, depending on the velocity of the particle and its dispersion in space.

But if it also has a dispersion in time, then part of the wave function will reach the detector – thanks to the fuzziness in time – a bit sooner and also a bit later than otherwise expected. There will be an additional dispersion in the time-of-arrival due to the dispersion in time.

As we will see (subsection 4.3), at non-relativistic speeds, the dispersion in the time-of-arrival is dominated by the dispersion in space, so this effect may be hard to pick out. At relativistic speeds, the contributions of the space and time dispersions can be comparable.

1.1.2. Uncertainty principle for time and energy Of particular significance for this work are differences in the treatment of the uncertainty principle for time/energy as opposed to that for space/momentum.

In the early days of quantum mechanics, these were treated on same basis. See for instance the discussions between Bohr and Einstein of the famous clock-in-a-box experiment [8] or the comments of Heisenberg in [9].

In later work this symmetry was lost. As Busch [10] puts it "... different types of time energy uncertainty can indeed be deduced in specific contexts, but ... there is no unique universal relation that could stand on equal footing with the position-momentum uncertainty relation." See also Pauli, Dirac, and Muga [11, 12, 13, 14].

There does not appear to be any experimental test of this or observational evidence for it; it is merely the way the field has developed.

IOP Conf. Series: Journal of Physics: Conf. Series **1239** (2019) 012015 doi:10.1088/1742-6596/1239/1/012015

That is not to say that there are not uncertainties with respect to time, but they are side effects of other uncertainties in quantum mechanics. For instance, if a particle is spread out in space, moving to the right, and going towards a detector at a fixed position, its time-of-arrival will have a dispersion in time. But this is a side-effect of the dispersion in space.

Now consider a particle going through a narrow slit in time, for instance a camera shutter. Its wave function will be clipped in time. If the wave function is not extended in time, then the wave function will merely be clipped: the resulting dispersion in time at detector will be *reduced.*

But if the wave function is extended in time and the Heisenberg uncertainty principle applies in time/energy on the same basis as with space/momentum, then an extremely fast camera shutter will give a small uncertainty in time at the gate:

$$\Delta t \to 0 \tag{9}$$

causing the uncertainty in energy to become arbitrarily great:

$$\Delta E \geq \frac{1}{\Delta t} \Rightarrow \Delta E \to \infty \tag{10}$$

which will in turn cause the wave function to be diffracted, to fan out in time, and the dispersion in time-of-arrival to become arbitrarily great.

1.1.3. A necessary hypothesis This question does not appear to have been attacked directly. As noted, the assumption that the wave function is not extended in time seems to have crept into the literature of its own, without experimental test or observational evidence.

To make an experimental test of this question we have to develop predictions for both branches:

(i) Assume the wave function is not extended in time. Make predictions about time-of-arrival and the like.
(ii) Assume the wave function is extended in time. Make equivalent predictions.
(iii) Compare.

We have to develop both branches in a way that makes the comparison straightforward.

Further, to make the results falsifiable we have to develop the extended-in-time branch in a way that is clearly correct. A null result should show that the wave function is *not* extended in time.

These objectives drive what follows.

1.1.4. Literature The literature for special relativity and for quantum mechanics is vast. Our focus is on the critical intersection of the two. References of particular interest here include:

- Stueckelberg & Feynman's original papers: [15, 16, 17, 18, 19, 20].
- Reviews of the role of time, Schulman, Zeh, Muga, Callender: [21, 22, 13, 14, 23].
- Block universe picture: Parmenides, Barbour, Price: [24, 25, 26].
- Cramer's transactional interpretation: [27, 28, 29, 30].
- The time symmetric quantum mechanics of Aharonov and Reznik: [31, 32].
- The relativistic dynamics program of Horwitz, Piron, Land, Collins, and others: [33, 34, 35, 36, 37, 38, 39, 40]. This program is a natural outgrowth of Stueckelberg & Feynman's work.

Our approach here may be understood as falling within the relativistic dynamics framework, but with the emphasis placed on the the "coordinate time" rather than the "evolution parameter" aspects of that program. We will discuss this in detail once an appropriate foundation has been laid.

IOP Conf. Series: Journal of Physics: Conf. Series **1239** (2019) 012015 doi:10.1088/1742-6596/1239/1/012015

1.2. Order of magnitude estimate

Has this hypothesis has already been falsified? Quantum mechanics has been tested with extraordinary precision. Should associated effects have been seen already, even if not looked for?

Consider the atomic scale given by the Bohr radius $5.3\ 10^{-11}m$. We take this as an estimate of the uncertainty in space.

We assume the maximum symmetry possible between time and space. We therefore infer that the uncertainty in time should be of order the uncertainty in space (in units where $c = 1$).

Dividing the Bohr radius by the speed of light we get the Bohr radius in time $a_0 = .177\ 10^{-18}s$, or less than an attosecond. $.177as$ is therefore our starting estimate of the uncertainty in time.

Therefore from strictly dimensional and symmetry arguments, the effects will be small, of order attoseconds. This is sufficient to explain why such effects have not been seen.

At the same time, the time scales we can look at experimentally are now getting down to the attosecond range. A recent paper by Ossiander et al [41] reports results at the sub-attosecond level.

Therefore if we can provide the experimentalists with a sufficiently well-defined target, the hypothesis should be falsifiable in practice.

1.3. Plan of attack

> *Look, I don't care what your theory of time is. Just give me something I can prove wrong.* – experimentalist at the 2009 Feynman Festival in Olomouc

1.3.1. Primary objective is falsifiability It is not enough to extend quantum mechanics to include time. It is necessary to do so in a way that can be proved wrong. The approach has to be so strongly and clearly constrained that if it is proved wrong, the whole project of extending quantum mechanics to include time is falsified.

Our requirements are therefore that we have:

(i) the most complete possible equivalence in the treatment of time and space – manifest covariance at every point at a minimum,

(ii) consistency with existing experimental and observation results,

(iii) and consistency between the single particle and multiple particle domains.

These requirements leave us with no free parameters. And having no free parameters means in turn that our hypothesis is falsifiable in principle.

To get to falsifiable in practice, we will look for the simplest cases that make a direct comparison possible.

We will also look at points of principle that need to be addressed, to ensure that the approach is not ruled out by, say, violations of unitarity.

We will use the acronym SQM for standard quantum mechanics. We will use the acronym TQM for temporal quantum mechanics. By TQM we mean SQM with time treated on the system basis as space: time just as much an observable as the three space dimensions.

We do *not* mean by "temporal quantum mechanics" that time itself comes in small chunks or quanta! For instance, there has been speculation that time is granular at the scale of the Planck time: $t_{Planck} \equiv \sqrt{\frac{\hbar G}{c^5}} \approx 5.39116x10^{-44}s = 5.39116x10^{-26}as$. Maybe it is, maybe it isn't. But as this is 25 orders of magnitude smaller than the times we are considering here, it is reasonable for us to take time as continuous. And since space is treated by SQM as continuous, and since the defining assumption of TQM is the maximum symmetry between time and space, we are required to take time in TQM continuous.

IOP Conf. Series: Journal of Physics: Conf. Series **1239** (2019) 012015 doi:10.1088/1742-6596/1239/1/012015

1.3.2. Extruding quantum mechanics in time If you have worked with a CAD/CAM or 3D drawing program you are familiar with the "extrude" operation, where you take a circle or square and extrude it into the third dimension[1]. That is all we are going to do here. We are going to use manifest covariance as the rule for going from three dimensions to four – so that $s^2 = t^2 - x^2 - y^2 - z^2$ is the invariant rather than $r^2 = x^2 + y^2 + z^2$ – but the principle is the same.

The advantage is that with this approach we have no free parameters. The moment we introduce a free parameter we put falsifiability at risk since we are thereby creating a "fudge factor". And any fudge factor might let us say "well perhaps no effect was seen at scale x, but at smaller scale y, then we shall see!". We want our experimentalists to be able to say, "not seen, therefore not there!"

It might be that we should have done the extrude at a slight angle, extruding our circle into, say, an oblate or prolate spheroid rather than a perfect one. But this does not seriously impact falsifiability – such an oblate or prolate spheroid should still have the same overall scale of a Bohr radius in time, it will still give our experimentalist a fixed target. (We discuss possible meanings of "a slight angle" in subsection 3.5 and again in the discussion).

We will use path integrals as our defining formalism. With the path integral approach we can extrude the paths from three to four dimensions in a straightforward way, while leaving the rest of the machinery essentially untouched. With other formalisms, it is less clear what "extrude" means. But once we have the meaning of extrude worked out for path integrals, we can, as it were, rotate our path integral formalism into other formalisms, seeing what "extruding quantum mechanics in time" looks like as a Schrödinger equation or in quantum field theory.

Our plan:

1.3.3. Single particle case In the single particle case we will:

(i) Generalize path integrals to include time as an observable.

(ii) Derive the corresponding Schrödinger equation as the short time limit of the path integral.

(iii) Develop the free solutions. We will estimate the initial wave function, let it evolve in time, and detect it. We will compute the dispersions of time-of-arrival measurements in SQM and in TQM. In general the differences are real but small.

(iv) Analyze the single and double slit experiments. The single slit in time experiment provides the decisive test of temporal quantum mechanics. In SQM, the narrower the slit, the *less* the dispersion in subsequent time-of-arrival measurements. In TQM, the narrower the slit, the *greater* the dispersion in subsequent time-of-arrival measurements. *In principle, the difference may be made arbitrarily great.*

1.3.4. Multiple particle case We will then extend TQM to include the multiple particle case, i.e. field theory. We will show, using a toy model, that:

(i) We can extend the usual path integral approach to include time as an observable. The basis functions in Fock space extend in a natural way from three to four dimensions, the Lagrangian is unchanged, and the usual Feynman diagram expansions appear.

(ii) For each Feynman diagram in SQM we can compute the TQM equivalent. Therefore any problem that can be solved using Feynman diagrams in SQM can be solved in TQM.

(iii) *The usual ultraviolet divergences do not appear* (the combination of dispersion in time and entanglement in time contain them).

(iv) And that there are a large number of additional experimental effects to be seen, including:

[1] See Abbott's classic Flatland [42] for a delightful example.

(a) wave functions anti-symmetric in time,
(b) correlations, entanglement, and interference in time,
(c) and forces of anticipation and regret.

1.3.5. Overall conclusions With this done, we will argue in the discussion:

(i) that TQM is not ruled out a priori.
(ii) that TQM is falsifiable. And given experimental work like Ossiander's, probably with current technology.
(iii) that TQM is a source of interesting experiments. Every foundational experiment in SQM has an "in time" variant.
(iv) that TQM is a potential starting point for attacks on the quantum gravity problem, since TQM is manifestly covariant and untroubled by the ultraviolet divergences.
(v) that as TQM is a straight-forward extrapolation of quantum mechanics and special relativity, experiments that falsify TQM are likely to require modification of our understanding of either quantum mechanics or special relativity or both. Something will have to break.

1.3.6. Limits of the current investigation We do no more here than is required to establish that TQM is well-defined, self-consistent, and falsifiable.

It will be clear from the development that the basic idea could be extended in a number of directions; we review some of these in the concluding discussion.

The conventions used in this work are spelled out in Appendix A.

2. Path integrals

2.1. Overview

To extend quantum mechanics to include time we will take as our starting point Feynman's path integral approach to quantum mechanics [43, 44, 45, 46, 47, 48, 49, 50, 51, 52].

With the path integral approach, the only change we will need to make is to generalize the paths from varying in three dimensions to varying in four.

To make clear what this means, consider the case of Alice walking her dog, say from her front door to Bob's.

Alice will take the shortest (classical) path from door to door.

But her dog will dart from side to side, now investigating a mailbox to the left, now checking out a lamppost to the right. In fact, as a quantum dog he will investigate all such paths simultaneously. While he will start at the same time and place as Alice, and finish at the same time and place as Alice, in between he will travel simultaneously along all possible paths.

But – in SQM – only along paths in space. At each tick of Alice's digital watch, her dog will be found off to the left or right, jumping up or digging down, further along the path to Bob's, or holding back for an important investigation.

But in TQM, the quantum dog can – and therefore will – advance into the future and drop back into the past. So that tick by tick of Alice's watch, her dog's paths will have to tracked in four dimensions rather than three.

This is harder to visualize, being out of our normal experience. So we develop the analysis a bit formally, letting math take the place of an as yet undeveloped intuition.

Path integrals, as the name suggests, are done by summing over all paths from starting point to finish, weighting each path by the integral of the Lagrangian (the action) along it:

IOP Conf. Series: Journal of Physics: Conf. Series **1239** (2019) 012015 doi:10.1088/1742-6596/1239/1/012015

$$\psi_T(x_T) = \int \mathcal{D}x_\tau \exp\left(\imath \int_0^T d\tau L[x,\dot{x}]\right)\psi_0(x_0) \tag{11}$$

Piece by piece:

(i) $\psi_0(x_0)$ is the initial wave function. We will be breaking these down into sums over Gaussian test functions using Morlet wavelet analysis.

(ii) τ is the clock time as given by Alice's digital watch. We will break up the paths into the bits from one clock tick to the next.

(iii) $\mathcal{D}x_\tau$ represents the paths. Each path is defined by its coordinates at each clock tick. In SQM, these are the values of x_τ, y_τ, z_τ at each clock tick. In TQM these are values of $t_\tau, x_\tau, y_\tau, z_\tau$ at each clock tick.

(iv) $L[x,\dot{x}]$ is a suitable Lagrangian. We will be using one that works equally well for both SQM and TQM.

(v) $\imath \int_0^T d\tau L[x,\dot{x}]$ is the action, the integral over the Lagrangian taken path by path.

(vi) And $\psi_T(x_T)$ is the final wave function, the amplitude for the dog to arrive at Bob's door step.

We will look at:

(i) What do we mean by τ the clock time?

(ii) What do we mean by the coordinate time t in t,x,y,z?

(iii) How do we define the initial wave function in a way that does not potentially bias the outcome?

(iv) What Lagrangian shall we use?

(v) How do we get the sums to converge?

(vi) Having gotten the sums to converge, how do we normalize them?

(vii) And what do all the pieces look like when we put them back together?

2.2. Laboratory time

We consider the action, the integral of the Lagrangian over time:

$$\int_0^T d\tau \mathcal{L}[x,\dot{x}] \tag{12}$$

In classical mechanics, we are free to take the parameter τ as any monotonically increasing variable. We will get the same classical equations of motion in any case.

A typical choice is to select τ as the proper time of the particle in question. However this makes it difficult to extend the work to the multiple particle case, where there are many particles and therefore many proper times in play.

Here we choose to use the time as shown on a laboratory clock. We take the term laboratory time from Busch [10]. We will use the terms clock time and laboratory time interchangeably.

It is useful to visualize this clock as a metronome, breaking up the clock time into a series of N ticks each of length ϵ. If $\tau = 0$ at the source, and $\tau = T$ at the detector, we have:

$$\epsilon \equiv \frac{T}{N} \tag{13}$$

We will take the limit as $N \to \infty$ as the final step in the calculation.

IOP Conf. Series: Journal of Physics: Conf. Series **1239** (2019) 012015 doi:10.1088/1742-6596/1239/1/012015

2.3. Coordinate time

We visualize a four dimensional coordinate system coordinates t, x, y, z. We will refer to t as coordinate time by analogy with the three coordinate space dimensions: coordinate x, coordinate y, and coordinate z.

Paths are defined with reference to this coordinate system. If the time by Alice's watch is τ, then each path π will have a location at τ given by:

$$\pi_\tau (t, x, y, z) \tag{14}$$

It may help to think of the coordinates as laid out on a piece of four dimensional graph paper. At a specific clock tick n, a specific path π will be represented by a dot on a specific vertex on the four dimensional graph paper. If we want to see the progress of the path with respect to clock time, we can flip the series of pieces of graph paper like one of those old time flip movies.

If our graph paper has M grid lines in each direction, the number of vertices on a page is M^4, and the number of paths total is M^{4N}. Each different sequence of grid points counts as a distinct path.

The path integral measure $\mathcal{D}x$ is usually defined by assigning a weight of one to each distinct path, and then taking the limit as the spacing goes to zero.

Since coordinate time is on the same footing as the three space coordinates it has a corresponding energy operator:

$$p_x \equiv -\imath\frac{\partial}{\partial x} \Rightarrow E \equiv \imath\frac{\partial}{\partial t} \tag{15}$$

We will refer to this as coordinate energy. It is not positive definite or bounded from below. Since p_x can be positive or negative, by our controlling requirement of covariance E can be positive or negative.

We discuss the relationship between clock time and coordinate time in Appendix F.

2.4. Initial wave function

We need a starting set of wave functions ψ_0 at clock time $\tau = 0$. We will need wave functions that extend in both coordinate time and space. The usual choices would be δ functions or plane waves.

In coordinate time these might be:

$$\delta(t - t_0) \tag{16}$$

$$e^{-\imath E(t-t_0)} \tag{17}$$

or in space:

$$\delta(x - x_0) \tag{18}$$

$$e^{\imath p(x-x_0)} \tag{19}$$

But neither δ functions nor plane waves are physical. Their use creates a risk of artifacts.

More physical would be Gaussian test functions, for instance in coordinate time:

$$\varphi(t) \equiv \sqrt[4]{\frac{1}{\pi\sigma_t^2}}e^{-\imath E(t-t_0)-\frac{(t-t_0)^2}{2\sigma_t^2}} \tag{20}$$

or in space:

$$\varphi(x) \equiv \sqrt[4]{\frac{1}{\pi\sigma_x^2}}e^{\imath p(x-x_0)-\frac{(x-x_0)^2}{2\sigma_x^2}} \tag{21}$$

But while Gaussian test functions are physically reasonable they are not completely general.

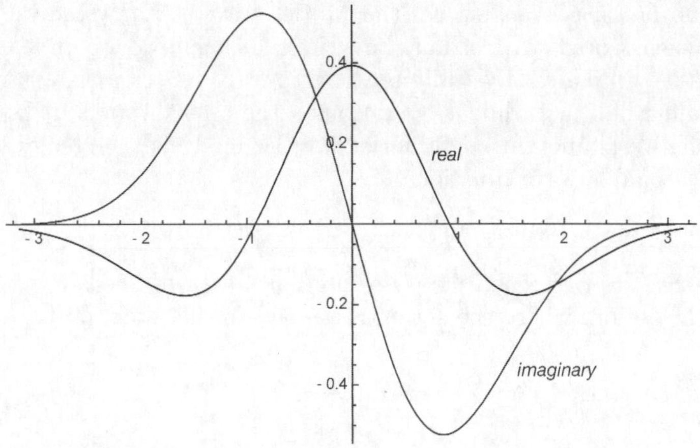

Figure 2. Mother Morlet wavelet

We can achieve both generality and physical reasonableness by using a basis of Morlet wavelets [53, 54, 55, 56, 57, 58, 59, 60, 61, 62].

Morlet wavelets are derived by starting with a "mother" wavelet:

$$\phi^{(mother)}(t) \equiv \left(e^{-\imath t} - \frac{1}{\sqrt{e}} \right) e^{\left(-\frac{t^2}{2}\right)} \tag{22}$$

and scaling and displacing it with the replacement $t \to \frac{t-l}{s}$:

$$\phi_{sl}(t) = \frac{1}{\sqrt{|s|}} \left(e^{-\imath\left(\frac{t-l}{s}\right)} - \frac{1}{\sqrt{e}} \right) e^{-\frac{1}{2}\left(\frac{t-l}{s}\right)^2} \tag{23}$$

Any normalizable function f can be broken up into wavelet components $\hat{f}$ using:

$$\hat{f}_{sl} = \int_{-\infty}^{\infty} dt \phi^*_{sl}(t) f(t) \tag{24}$$

And then recovered using the inverse Morlet wavelet transform:

$$f(t) = \frac{1}{C} \int_{-\infty}^{\infty} \frac{ds dl}{s^2} \phi_{sl}(t) \hat{f}_{sl} \tag{25}$$

The value of C is worked out in [62].

Therefore we can write any physically reasonable wave function f in time in terms of Morlet wavelets.

We may include space by using products of Morlet wavelets:

$$\begin{aligned} \hat{f}_{s_t l_t s_x l_x} &= \int dt dx \phi^*_{s_t l_t}(t) \phi^*_{s_x l_x}(x) f(t,x) \\ f(t,x) &= \frac{1}{C^2} \int \frac{ds_t dl_t}{s_t^2} \frac{ds_x dl_x}{s_x^2} \phi_{s_t l_t}(t) \phi_{s_x l_x}(x) \hat{f}_{s_t l_t s_x l_x} \end{aligned} \tag{26}$$

Clearly it would be cumbersome to track four dimensional Morlet wavelets at every step.

IOP Conf. Series: Journal of Physics: Conf. Series **1239** (2019) 012015 doi:10.1088/1742-6596/1239/1/012015

Fortunately we do not need to perform the Morlet wavelet analyses; we merely need the ability to do so. As each Morlet wavelet may be written as a sum of a pair of Gaussians, Morlet wavelet analysis lets us write any physically reasonable wave function as a sum over Gaussians. Provided we are dealing only with linear operations – the case throughout here – we can work directly with Gaussian test functions. By Morlet wavelet analysis the results will then be valid for any physically reasonable wave functions.

2.5. Lagrangian

To sum over the paths – to construct the path integral – we will need to weight each path by the exponential of the action, where the action is defined as the integral of the Lagrangian over the laboratory time:

$$e^{\imath\int\limits_0^T d\tau L(x^\mu,\dot{x}^\mu)} \tag{27}$$

We require a Lagrangian which:

(i) Is manifestly covariant,

(ii) Produces the correct classical equations of motion,

(iii) And gives the correct Schrödinger equation.

We would further prefer a Lagrangian which is the same for both SQM and TQM. This will let us argue that we are treating SQM and TQM with the most complete possible equality.

Somewhat surprisingly such a Lagrangian exists. In Goldstein's well-known text on classical mechanics [63] we find:

$$L\left(x^\mu,\dot{x}^\mu\right) = -\frac{1}{2}m\dot{x}^\mu\dot{x}_\mu - q\dot{x}^\mu A_\mu\left(x\right) \tag{28}$$

The potentials are not themselves functions of the laboratory time τ. The mass m is the rest mass of the particle, an invariant.

This Lagrangian is unusual in that it uses four independent variables (the usual three space coordinates plus a time variable) but still gives the familiar classical equations of motion (see Appendix B).

This Lagrangian therefore provides a natural bridge from a three to a four dimensional picture.

The classical equations of motion are still produced if we add a dimensionless scale a and an additive constant b to the Lagrangian:

$$-\frac{1}{2}am\dot{x}^\mu\dot{x}_\mu - aq\dot{x}^\mu A_\mu\left(x\right) - ab\frac{m}{2} \tag{29}$$

The Lagrangian is therefore only determined up to a and b. The requirement that we match the SQM results will fix a and b (subsection 3.3).

2.6. Convergence

How do we get the integrals over the paths to converge without breaking covariance?

We compute the path integral for the kernel by slicing the clock time into an infinite number of intervals and integrating over each:

$$K_{BA} = \lim_{N\to\infty} C_N \int \prod_{n=1}^{N-1} dt_n d\vec{x}_n e^{\imath\varepsilon\sum\limits_{j=1}^{N} L_j} \tag{30}$$

with C_N an appropriate normalization factor.

IOP Conf. Series: Journal of Physics: Conf. Series **1239** (2019) 012015 doi:10.1088/1742-6596/1239/1/012015

Consider the discrete form of the Lagrangian. We use a tilde to mark the coordinate time part and an overbar to mark the space part:

$$L_j \equiv \tilde{L}_j^t + \bar{L}_j^{\vec{x}} + L_j^m \tag{31}$$

$$\tilde{L}_j^t \equiv -a\frac{m}{2}\left(\frac{t_j - t_{j-1}}{\varepsilon}\right)^2 - qa\frac{t_j - t_{j-1}}{\varepsilon}\frac{\Phi(x_j) + \Phi(x_{j-1})}{2} \tag{32}$$

$$\bar{L}_j^{\vec{x}} \equiv a\frac{m}{2}\left(\frac{\vec{x}_j - \vec{x}_{j-1}}{\varepsilon}\right)^2 + qa\frac{\vec{x}_j - \vec{x}_{j-1}}{\varepsilon}\cdot\frac{\vec{A}(x_j) + \vec{A}(x_{j-1})}{2} \tag{33}$$

$$L_j^m \equiv -ab\frac{m}{2} \tag{34}$$

We are using the mid-point rule, averaging the scalar and the vector potentials over the start and end points of the step, by analogy with the rule for three dimensions (Schulman [44], Grosche and Steiner [49]).

Now look at a single step for the free case, vector potential A_μ zero:

$$\int_{-\infty}^{\infty}\int_{-\infty}^{\infty} dt_j d\vec{x} e^{-\imath am\frac{\left(t_j - t_{j-1}\right)^2}{2\varepsilon} + \imath am\frac{\left(\vec{x}_j - \vec{x}_{j-1}\right)^2}{2\varepsilon}} \tag{35}$$

The formal tricks normally used to ensure convergence do not work here (e.g. Kashiwa or Zinn-Justin [48, 50]). Perhaps the most popular of these is the use of Wick rotation to shift to a Euclidean time:

$$\tau \to -\imath\tau \tag{36}$$

This causes integrals to converge rapidly going into the future, but makes the past inaccessible. For instance, factors of $\exp(-\imath\omega\tau)$ – which spring up everywhere in path integrals – converge going into the future, but blow up going into the past. If we are to treat time on the same footing as space – our central assumption – then past and future must be treated as symmetrically as left and right.

Another approach is to add a small convergence factor at a cleverly chosen spot in the arguments of the exponentials. But if we attach a convergence factor to t and x separately, we break manifest covariance. If we attach our convergence factor to both, the fact that the t and x parts enter with opposite sign means any convergence factor that works for one will fail for the other. We could try attaching one to the mass m, but this also fails. For instance if $a > 0$ and we subtract a small factor of $\imath\delta$ from the mass:

$$m \to m - \imath\delta \tag{37}$$

the t integral converges but the x integral diverges.

We recall the kernel has meaning only when applied to a specific physical wave function. If we break the incoming wave up into Morlet wavelets and then into Gaussian test functions, we see that each integral converges by inspection, the factor $e^{-\frac{1}{2}\left(\frac{t-l}{s}\right)^2}$ ensures this.

So for physically significant wave functions, there is no problem in the first place. Effectively we are taking seriously the point that the path integral kernel is a distribution, only meaningful with respect to specific wave functions.

IOP Conf. Series: Journal of Physics: Conf. Series **1239** (2019) 012015 doi:10.1088/1742-6596/1239/1/012015

2.7. Normalization

Now that we have our path integrals converging, we have to normalize them. If we start from the Schrödinger equation, the normalization is wired in. But in path integrals we are a bit at sea.

We will here deal with the free case, verifying the normalization is correct in the general case in Appendix C.

The normalization factor for N steps we will call C_N. The defining requirement is that, if the initial wave function is normalized to one, then with the inclusion of C_N, the final wave function will be normalized to one as well:

$$\int dt_0 d\vec{x}_0 \left|\psi_0(t_0,\vec{x}_0)^2\right| = 1 \rightarrow \int dt_N d\vec{x}_N \left|\psi_N(t_N,\vec{x}_N)^2\right| = 1 \tag{38}$$

If C_N depends on the particular ψ_0, then we have failed.

We now compute the factor of C_N.

2.7.1. Normalization in time We start with the coordinate time dimension only. Consider a Gaussian test function centered on an initial position in coordinate time $\bar{t}_0$:

$$\tilde{\varphi}_0(t_0) \equiv \sqrt[4]{\frac{1}{\pi\sigma_t^2}} e^{-\imath E_0 t_0 - \frac{(t_0-\bar{t}_0)^2}{2\sigma_t^2}} \tag{39}$$

We write the kernel for the time part as:

$$\tilde{K}_\tau(t_N;t_0) \sim \int \prod_{j=1}^{N-1} dt_j e^{-\imath am \sum\limits_{k=1}^{N} \frac{(t_k-t_{k-1})^2}{2\varepsilon}} \tag{40}$$

The wave function after the initial integral over t_0 is:

$$\tilde{\varphi}_\varepsilon(t_1) = \int dt_0 e^{-\imath \frac{am}{2\varepsilon}(t_1-t_0)^2} \tilde{\varphi}_0(t_0) \tag{41}$$

or:

$$\tilde{\varphi}_\varepsilon(t_1) = \sqrt{\frac{2\pi\varepsilon}{\imath am}} \sqrt[4]{\frac{1}{\pi\sigma_t^2}} \sqrt{\frac{1}{f_\varepsilon^{(t)}}} e^{-\imath E_0 t_1 + \imath \frac{E_0^2}{2am}\varepsilon - \frac{1}{2\sigma_t^2 f_\varepsilon^{(t)}}\left(t_1-\bar{t}_0-\frac{E_0}{am}\varepsilon\right)^2} \tag{42}$$

with the dispersion factor $f_\tau^{(t)} \equiv 1 - \imath \frac{\tau}{am\sigma_t^2}$.

The normalization requirement is:

$$1 = \int dt_1 \tilde{\varphi}_\varepsilon^*(t_1)\tilde{\varphi}_\varepsilon(t_1) \tag{43}$$

The first step normalization is correct if we multiply the kernel by a factor of $\sqrt{\frac{\imath am}{2\pi\varepsilon}}$. Since this normalization factor does not depend on the laboratory time the overall normalization for N infinitesimal kernels is the product of N of these factors:

$$C_N \equiv \sqrt{\frac{\imath am}{2\pi\varepsilon}}^N \tag{44}$$

Note also that the normalization does not depend on the specifics of the Gaussian test function (the values of E_0, σ_t^2, and $\bar{t}_0$) so it is valid for an arbitrary sum of Gaussian test functions as well. And therefore, by Morlet wavelet decomposition, for an arbitrary wave function.

As noted, the phase is arbitrary. If we were working the other way, from Schrödinger equation to path integral, the phase would be determined by the Schrödinger equation itself. The specific phase choice we are making here has been chosen to help ensure the four dimensional Schrödinger equation is manifestly covariant, see below. We may think of the phase choice as a choice of gauge (see Appendix D).

Therefore the expression for the free kernel in coordinate time is (with $t'' \equiv t_N$, $t' \equiv t_0$):

$$\tilde{K}_\tau\left(t'';t'\right) = \int dt_1 dt_2 \dots dt_{N-1} \sqrt{\frac{\imath am}{2\pi\varepsilon}}^N e^{-\imath \sum_{j=1}^{N} \left(\frac{am}{2\varepsilon}(t_j - t_{j-1})^2\right)} \tag{45}$$

Doing the integrals we get:

$$\tilde{K}_\tau\left(t'';t'\right) = \sqrt{\frac{\imath am}{2\pi\tau}} e^{-\imath am \frac{(t''-t')^2}{2\tau}} \tag{46}$$

and free wave functions in coordinate time:

$$\tilde{\varphi}_\tau(t) = \sqrt[4]{\frac{1}{\pi\sigma_t^2}} \sqrt{\frac{1}{f_\tau^{(t)}}} e^{-\imath E_0 t - \frac{1}{2\sigma_t^2 f_\tau^{(t)}}\left(t - \bar{t}_0 - \frac{E_0}{am}\tau\right)^2 + \imath \frac{E_0^2}{2am}\tau} \tag{47}$$

2.7.2. Normalization in space We redo the analysis for coordinate time for space. We use the correspondences:

$$t \to x, m \to -m, \bar{t}_0 \to \bar{x}_0, E_0 \to -p_0, \sigma_t^2 \to \sigma_x^2 \tag{48}$$

With these we can write down the equivalent set of results by inspection. Since we will need the results below, we do this explicitly. We get the initial Gaussian test function:

$$\bar{\varphi}_0(x_0) = \sqrt[4]{\frac{1}{\pi\sigma_x^2}} e^{\imath p_0 x_0 - \frac{(x_0 - \bar{x}_0)^2}{2\sigma_x^2}} \tag{49}$$

free kernel:

$$\bar{K}_\tau\left(x'';x'\right) \sim \int dx_1 dx_2 \dots dx_{N-1} e^{\imath \sum_{j=1}^{N} \frac{am}{2\varepsilon}(x_j - x_{j-1})^2} \tag{50}$$

and normalized kernel:

$$\bar{K}_\tau\left(x'';x'\right) = \sqrt{-\frac{\imath am}{2\pi\tau}} e^{\imath am \frac{(x''-x')^2}{2\tau}} \tag{51}$$

The kernel matches the usual (non-relativistic) kernel [64, 44] if $a = 1$.

The wave function is:

$$\bar{\varphi}_\tau(x) = \sqrt[4]{\frac{1}{\pi\sigma_x^2}} \sqrt{\frac{1}{f_\tau^{(x)}}} e^{\imath p_0 x - \frac{1}{2\sigma_x^2 f_\tau^{(x)}}\left(x - \bar{x}_0 - \frac{p_0}{am}\tau\right)^2 - \imath \frac{p_0^2}{2am}\tau} \tag{52}$$

with the definition of the dispersion factor $f_\tau^{(x)} = 1 + \imath \frac{\tau}{am\sigma_x^2}$ parallel to that for coordinate time (but with opposite sign for the imaginary part).

2.7.3. Normalization in time and space The full kernel is the product of the coordinate time kernel, the three space kernels, and the constant term $e^{-\imath \frac{abm}{2}\tau}$. We understand x to refer to coordinate time and all three space dimensions:

$$K_\tau\left(x'';x'\right) = -\imath \frac{a^2 m^2}{4\pi^2\tau^2} e^{-\frac{\imath am}{2\tau}(x''-x')^2 - \imath \frac{abm}{2}\tau} \tag{53}$$

We have done the analysis only for Gaussian test functions, but by Morlet wavelet decomposition it is completely general.

IOP Conf. Series: Journal of Physics: Conf. Series **1239** (2019) 012015 doi:10.1088/1742-6596/1239/1/012015

2.8. Formal expression
We now have the full path integral:

$$K_\tau\left(x'';x'\right) = \lim_{N\to\infty} \int \mathcal{D}x e^{\imath \sum\limits_{j=1}^{N} L_j \epsilon} \tag{54}$$

with the measure:

$$\mathcal{D}x \equiv \left(-\imath a^2 \frac{m^2}{4\pi^2\varepsilon^2}\right)^N \prod_{n=1}^{N-1} d^4 x_n \tag{55}$$

and the discretized Lagrangian at each step:

$$L_j \equiv -am\frac{\left(x_j - x_{j-1}\right)^2}{2\varepsilon^2} - aq\frac{x_j - x_{j-1}}{\epsilon}\frac{A\left(x_j\right) + A\left(x_{j-1}\right)}{2} - ab\frac{m}{2} \tag{56}$$

We will show that $a = b = 1$ in the next section.

3. Schrödinger equation

3.1. Overview
The path integral and Schrödinger equation views are complementary. We need both to fully understand either.

We derive the Schrödinger equation from the path integral by taking the short time limit of the path integral form.

By comparing the result to the Klein-Gordon equation – and making a reasonable assumption about the long time evolution of the wave functions – we are able to fix the additive and scale constants in the Lagrangian.

The resulting equation looks like the Klein-Gordon equation over short times but shows some drift over longer times. We use some heuristic arguments to estimate the scale of the long term drift as of order picoseconds, a million times longer than the attosecond scale of the time dispersions we are primarily concerned with here. We will therefore be able to largely ignore this drift.

With a, b defined, we look at a further problem. We have done the derivation of path integral and Schrödinger equation forms from Alice's perspective. But what of Bob, jetting around like a fusion powered mosquito?

We resolve this conflict by arguing that we can find a natural rest frame that both can use. Starting with an argument of Weinberg's, we argue we can associate an energy-momentum tensor with spacetime. This means we can associate a local rest frame with spacetime. And this local rest frame can provide the neutral and agreed defining frame for TQM.

This will complete the formal development of TQM.

Before turning to applications, we will then look at the relationship of TQM to the relativistic dynamics program. We will argue that TQM may be understood as a member of that program aggressively specialized to achieve falsifiability.

3.2. Derivation of the Schrödinger equation
Normally the path integral expression is derived from the Schrödinger equation. But because for us the path integral provides the defining formulation we need to run the analysis in the "wrong" direction.

Our starting point is a derivation of the path integral from the Schrödinger equation by Schulman [44]. We run his derivation in reverse and with one extra dimension[2].

[2] This derivation is done for the free case in Fanchi [36].

IOP Conf. Series: Journal of Physics: Conf. Series **1239** (2019) 012015 doi:10.1088/1742-6596/1239/1/012015

We start with the discrete form of the path integral. We consider a single step of length ϵ, taking $\epsilon \to 0$ at the end. Because of this, only terms first order in ϵ are needed.

Following Schulman, we define the coordinate difference:

$$\xi \equiv x_j - x_{j+1} \tag{57}$$

We rewrite the functions of x_j as functions of ξ and x_{j+1}. We expand the vector potential:

$$A_\nu (x_j) = A_\nu (x_{j+1}) + (\xi^\mu \partial_\mu) A_\nu (x_{j+1}) + \ldots \tag{58}$$

and the wave function:

$$\psi_\tau (x_j) = \psi_\tau (x_{j+1}) + (\xi^\mu \partial_\mu) \psi_\tau (x_{j+1}) + \frac{1}{2}\xi^\mu \xi^\nu \partial_\mu \partial_\nu \psi_\tau (x_{j+1}) + \ldots \tag{59}$$

giving:

$$\begin{gathered} \psi_{\tau+\varepsilon} (x_{j+1}) = \sqrt{\tfrac{\imath am}{2\pi\varepsilon}} \sqrt{-\tfrac{\imath am}{2\pi\varepsilon}}^3 \int d^4\xi e^{-\frac{\imath am\xi^2}{2\varepsilon} - \imath ab\frac{m}{2}\varepsilon} \\ \times e^{\imath aq\xi^\nu \left(A_\nu(x_{j+1}) + \frac{1}{2}(\xi^\mu\partial_\mu)A_\nu(x_{j+1}) + \ldots\right)} \\ \times \left(\psi_\tau (x_{j+1}) + (\xi^\mu \partial_\mu) \psi_\tau (x_{j+1}) + \tfrac{1}{2}\xi^\mu \xi^\nu \partial_\mu \partial_\nu \psi_\tau (x_{j+1}) + \ldots\right) \end{gathered} \tag{60}$$

We now expand in powers of $\xi \sim \sqrt{\epsilon}$. We do not need more than the second power:

$$\begin{gathered} \psi_{\tau+\varepsilon} (x_{j+1}) = \sqrt{\tfrac{\imath am}{2\pi\varepsilon}} \sqrt{-\tfrac{\imath am}{2\pi\varepsilon}}^3 \int d^4\xi e^{-\frac{\imath am\xi^2}{2\varepsilon}} \\ \times \left(1 + \imath aq\xi^\nu A_\nu (x_{j+1}) + \tfrac{\imath aq}{2}\xi^\nu \xi^\mu \partial_\mu A_\nu (x_{j+1}) - \tfrac{a^2q^2}{2}\xi^\mu A_\mu (x_{j+1}) \xi^\nu A_\nu (x_{j+1}) - \imath\tfrac{abm\varepsilon}{2}\right) \\ \times \left(\psi_\tau (x_{j+1}) + (\xi^\mu \partial_\mu) \psi_\tau (x_{j+1}) + \tfrac{1}{2}\xi^\mu \xi^\nu \partial_\mu \partial_\nu \psi_\tau (x_{j+1}) + \ldots\right) \end{gathered} \tag{61}$$

The term zeroth order in ξ gives:

$$\sqrt{\frac{\imath am}{2\pi\varepsilon}} \sqrt{-\frac{\imath am}{2\pi\varepsilon}}^3 \int d\xi^4 e^{-\frac{\imath am\xi^2}{2\varepsilon}} = 1 \tag{62}$$

This is not surprising; the normalization above was chosen to do this.

Terms linear in ξ give zero when integrated.

The terms second order in ξ (first in ϵ) are:

$$\left(\begin{array}{l} -\imath\frac{abm\varepsilon}{2} + \imath aq (\xi^\nu A_\nu) (\xi^\mu \partial_\mu) + \frac{\imath aq}{2}\xi^\nu \xi^\mu (\partial_\mu A_\nu) \\ -\frac{a^2q^2}{2} (\xi^\mu A_\mu) (\xi^\nu A_\nu) + \frac{1}{2}\xi^\mu \xi^\nu \partial_\mu \partial_\nu \end{array}\right) \psi_\tau \tag{63}$$

Integrals over off-diagonal powers of order ξ^2 give zero. Integrals over diagonal ξ^2 terms give:

$$\begin{gathered} \sqrt{\tfrac{\imath am}{2\pi\varepsilon}} \int d\xi_0 e^{-\frac{\imath am\xi_0^2}{2\varepsilon}} \xi_0^2 = \tfrac{\varepsilon}{\imath am} \\ \sqrt{-\tfrac{\imath am}{2\pi\varepsilon}} \int d\xi_i e^{\frac{\imath am\xi_i^2}{2\varepsilon}} \xi_i^2 = -\tfrac{\varepsilon}{\imath am} \end{gathered} \tag{64}$$

The expression for the wave function is therefore:

$$\psi_{\tau+\varepsilon} = \psi_\tau - \frac{\imath abm\varepsilon}{2}\psi_\tau + \frac{q\varepsilon}{m} (A^\mu \partial_\mu) \psi_\tau + \frac{q}{2m}\varepsilon (\partial^\mu A_\mu) \psi_\tau + \frac{\imath aq^2\varepsilon}{2m} A^2 \psi_\tau - \frac{\imath\varepsilon}{2am}\partial^2 \psi_\tau \tag{65}$$

Taking the limit $\epsilon \to 0$ and multiplying by $\imath$, we get the Schrödinger equation for TQM:

$$\imath\frac{\partial\psi_\tau}{\partial\tau} = ab\frac{m}{2}\psi_\tau + \frac{\imath q}{m} (A^\mu \partial_\mu) \psi_\tau + \frac{\imath q}{2m} (\partial^\mu A_\mu) \psi_\tau - \frac{aq^2}{2m} A^2 \psi_\tau + \frac{1}{2ma}\partial^2 \psi_\tau \tag{66}$$

IOP Conf. Series: Journal of Physics: Conf. Series **1239** (2019) 012015 doi:10.1088/1742-6596/1239/1/012015

or:

$$\imath\frac{\partial\psi_\tau}{\partial\tau}(t,\vec{x}) = -\frac{1}{2ma}\left((\imath\partial_\mu - aqA_\mu(t,\vec{x}))(\imath\partial^\mu - aqA^\mu(t,\vec{x})) - a^2bm^2\right)\psi_\tau(t,\vec{x}) \tag{67}$$

If we make the customary identifications $\imath\frac{\partial}{\partial t} \to E$, $-\imath\vec{\nabla} \to \hat{\vec{p}}$ or $\imath\partial_\mu \to p_\mu$ we have:

$$\imath\frac{\partial\psi_\tau}{\partial\tau} = -\frac{1}{2ma}\left((p_\mu - aqA_\mu)(p^\mu - aqA^\mu) - a^2bm^2\right)\psi_\tau \tag{68}$$

3.3. Long, slow approximation

We can now fix the scale and additive constants by looking at the behavior of the Schrödinger equation over longer times.

In his development of quantum mechanics from a time-dependent perspective [65], Tannor used a requirement of constructive interference in time to derive the Bohr condition for the allowed atomic orbitals. We use a similar approach here.

If we average over a sufficiently long period of time, the results will be dominated by the components with:

$$\imath\frac{\partial\psi_\tau(x)}{\partial\tau} = 0 \tag{69}$$

The argument here is not that the typical variation from the long, slow solution is small, but rather that over time interactions with the system in question will tend to be dominated by interactions with the stabler, slower moving components. Interactions with more rapidly varying components will tend to average to zero.

Accepting this, then the right side looks like the Klein-Gordon equation. To complete this identification, first look at the case with the vector potential A zero:

$$\left(\hat{p}^2 - a^2bm^2\right)\psi = 0 \to a^2b = 1 \tag{70}$$

Now when A is not zero we have:

$$\left((\hat{p} - aqA)^2 - m^2\right)\psi = 0 \to a = 1 \to b = 1 \tag{71}$$

We will refer to this as the "long, slow approximation".

In the free case, the long, slow approximation picks out the on-shell components:

$$\left(\hat{p}^2 - m^2\right)\psi = 0 \tag{72}$$

And more generally the solutions of the Klein-Gordon equation with the minimal substitution $p \to p - qA$:

$$\left((\hat{p} - qA)^2 - m^2\right)\psi = 0 \tag{73}$$

The two constants are now fixed. The full Schrödinger equation is:

$$\imath\frac{\partial\psi_\tau}{\partial\tau}(t,\vec{x}) = -\frac{1}{2m}\left((\imath\partial_\mu - qA_\mu(t,\vec{x}))(\imath\partial^\mu - qA^\mu(t,\vec{x})) - m^2\right)\psi_\tau(t,\vec{x}) \tag{74}$$

and in momentum space:

$$\imath\frac{\partial\psi_\tau}{\partial\tau} = -\frac{1}{2m}\left((p_\mu - qA_\mu)(p^\mu - qA^\mu) - m^2\right)\psi_\tau \tag{75}$$

IOP Conf. Series: Journal of Physics: Conf. Series **1239** (2019) 012015 doi:10.1088/1742-6596/1239/1/012015

The free Schrödinger equation is:

$$2m\imath\frac{\partial\psi_\tau}{\partial\tau}(t,\vec{x}) = \left(\partial_\mu\partial^\mu + m^2\right)\psi_\tau(t,\vec{x}) \tag{76}$$

and in momentum space:

$$2m\imath\frac{\partial\psi_\tau}{\partial\tau} = -\left(p_\mu p^\mu - m^2\right)\psi_\tau(t,\vec{x}) \tag{77}$$

We establish in Appendix C this equation is unitary. Therefore if a wave function is normalized at $\tau = 0$:

$$1 = \int dtd\vec{x}\psi_0^*(t,\vec{x})\psi_0(t,\vec{x}) \tag{78}$$

then at any later clock time $\tau > 0$ we will have:

$$1 = \int dtd\vec{x}\psi_\tau^*(t,\vec{x})\psi_\tau(t,\vec{x}) \tag{79}$$

We next estimate the time scales over which we expect the long, slow approximation to be valid.

3.4. *How long and how slow?*

We have argued that we can fix the scaling and additive constants by looking at the behavior of the Schrödinger equation over long times. What do we mean by long times?

To see the relevant scale, we estimate the clock frequency f:

$$f \sim -\frac{E^2 - \vec{p}^2 - m^2}{2m} \tag{80}$$

In the non-relativistic case E is of order mass plus kinetic energy:

$$E \sim m + \frac{\vec{p}^2}{2m} \tag{81}$$

So we have:

$$E^2 - \vec{p}^2 - m^2 \sim \left(m + \frac{\vec{p}^2}{2m}\right)^2 - \vec{p}^2 - m^2 = \left(\frac{\vec{p}^2}{2m}\right)^2 \tag{82}$$

This is just the kinetic energy, squared. In an atom the kinetic energy is of order the binding energy:

$$\frac{\vec{p}^2}{2m} \sim eV \tag{83}$$

So the numerator is of order eV squared. But the denominator is of order MeV. So we can estimate the clock frequency f as:

$$f \sim \frac{eV^2}{MeV} \sim 10^{-6}eV \tag{84}$$

Energies of millionths of an electron volt $10^{-6}eV$ correspond to times of order millions of attoseconds $10^6 as$ or picoseconds, a million times longer than the natural time scale of the effects we are looking at. So the long, slow approximation is reasonable.

The physical picture that emerges is of a particle that is extended in time on the same basis as space, with its wave function extend in all four dimensions, and – if we think in terms of path integrals – the associated paths wandering around in all four dimensions, not just the traditional three.

It may be amusing to note that from this perspective there is no such thing as an onshell particle; in momentum space the onshell part of the wave function is always a set of measure zero. It is only over scales of picoseconds and greater that an onshell description of the particle may give an acceptable approximation.

IOP Conf. Series: Journal of Physics: Conf. Series **1239** (2019) 012015 doi:10.1088/1742-6596/1239/1/012015

3.5. Observer independent choice of frame

There remains one piece of the puzzle; we need to establish that the use of the clock time τ does not itself violate covariance.

If we did the above derivation for Bob rather than Alice, we would see Alice's clock time replaced by Bob's clock time $\tau \to \gamma\tau$ ($\gamma \equiv \frac{1}{\sqrt{1-v^2}}$) where v is his velocity relative to her.

We therefore have one free parameter left to fix before we can declare our analysis free of free parameters.

If Bob is not going that quickly (relative to Alice) the errors created by ambiguities with respect to τ will introduce only small corrections; of only second order and therefore not relevant for falsifiability.

Even if we were prepared to accept that, establishing frame independence is interesting as a point of principle.

This may be done in a natural way by making use of an observation from Weinberg [66]. Per Weinberg, we may treat the Einstein field equation for general relativity as representing conservation of energy-momentum when exchanges of energy momentum with spacetime are included.

Consider the Einstein field equations:

$$G_{\mu\nu} \equiv R_{\mu\nu} - \frac{1}{2} g_{\mu\nu} R = -8\pi G T_{\mu\nu} \tag{85}$$

Rewrite as:

$$\left(G^{\mu\nu} + 8\pi G T^{\mu\nu}\right)_{;\nu} = 0 \tag{86}$$

We may use this to associate an energy momentum tensor ($t_{\mu\nu}$ in Weinberg's notation) with local space time. Define:

$$g_{\mu\nu} = \eta_{\mu\nu} + h_{\mu\nu} \tag{87}$$

where $h_{\mu\nu}$ vanishes at infinity but is not assumed small. The part of the Ricci tensor linear in h is:

$$R^{(1)}_{\mu\nu} \equiv \frac{1}{2}\left(\frac{\partial^2 h^\lambda_\lambda}{\partial x^\mu \partial x^\nu} - \frac{\partial^2 h^\lambda_\mu}{\partial x^\lambda \partial x^\nu} - \frac{\partial^2 h^\lambda_\nu}{\partial x^\lambda \partial x^\mu} + \frac{\partial^2 h_{\mu\nu}}{\partial x^\lambda \partial x_\lambda}\right) \tag{88}$$

The exact Einstein equations may be written as:

$$R^{(1)}_{\mu\nu} - \frac{1}{2}\eta_{\mu\nu} R^{(1)\lambda}_{\lambda} = -8\pi G \left(T_{\mu\nu} + t_{\mu\nu}\right) \tag{89}$$

where $t_{\mu\nu}$ is defined by:

$$t_{\mu\nu} \equiv \frac{1}{8\pi G}\left(R_{\mu\nu} - \frac{1}{2} g_{\mu\nu} R^\lambda_\lambda - R^{(1)}_{\mu\nu} + \frac{1}{2}\eta_{\mu\nu} R^{(1)\lambda}_{\lambda}\right) \tag{90}$$

Weinberg then argues we may interpret $t_{\mu\nu}$ as the energy-momentum of the gravitational field itself.

Accepting this, we change to a coordinate frame in which $t_{\mu\nu}$ is diagonalized. We will refer to this as the rest frame of the vacuum or the V frame. We can treat this V frame as the defining frame for the four dimensional Schrödinger equation. As the V frame is invariant (up to rotations in three-space) we now have an invariant definition of the four dimensional Schrödinger equation.

This is obviously going to be a free-falling frame. So Alice and Bob – if they are working in a terrestrial laboratory – will have to adjust their calculations to include a correction for the upwards force the laboratory floor exerts against them. If their colleague Vera is working in an orbiting laboratory, she will be able to calculate without correction.

Here after, unless stated to contrary, we will assume we are working in the "rest frame of the vacuum".

3.6. Relationship between TQM and the Relativistic Dynamics program

"In other words, we are trying to prove ourselves wrong as quickly as possible, because only in that way can we find progress." – Richard P. Feynman [2]

We have now fully defined the path integral expression and the Schrödinger equation.

It is appropriate therefore to pause to look at relationship of TQM to the relativistic dynamics program. (A helpful overview of the relativistic dynamics program is provided in Fanchi [35]).

3.6.1. TQM as part of the Relativistic Dynamics program We start with Feynman [17], who uses the Lagrangian:

$$L_{Feynman}\left(x^{\mu},\dot{x}^{\mu}\right)=-\dot{x}^{\mu}\dot{x}_{\mu}-q\dot{x}^{\mu}A_{\mu}\left(x\right) \tag{91}$$

giving:

$$\imath\frac{\partial}{\partial s}\psi_s=\frac{\left(p^{\mu}-qA^{\mu}\right)\left(p_{\mu}-qA_{\mu}\right)}{2}\psi_s \tag{92}$$

However with this approach, as Feynman notes, each particle would need its own evolution parameter s. This would make extension to the multiple particle case problematic.

There are many variations on this in the literature. Fanchi [36] derives:

$$\imath\frac{\partial}{\partial s}\psi_s=\frac{\left(p^{\mu}-qA^{\mu}\right)\left(p_{\mu}-qA_{\mu}\right)}{2m}\psi_s \tag{93}$$

This differs from Feynman's by a factor of $1/m$. Fanchi's evolution parameter s is for all practical purposes the negative of our clock time τ.

Land and Horwitz [37] give:

$$\left(\imath\frac{\partial}{\partial s}+qa_5\right)\psi_s=\frac{\left(p^{\mu}-qA^{\mu}\right)\left(p_{\mu}-qA_{\mu}\right)}{2m}\psi_s \tag{94}$$

This is Fanchi's with the addition of the a_5, a gauge term.

If we take $s\to-\tau$ and make the choice of gauge (see Appendix D):

$$a_5\Rightarrow\frac{m}{2q} \tag{95}$$

we get:

$$\left(\imath\frac{\partial}{\partial\tau}\right)\psi_s=-\frac{\left(p^{\mu}-qA^{\mu}\right)\left(p_{\mu}-qA_{\mu}\right)-m^2}{2m}\psi_s \tag{96}$$

which is the same as our Schrödinger equation (equation 74 on page 17).

At this point we may argue that as the right hand-side is the Klein-Gordon equation, and the Klein-Gordon equation is strongly confirmed by experimental evidence we have:

$$0\approx\frac{\left(p^{\mu}-qA^{\mu}\right)\left(p_{\mu}-qA_{\mu}\right)-m^2}{2m}\psi_s \tag{97}$$

So the combination of the mass gauge ($a_5=\frac{m}{2q}$) and the comparison with the Klein-Gordon equation give us the long slow approximation.

We have therefore placed our Schrödinger equation within the context of the relativistic program. This will let us use results from that program in the appropriate limit.

IOP Conf. Series: Journal of Physics: Conf. Series **1239** (2019) 012015 doi:10.1088/1742-6596/1239/1/012015

3.6.2. TQM and falsifiability TQM may be thought of as a specialization of the relativistic dynamics program aimed at making the hypothesis "time should be treated on the same basis as space in quantum mechanics" falsifiable. This means in turn that we have to be able to rule out the *class* of such theories, not just one in particular.

To do this we need to extend quantum mechanics to include time in a way that depends only on covariance. We can admit no free or adjustable parameters; any such risk falsifiability.

We may understand each step of the program here from this point of view.

(i) We used clock time as the "evolution parameter". This is defined operationally – by the laboratory clock – and therefore introduces no free parameters.

(ii) We used path integrals as the defining representation; with these the only change we need to make is to have the paths extend in four rather than three dimensions.

(iii) We could not use plane waves or delta functions as the initial wave functions: as they are not physical, they introduce the risk of inducing mathematical artifacts and again threatening falsifiability.

(iv) We needed to treat general wave functions, not just a specialized subset. Otherwise the results might be conditional on the subset chosen.

(v) We needed to ensure that our path integrals are convergent without the use of tricks. Otherwise the tricks can bring the results into question. The use of Morlet wavelet decomposition solved this and the previous two problems.

(vi) The resulting Schrödinger equation depends on the clock time. To eliminate such dependence (to lowest order) we showed that the time scales associated with the clock time are of order picoseconds, while the scales associated with dispersion in time are of order attoseconds – a million times smaller. So long as we work at the attosecond scale we should be able to ignore effects associated with clock time.

(vii) The clock time in turn depends on the choice of the laboratory frame. To further control for effects of the specific choice of laboratory frame, we showed we can define an invariant frame (the "rest frame of the vacuum"). By defining all calculations with reference to this invariant frame we eliminate the dependence of the clock time on the choice of a specific laboratory frame.

While we have fully defined the path integral expression and the Schrödinger equation, there are still some further questions that need to be addressed if we are to achieve falsifiability:

(i) How do we estimate the initial wave functions in time in a robust and frame independent way (subsection 4.1)?

(ii) How do we understand the evolution of the wave function with respect to clock time: with the long, slow approximation we appear to be saying the wave function is frozen in time, that $\imath\frac{\partial}{\partial\tau}\psi \approx 0$, which is clearly far from the case (subsection 4.2)?

(iii) How do we define the rules for detection in a way which is manifestly covariant (subsection 4.3)?

(iv) How do we produce at least one experiment with which we can unambiguously show that the wave function should not be extended in time (section 5)?

(v) Realistic experiments will need to run at ultra-short times and therefore high energies. At these energies, particles may spring into existence from the vacuum. How do we extend this analysis to the multiple particle case in a way which is consistent with the work done in the single particle case (section 6)?

At attosecond times, we expect that the effects of dispersion in time will dominate. Over picosecond and longer times we expect that the effects of the clock time will need to be included.

IOP Conf. Series: Journal of Physics: Conf. Series **1239** (2019) 012015 doi:10.1088/1742-6596/1239/1/012015

Therefore, at attosecond times, we expect TQM will function as a kind of lowest common denominator for the relativistic dynamics program. At picosecond times, we may need to include contributions from the evolution parameter and therefore be able to discriminate among various branches of the relativistic dynamics program.

4. Free solutions

We now have the Schrödinger equation. What do its free solutions look like?

We examine in turn the birth, life, and death of a free particle. The calculations are straightforward. But each stage will present problems specific to TQM.

For convenience, we assemble the solutions of the free Schrödinger's equation in Appendix E.

4.1. Initial wave function

What do our wave functions look like at start?

We have a chicken and egg problem here. Any initial wave function had itself to come from somewhere. How can we estimate the initial wave functions without first knowing them?

We look specifically at the Klein-Gordon equation for a static electric potential with no magnetic field:

$$A_0 = \Phi\left(\vec{x}\right), \vec{A} = 0 \tag{98}$$

with potential:

$$V \equiv q\Phi \tag{99}$$

This includes attractive potentials, scattering potentials, and the free case as the special case when $\Phi = 0$.

In SQM the Klein-Gordon equation is:

$$\left(\left(\imath\partial_\tau - V\left(\vec{x}\right)\right)^2 - \nabla^2 - m^2\right)\bar{\psi}_\tau\left(\vec{x}\right) = 0 \tag{100}$$

In TQM the equivalent is the four dimensional Schrödinger equation:

$$\imath\frac{\partial}{\partial\tau}\psi_\tau\left(t,\vec{x}\right) = -\frac{1}{2m}\left(\left(\imath\partial_t - V\left(\vec{x}\right)\right)^2 - \nabla^2 - m^2\right)\psi_\tau\left(t,\vec{x}\right) \tag{101}$$

The long, slow approximation picks out the solutions with:

$$\left(\left(\imath\partial_t - V\left(\vec{x}\right)\right)^2 - \nabla^2 - m^2\right)\psi_\tau\left(t,\vec{x}\right) = 0 \tag{102}$$

We will assume we are already in possession of solutions to the SQM version of the problem:

$$\left(\left(E - V\left(\vec{x}\right)\right)^2 - \vec{p}^2 - m^2\right)\bar{\psi}_E\left(\vec{x}\right) = 0 \tag{103}$$

This is necessarily in a specific frame, since SQM solutions are always (from our perspective) taken with reference to a specific frame.

We leverage the SQM solutions in two different ways[3] to get the TQM solution:

(i) *Separation of variables.* Each SQM solution induces a TQM solution, which is the SQM solution with a plane wave bolted on. This is technically correct, but unphysical.

(ii) *Maximum entropy.* We can use the long, slow approximation to estimate the mean and uncertainty of the coordinate energy. With these we can use the method of Lagrange multipliers to get the maximum entropy solution. Maximum entropy solutions tend to be robust: even if we are wrong about the details, the order of magnitude should be correct. This will provide our preferred starting point.

[3] This problem is also solved in Fanchi [36], using different methods.

IOP Conf. Series: Journal of Physics: Conf. Series **1239** (2019) 012015 doi:10.1088/1742-6596/1239/1/012015

As a quick check on the sanity of all this we will use the virial theorem to estimate the TQM version of the atomic wave functions. The width in time/energy of these matches the initial order of magnitude estimate we gave in the introduction.

We will also show that while we have chosen a specific frame in which to estimate the dispersion in time/energy we can get the estimate in an invariant way.

4.1.1. Solution by separation of variables We solve the TQM Klein-Gordon equation using separation of variables, looking for solutions of the form:

$$\varphi_n(t,\vec{x}) = \tilde{\phi}_n(t)\,\bar{\varphi}_n(\vec{x}) \tag{104}$$

where $\bar{\varphi}_n(\vec{x})$ is a solution of the SQM Klein-Gordon equation:

$$\left(\left(\bar{E}_n - V(\vec{x})\right)^2 - \vec{p}^2 - m^2\right)\bar{\varphi}_n(\vec{x}) = 0 \tag{105}$$

Then the coordinate time part is:

$$\tilde{\phi}_n(t) = \frac{1}{\sqrt{2\pi}} e^{-\imath \bar{E}_n t} \tag{106}$$

Because the potential is constant in time each use of the operator E turns into a constant E_n via:

$$E \to \imath\frac{\partial}{\partial t} \to \bar{E}_n \tag{107}$$

We get immediately:

$$\left(\left(\bar{E}_n - V(\vec{x})\right)^2 - \vec{p}^2 - m^2\right)\varphi_n(t,\vec{x}) = 0 \tag{108}$$

So every solution of the Klein-Gordon equation in SQM generates a corresponding solution in TQM.

We could accomplish the same thing, formally, by taking $\tau \to t$. Since we expect in general that $\langle t \rangle \approx \tau$ (discussed further in the next section) this will often a give reasonable first approximation.

However this is not entirely satisfactory. We have a solution which is "fuzzy" in space, but "crisp" in time. A more realistic, if more complex solution, would include off-shell components. Even if our wave function started out as a simple plane wave in time, internal decoherence would rapidly turn it into something a bit more cloud-like. While mathematically acceptable, our solution is not physically plausible.

4.1.2. Solution by maximum entropy The long, slow approximation picks out a single solution. But in practice we expect there would be a great number of solutions, with the one given by the long slow approximation merely the most typical.

Such a sum will have an associated probability density function. We can get a reasonable first estimate of this by defining appropriate constraints and then using the method of Lagrange multipliers to pick out the distribution with maximum entropy.

From the probability density, we will infer the wave function.

Estimate of the probability density We start with the free case, as representing the simplest case of a constant potential. We treat the bound case below.

IOP Conf. Series: Journal of Physics: Conf. Series **1239** (2019) 012015 doi:10.1088/1742-6596/1239/1/012015

We assume we are given the SQM wave function in three dimensions. We use this to compute the the expectation of the energy, and the expectation of the energy squared. These give us our constraints:

$$\begin{aligned} \langle 1 \rangle &= 1 \\ \bar{E} \equiv \langle E \rangle &= \sqrt{m^2 + \langle \vec{p} \rangle^2} \\ \langle E^2 \rangle &= \langle m^2 + \vec{p}^2 \rangle = m^2 + \langle \vec{p}^2 \rangle \end{aligned} \tag{109}$$

The uncertainty in energy is defined as:

$$\Delta E \equiv \sqrt{\langle E^2 \rangle - \langle E \rangle^2} \tag{110}$$

The expectations are defined as integrals over the probability density:

$$\langle f \rangle \equiv \int d\vec{p} \bar{\rho} \left(\vec{p}\right) f \left(\vec{p}\right) \tag{111}$$

The constraints imply $\Delta E = \Delta p$.

We will work with box normalized energy eigenfunctions:

$$\phi_n \left(t\right) \equiv \frac{1}{\sqrt{2T}} e^{-\imath E_0 n t} \tag{112}$$

$$E_0 \equiv \frac{\pi}{T} \tag{113}$$

Here n is an integer running from negative infinity to positive and the eigenfunctions are confined to a box extending T seconds into the future and T seconds into the past, where T is much larger than any time of interest to us. (We will use a similar approach in developing the four dimensional Fock space below, subsection 6.3.2).

A general wave function can be written as:

$$\psi \left(t\right) = \sum_{n=-\infty}^{\infty} c_n \phi_n \left(t\right) \tag{114}$$

The coefficients c only appear as the square:

$$\rho_n \equiv c_n^* c_n \tag{115}$$

Expressed in this language we have the constraints:

$$\begin{aligned} C_0 &\equiv \sum_{n=-\infty}^{\infty} \rho_n - 1 = 0 \\ C_1 &\equiv E_0 \sum_{n=-\infty}^{\infty} n \rho_n - \langle E \rangle = 0 \\ C_2 &\equiv E_0^2 \sum_{n=-\infty}^{\infty} n^2 \rho_n - \langle E^2 \rangle = 0 \end{aligned} \tag{116}$$

We would like to find the solution that maximizes the entropy:

$$S \equiv \sum_{n=-\infty}^{\infty} -\rho_n \ln \left(\rho_n\right) \tag{117}$$

We form the Lagrangian from the sum of the entropy and the constraints, with Lagrange multipliers λ_0, λ_1, and λ_2:

$$L \equiv S + \lambda_0 C_0 + \lambda_1 C_1 + \lambda_2 C_2 \tag{118}$$

To locate the configuration of maximum entropy, we take the derivative with respect to the ρ_n:

$$\frac{\partial L}{\partial \rho_n} = 0 \tag{119}$$

getting:

$$-\ln \rho_n - 1 + \lambda_0 + E_0 n \lambda_1 + E_0^2 n^2 \lambda_2 = 0 \tag{120}$$

Therefore the distribution of the ρ_n is given by an exponential with zeroth, first, and second powers of the energy:

$$\rho(E) \sim e^{-a-bE-cE^2} \tag{121}$$

The constraints force the constants:

$$\rho(E) = \frac{1}{\sqrt{2\pi\Delta E^2}} e^{-\frac{(E-\bar{E})^2}{2(\Delta E)^2}} \tag{122}$$

Estimate of the wave function We therefore have the probability density in energy; now we wish to estimate the corresponding wave function in energy (or equivalently time).

The simplest – and therefore best – way to do this is to write the total wave function as the direct product of a wave function in energy times the postulated wave function in momentum:

$$\hat{\varphi}(E, p) \sim \hat{\tilde{\varphi}}(E)\, \hat{\bar{\varphi}}(p) \tag{123}$$

We then use the Gaussian test function that matches the derived density function for the energy factor:

$$\hat{\tilde{\varphi}}_0(E) = \sqrt[4]{\frac{1}{\pi\sigma_E^2}} e^{\imath(E-E_0)t_0 - \frac{(E-E_0)^2}{2\sigma_E^2}} \tag{124}$$

with values:

$$\sigma_E^2 = \sigma_p^2 \tag{125}$$

and:

$$E_0 = \sqrt{m^2 + \bar{p}^2} \tag{126}$$

Taking the Fourier transform of the energy part we have:

$$\tilde{\varphi}_0(t) = \sqrt[4]{\frac{1}{\pi\sigma_t^2}} e^{-\imath E_0(t-t_0) - \frac{t^2}{2\sigma_t^2}} \tag{127}$$

$$\sigma_t^2 = \frac{1}{\sigma_E^2} \tag{128}$$

We set $t_0 = 0$ as the overall phase is already supplied by the space/momentum part.

The full wave functions are the products of the coordinate time and space (or energy and momentum) parts:

$$\varphi_0(t, x) = \tilde{\varphi}_0(t)\, \bar{\varphi}_0(x) \tag{129}$$

$$\hat{\varphi}_0(E, p) = \hat{\tilde{\varphi}}_0(E)\, \hat{\bar{\varphi}}_0(p) \tag{130}$$

IOP Conf. Series: Journal of Physics: Conf. Series **1239** (2019) 012015 doi:10.1088/1742-6596/1239/1/012015

4.1.3. Bound state wave functions We extend this approach to estimate the dispersion of a bound wave function in time.

In the case of a Coulomb potential we can estimate Δp from the virial theorem:

$$\left\langle \frac{\vec{p}^2}{2m} \right\rangle = -\frac{1}{2} \langle V \rangle \tag{131}$$

which implies:

$$\left\langle \frac{\vec{p}^2}{2m} \right\rangle + \langle V \rangle = \bar{E}_n \rightarrow \left\langle \frac{\vec{p}^2}{2m} \right\rangle = -\bar{E}_n \tag{132}$$

Since the average momentum is zero, we have the estimate of the uncertainty in energy as:

$$\Delta E_n = \sqrt{-2m\bar{E}_n} \tag{133}$$

Substituting the mass of the electron and the Rydberg constant, we have for the hydrogen ground state:

$$\Delta E_1 = \sqrt{2 \cdot 13.6\text{eV} \cdot (0.511 \cdot 10^6)\,\text{eV}} = 3728 \text{ eV} \tag{134}$$

And the corresponding dispersion in coordinate time is:

$$\Delta t = \frac{\hbar}{\Delta E_n} = .1766as \tag{135}$$

This matches the order of magnitude estimate we started with (subsection 1.2). The numerical closeness is coincidental, but does increase confidence the order of magnitude is correct.

We have therefore as the initial estimate of the TQM wave function for a hydrogen atom:

$$\begin{aligned} \hat{\psi}_n(E, \vec{p}) &= \hat{\varphi}_n(E)\,\hat{\hat{\psi}}_n(\vec{p}) \\ \hat{\hat{\varphi}}_n(E) &= \sqrt[4]{\frac{1}{2\pi\hat{\sigma}_n^2}}\, e^{\imath(E-\bar{E}_n)t_0 - \frac{(E-\bar{E}_n)^2}{2\hat{\sigma}_n^2}} \\ \hat{\sigma}_n^2 &= -4m\bar{E}_n \end{aligned} \tag{136}$$

4.1.4. Frame independence of the estimate We have given a reasonable estimate of the initial wave function in time/energy. However, we have not yet established that the estimate is independent of the initial choice of frame.

For both the bound and free cases, we choose our initial frame as the rest frame. We make our estimate, via maximum entropy in this frame, then to get the wave function in a arbitrary frame Lorentz transform to that frame:

$$\psi'(x') = \psi(\Lambda x) \tag{137}$$

Since the rest frame is an invariant, the resulting initial wave functions are well-defined.

4.2. Evolution of the free wave function

Now that we know what our starting wave functions look like, how do they evolve over time?

For a first examination we work with the non-relativistic case; we will extend to the relativistic case below (section 6). We will first look at the SQM case, then the TQM case, then compare the two.

4.2.1. Evolution in SQM We start with the familiar problem of the evolution of the non-relativistic wave function with respect to clock time. We work with two dimensions – τ, x – since the extension to y, z is straightforward.

The non-relativistic Schrödinger equation is:

$$\imath \frac{\partial}{\partial \tau} \bar{\psi} = \frac{p^2}{2m} \bar{\psi} = -\frac{\partial_x^2}{2m} \bar{\psi} \tag{138}$$

At clock time zero we start with a Gaussian test function in momentum with average position x_0, average momentum p_0, and dispersion in momentum σ_p. To reduce clutter we use p for p_x:

$$\hat{\bar{\varphi}}_0 (p) = \sqrt[4]{\frac{1}{\pi \sigma_p^2}} e^{-\imath p x_0 - \frac{(p-p_0)^2}{2\sigma_p^2}} \tag{139}$$

In momentum space the problem is trivial. The solution is:

$$\hat{\bar{\varphi}}_\tau (p) = \sqrt[4]{\frac{1}{\pi \sigma_p^2}} e^{-\imath p x_0 - \frac{(p-p_0)^2}{2\sigma_p^2} - \imath \frac{p^2}{2m}\tau} \tag{140}$$

In coordinate space we get:

$$\bar{\varphi}_\tau (x) = \sqrt[4]{\frac{1}{\pi \sigma_x^2}} \sqrt{\frac{1}{f_\tau^{(x)}}} e^{\imath p_0 x - \frac{1}{2\sigma_x^2 f_\tau^{(x)}} \left(x - x_0 - \frac{p_0}{m}\tau\right)^2 - \imath \frac{p_0^2}{2m}\tau} \tag{141}$$

with:

$$\sigma_x = \frac{1}{\sigma_p} \tag{142}$$

$$f_\tau^{(x)} = 1 + \imath \frac{\tau}{m \sigma_x^2} \tag{143}$$

4.2.2. Evolution in TQM In two dimensions the Schrödinger equation for TQM is:

$$\imath \frac{\partial \psi_\tau}{\partial \tau} (t, \vec{x}) = -\frac{E^2 - p^2 - m^2}{2m} \psi_\tau (t, \vec{x}) = \left(\frac{\partial_t^2}{2m} - \frac{\partial_x^2}{2m} + \frac{m}{2} \right) \psi_\tau (t, \vec{x}) \tag{144}$$

We start in energy momentum space. The momentum part is as above. We start with a Gaussian test function in energy, with average time at start t_0, average energy E_0, and dispersion in energy σ_E:

$$\hat{\tilde{\varphi}}_0 (E) \equiv \sqrt[4]{\frac{1}{\pi \sigma_E^2}} e^{\imath E t_0 - \frac{(E-E_0)^2}{2\sigma_E^2}} \tag{145}$$

Usually we will take $t_0 = \tau_0 = 0$.

As a function of clock time we get:

$$\hat{\psi}_\tau (E, p) = \hat{\tilde{\varphi}}_0 (E) \hat{\bar{\varphi}}_0 (p) \exp\left(-\imath \frac{E^2 - p^2 - m^2}{2m} \tau \right) \tag{146}$$

We divide up the pieces of the clock time part, assigning the $\frac{E^2}{2m}$ to the energy part, the $\frac{p^2}{2m}$ to the momentum part, and keeping the third part outside:

$$\hat{\psi}_\tau (E, p) = \hat{\tilde{\varphi}}_\tau (E) \hat{\bar{\varphi}}_\tau (p) \exp\left(\imath \frac{m}{2} \tau \right) \tag{147}$$

Now the energy part works in parallel to the momentum part:

$$\hat{\varphi}_\tau(E) \equiv \sqrt[4]{\frac{1}{\pi\sigma_E^2}} e^{\imath E t_0 - \frac{(E-E_0)^2}{2\sigma_E^2} - \imath \frac{E_0^2}{2m}\tau} \tag{148}$$

And in coordinate space:

$$\tilde{\varphi}_\tau(t) = \sqrt[4]{\frac{1}{\pi\sigma_t^2}} \sqrt{\frac{1}{f_\tau^{(t)}}} e^{-\imath E_0 t + \imath \frac{E_0^2}{2m}\tau - \frac{1}{2\sigma_t^2 f_\tau^{(t)}}\left(t - t_0 - \frac{E_0}{m}\tau\right)^2} \tag{149}$$

with the usual ancillary definitions:

$$\sigma_E = \frac{1}{\sigma_t} \tag{150}$$

$$f_\tau^{(t)} \equiv 1 - \imath \frac{\tau}{m\sigma_t^2} \tag{151}$$

and with the expectation for coordinate time:

$$\bar{t}_\tau = t_0 + \frac{E_0}{m}\tau \tag{152}$$

implying a velocity for coordinate time with respect to laboratory time:

$$\gamma = \frac{E_0}{m} \tag{153}$$

In the non-relativistic case, $\gamma \approx 1$. So the expectation of the coordinate time advances at the traditional one-second-per-second rate relative to the clock time.

4.2.3. Comparison of TQM to SQM The TQM and SQM approaches develop in close parallel. In both, the wave functions are centered on the classical trajectory as given by the solution of the Euler-Lagrange equations.

There is however one peculiarity.

We are relying on the long, slow approximation. Especially over short times, this means that the total wave function appears to be relatively static with respect to evolution in clock time:

$$\imath \frac{\partial}{\partial \tau}\psi \approx 0 \tag{154}$$

In momentum space:

$$f_p \equiv -\frac{E^2 - p^2 - m^2}{2m} \approx 0 \tag{155}$$

However, real wave functions are not static with respect to clock time.

The resolution is that most of the clock time dependence is carried by the coordinate time:

$$\frac{d}{d\tau}\psi = \frac{\partial}{\partial \tau}\psi + \frac{dt}{d\tau}\frac{\partial}{\partial t}\psi \approx \frac{dt}{d\tau}\frac{\partial}{\partial t}\psi \tag{156}$$

And in the non-relativistic case we have:

$$\frac{dt}{d\tau} = \frac{E}{m} \approx 1 \tag{157}$$

So we get as a rough approximation:

$$\frac{d}{d\tau}\psi \approx \frac{\partial}{\partial t}\psi \tag{158}$$

We see that the expectation of the coordinate time is about equal to the clock time. While Alice's dog is always getting ahead of and behind her, on average his position is about equal to hers:

$$\langle t \rangle \approx \tau \tag{159}$$

The Schrödinger equation gives the partial derivative with respect to clock time; the total derivative behaves as expected.

4.3. Time of arrival measurements

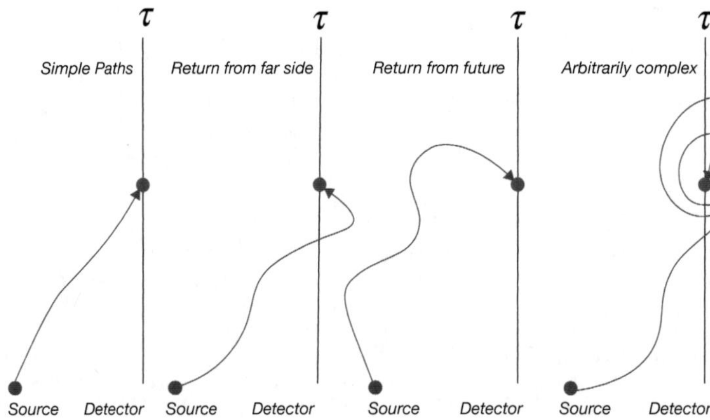

Figure 3. Complex paths near the detector

So we know what our wave function looks like at start and how it evolves with time. To complete the analysis of the free case we look at how it is detected.

We look specifically at the measurement of time-of-arrival. We assume we have a particle going left to right, starting at $x = 0$. We place a detector at position $x = L$. It records when it detects the particle. The metric we are primarily interested in is the dispersion in time-of-arrival at the detector.

In SQM, if a detector located at position X registers a hit by a particle we take the particle's position in space as also $x = X$. Therefore in TQM, if detector active at laboratory time T registers a hit by a particle we *must* take the particle's position in time as $t = T$. This is required by our principle of maximum symmetry between time and space.

By the same token, in SQM if an emitter located at position X emits a particle, we take the start position of the path as $x = X$. Therefore in TQM, if an emitter active at laboratory time T emits a particle, we *must* take the start position in coordinate time as $t = T$.

In a practical treatment we would replace the phrases "at X" or "at T" with "within the range $X \pm \frac{\Delta X}{2}$" and "within the range $T \pm \frac{\Delta T}{2}$".

If we know both source and detector positions in space and time then all corresponding paths are clamped at both ends. In between source and detector the paths can examine all sorts of interesting times and spaces but each path is clamped at the endpoints. Alice and her dog leave from the same starting point in space time and arrive at the same ending point in space time, but while classical Alice takes the shortest path between the start and end points, *the quantum dog explores all paths.*

IOP Conf. Series: Journal of Physics: Conf. Series **1239** (2019) 012015 doi:10.1088/1742-6596/1239/1/012015

Meaning of "all paths" Paths in TQM are much more complex than those in SQM. If a detector is a camera shutter, open for a fraction of a second, then any paths that arrive early or late will merely be "eaten" by the closed shutter. But what if our apparatus can somehow be toggled from transparent to absorptive and back, as via "electromagnetically induced transparency" [67]? Then the paths can arrive early but then circle back, or arrive late but circle forward, or even perform a drunkard's walk around the detector till they choose to fall into it.

Since we are primarily interested in comparisons of TQM to SQM, rather than in fully exploring the elaborations of TQM, we will focus on the camera shutter model. Paths that arrive before the shutter is open or after the shutter is closed again will be silently absorbed by the camera itself. We defer to a later investigation examination of more complex paths.

We employ the rest frame of the camera.

4.3.1. Metrics With that dealt with, to compute the dispersion in time, we log how many hits we get in each time interval ("clicks per tick"):

$$\rho(\tau) \tag{160}$$

then calculate the average:

$$\langle\tau\rangle \equiv \int_{-\infty}^{\infty} d\tau\rho(\tau) \tag{161}$$

and the uncertainty:

$$\langle\Delta\tau\rangle^2 \equiv \int_{-\infty}^{\infty} d\tau\tau^2\rho(\tau) - \langle\tau\rangle^2 \tag{162}$$

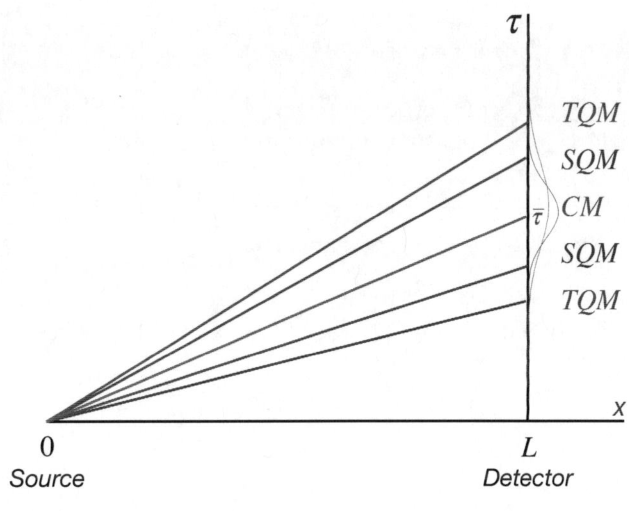

Figure 4. Time of arrival

4.3.2. Time of arrival in SQM We start with a particle with initial position $x = 0$ and with average momentum (in the x direction) of p_0. We will assume that the initial dispersion in momentum is small. We have the wave function from above. The probability density is then:

$$\bar{\rho}_\tau(x) \equiv |\bar{\varphi}_\tau(x)|^2 = \sqrt{\frac{1}{\pi\sigma_x^2 \left|f_\tau^{(x)}\right|^2}} e^{-\frac{\left(x-\frac{p_0}{m}\tau\right)^2}{\sigma_x^2\left|f_\tau^{(x)}\right|^2}} \tag{163}$$

We assume there is a detector at position $x = L$. If the particle is released at time $\tau = 0$ the average time of arrival is:

$$\bar{\tau}_D \equiv \frac{mL}{p_0} \tag{164}$$

We can write the difference from the average time as $\delta\tau \equiv \tau - \bar{\tau}$ We are interested in the uncertainty in time of arrival at the detector or the expectation of $\delta\tau^2$.

If we were looking at measurements of the x position we would know how to proceed. We would compute:

$$\Delta x = \sqrt{\int dx (x-\bar{x})^2 \rho_\tau(x)} \tag{165}$$

We would like something not much more complex for the uncertainty in time. And which makes sense for both SQM and TQM.

We will take an ad hoc approach here but then check it against the results of a more detailed analysis by Muga and Leavens[68].

Normally we think of a wave function as something that evolves in time; it is first a function of τ, then of x. But here we are not interested in the probability to be at x at a specific time τ; we are interested in the probability to be at a specific clock time τ for a fixed $x = L$. So we will rewrite the probability density as a function of τ.

We will assume that we are dealing with a reasonably well-focused particle so we may use a paraxial approximation. We can write $x = v\tau$ or $x = L + \delta x$. We use this to rewrite the density function as a function of $\delta\tau$. We take:

$$L + \delta x = v(\bar{\tau}_D + \delta\tau) \to \delta x = v\delta\tau \tag{166}$$

We keep only terms up through second order in $\delta\tau$.

We rewrite the numerator in the density function in terms of $\delta\tau$:

$$\bar{\rho}_\tau(L) \approx \sqrt{\frac{1}{\pi\sigma_x^2 \left|f_\tau^{(x)}\right|^2}} e^{-\frac{(v\delta\tau)^2}{\sigma_x^2\left|f_\tau^{(x)}\right|^2}} \tag{167}$$

Since the numerator is already only of second order in $\delta\tau$ we need only keep the zeroth order in $\delta\tau$ in the denominator:

$$\sigma_x^2\left|f_\tau^{(x)}\right|^2 = \sigma_x^2 + \frac{\tau^2}{m^2\sigma_x^2} = \sigma_x^2 + \frac{(\bar{\tau}+\delta\tau)^2}{m^2\sigma_x^2} \approx \frac{(\bar{\tau}+\delta\tau)^2}{m^2\sigma_x^2} \approx \frac{\bar{\tau}^2}{m^2\sigma_x^2} \tag{168}$$

giving:

$$\bar{\rho}_{\delta\tau} \approx \sqrt{\frac{v^2m^2\sigma_x^2}{\pi\bar{\tau}^2}} e^{-\frac{v^2m^2\sigma_x^2}{\bar{\tau}^2}(\delta\tau)^2} \tag{169}$$

We define an effective dispersion in time:

$$\bar{\sigma}_\tau \equiv \frac{1}{mv\sigma_x}\bar{\tau} \tag{170}$$

And the probability of detection as:

$$\bar{\rho}_{\delta\tau} = \sqrt{\frac{1}{\pi\bar{\sigma}_\tau^2}} e^{-\frac{(\delta\tau)^2}{\bar{\sigma}_\tau^2}} \tag{171}$$

This is normalized to one, centered on $\tau = \bar{\tau}$, and with uncertainty:

$$\Delta\tau = \frac{1}{\sqrt{2}}\bar{\sigma}_\tau \tag{172}$$

Particularly important is the inverse dependence on the velocity. Intuitively if we have a slow moving (non-relativistic) particle, it will take a long time to pass through by the x position of the detector, causing the associated uncertainty in time to be relatively large.

Comparison to a time-of-arrival operator As a cross-check, we compare our treatment to the time-of-arrival operator analysis in Muga and Leavens (who are following Kijowski [69]). They give a probability density in time of:

$$\rho(\tau) = \left| \int_0^\infty dp \sqrt{\frac{p}{m}} e^{-\imath \frac{p^2 \tau}{2m}} \hat{\varphi}(p) \exp(\imath pL) \right|^2 + \left| \int_{-\infty}^0 dp \sqrt{\frac{-p}{m}} e^{-\imath \frac{p^2 \tau}{2m}} \hat{\varphi}(p) \exp(-\imath pL) \right|^2 \tag{173}$$

where $\hat{\varphi}$ is an arbitrary momentum space wave function normalized to one.

Assume:

$$p_0 \gg \sigma_p \tag{174}$$

so:

$$p \approx p_0 \tag{175}$$

If our wave functions are closely centered on p, the term with negative momentum can be dropped or even flipped in sign without effect on the value of the integral. Further, by comparison to the exponential part, the term under the square root is roughly constant:

$$\sqrt{\frac{p}{m}} \approx \sqrt{\frac{p_0}{m}} \tag{176}$$

So we may replace Muga and Leaven's expression by the simpler:

$$\rho(\tau) \approx \frac{p_0}{m} \left| \int_{-\infty}^\infty dp e^{-\imath \frac{p^2 \tau}{2m}} \hat{\varphi}(p) \exp(\imath pL) \right|^2 \tag{177}$$

As the contents of the integral are the Fourier transform of the (clock) time dependent momentum space wave function, it is the clock time dependent space wave function at $x = L$:

$$\rho(\tau) \approx \frac{p_0}{m} |\bar{\varphi}_\tau(L)|^2 \tag{178}$$

We are now in x space. To get to τ space, we take:

$$\int \frac{p_0}{m} dx = \int v dx = \int d\tau \tag{179}$$

and the probability density from Muga and Leavens is now identical to ours.

IOP Conf. Series: Journal of Physics: Conf. Series **1239** (2019) 012015 doi:10.1088/1742-6596/1239/1/012015

4.3.3. Time of arrival in TQM It is striking that there is considerable uncertainty in time even when time is treated classically. Our hypothesized uncertainty in time will be added to this pre-existing uncertainty.

Using the time wave function from equation 149 we have for the probability density in time:

$$\tilde{\rho}_\tau(t) = \sqrt{\frac{1}{\pi\sigma_t^2\left|f_\tau^{(t)}\right|^2}} e^{-\frac{1}{\sigma_t^2\left|f_\tau^{(t)}\right|^2}(t-\tau)^2} \tag{180}$$

We multiply by the space part from above to get the full probability density:

$$\rho_D(t, L) = \tilde{\rho}_D(t)\,\bar{\rho}_D(L) \tag{181}$$

If t were replaced by space dimension y, we would have no doubt as to how to proceed. To get the overall uncertainty in y we would integrate over clock time:

$$(\Delta y)^2 = \int dy\,(y-\bar{y})^2 \int d\tau\,\rho_\tau(y)\,\bar{\rho}_\tau(L) \tag{182}$$

Therefore we write (taking $y \to t$):

$$(\Delta t)^2 = \int dt\,(t-\bar{\tau})^2 \int d\tau\,\tilde{\rho}_\tau(t)\,\bar{\rho}_\tau(L) \tag{183}$$

This is a convolution of clock time with coordinate time. To solve we first invoke the same approximations as above:

$$\begin{gathered}\sigma_t^2\left|f_\tau^{(t)}\right|^2 = \sigma_t^2 + \frac{\tau^2}{m^2\sigma_t^2} = \sigma_t^2 + \frac{(\bar{\tau}+\delta\tau)^2}{m^2\sigma_t^2} \approx \frac{(\bar{\tau}+\delta\tau)^2}{m^2\sigma_t^2} \approx \frac{\bar{\tau}^2}{m^2\sigma_t^2}\\ \tilde{\sigma}_\tau \equiv \frac{\bar{\tau}}{m\sigma_t}\\ \tilde{\rho}_\tau(t) \approx \sqrt{\frac{1}{\pi\tilde{\sigma}_\tau^2}} e^{-\frac{1}{\tilde{\sigma}_\tau^2}(t-\tau)^2}\end{gathered} \tag{184}$$

So we have for the full probability distribution in t:

$$\begin{gathered}\rho_{\bar{\tau}}(t) \equiv \int d\tau \sqrt{\frac{1}{\pi\tilde{\sigma}_\tau^2}} e^{-\frac{1}{\tilde{\sigma}_\tau^2}(t-\tau)^2} \sqrt{\frac{1}{\pi\bar{\sigma}_\tau^2}} e^{-\frac{1}{\bar{\sigma}_\tau^2}(\tau-\bar{\tau})^2}\\ (\Delta t)^2 = \int dt\,(t-\bar{\tau})^2\,\rho_{\bar{\tau}}(t)\end{gathered} \tag{185}$$

The convolution over τ is trivial giving:

$$\rho_{\bar{\tau}}(t) = \sqrt{\frac{1}{\pi\sigma_\tau^2}} e^{-\frac{(t-\bar{\tau})^2}{\sigma_\tau^2}} \tag{186}$$

with the total dispersion in clock time being the sum of the dispersions in coordinate time and in space:

$$\sigma_\tau^2 \equiv \tilde{\sigma}_\tau^2 + \bar{\sigma}_\tau^2 \tag{187}$$

This is intuitively reasonable.

The uncertainty is:

$$\Delta\tau = \frac{1}{\sqrt{2}}\sqrt{\tilde{\sigma}_\tau^2 + \bar{\sigma}_\tau^2} \tag{188}$$

We collect the definitions for the two dispersions:

$$\begin{gathered}\bar{\sigma}_\tau^2 = \frac{\bar{\tau}^2}{m^2 v^2 \sigma_x^2}\\ \tilde{\sigma_t}^2 \approx \frac{\bar{\tau}^2}{m^2\sigma_t^2}\end{gathered} \tag{189}$$

IOP Conf. Series: Journal of Physics: Conf. Series **1239** (2019) 012015 doi:10.1088/1742-6596/1239/1/012015

From the long, slow approximation, we would expect particle wave functions to have initial dispersions in energy/time comparable to their dispersions in momentum/space. $\sigma_t \sim \sigma_x$. But the conventional contribution has an additional $\frac{1}{v}$ in it. Since in the non-relativistic case, $v \ll 1$ the total uncertainty will be dominated by the space part.

This helps to explain why dispersion in time has not been seen by accident. It also motivates an exploration of the relativistic case, where the effects of dispersion in time should be at least comparable to the effects of dispersion in space (section 6).

5. Single and double slit experiments

There is already a significant literature on the "in time" versions of the single and double slit experiments. The investigation of this problem started over sixty years ago with Moshinsky [70, 71] and continues, with a recent review by Gerhard and Paulus [72]. Particularly interesting for our purposes are treatments of scattering of wave functions, as Umul [73] and Marchewka and Schuss [74, 75].

Neither single nor double slit has, to the best of our knowledge, been solved exactly. Approximations are necessary. Further the SQM and TQM branches have to be approximated in a way that lets us compare the two branches directly. We will treat the single slit first, then the double.

5.1. Single slit in time

The single slit in time experiment provides the decisive test of temporal quantum mechanics.

In SQM, the narrower the slit, the *less* the dispersion in subsequent time-of-arrival measurements. In TQM, the narrower the slit, the *greater* the subsequent dispersion in subsequent time-of-arrival measurements. *In principle, the difference may be made arbitrarily great.*

This distinction follows directly from the fundamental principles of quantum mechanics. Picture a quantum wave function going through a gate in space. If the gate is wide, diffraction by the edges is minimal and the subsequent broadening of the wave function minimal. The gate will clip the beam around the edges and that will be about it. But if the gate is narrow, then the wave function will spread in a nearly circular pattern and the subsequent broadening will be arbitrarily great.

In terms of the uncertainty principle, the gate represents a measurement of the position. The narrower the gate, the less the uncertainty at the gate. If Δy is small, then Δp_y must be correspondingly large and the resulting spread greater at the detector. As $\Delta y \to 0 \Rightarrow \Delta p_y \to \infty$. But a large Δp_y implies – with a bit of time – a large spread at the detector.

We translate this from space to time. We will start with a beam moving from left to right in the x direction, going through an extremely fast camera shutter, and arriving at a detector. In both SQM and TQM, the faster the shutter the smaller Δt. But in SQM the beam is clipped and the dispersion at the detector correspondingly reduced. While in TQM the smaller the Δt the correspondingly greater the ΔE. The greater the ΔE, the greater the dispersion in velocities and the greater the dispersion in time-of-arrival at the detector.

The distribution of detections in clock time will give us the dispersion in time-of-arrival, which is the key measurement.

We will ignore paths that loop back and forth through the gate. Most elementary treatments make this assumption of a single passage (for an analysis of the effects of multiple passages of a gate see Yabuki, Raedt, and Sawant [76, 77, 78]).

We will also, following Feynman and Hibbs [64], take the gate as having a Gaussian shape, rather then turning on and off instantly. This is more realistic and avoids some distracting mathematical complexities that result from using hard-edged gates.

IOP Conf. Series: Journal of Physics: Conf. Series **1239** (2019) 012015 doi:10.1088/1742-6596/1239/1/012015

Note also, in space we can make the gate entirely perpendicular to the beam. Beam traveling in x, gate in y. The beam can start with zero momentum in the y direction, letting the x momentum act as a carrier. But there is no such thing as a particle that does not have at least some momentum in time, i.e. energy. Therefore it is difficult to achieve a complete separation between the measurement of p_x and E. Here we will use increased dispersion in the time-of-arrival measurements as the test of TQM. (Below we will discuss briefly a way to separate the measurements of uncertainty in space and in time: subsubsection 6.8.4.)

As usual we look first at SQM, then TQM.

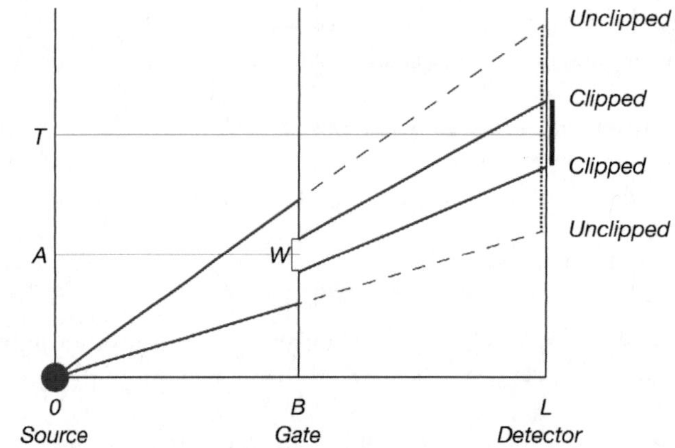

Figure 5. Single slit in SQM

5.1.1. Single slit in SQM We start by calculating the effects of a single slit in time on the dispersion of time-of-arrival measurements in SQM.

The gate is located at $x = B$, centered on clock time A, with width in time W:

$$\overline{G}_\tau = e^{-\frac{(\tau-A)^2}{2W^2}} \tag{190}$$

We start with a wave function in x:

$$\bar{\varphi}_\tau(x) = \sqrt[4]{\frac{1}{\pi\sigma_x^2}}\sqrt{\frac{1}{f_\tau^{(x)}}}e^{\imath p_0 x - \frac{1}{2\sigma_x^2 f_\tau^{(x)}}\left(x - x_0 - \frac{p_0}{m}\tau\right)^2 - \imath\frac{p_0^2}{2m}\tau} \tag{191}$$

or in p:

$$\hat{\bar{\varphi}}_\tau(p) = \sqrt[4]{\frac{1}{\pi\sigma_p^2}}e^{-\imath p x_0 - \frac{(p-p_0)^2}{2\sigma_p^2} - \imath\frac{p^2}{2m}\tau} \tag{192}$$

We take the initial position $x_0 = 0$, initial velocity $v \equiv \frac{p_0}{m}$. We assume the particle is non-relativistic so that $v \ll 1$.

We assume that the incoming wave function can be treated as a sum of p rays. This worked in the time-of-arrival case, has the merit of simplicity, and lets us make a direct comparison between TQM and SQM.

The evolution of the wave function is simplest in the p basis, but the gate is defined in the τ basis. Therefore we will need to shift back and forth between the p and the τ basis. To do this we use:

$$p = m\frac{x}{\tau} \tag{193}$$

IOP Conf. Series: Journal of Physics: Conf. Series **1239** (2019) 012015 doi:10.1088/1742-6596/1239/1/012015

with the ancillary definitions of δp and $\delta\tau$:

$$\begin{aligned} p &= p_0 + \delta p = mv + \delta p \\ \tau &= \bar{\tau} + \delta\tau \end{aligned} \tag{194}$$

For simplicity we center the particle beam on the gate. If $\bar{\tau}_G$ is the average time at which the particle reaches the gate, we arrange it so that:

$$\bar{\tau}_G = A \tag{195}$$

With the detector at position $x = L$, we define T as the average time at which the particle is detected. This gives:

$$v = \frac{B}{A} = \frac{L}{T} \tag{196}$$

We assume the beam is well-focused and use a paraxial approximation. To quadratic order:

$$\begin{aligned} \delta p &= m\tfrac{x}{\bar{\tau}+\delta\tau} - m\tfrac{x}{\bar{\tau}} \approx mv\left(-\tfrac{\delta\tau}{\bar{\tau}} + \tfrac{\delta\tau^2}{\bar{\tau}^2}\right) \\ \delta\tau &= m\tfrac{x}{p_0+\delta p} - m\tfrac{x}{p_0} \approx \tfrac{x}{v}\left(-\tfrac{\delta p}{p_0} + \tfrac{\delta p^2}{p_0^2}\right) \end{aligned} \tag{197}$$

To first order:

$$\frac{\delta p}{p_0} \approx -\frac{\delta\tau}{\bar{\tau}} \tag{198}$$

δp and $\delta\tau$ have opposite signs; faster means earlier, slower means later.

If the gate is much wider than the wave function then the particle will pass through unscathed. If the gate is narrower than the wave function then the particle will clipped to the width of the gate.

The wave function just before the gate (using $p = \frac{mB}{A+\delta\tau}$) is:

$$\bar{\varphi}_{G(pre)}(\delta\tau) = \sqrt[4]{\frac{1}{\pi\sigma_G^2}}\, e^{-\frac{\delta\tau^2}{2\sigma_G^2} - \imath\frac{1}{2m}\left(\frac{mB}{A+\delta\tau}\right)^2(A+\delta\tau)} \tag{199}$$

with σ_G defined by:

$$\sigma_G \equiv A\frac{\sigma_p}{p_0} \tag{200}$$

Note that the wave function is correctly normalized for integration over τ. And that to lowest order:

$$\frac{\delta\tau}{\sigma_G} = -\frac{\delta p}{\sigma_p} \tag{201}$$

σ_G is the measure of how wide the beam is in clock time when it reaches the gate. It is linear in A and proportional to σ_p.

For the SQM case, we assume no diffraction in time: the wave function will be clipped by the gate, not diffracted. This means that to get the wave function post-gate we must multiply the wave function pre-gate by the gate function:

$$\bar{\varphi}_{G(post)}(\delta\tau) = e^{-\frac{(\delta\tau)^2}{2W^2}}\bar{\varphi}_{G(pre)}(\delta\tau) \tag{202}$$

Now we convert back to p space, but on the far side of the gate:

$$\hat{\bar{\varphi}}_{G(post)}(p) = \sqrt[4]{\frac{1}{\pi\sigma_p^2}}\, e^{-\frac{\delta p^2}{2\sigma_p^2} - \frac{A^2}{2W^2}\frac{\delta p^2}{p_0^2} - \imath\frac{p^2}{2m}\tau_G} \tag{203}$$

IOP Conf. Series: Journal of Physics: Conf. Series **1239** (2019) 012015 doi:10.1088/1742-6596/1239/1/012015

The evolution from gate to detector only adds a phase:

$$\hat{\bar{\varphi}}_D(p) = \sqrt[4]{\frac{1}{\pi\sigma_p^2}} e^{-\frac{\delta p^2}{2\sigma_p^2} - \frac{\delta p^2}{2p_0^2}\frac{A^2}{W^2} - \imath\frac{p^2}{2m}\tau_D} \tag{204}$$

We again switch to τ space at the detector. The significant change is from $\sigma_p \to \sigma'_p$ where the primed dispersion in p is:

$$\frac{1}{(\sigma'_p)^2} = \frac{1}{\sigma_p^2} + \frac{A^2}{W^2 p_0^2} \tag{205}$$

With this we can write out the wave function at the detector by inspection. In p space:

$$\hat{\bar{\varphi}}_D(p) = \sqrt[4]{\frac{1}{\pi\sigma_p^2}} e^{-\frac{\delta p^2}{(\sigma'_p)^2} - \imath\frac{p^2}{2m}\tau_D} \tag{206}$$

In τ space:

$$\bar{\varphi}_D(\delta\tau) = \sqrt[4]{\frac{1}{\pi\sigma_G^2}} e^{-\frac{\delta\tau^2}{(\sigma'_G)^2} - \imath\frac{mL^2}{2T}\left(1 - \frac{\delta\tau}{T} + \frac{\delta\tau^2}{T^2}\right)} \tag{207}$$

The primed dispersion in clock time at the detector is a scaled version of the dispersion in momentum post gate:

$$\sigma'_G \equiv T\frac{\sigma'_p}{p_0} \tag{208}$$

Note the factor $\sqrt[4]{\frac{1}{\pi\sigma_G^2}}$ is unchanged. The ratio:

$$\left(\frac{\sigma'_G}{\sigma_G}\right)^2 \tag{209}$$

tells us what percentage of the particles get through the gate. But this does not affect the calculation of the dispersion, which is normalized.

The effective dispersion of the wave function is proportional to T. It is useful to scale that out. The result is controlled by the ratio of the angular width of the beam in momentum space $\theta_p \equiv \frac{\sigma_p}{p_0}$ to the angular width of the gate in time $\theta_G \equiv \frac{W}{A}$:

$$\frac{(\sigma'_G)^2}{T^2} = \frac{\theta_p^2\theta_G^2}{\theta_p^2 + \theta_G^2} \tag{210}$$

When the gate is wide open, the dispersion at the detector is proportional to the initial dispersion of the beam:

$$W \to \infty \Rightarrow \frac{\sigma'_G}{T} \to \frac{\sigma_p}{p_0} \tag{211}$$

But when the gate is much narrower than the beam, the dispersion at the detector proportional to the width of the gate:

$$W \to 0 \Rightarrow \frac{\sigma'_G}{T} \to \frac{W}{A} \tag{212}$$

The narrower the gate, the narrower the beam.

While the approximations we have used have been simple, this is a fundamental implication of the SQM view of time, of time as classical. In SQM, wave functions are not, by assumption, diffracted by a gate, they are clipped. And therefore their dispersion in time must be reduced by the gate rather than increased by it.

And we have therefore the uncertainty in time-of-arrival:

$$\frac{\Delta\tau}{T} = \frac{1}{\sqrt{2}}\frac{\sigma'_G}{T} \tag{213}$$

5.1.2. Single slit in TQM We now have a baseline from SQM; we turn to the single slit in time in TQM using the SQM treatment as a starting point.

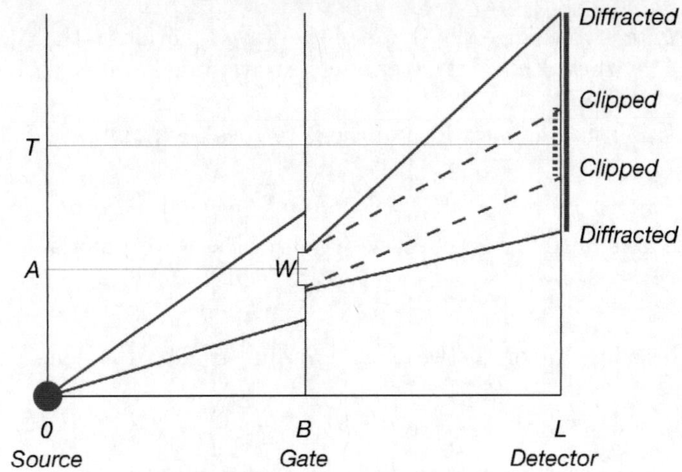

Figure 6. Single slit in TQM

By assumption, the gate will act only on the time part:

$$\widetilde{G}_t = e^{-\frac{(t-A)^2}{2W^2}} \tag{214}$$

For TQM, we start with a wave function factored in time and momentum:

$$\psi_\tau(t) = \tilde{\varphi}_\tau(t)\,\hat{\varphi}_\tau(p) \tag{215}$$

We are treating the p part as a carrier, using the wave function from above. The momentum part will be unchanged throughout.

We take the particle as starting with $t_0 = 0$. The wave function pre-gate is:

$$\tilde{\varphi}_\tau(t) = \sqrt[4]{\frac{1}{\pi\sigma_t^2}}\sqrt{\frac{1}{f_\tau^{(t)}}}\,e^{-\imath E_0 t - \frac{1}{2\sigma_t^2 f_\tau^{(t)}}\left(t-\frac{E_0}{m}\tau\right)^2 + \imath\frac{E_0^2}{2m}\tau} \tag{216}$$

We will again take the particle as non-relativistic: E_0 such that $\frac{E_0}{m} \approx 1$.

On arrival at the gate the time part of the wave function is:

$$\tilde{\varphi}_{G^{(pre)}}(t_G) = \sqrt[4]{\frac{1}{\pi\sigma_t^2}}\sqrt{\frac{1}{f_G^{(t)}}}\exp\left(-\imath E_0 t_G - \frac{(t_G-A)^2}{2\sigma_t^2\left(1-\imath\frac{A}{m\sigma_t^2}\right)} + \imath\frac{E_0^2}{2m}A\right) \tag{217}$$

Post gate:

$$\tilde{\varphi}_{G^{(post)}}(t_G) = e^{-\frac{(t_G-A)^2}{2W^2}}\,\tilde{\varphi}_{G^{(pre)}}(t_G) \tag{218}$$

The effect of the gate on the wave function is to rescale it:

$$\frac{1}{2\sigma_t^2\left(1-\imath\frac{A}{m\sigma_t^2}\right)} \to \frac{1}{2W^2} + \frac{1}{2\sigma_t^2\left(1-\imath\frac{A}{m\sigma_t^2}\right)} \tag{219}$$

We define rescaling constants σ_t^{*2}, τ^* by:

$$\frac{1}{W^2} + \frac{1}{\sigma_t^2 - \imath\frac{A}{m}} = \frac{1}{\sigma_t^{*2} - \imath\frac{\tau^*}{m}} \tag{220}$$

The effect of the gate is to change the shape of the wave function so that it looks *as if* it had started at time $A - \tau^*$ with width σ^*. This parameterization makes it easy to see what the gate is doing to the wave function.

We have by inspection that as the width of the gate goes to infinity, it lets the wave function through unchanged:

$$W \to \infty \Rightarrow \sigma_t^* \to \sigma_t, \tau^* \to A \tag{221}$$

While as the gate gets narrow, the effective width of the wave function in time goes to the width of the gate:

$$W \to 0 \Rightarrow \sigma_t^* \to W, \tau^* \to 0 \tag{222}$$

To compute the rescaling factors between these limits we clear the denominators:

$$\left(\sigma_t^2 - \imath\frac{A}{m}\right)\left(\sigma_t^{*2} - \imath\frac{\tau^*}{m}\right) + W^2\left(\sigma_t^{*2} - \imath\frac{\tau^*}{m}\right) = W^2\left(\sigma_t^2 - \imath\frac{A}{m}\right) \tag{223}$$

And equate the real and imaginary parts:

$$\begin{aligned} \left(\sigma_t^2 + W^2\right)\sigma_t^{*2} - \frac{A}{m}\frac{\tau^*}{m} &= W^2\sigma_t^2 \\ -\frac{A}{m}\sigma_t^{*2} - \left(\sigma_t^2 + W^2\right)\frac{\tau^*}{m} &= -W^2\frac{A}{m} \end{aligned} \tag{224}$$

This gives us a two by two matrix equation for σ_t^{*2}, τ^* which we invert and apply to the right hand side:

$$\begin{pmatrix} \sigma_t^{*2} \\ \frac{\tau^*}{m} \end{pmatrix} = \frac{W^2}{D}\begin{pmatrix} \sigma_t^2 + W^2 & -\frac{A}{m} \\ -\frac{A}{m} & -\left(\sigma_t^2 + W^2\right) \end{pmatrix}\begin{pmatrix} \sigma_t^2 \\ -\frac{A}{m} \end{pmatrix} \tag{225}$$

with determinant:

$$D = \left(\sigma_t^2 + W^2\right)^2 + \frac{A^2}{m^2} \tag{226}$$

getting:

$$\begin{aligned} \sigma_t^{*2} &= \frac{W^2}{D}\left(\left(\sigma_t^2 + W^2\right)\sigma_t^2 + \frac{A^2}{m^2}\right) \\ \tau^* &= \frac{A}{D}W^4 \end{aligned} \tag{227}$$

As a double check, we see we get the expected limits for σ_t^{*2} and τ^* as $W \to \infty, W \to 0$.

σ_t^{*2} and τ^* in turn gives us the t wave function at the detector. We replace σ_t and τ_G in the initial wave function by σ_t^* and τ^*:

$$\tilde{\varphi}_D\left(t_D\right) = N\sqrt[4]{\frac{1}{\pi\sigma_t^{*2}}}\sqrt{\frac{1}{f'_G}}e^{-\imath E_0 t_D - \frac{1}{2\sigma_t^{*2}f'_D}\left(t_D - \bar{t}_D\right)^2 + \imath\frac{E_0^2}{2m}\tau_D} \tag{228}$$

with:

$$\begin{aligned} f'_D &\equiv 1 - \imath\frac{\tau^* + (T - A)}{m\sigma_t^{*2}} \\ N &\equiv \sqrt[4]{\frac{\sigma_t^{*2}}{\sigma_t^2}}\sqrt{\frac{f'_G}{f_G}} \end{aligned} \tag{229}$$

The value of N drops out in the normalization when we calculation the uncertainty.

The use of rescaling constants makes clear that the effect of the gate is to reset the wave function, changing its shape and leaving no other memory of the wave function before the gate. The wave function gets a fresh start with the gate.

Of particular interest here is the limit $W \to 0$: an extremely narrow gate – an extremely fast camera shutter – resets the wave function to have a width of order width of the gate.

And therefore, given the Heisenberg uncertainty principle, the uncertainty in energy post-gate goes as $\frac{1}{W}$. And this in turn forces the wave to diverge more rapidly in time than would otherwise have been the case.

At large distances from the gate the dispersion of the wave function in time $\tilde{\sigma}_\tau^2 \equiv \sigma_t^{*2} f'_\tau$ goes as:

$$\tilde{\sigma}_\tau \sim \frac{\tau}{m}\sigma_E = \frac{\tau}{mW} \tag{230}$$

We can reapply the analysis from above (subsection 4.3.2) to get:

$$\Delta\tau = \frac{1}{\sqrt{2}}\sqrt{\tilde{\sigma}_\tau^2 + \bar{\sigma}_\tau^2} = \frac{1}{\sqrt{2}}\frac{T}{m}\sqrt{\frac{1}{W^2} + \frac{1}{v^2\sigma_x^2}} \tag{231}$$

So in principle by making W small enough we can make the uncertainty in time of arrival as large as we wish.

Therefore:

(i) In SQM the uncertainty in clock time is *directly* proportional to the width of the gate.

(ii) In TQM the uncertainty in clock time is *inversely* proportional to the width of the gate.

These effects are results of the fundamental assumptions. In SQM, time is a parameter and the gate must *clip* the incoming wave function. In TQM, time is an operator and the gate must *diffract* the incoming wave function.

Therefore the observable effect may be made, in principle, *arbitrarily* large.

To go from observable in principle to observable in practice will require us to resolve two difficulties.

First we have to include the relativistic case: it is clear that the $\frac{1}{v}$ factor will make it hard for the effects of uncertainty in time to compete with the effects of uncertainty in space. This is a large part of the motivation for looking at the multiple particle case in the next section: relativistic velocities imply particle creation and therefore a need to understand what TQM looks like when there is more than one particle in play.

Secondly we need a way to separate measurements of uncertainty in time from measurements of uncertainty in space; an approach is suggested below (subsubsection 6.8.4).

5.2. Double slit in time

The two principle effects we have been focused on are additional dispersion in time and the uncertainty principle in time. We have not looked at effects related to interference in time, mostly notably the double slit in time experiment.

The double slit experiment has been described by Feynman [79] as the "only mystery" of quantum mechanics. We therefore take a look at it in the TQM context.

There have been many tests of the double slit experiment in space. The double slit in time variation has been performed by Lindner et al [80]. In the Linder et al experiment argon atoms are ionized using few-cycle electric pulses. If the electron is ionized by a single peak, that scores as a single slit in time; if by two, that scores as a double slit in time. The times are of order 500 attoseconds, much greater then the times we have been concerned with. But the principles of the experiment are of great interest here.

Horwitz [38] argues that the analysis in Lindner et al is not properly characterized as interference in time, since the analysis in Lindner et al is based on SQM, in which time is not an operator.

But Horwitz then shows you get the same diffraction spacing using as a starting point the relativistic dynamics equation[4]:

$$\imath \frac{\partial}{\partial s} \psi_s = -\frac{E^2 - \vec{p}^2}{2m} \psi_s \tag{232}$$

As noted above, this differs from our Schrödinger equation by only a gauge term. As gauge terms are not physically meaningful, Horwitz has provided us with the diffraction pattern for TQM.

And as Horwitz gets the same spacing as Lindner did, the double slit in time experiment does not help us differentiate between SQM and TQM, at least not to first order[5]. And therefore – somewhat surprisingly – the double slit in time experiment does not help us to falsify TQM

And it follows in turn that it is not the double slit but the single slit experiment which provides the decisive test of TQM.

6. Multiple particles

> "In what follows we assume that even though $A = \delta m$ is formally divergent, it is still `small´ in the sense that it is of the order of 1/137 times the electron mass." – J. J. Sakurai [81]

6.1. Why look at the multiple particle case?

In principle, we have enough to falsify TQM. Why then look at the multiple particle case?

(i) From the above we see the effects of TQM will be largest at relativistic velocities and short times. This implies we need to look at high energies and, since high energies imply particle creation, at the multiple particle case. (We thank Dr. Steve Libby for bringing this point to our attention.)

(ii) There is the further concern that TQM may not be renormalizable. If the loop integrals in SQM are barely renormalizable, with the addition of one more dimension the loop integrals in TQM may well become completely intractable.

(iii) Extending TQM to the multiple particle case opens up some new effects.

(iv) And then there is the intrinsic interest of the question.

At the same time, there are formidable difficulties: the literature on quantum field theory is vast and complex (among the references we have found helpful are [82, 83, 84, 45, 85, 86, 87, 88, 89, 90, 91, 92, 52, 93, 94, 95, 40]).

We will focus on analyzing a toy model in a series of toy situations – but in a way that will make clear how to extend the toy model to more realistic and useful cases.

Since the complications of spin are inessential here, we will look at a simple model with three massive spinless particles A, B, and C (correspondingly very loosely to the electron, photon, and proton).

As in the single particle case, we will first work out the rules for the various pieces of the path integral, then apply these rules to a few simple cases. There is no possibility of covering all cases, but we will do enough to make clear how to extend the approach to an arbitrary problem.

[4] Note the sign of the evolution parameter s here is opposite from the sign used in Land and Horwitz [37].

[5] To higher orders, a preliminary examination suggests that while the spacing peak-to-peak is unaffected, the widths of the individual peaks are increased.

We will require, as before, manifest covariance and consistency with established results. We will also require that the treatment of the single and multiple cases be consistent; the single particle case should appear at the end to be a specialized version of the multiple particle case.

The analysis divides into two parts:

First, we use a Lagrangian approach to derive the the appropriate Feynman path integrals. This approach will be loosely parallel to the approach in the single particle case:

(i) Extend Fock space from 3D to 4D.
(ii) Reuse the existing field theory Lagrangian densities by interpreting the time as the coordinate time.
(iii) Compute the action by integrating over both clock and coordinate times, so that the integrals over the Lagrangian density goes from $\int d\tau d\vec{x} \rightarrow \int d\tau dt d\vec{x}$.
(iv) Verify that the free particle propagator computed with this approach matches the single particle free particle propagator.

The result is that we get the TQM Feynman diagrams from their SQM equivalents by keeping the topology of lines and vertexes the same, but replacing the SQM parts with their TQM equivalents. We replace:

(i) SQM wave functions with TQM,
(ii) SQM propagators with TQM,
(iii) and SQM integrals over three space dimensions and clock time with TQM integrals over the three space dimensions, coordinate time, and clock time.

In principle, we could simply have postulated these substitutions; the derivation from first principles adds depth to our understanding, and also makes possible the extension of TQM to problems not amenable to attack via the Feynman diagram approach.

In the second part, we apply the derived Feynman rules to a few simple cases:

(i) Free particle,
(ii) Emission of a particle,
(iii) Absorption of a particle,
(iv) Exchange of a particle,
(v) Loop correction to the mass.

By combining these elements we can in principle calculate any Feynman diagram.

The loop calculation provides a preliminary success for TQM. If treated naively, the loop diagrams in TQM are divergent. But the combination of Morlet wavelet decomposition and entanglement in time – neither alone sufficient – causes the loop integrals to converge.

As we develop the multiple particle case, we will see a number of additional effects:

(i) anti-symmetry in time,
(ii) forces of anticipation and regret,
(iii) interference and entanglement in time.

With the approach taken here we have no free or adjustable parameters. We have a simple transition from the single to the multiple particle treatment. We do not have the familiar ultraviolet divergences. And we have a large number of opportunities for experimental test.

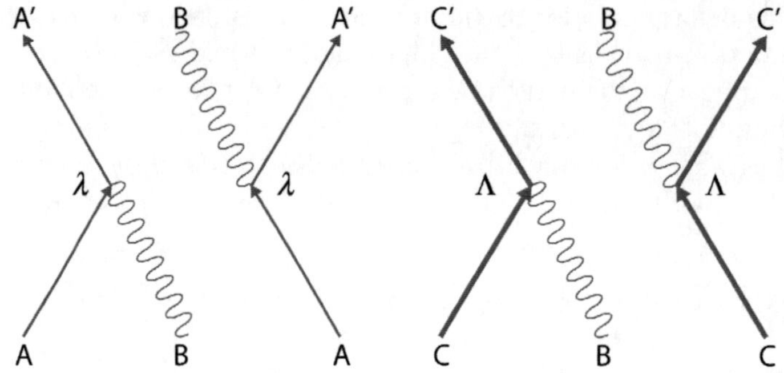

Figure 7. Interactions in the *ABC* model

6.2. ABC model

The simplest model we can find that lets us cover the basic interactions has three spinless, massive particle species A, B, and C. We assign them non-zero masses m, μ, and M respectively. We will take μ as small as it needs to be. They are real fields. A's emit and absorb B's with amplitude λ. C's emit and absorb B's with amplitude Λ. A's and C's do not talk with each other directly. There are no other interactions.

Loosely, A is a spinless model of an electron, B of a photon, and C of a proton.

We will focus primarily on the A and B particles. We will need the C for the discussion of particle exchange.

6.3. Fock space

In the single particle case we generalized the wave functions from three dimensions to four. Here we generalize Fock space from three dimensions to four. As paths may be seen as a series of wave functions, one wave function per clock tick, this implicitly generalizes the associated paths from three dimensions to four as well.

We use box normalization. The box runs from $-L \to L$ in all four coordinates: coordinate time and the three space dimensions. It is taken to extend well past the wavelets we are working with.

Using box normalization means that the Fourier transforms will be discrete. Discrete Fourier transforms are convenient for discussing various points of principle and to help in visualizing the field theory calculations. For the actual calculations we will use continuous wave functions.

In general the extrusion from three to four dimensions is straight forward; a few specific points require attention.

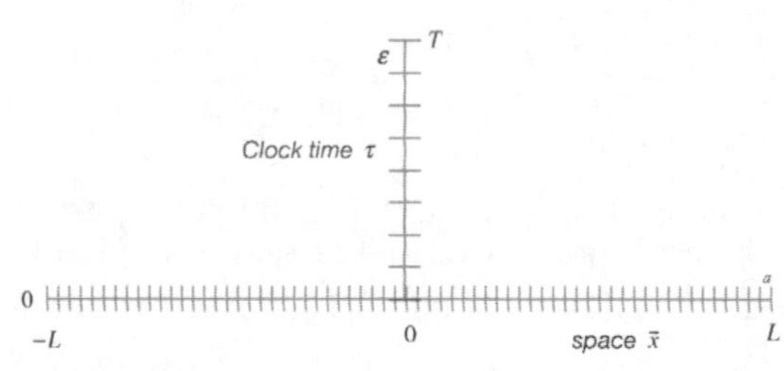

Figure 8. Fock space in three dimensions

IOP Conf. Series: Journal of Physics: Conf. Series **1239** (2019) 012015 doi:10.1088/1742-6596/1239/1/012015

6.3.1. *Fock space in three dimensions*

Single particle basis wave functions We will focus on the x coordinate here; y and z are the same. We break our box into 2M pieces, implying a lattice spacing $a \equiv \frac{L}{M}$. a has dimensions of length. This lets us replace the smoothly varying x with the values of x at a series of points:

$$x \in [-L, L] \rightarrow x = (-aM, \ldots, 0 \ldots, aM) \tag{233}$$

The continuous/discrete translation table is:

$$(x, y, z) \leftrightarrow (ai, aj, ak) \tag{234}$$

Integrals over space go to sums over i, j, k:

$$\int_{-L}^{L} dxdydz \leftrightarrow a^3 \sum_{i,j,k=-M}^{M} \tag{235}$$

The coordinate basis is trivial, just Kronecker δ functions:

$$\phi_{\vec{x}'}(\vec{x}) \equiv \delta^3(\vec{x} - \vec{x}') \leftrightarrow \delta_{ii'}\delta_{jj'}\delta_{kk'} \tag{236}$$

We assume the wave functions are periodic in $2L$. The periodic condition is not important; L will be chosen large enough that all interesting wave functions are well inside of it. They will be trivially periodic because they are zero on both sides of each dimension.

We normalize the basis wave functions to one:

$$\int_{-L,-L,-L}^{L,L,L} d\vec{x}\phi^*_{\vec{k}}(\vec{x})\phi_{\vec{k}'}(\vec{x}) = \delta_{\vec{k}\vec{k}'} \tag{237}$$

giving:

$$\phi_{\vec{k}}(\vec{x}) = \frac{1}{\sqrt{2L}^3} \exp\left(\imath\vec{k} \cdot \vec{x}\right) \tag{238}$$

Now we can expand an arbitrary wave function in terms of the basis functions:

$$\phi(\vec{x}) = \sum_{\vec{k}} c_{\vec{k}}\phi_{\vec{k}}(\vec{x}) \tag{239}$$

The measure in the path integrals is in terms of the c's:

$$\mathcal{D}\phi \equiv \prod_{n=0}^{N} \mathcal{D}_n\phi, \mathcal{D}_n\phi \equiv \prod_{\vec{k}} dc_{\vec{k}} \tag{240}$$

so there is one set of space integrals at each clock tick.

At the end of the discrete part of the calculation we will be letting M, N, and L go to infinity. As the effects of TQM are averaged out over larger times we will *not* be letting T go to infinity, even in the SQM case.

In the continuum limit we have:

$$\phi_{\vec{k}}(\vec{x}) \rightarrow \frac{1}{\sqrt{2\pi}^3} \exp\left(\imath\vec{k} \cdot \vec{x}\right) \tag{241}$$

IOP Conf. Series: Journal of Physics: Conf. Series **1239** (2019) 012015 doi:10.1088/1742-6596/1239/1/012015

Multi-particle wave functions Things become interesting when we go to multiple particle wave functions. For two particles:

$$\phi_{\vec{k}\vec{k}'}(1,2) \equiv \frac{1}{\sqrt{2}}\left(\phi_{\vec{k}}(1)\,\phi_{\vec{k}'}(2) + \phi_{\vec{k}}(2)\,\phi_{\vec{k}'}(1)\right) \tag{242}$$

As we get to larger and larger numbers of particles, these wave functions become tricky to write out and manage. To simplify, we use the familiar annihilation and creation operators, defined by:

$$\begin{aligned} a^{\dagger}_{\vec{k}}\left|n_{\vec{k}}\right\rangle &= \sqrt{n_{\vec{k}}+1}\left|n_{\vec{k}}+1\right\rangle \\ a_{\vec{k}}\left|n_{\vec{k}}\right\rangle &= \sqrt{n_{\vec{k}}}\left|n_{\vec{k}}-1\right\rangle \end{aligned} \tag{243}$$

with the usual commutation operators:

$$\left[a_{\vec{k}}, a^{\dagger}_{\vec{k}'}\right] = \delta_{\vec{k}\vec{k}'} \tag{244}$$

We define the single particle operator:

$$\bar{\phi}(\vec{x}) = \sum_{\vec{k}} a_{\vec{k}}\bar{\phi}^{\dagger}_{\vec{k}}(\vec{x}) + a^{\dagger}_{\vec{k}}\bar{\phi}_{\vec{k}}(\vec{x}) \tag{245}$$

An arbitrary multiple particle basis state may be built up as products of these:

$$\left|\{n_{\vec{k}}\}\right\rangle = \frac{1}{\sqrt{\prod\limits_{\vec{k}} n_{\vec{k}}!}}\prod_{\vec{k}}\left(a^{\dagger}_{\vec{k}}\right)^{n_{\vec{k}}}|0\rangle \tag{246}$$

where $\{n_{\vec{k}}\}$ is a specific set of occupation numbers.

We define the general wave function in the occupation number basis:

$$\sum_{\{n_{\vec{k}}\}} c_{\{n_{\vec{k}}\}}\left|\{n_{\vec{k}}\}\right\rangle \tag{247}$$

with normalization:

$$1 = \sum_{\{n_{\vec{k}}\}} c^2_{\{n_{\vec{k}}\}} \tag{248}$$

Note we are *not* defining the creation and annihilation operators in terms of an infinite set of harmonic oscillators; we are defining them by their effects on Fock space. We can think of movement in Fock space as a giant game of snakes and ladders, with the creation operators as the ladders, the annihilation operators as the snakes.

6.3.2. Fock space in four dimensions

Single particle basis wave functions The development in TQM works along the same lines but with one more dimension t and a fourth index h. We are treating coordinate time essentially like a fourth spatial dimension, much like the $x_4 \equiv \imath ct$ trick of earlier works on special relativity. Paths in TQM will typically start at $\tau = 0$ and finish at some defined clock time $\tau = T$. But their paths in coordinate time may well dive before $t = 0$ and climb past $t = T$. To make sure that all of the relevant paths are included in our box, we require that $-L \ll 0$ and $L \gg T$. Using the same L for time as for space, and requiring that $L \to \infty$ accomplishes this.

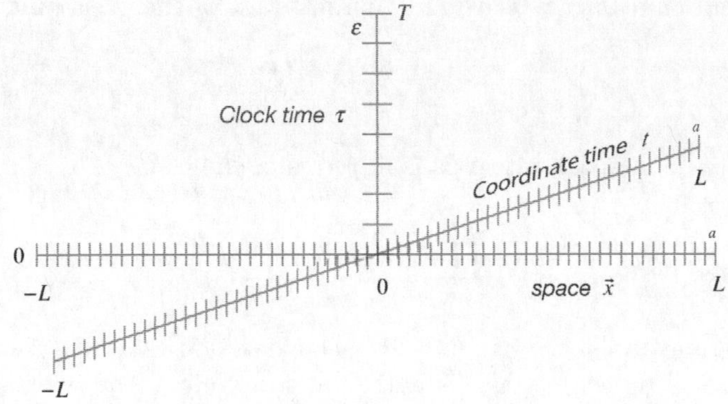

Figure 9. Fock space in four dimensions

The continuous/discrete translation table is now:

$$(t, x, y, z) \leftrightarrow (ah, ai, aj, ak) \tag{249}$$

So we have:

$$t = ah \tag{250}$$

Note there is no requirement that the lattice spacing a in coordinate time t match the step spacing ϵ in clock time τ; In general we will have: $a \neq \varepsilon$.

We promote integrals and sums over three dimensions to integrals and sums over four:

$$\int d\vec{x} \to \int dt d\vec{x}, \sum_{ijk} f_{ijk} \to \sum_{hijk} f_{hijk} \tag{251}$$

$$\phi_{x'}(x) \equiv \delta^4(x - x') \leftrightarrow \delta_{hh'}\delta_{ii'}\delta_{jj'}\delta_{kk'} \tag{252}$$

The k's are again periodic. We need to think for a moment about how our wave functions get their start in life. At clock time zero, the TQM wave function will look something like:

$$\tilde{\varphi}_0(t) \sim \exp\left(-\frac{(t - t_0)^2}{2\sigma_t^2}\right) \tag{253}$$

One subtlety: we always develop our wave functions from $\tau = 0 \to \tau = T$. Consider the wave function at $\tau = 0$. How did it get there? what of its past? Won't at some time the wave function have existed before $-L$, before the box appeared? The working answer is that the shape of the wave function at $\tau = 0$ tells us all we need to know of its history before that time. If the wave function *at* $\tau = 0$ is well within the box, if $\sigma_t \ll L$, we have what we need. What happened in the past, stays in the past.

We normalize the basis wave functions to one:

$$\int_{-L,-L,-L,-L}^{L,L,L,L} dt d\vec{x} \phi_k^*(t, \vec{x}) \phi_{k'}(t, \vec{x}) = \delta_{kk'} \tag{254}$$

giving:

$$\phi_k(x) = \frac{1}{4L^2} \exp(-\imath kx) \tag{255}$$

IOP Conf. Series: Journal of Physics: Conf. Series **1239** (2019) 012015 doi:10.1088/1742-6596/1239/1/012015

Now we can expand an arbitrary wave function in terms of the basis functions:

$$\phi(x) = \sum_k c_k \phi_k(x) \tag{256}$$

Again, the measure in the path integrals is in terms of the c's::

$$\mathcal{D}\phi \equiv \prod_{n=0}^{N} \mathcal{D}_n\phi, \mathcal{D}_n\phi \equiv \prod_k dc_k \tag{257}$$

so there is one set of time and space integrals at each clock tick. At the end of the discrete part of the calculation we will be letting M, N, and L go to infinity. And we will *not* be letting T go to infinity.

In the continuum limit we have:

$$\phi_k(x) \to \frac{1}{4\pi^2} \exp\left(-\imath wt + \imath \vec{k} \cdot \vec{x}\right) \tag{258}$$

Multiple particle wave functions Symmetrization in four dimensions works exactly as in three. For two particles we have:

$$\phi_{kk'}(1,2) \equiv \frac{1}{\sqrt{2}} \left(\phi_k(1)\,\phi_{k'}(2) + \phi_k(2)\,\phi_{k'}(1)\right) \tag{259}$$

And the creation and annihilation operators work in the same way:

$$\begin{aligned} a_k^\dagger \left|n_k\right\rangle &= \sqrt{n_k+1}\left|n_k+1\right\rangle \\ a_k \left|n_k\right\rangle &= \sqrt{n_k}\left|n_k-1\right\rangle \end{aligned} \tag{260}$$

With the commutators:

$$\left[a_k, a_{k'}^\dagger\right] = \delta_{kk'} \tag{261}$$

Single particle operator:

$$\phi(x) = \sum_k a_k \phi_k^\dagger(x) + a_k^\dagger \phi_k(x) \tag{262}$$

Arbitrary multiple particle:

$$\left|\{n_k\}\right\rangle = \frac{1}{\sqrt{\prod\limits_k n_k!}} \prod_k \left(a_k^\dagger\right)^{n_k} \left|0\right\rangle \tag{263}$$

where $\{n_k\}$ is a specific set of occupation numbers, now over all possibilities in four dimensions.

And we again define the general wave function as a sum over all possible $\{n_k\}$:

$$\sum_{\{n_k\}} c_{\{n_k\}} \left|\{n_k\}\right\rangle \tag{264}$$

with normalization:

$$1 = \sum_{\{n_k\}} c_{\{n_k\}}^2 \tag{265}$$

This defines the Fock space in four dimensions, along the same lines as the one in three. We are again playing snakes and ladders, but with four dimensional snakes and ladders rather than three.

IOP Conf. Series: Journal of Physics: Conf. Series **1239** (2019) 012015 doi:10.1088/1742-6596/1239/1/012015

6.3.3. Anti-symmetry in time We assume the same overall symmetry properties are required in four dimensions as in three. This implies that wave functions can use the coordinate time to help meet their symmetry responsibilities, with potentially amusing implications. In particular if the wave function is anti-symmetric in time, it will have the "wrong" symmetry properties in space.

This is testable, at least in principle.

Say we have wide wave functions in time and space $A(t)$ and $B(x)$, and narrow wave functions in time and space $a(t)$ and $b(x)$. The particles are identified as 1 and 2.

An acceptable initial wave function is:

$$\varphi_{sym}(1,2) = \frac{1}{\sqrt{2}}\left(A(t_1)B(x_1)a(t_2)b(x_2) + A(t_2)B(x_2)a(t_1)b(x_1)\right) \tag{266}$$

This clearly has the right symmetry between particles 1 and 2. We wish to break this down into sums over products of wave functions in time and space.

The symmetrical basis functions in time and space are:

$$\begin{aligned}\tilde{\varphi}_{sym}(1,2) &= \tfrac{1}{\sqrt{2}}\left(A(t_1)a(t_2) + A(t_2)a(t_1)\right)\\ \bar{\varphi}_{sym}(1,2) &= \tfrac{1}{\sqrt{2}}\left(B(x_1)b(x_2) + B(x_2)b(x_1)\right)\end{aligned} \tag{267}$$

If we use these as a product we get:

$$\tilde{\varphi}_{sym}(1,2)\,\bar{\varphi}_{sym}(1,2) = \frac{1}{2}\begin{pmatrix} A(t_1)\,B(x_1)\,a(t_2)\,b(x_2)\\ +A(t_1)\,B(x_2)\,a(t_2)\,b(x_1)\\ +A(t_2)\,B(x_1)\,a(t_1)\,b(x_2)\\ +A(t_2)\,B(x_2)\,a(t_1)\,b(x_1)\end{pmatrix} \tag{268}$$

where the two middle terms do not belong.

The anti-symmetric basis functions in time and space are:

$$\begin{aligned}\tilde{\varphi}_{anti}(1,2) &= \tfrac{1}{\sqrt{2}}\left(A(t_1)a(t_2) - A(t_2)a(t_1)\right)\\ \bar{\varphi}_{anti}(1,2) &= \tfrac{1}{\sqrt{2}}\left(B(x_1)b(x_2) - B(x_2)B(x_1)\right)\end{aligned} \tag{269}$$

and their product is:

$$\tilde{\varphi}_{anti}(1,2)\,\bar{\varphi}_{anti}(1,2) = \frac{1}{2}\begin{pmatrix} A(t_1)\,B(x_1)\,a(t_2)\,b(x_2)\\ -A(t_1)\,B(x_2)\,a(t_2)\,b(x_1)\\ -A(t_2)\,B(x_1)\,a(t_1)\,b(x_2)\\ +A(t_2)\,B(x_2)\,a(t_1)\,b(x_1)\end{pmatrix} \tag{270}$$

Therefore the sum of the completely symmetric and the completely anti-symmetric gives the target wave function:

$$\varphi_{sym}(1,2) = \frac{1}{2}\left(\tilde{\varphi}_{sym}(1,2)\,\bar{\varphi}_{sym}(1,2) + \tilde{\varphi}_{anti}(1,2)\,\bar{\varphi}_{anti}(1,2)\right) \tag{271}$$

To get a wave function which is completely symmetric in time and space together we need to use both the symmetric and the anti-symmetric basis functions.

6.4. Lagrangian

What should we use as a Lagrangian? We will look at this first from a classical perspective.

In SQM the Lagrangian for a massive spin 0 free particle is given by:

$$\bar{\mathcal{L}}^{free}\left[\phi,\dot{\phi}\right]=\frac{1}{2}\frac{\partial\phi}{\partial\tau}\frac{\partial\phi}{\partial\tau}-\frac{1}{2}\nabla\phi\nabla\phi-\frac{m^2}{2}\phi^2 \tag{272}$$

In classical mechanics, the wave functions may be written as sums over the basis plane waves:

$$\phi_\tau\left(\vec{x}\right)\sim\sum_{\vec{k}}c_{\tau,\vec{k}}\phi_{\tau,\vec{k}}\left(\vec{x}\right) \tag{273}$$

The action is the integral of this over space and clock time:

$$\int_0^T d\tau d\vec{x}\bar{\mathcal{L}}^{free}\left[\phi,\dot{\phi}\right] \tag{274}$$

Typically we let the limits in clock time go to $\pm\infty$, usually somewhere near the end of the analysis. Here that would average out the effects of any dispersion in time. So just as in the definition of Fock space, we keep the total clock time finite.

How to extend this Lagrangian to include coordinate time?

By our first requirement, x and t have to rotate into each other under a Lorentz transformation. The only way to do this is to change clock time to coordinate time: $\tau\rightarrow t$. So we start with:

$$\mathcal{L}^{free}\left[\phi\right]=\frac{1}{2}\partial_t\phi\partial_t\phi-\frac{1}{2}\nabla\phi\nabla\phi-\frac{m^2}{2}\phi^2 \tag{275}$$

The wave functions may be written as sums over the basis plane waves, with all the clock time dependence in the coefficients c.

$$\phi_\tau\left(t,\vec{x}\right)\ \sum_k c_\tau^{(k)}\phi_k\left(t,\vec{x}\right) \tag{276}$$

Neither basis functions nor operators are functions of clock time. Therefore the Lagrangian is not. To include the dependence on clock time we will need to include the integral over clock time from 0 to T

$$S_0\sim\int_0^T d\tau\int dtd\vec{x}\mathcal{L}^{free}\left[\varphi\right] \tag{277}$$

We can write the Lagrangian in momentum space as:

$$\hat{\mathcal{L}}^{free}\sim\frac{w^2-\vec{k}^2-m^2}{2} \tag{278}$$

The integral over clock time gives:

$$\imath\int d\tau\rightarrow\imath\frac{w^2-\vec{k}^2-m^2}{2} \tag{279}$$

This has two problems: it is not dimensionless and it does not match the results for the single particle propagator. We can fix both by adding a factor of $\frac{1}{m}$:

$$\mathcal{L}^{(free)}\rightarrow\frac{1}{2m}\left(\partial_t\phi\partial_t\phi-\nabla\phi\nabla\phi-m^2\phi^2\right) \tag{280}$$

IOP Conf. Series: Journal of Physics: Conf. Series **1239** (2019) 012015 doi:10.1088/1742-6596/1239/1/012015

With that noted, we will get the single particle propagator:

$$\hat{K} \sim \exp\left(\imath \frac{w^2 - \vec{k}^2 - m^2}{2m} \tau \right) \tag{281}$$

Our recipe for going from an existing SQM Lagrangian and action to the equivalent TQM Lagrangian and action is therefore:

(i) replace the time in the SQM Lagrangian with coordinate time,

(ii) replace the integral over three space dimensions with an integral over coordinate time and the three space dimensions,

(iii) add an overall integral over clock time,

(iv) divide the Lagrangian by the mass of the particle,

(v) and in the case of a massless particle – not covered here – use the familiar trick of taking the limit as the mass goes to zero

Spelling this out for the free Lagrangian we get:

$$S = \int_0^T d\tau dt d\vec{x} \frac{1}{2m} \left(\partial_t \phi \partial_t \phi - \nabla \phi \nabla \phi - m^2 \phi^2 \right) \tag{282}$$

The extension to include B, C particles is:

$$\mathcal{L}_{AB} [A, B, C] = \mathcal{L}^{free} [A] + \mathcal{L}^{free} [B] + \mathcal{L}^{free} [C] - \frac{\lambda}{2} ABA - \frac{\Lambda}{2} CBC \tag{283}$$

Since there is no longer any explicit dependence on clock time in the TQM Lagrangian, the corresponding Hamiltonian is merely $-\mathcal{L}$, with the slightly disconcerting result that there are no non-trivial canonical momenta.

In SQM the next step is to promote the classical fields to operators:

$$\bar{\phi} (\vec{x}) \to \sum_{\vec{k}} a_{\vec{k}} \bar{\phi}^{\dagger}_{\vec{k}} (\vec{x}) + a^{\dagger}_{\vec{k}} \bar{\phi}_{\vec{k}} (\vec{x}) \tag{284}$$

so in TQM we do the same:

$$\phi (x) \to \sum_{k} a_k \phi^{\dagger}_k (x) + a^{\dagger}_k \phi_k (x) \tag{285}$$

6.5. Path integrals

With the Lagrangian defined, we can write the full kernel as:

$$K_T \equiv \int \mathcal{D}\phi \exp \left(\imath \int_0^T d\tau \int d^4 x \mathcal{L} [\phi] \right) \tag{286}$$

This notation conceals much complexity. We start with the zero dimensional free case.

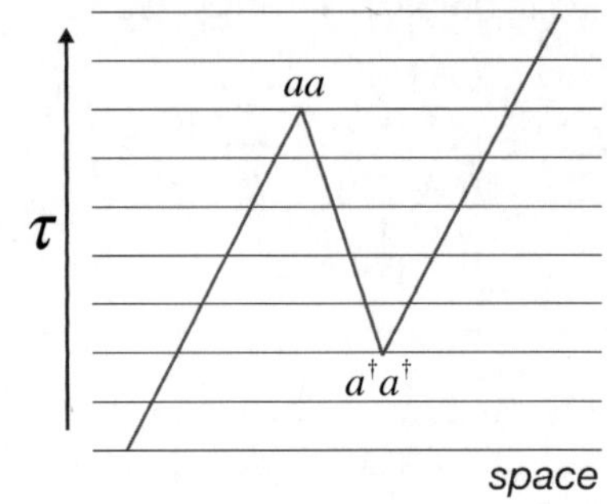

Figure 10. Zigzag paths in clock time

6.5.1. Zero dimensional free case The basis of wave functions for the zero dimensional case is the set of possible occupation numbers from 0 to infinity. Any wave function may be written as a sum over these:

$$\psi = \sum_{l=0}^{\infty} c_l \left| l \right\rangle \tag{287}$$

with the normalization condition that:

$$1 = \sum_{l=0}^{\infty} c_l^2 \tag{288}$$

We take the zero single particle wave function as simply the number one. For multiple particles we have:

$$\begin{aligned} \phi_2 &= \frac{1}{\sqrt{2!}} \left(\phi_1 (1) \phi_1 (2) + \phi_1 (2) \phi_1 (1) \right) \\ \phi_3 &= \frac{1}{\sqrt{3!}} \left(\phi_1 (1) \phi_1 (2) \phi_1 (3) + \ldots \right) \\ \ldots \end{aligned} \tag{289}$$

The number states are orthogonal:

$$\left\langle l \mid l' \right\rangle = \delta_{ll'} \tag{290}$$

The amplitude to go from one wave function to another is given by a Fock space sandwich:

$$\left\langle \psi' \right| K_T \left| \psi \right\rangle = \left\langle \sum_{l'=0}^{\infty} c_{l'} \phi_{l'} \middle| \int \mathcal{D}\phi \exp \left(\imath \int_0^T d\tau \mathcal{L} \right) \middle| \sum_{l=0}^{\infty} c_l \phi_l \right\rangle \tag{291}$$

with measure:

$$\mathcal{D}\phi = \prod_{n=0}^{n=N} \mathcal{D}\phi^{(n)}, \mathcal{D}\phi^{(n)} = \prod_{l=0}^{\infty} dc_l^{(n)} \tag{292}$$

and Lagrangian:

$$\mathcal{L} = -\frac{m}{2} \left(aa + aa^\dagger + a^\dagger a + a^\dagger a^\dagger \right) \tag{293}$$

Each term is a pair of operators. Two terms change the particle numbers; two do not. The term with two annihilation operators will reduce the number of particles by two, the term with two creation operators will increase the number of particles by two.

This means that even the simple free case has us moving up and down in Fock space. We have to deal with populations of particles.

IOP Conf. Series: Journal of Physics: Conf. Series **1239** (2019) 012015 doi:10.1088/1742-6596/1239/1/012015

We proceed discretely:

$$\exp\left(\imath\int_0^T d\tau\mathcal{L}\right)\to\exp\left(\imath\varepsilon\sum_{n=1}^{N}\mathcal{L}\right) \tag{294}$$

and one step at a time:

$$\exp\left(\imath\varepsilon\mathcal{L}\right)\approx 1+\imath\varepsilon\mathcal{L} \tag{295}$$

We will start with a common tactic: we will throw out all disconnected diagrams. For instance at one clock tick an $a^\dagger a^\dagger$ term could create a virtual particle/anti-particle pair which a few clock ticks later an aa then deletes. These represent self-interactions of the vacuum and are mere background noise, in common across all diagrams.

We will also use normal ordering ("always annihilate before you create"):

$$\frac{1}{2}\left(aa^\dagger+a^\dagger a\right)\to a^\dagger a+\frac{1}{2}=\hat{n}+\frac{1}{2} \tag{296}$$

Here $\hat{n}$ is the number operator. The $\frac{1}{2}$ term gives us an overall constant, which we can also ignore, since it also is present for all diagrams.

There is a third problem: sometimes a particle can interact with a virtual pair. The particle is evolving in clock time and encounters a term with two annihilation operators. By chance, a particle in the vacuum encounters the same term at the same clock tick. To an outside observer, it looks as if our particle has reversed direction in clock time and is now headed backwards. This is now a connected diagram, however, so the previous rules do not exclude it.

This is however implicitly included in our analysis of particle exchange below – the middle part of the diagram is the exchanged particle. We will drop these terms unless we explicitly need them, e.g. if we are looking at pair creation or annihilation.

With this the Lagrangian reduces to a sum over number operators:

$$\mathcal{L}\to -\imath m\frac{\hat{n}}{2}=-\imath\frac{m}{2}\sum_{l=0}^{\infty}l\delta_{ll'} \tag{297}$$

We return the infinitesimal Lagrangian to the exponential:

$$\exp\left(-\imath\varepsilon\sum_{n=1}^{N}\frac{m}{2}\sum_{l=0}^{\infty}l\delta_{ll'}\right) \tag{298}$$

and get:

$$K_\tau\left(l;l'\right)=\exp\left(-\imath\frac{m}{2}l\tau\right)\delta_{ll'} \tag{299}$$

which makes sense as the zero dimensional matrix element: if there are l particles present, they oscillate l times as quickly as one.

6.5.2. Four dimensional free case We now extend the zero dimensional treatment to four dimensions. Free Lagrangian:

$$\mathcal{L}^{free}\left[\phi\right]=\frac{1}{2m}\frac{\partial\phi}{\partial t}\frac{\partial\phi}{\partial t}-\frac{1}{2m}\nabla\phi\nabla\phi-\frac{m^2}{2m}\phi^2 \tag{300}$$

where ϕ is an operator:

$$\phi\left(x\right)=\int d^4k\hat{a}_k\phi_k^\dagger\left(x\right)+\hat{a}_k^\dagger\phi_k\left(x\right) \tag{301}$$

IOP Conf. Series: Journal of Physics: Conf. Series **1239** (2019) 012015 doi:10.1088/1742-6596/1239/1/012015

Again the amplitude to go from one state to another is computed by constructing a Fock space sandwich:

$$A_\varepsilon = \langle \{n_{k'}\} | \exp\left(\imath\varepsilon \int d^4x \mathcal{L}[\phi, \partial\phi]\right) | \{n_k\} \rangle \tag{302}$$

We will use as the single particle wave functions the exponentials $\phi_k(x) \equiv \frac{1}{4\pi^2} \exp(-\imath kx)$. In momentum space the partial derivatives ∂_x turn into powers of k. We have the overall integral over x and (from the transition to momentum space) integrals over k and k' for each of the two operators in each term. The basis functions integrated over x give δ functions in k, k'. The integral over k' gives an integral over k. We are left with:

$$\int d^4x\phi\phi \to \int d^4k \left(a_k a_{-k} + a_k a_k^\dagger + a_k^\dagger a_k + a_k^\dagger a_{-k}^\dagger\right) \tag{303}$$

This is the zero dimensional case with an index k. All four terms conserve momentum, but as before two terms change the particle numbers, two do not.

The momentum space integral of the Lagrangian is:

$$-\imath \int_0^\tau d\tau' f_k n_k = -\imath f_k \tau \tag{304}$$

giving the kernel for one frequency as:

$$K_T \sim \exp(-\imath f_k n_k \tau) \tag{305}$$

If $n_k = 1$ we have the TQM single particle propagator:

$$\hat{K}_\tau(k; k') = \exp(-\imath f_k \tau)\, \delta(k - k') \tag{306}$$

6.5.3. Measure Fock space consists of products of basis wave functions:

$$|\{n_k\}\rangle \to |n_0 k_0\rangle |n_1 k_1\rangle |n_2 k_2\rangle \ldots \tag{307}$$

A path is a series of positions in this Fock space. The measure weights each possible position equally, so the measure is:

$$\mathcal{D}\phi \equiv \prod_{n=0}^{N} \mathcal{D}_n\phi, \mathcal{D}_n\phi \equiv \prod_{k,n_k} dc_{k,n_k}^{(n)} \tag{308}$$

6.5.4. Interaction terms The coupling term is represented by a factor of the form:

$$\exp\left(-\imath\varepsilon\lambda \int d^4x \frac{ABA}{2}\right) \tag{309}$$

Assume we are looking at the case where this happens at clock tick number n. Note – in striking contrast to the SQM case – the interaction term has no dependence on n, on the clock time.

As $N \to \infty$ $\epsilon \to 0$ so we can approximate the exponential by:

$$1 - \imath\varepsilon\lambda \int d^4x \frac{ABA}{2} \tag{310}$$

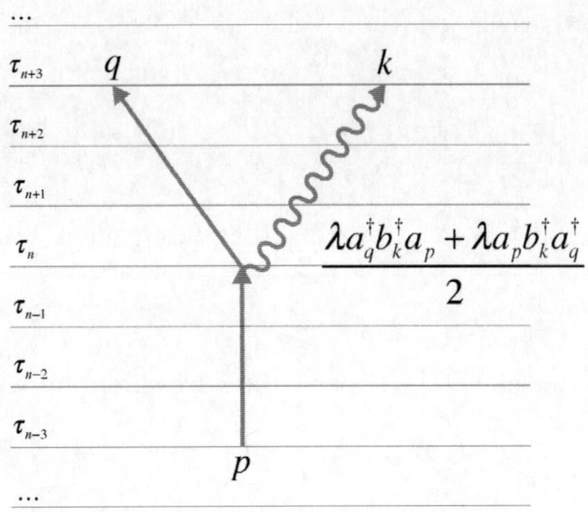

Figure 11. Interaction term in ABC model

The path integral is formed by doing the integrals from $0 \to n-1$, then the integral over the interaction term at n, then the integrals from $n+1 \to N$. The terms before n are included in the free propagator(s) from 0 to n; the terms after n are included in the free propagators from n to N, we have the interaction term to consider here.

The integral over space at step n will give us a δ function in momentum at step n: $\delta(k+q-p)$. Notice that four momentum is conserved at the vertex. This is another point of difference with SQM. In SQM only the three momentum is conserved at a vertex, the conservation of energy comes from the integral over the clock time.

For the nth time in our path integral from 0 to T, we role the dice in our game of snakes and ladders. Spelled out in terms of a and b operators we have:

$$\begin{aligned} A_p &= a_p \phi_p^\dagger + a_p^\dagger \phi_p \\ B_k &= b_k \phi_k^\dagger + b_k^\dagger \phi_k \\ A_q &= a_q \phi_q^\dagger + a_q^\dagger \phi_q \end{aligned} \tag{311}$$

We might drop down one step in terms of A particles with momentum p while going up one step for A particles with momentum q and one step for B particles with momentum k. This would be accomplished by a term of the form $\lambda a_q^\dagger b_k^\dagger a_p$. There are two such terms in the interaction term, neatly canceling out the factor of $\frac{1}{2}$. To contribute to the first order perturbation diagram the interaction must hit exactly once on the way from 0 to N. Result:

$$\hat{\psi}_T(q,k) = -\imath\lambda \int_0^T d\tau \int dp \hat{K}_{T\tau}^{(m)}(q)\, \hat{K}_{T\tau}^{(\mu)}(k)\, \delta(q+k-p)\, \hat{K}_{\tau}^{(m)}(p)\, \hat{\varphi}(p) \tag{312}$$

6.5.5. Full propagator We now have what is required to compute an arbitrary propagator.

The topology of the diagrams is unchanged from SQM: we get exactly the same set of diagrams, but with the intermediate integrals and initial wave function(s) over four dimensions rather than three.

The general propagator is given by:

$$\langle\{n_{k'}\}|\int\mathcal{D}\phi\exp\left(\imath\int_0^T d\tau\mathcal{L}^{free}[\phi]-V[\phi]\right)|\{n_k\}\rangle \tag{313}$$

The Feynman diagrams are generated by expanding this expression in powers of the coupling constant.

6.6. Free particles

How does the free propagator in TQM compare to the free propagator in SQM?

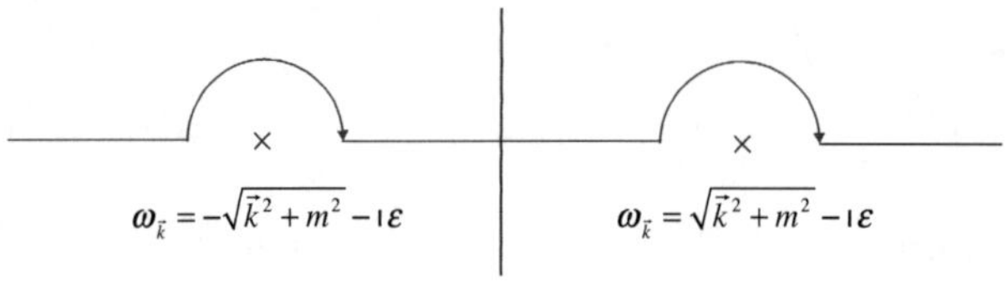

Figure 12. Contour for retarded propagator

We compute the free propagators for SQM and then TQM. We work these out for an A particle; the B and C are the same.

Our goal here to establish clearly the relationship between the SQM and TQM propagators; to make an apples-to-apples comparison between the two. The best way to do this is to look not just at the propagators but – as usual – at their effects on Gaussian test functions.

We will take as the starting point the respective differential equations for the SQM and TQM propagators. These may be derived from the path integral approach using the powerful generating approaches described in for instance Kashiwa [48] and Zee [52]. Here we take them as a given of the analysis.

6.6.1. Free particle in SQM We start with the Klein-Gordon equation. We define the propagator by:

$$\left(-\frac{\partial^2}{\partial\tau^2}+\nabla^2-m^2\right)\bar{K}_\tau\left(\vec{x};\vec{x}'\right)=\imath\delta\left(\tau\right)\delta^3\left(\vec{x}-\vec{x}'\right) \tag{314}$$

In momentum space we have:

$$\frac{-\imath}{k_0^2-\vec{k}^2-m^2}=\frac{1}{(2\pi)^4} \tag{315}$$

We choose retarded boundary conditions. This implies that both poles have a small negative imaginary part:

$$k_0=\pm\omega_{\vec{k}}-\imath\varepsilon,\omega_{\vec{k}}\equiv\sqrt{m^2+\vec{k}^2} \tag{316}$$

and the inverse Fourier transform is:

$$\bar{K}_\tau\left(\vec{x},\vec{x}'\right)=-\imath\lim_{\varepsilon\to0}\frac{1}{(2\pi)^4}\int dk_0 d\vec{k}\frac{\exp\left(-\imath k_0\tau+\imath\vec{k}\cdot\left(\vec{x}-\vec{x}'\right)\right)}{\left(k_0+\imath\varepsilon\right)^2-\vec{k}^2-m^2} \tag{317}$$

IOP Conf. Series: Journal of Physics: Conf. Series **1239** (2019) 012015 doi:10.1088/1742-6596/1239/1/012015

Doing the k_0 integral explicitly we get:

$$\bar{K}_\tau\left(\vec{x},\vec{x}'\right)=\frac{1}{(2\pi)^3}\int d\vec{k}\frac{\exp\left(-\imath\omega_{\vec{k}}\tau\right)-\exp\left(\imath\omega_{\vec{k}}\tau\right)}{2\omega_{\vec{k}}}\exp\left(\left(\imath\vec{k}\cdot\left(\vec{x}-\vec{x}'\right)\right)\right) \tag{318}$$

This corresponds to the basis (see for instance section 2.3 of Peskin [87]):

$$\frac{\sqrt{2\omega_{\vec{k}}}}{\sqrt{2\pi}^3}\exp\left(\imath\vec{k}\cdot\vec{x}\right) \tag{319}$$

To make a closer comparison to TQM we shift to the basis functions:

$$\frac{1}{\sqrt{2\pi}^3}\exp\left(\imath\vec{k}\cdot\vec{x}\right) \tag{320}$$

This gives:

$$\bar{K}_\tau\left(\vec{x},\vec{x}'\right)=\int d\vec{k}\frac{\exp\left(-\imath\omega_{\vec{k}}\tau\right)-\exp\left(\imath\omega_{\vec{k}}\tau\right)}{(2\pi)^3}\exp\left(\left(\imath\vec{k}\cdot\left(\vec{x}-\vec{x}'\right)\right)\right) \tag{321}$$

We apply this to a Gaussian test function. We choose one centered on $\vec{k}_0$, initial position $\vec{x}_0$. We choose one which is separable in the three space directions. We use the obvious definition $\delta\vec{k}\equiv\vec{k}-\vec{k}_0$:

$$\hat{\varphi}_0\left(\vec{k}\right)=\hat{\varphi}_0^{(y)}\left(k_y\right)\hat{\varphi}_0^{y}\left(k_y\right)\hat{\varphi}_0^{(z)}\left(k_z\right) \tag{322}$$

where the x Gaussian test function is:

$$\hat{\varphi}_0^{(x)}\left(k_x\right)=\sqrt[4]{\frac{1}{\pi\hat{\sigma}_x^2}}e^{-\imath k_x x_0-\frac{\left(k_x-k_x^{(0)}\right)^2}{2\hat{\sigma}_x^2}}=\sqrt[4]{\frac{1}{\pi\hat{\sigma}_x^2}}e^{-\imath k_x x_0-\frac{\delta k_x^2}{2\hat{\sigma}_x^2}} \tag{323}$$

and y,z the same.

We include both positive and negative frequencies for each wave vector.

We now take as a first working assumption that our incoming wave function is dominated by the positive frequency part. (This is the same trick we used in the analysis of the time-of-arrival measurements in subsection 4.3). If we were going to examine phenomena like Zitterbewegung we would need to relax this assumption.

As a second assumption, we will assume our incoming Gaussian test function is reasonably well-focused, $\hat{\sigma_k}$ small, so that we can replace the factor overall factor by its expectation:

$$\frac{1}{\sqrt{2\omega_{\vec{k}}}}\to\frac{1}{\sqrt{2\omega_{\vec{k}_0}}} \tag{324}$$

We therefore simplify our kernel to:

$$\hat{K}_\tau\left(\vec{k};\vec{k}'\right)=\frac{1}{2\omega_{\vec{k}_0}}\exp\left(-\imath\omega_{\vec{k}}\tau\right)\delta^3\left(\vec{k}-\vec{k}'\right) \tag{325}$$

We next expand $\omega_{\vec{k}}$ in powers of the kinetic energy:

$$\omega_{\vec{k}}\approx m+\frac{\vec{k}^2}{2m}-\frac{\vec{k}^4}{8m^3}+O\left(\vec{k}^6\right) \tag{326}$$

Applied to the Gaussian test function we get:

$$\hat{\hat{\varphi}}_\tau\left(\vec{k}\right) = \int d\vec{k}' \hat{\hat{K}}_\tau\left(\vec{k};\vec{k}'\right)\hat{\hat{\varphi}}_0\left(\vec{k}\right) \tag{327}$$

or:

$$\hat{\hat{\varphi}}_\tau\left(\vec{k}\right) = \exp\left(-\imath m\tau - \imath\frac{\vec{k}^2}{2m}\tau + \imath\frac{\vec{k}^4}{8m^3}\tau\right)\hat{\hat{\varphi}}_0\left(\vec{k}\right) \tag{328}$$

We see on the SQM side the relationship between the various levels of analysis: the particle at rest (m term), the particle moving slowly ($\vec{k}^2$) term, and the particle moving relativistically ($\vec{k}^4$ and higher corrections).

We summarize the SQM propagator (as applied to a Gaussian test function) as:

$$\hat{\hat{K}}_\tau\left(\vec{k}\right) = \exp\left(-\imath m\tau - \imath\frac{\vec{k}^2}{2m}\tau + \imath\frac{\vec{k}^4}{8m^3}\tau\right) \tag{329}$$

6.6.2. Free particle in TQM We have the free propagator from above:

$$\hat{K}_\tau\left(\vec{k}\right) = \exp\left(-\imath\frac{m}{2}\tau + \imath\frac{w^2}{2m}\tau - \imath\frac{\vec{k}^2}{2m}\tau\right) \tag{330}$$

There are three differences between the TQM and SQM propagators. The first two are not that interesting for our purposes; the third is critical.

First we have an overall factor of:

$$\exp\left(-\imath\frac{m}{2}\tau\right) \tag{331}$$

as compared to the factor of:

$$\exp\left(-\imath m\tau\right) \tag{332}$$

in the SQM propagator.

The frequency associated with, say, the mass of an electron is:

$$\frac{m_e}{\hbar} \approx \frac{.51\cdot 10^6 eV}{6.6\cdot 10^{-16} eV\,\mathrm{sec}} = 7.7\cdot 10^{-24}\mathrm{sec}^{-1} \tag{333}$$

This is of order the frequencies associated with Zitterbewegung, about one million times the frequencies we are dealing with here. As Zitterbewegung has not itself been measured, we would have no comparison point on the SQM side. We will therefore ignore this factor.

The relativistic correction factor:

$$\exp\left(+\imath\frac{\vec{k}^4}{8m^3}\tau\right) \tag{334}$$

is present in SQM but not in TQM.

These represent higher order corrections. As TQM is predicting significant differences from SQM in even the non-relativistic case, relativistic corrections to SQM are not needed (and clutter up the analysis).

The key difference is the factor of:

$$\exp\left(+\imath\frac{w^2}{2m}\tau\right) \tag{335}$$

IOP Conf. Series: Journal of Physics: Conf. Series **1239** (2019) 012015 doi:10.1088/1742-6596/1239/1/012015

This represents the extension of the function in time/energy. Basically the TQM propagator is the SQM propagator with additional fuzziness in time/energy. This difference is the focus of attention in this work.

So we will take the SQM propagator as:

$$\hat{\hat{K}}_\tau\left(\vec{k}\right) = \exp\left(-\imath\frac{\vec{k}^2}{2m}\tau\right) \tag{336}$$

the TQM propagator as:

$$\hat{K}_\tau\left(\vec{k}\right) = \exp\left(\imath\frac{w^2}{2m}\tau - \imath\frac{\vec{k}^2}{2m}\tau\right) = \exp\left(\imath\frac{w^2}{2m}\tau\right)\hat{\hat{K}}_\tau\left(\vec{k}\right) \tag{337}$$

We apply this to a Gaussian test function. We use a Gaussian test function in energy times the previous Gaussian test function in three momentum.

$$\hat{\varphi}_0\left(w,\vec{k}\right) = \hat{\tilde{\varphi}}_0\left(w\right)\hat{\hat{\varphi}}_0\left(\vec{k}\right) \tag{338}$$

with energy part:

$$\hat{\tilde{\varphi}}_0\left(w\right) \equiv \sqrt[4]{\frac{1}{\pi\hat{\sigma}_t^2}}e^{\imath w t_0 - \frac{(w-w_0)^2}{2\hat{\sigma}_t^2}} \tag{339}$$

From the entropic analysis above (subsection 4.1):

$$\begin{aligned} w_0 &\sim \omega_{\vec{k}} \\ \hat{\sigma}_t^2 &\sim \hat{\sigma}_x^2 + \hat{\sigma}_y^2 + \hat{\sigma}_t^2 \end{aligned} \tag{340}$$

Application of the TQM kernel to the Gaussian test function is trivial:

$$\hat{\varphi}_\tau\left(k\right) = \exp\left(\imath\frac{k^2 - m^2}{2m}\tau\right)\hat{\varphi}_0\left(k\right) \tag{341}$$

We have therefore recovered, at the cost of a few approximations, the single particle propagator for TQM.

We turn now to specific applications.

6.7. Emission of a particle

What does the emission of a particle look like in TQM?

6.7.1. Overview We look at the case where an A particle emits a B. The initial particle expectation and dispersion are given; we wish to compute the outgoing particle expectations and dispersions.

We expect that this emission diagram will be part of a larger calculation. We will assume for convenience here that the momentum space expectations of the incoming particle are on-shell, but that need not be true in general.

We start with a Gaussian test function A, extended in space for the SQM case, in time and space for TQM.

The initial expectation and dispersion are given at clock time τ_0. We wish to compute the final expectations and dispersions at clock time τ_2. The B has an amplitude λ to be emitted at each intermediate clock time τ_1.

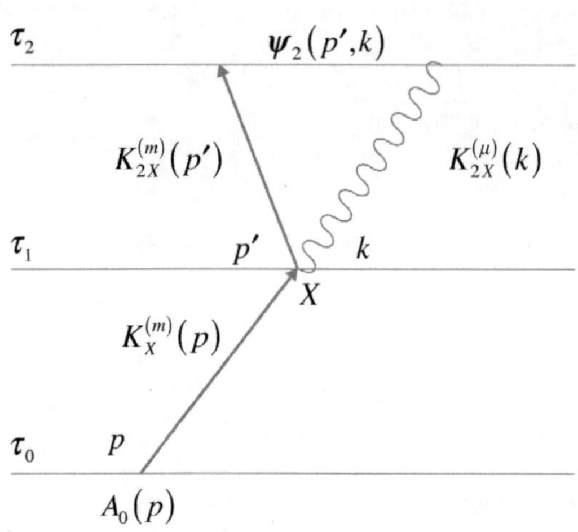

Figure 13. An A particle emits a B particle

In a first order perturbation expansion we would integrate over the intermediate clock time τ_1. But here we will focus on a smaller piece of the puzzle, looking at the contribution to the final wave function from a single point in clock time. We look at:

$$\psi_2(p,k) = -\imath\lambda \int dp K_{21}^{(m)}(p') K_{21}^{(\mu)}(k)\, \delta(p'+k-p)\, K_{10}^{(m)}(p)\, A_0(p) \tag{342}$$

We look at this first in SQM then in TQM.

6.7.2. Emission of a particle in SQM We start our A particle as a Gaussian test function in momentum space:

$$\hat{\hat{A}}_0(\vec{p}) = \hat{\hat{\varphi}}_a^{(x)}(p_y)\, \hat{\hat{\varphi}}_a^{(y)}(p_x)\, \hat{\hat{\varphi}}_a^{(z)}(p_x) \tag{343}$$

For the x component we have:

$$\hat{\hat{\varphi}}_a^{(x)}(p_x) \equiv \sqrt[4]{\frac{1}{\pi\hat{\sigma}_x^2}}\, e^{-\imath p_x x_a - \frac{\left(p_x - p_x^{(a)}\right)^2}{2\hat{\sigma}_x^2}} \tag{344}$$

with y, z in parallel. The full wave function at τ_0 is:

$$\hat{\hat{A}}_0(\vec{p}) = \sqrt[4]{\frac{1}{\pi^3 det\left(\hat{\hat{\Sigma}}\right)}}\, e^{-\imath\vec{p}\cdot\vec{x}_a - \frac{1}{2}\delta\vec{p}\cdot\hat{\hat{\Sigma}}^{-1}\cdot\delta\vec{p}} \tag{345}$$

with ancillary definitions:

$$\begin{aligned} \delta\vec{p} &\equiv \vec{p} - \vec{p}_a \\ \hat{\hat{\Sigma}} &\equiv \begin{pmatrix} \hat{\sigma}_x^2 & 0 & 0 \\ 0 & \hat{\sigma}_x^2 & 0 \\ 0 & 0 & \hat{\sigma}_x^2 \end{pmatrix} \end{aligned} \tag{346}$$

We are dropping the parts of the kernel that depend on the rest mass. For the A particle these are independent of the interaction and therefore irrelevant. For the B particle the $exp(-\imath\mu\tau)$ is not independent of the interaction, but in the limit as $\mu \to 0$ this factor is constant and therefore also irrelevant.

The kernel that carries A from $\tau_0 \to \tau_X$ is:

$$\hat{\hat{K}}_X^{(m)}(\vec{p}) = \exp\left(-\imath\frac{\vec{p}^2}{2m}\tau_X\right) \tag{347}$$

So the A wave function at X is:

$$\hat{\hat{A}}_X(\vec{p}) = \sqrt[4]{\frac{1}{\pi^3 det\left(\hat{\hat{\Sigma}}\right)}} e^{-\imath\vec{p}\cdot\vec{x}_a - \frac{1}{2}\delta\vec{p}\cdot\hat{\hat{\Sigma}}^{-1}\cdot\delta\vec{p} - \imath\frac{\vec{p}^2}{2m}\tau_X} \tag{348}$$

The integral over the δ function at τ_X gives:

$$\hat{\hat{\psi}}_2\left(\vec{p}',\vec{k}\right) = -\imath\lambda\hat{\hat{K}}_{2X}^{(m)}(\vec{p}')\,\hat{\hat{K}}_{2X}^{(\mu)}\left(\vec{k}\right)\hat{\hat{A}}_X\left(\vec{p}'+\vec{k}\right) \tag{349}$$

We have replaced the initial momentum with the sum of the final momenta.

The post-vertex kernels are:

$$\hat{\hat{K}}_{2X}^{(m)}(\vec{p}') = \exp\left(-\imath\frac{(\vec{p}')^2}{2m}\tau_{2X}\right), \hat{\hat{K}}_{2X}^{(\mu)}\left(\vec{k}\right) = \exp\left(-\imath\frac{\vec{k}^2}{2\mu}\tau_{2X}\right) \tag{350}$$

with $\tau_{2X} \equiv \tau_2 - \tau_X$.

The joint wave function at τ_2 is therefore:

$$\hat{\hat{\psi}}_2\left(\vec{p}',\vec{k}\right) = \sqrt[4]{\frac{1}{\pi^3 det\left(\hat{\hat{\Sigma}}\right)}} e^{\left(-\imath(\vec{p}'+\vec{k})\cdot\vec{x}_a - \frac{1}{2}\delta\vec{p}'\cdot\hat{\hat{\Sigma}}^{-1}\cdot\delta\vec{p}' - \imath\Omega_X\tau_{2X} - \imath\Omega_0\tau_X\right)} \tag{351}$$

with change in momentum:

$$\delta\vec{p}' \equiv \vec{p}' + \vec{k} - \vec{p}_a \tag{352}$$

and initial and final energies:

$$\Omega_0 \equiv \frac{\left(\vec{p}'+\vec{k}\right)^2}{2m}, \Omega_X \equiv \frac{(\vec{p}')^2}{2m} + \frac{\vec{k}^2}{2\mu} \tag{353}$$

We can see that an integral over τ_X would tend to subtract out components where $\Omega_0 \neq \Omega_X$, giving us an effective δ function in the SQM energy.

The final wave function is strongly correlated between left and right. The conservation condition at the vertex means A', B are each sharing part of the same initial momentum. They are like Siamese twins – separated at birth but still connected. This is the source of the mysterious spooky action at a distance complained of in the initial EPR paper [96].

6.7.3. Emission of a particle in TQM We take the same basic approach, but now with the coordinate energy/coordinate time included, and with the conservation condition at the vertex being for four momentum rather than three momentum.

We write the initial wave function as direct product of time and space parts: $A = \tilde{A}\bar{A}$.

$$\hat{A}_0(p) = \hat{\tilde{A}}_a(E)\,\hat{\bar{A}}_a(\vec{p}) \tag{354}$$

IOP Conf. Series: Journal of Physics: Conf. Series **1239** (2019) 012015 doi:10.1088/1742-6596/1239/1/012015

The three momentum part is the same as in SQM. For the energy part we have:

$$\hat{\tilde{A}}_a(E) \equiv \sqrt[4]{\frac{1}{\pi\hat{\sigma}_t^2}} e^{\imath E t_a - \frac{(E-E_a)^2}{2\hat{\sigma}_t^2}} \tag{355}$$

We take:

$$\begin{array}{l} t_a \approx \tau_0 = 0 \\ E_a \approx \sqrt{m^2 + \vec{p}_a^2} \\ \hat{\sigma}_E^2 \approx \hat{\sigma}_x^2 + \hat{\sigma}_y^2 + \hat{\sigma}_z^2 \end{array} \tag{356}$$

We write the entire wave function function as:

$$\hat{A}_0(p) = \sqrt[4]{\frac{1}{\pi^4 \det\left(\hat{\Sigma}\right)}} e^{-\imath p x_a - \frac{1}{2}\delta p \hat{\Sigma}^{-1}\delta p} \tag{357}$$

with ancillary definitions:

$$\begin{array}{l} \delta p \equiv p - p_a \\ \hat{\Sigma} \equiv \begin{pmatrix} \hat{\sigma}_t^2 & 0 & 0 & 0 \\ 0 & \hat{\sigma}_x^2 & 0 & 0 \\ 0 & 0 & \hat{\sigma}_y^2 & 0 \\ 0 & 0 & 0 & \hat{\sigma}_z^2 \end{pmatrix} \end{array} \tag{358}$$

The kernel that carries A from $\tau_0 \to \tau_X$ is:

$$K_X^{(m)}(p) = \exp\left(-\imath f_p \tau_X\right) = \hat{\tilde{K}}_X^{(m)}(E)\, \hat{K}_X^{(m)}(\vec{p}) \exp\left(-\imath \frac{m}{2}\tau_X\right) \tag{359}$$

with $f_p \equiv -\frac{E^2 - \vec{p}^2 + m^2}{2m}$ and with energy part:

$$\hat{\tilde{K}}_X^{(m)}(E) \equiv \exp\left(\imath \frac{E^2}{2m}\tau_X\right) \tag{360}$$

So the A wave function at X is:

$$\hat{A}_X(p) = \sqrt[4]{\frac{1}{\pi^4 det\left(\hat{\Sigma}\right)}} \exp\left(-\imath p \cdot x_a - \frac{1}{2}\delta p \hat{\Sigma}^{-1}\delta p - \imath f_p \tau_X\right) \tag{361}$$

The energy part at A is now:

$$\hat{\tilde{A}}_X(E) = \sqrt[4]{\frac{1}{\pi\hat{\sigma}_t^2}} \exp\left(\imath E t_a - \frac{(E-E_a)^2}{2\hat{\sigma}_t^2} + \imath \frac{E_a^2}{2m}\tau_X\right) \tag{362}$$

The integral over the δ function at X gives:

$$\hat{\psi}_2(p', k) = -\imath\lambda \hat{K}_{2X}^{(m)}(p')\, \hat{K}_{2X}^{(\mu)}(k)\, \hat{A}_X(p' + k) \tag{363}$$

Again we are pushing the sum of the final momenta back into the initial wave function. Post vertex kernels:

$$\hat{K}_{2X}^{(m)}(p') = \exp\left(-\imath f_{p'} \tau_{2X}\right), \hat{K}_{2X}^{(\mu)}(k) = \exp\left(-\imath f_k \tau_{2X}\right) \tag{364}$$

So the wave function at τ_2 is:

$$\hat{\psi}_2(p',k) = \sqrt[4]{\frac{1}{\pi^4 \det\left(\hat{\Sigma}\right)}} \exp\left(\imath(p'+k)x_a - \frac{1}{2}\delta p'\hat{\Sigma}^{-1}\delta p' - \imath F_X \tau_{2X} - \imath F_0 \tau_X\right) \tag{365}$$

with change in four momentum:

$$\delta p' \equiv p' + k - p_a \tag{366}$$

and initial and final clock frequencies:

$$\begin{array}{l} F_0 \equiv -\frac{(E'+w)^2-(\vec{p}'+\vec{k})^2-m^2}{2m}, \\ F_X \equiv f_{p'} + f_k = -\frac{(E')^2-(\vec{p}')^2-m^2}{2m} - \frac{w^2-\vec{k}^2-\mu^2}{2\mu} \end{array} \tag{367}$$

Per the long, slow approximation, we expect that both F_0 and F_X will be small. And an integral over τ_X would make their difference still smaller. As noted in the free particle section most of the dependence on clock time will be carried by the coordinate time part of the wave function.

The parts dependent on the rest masses do not play a critical role, for the same reasons as in the SQM case.

The energy part of the joint wave function is:

$$\hat{\tilde{\psi}}_2(E',w) = \sqrt[4]{\frac{1}{\pi\hat{\sigma}_t^2}} e^{\imath(E'+w)t_a - \frac{(E'+w-E_a)^2}{2\hat{\sigma}_t^2} + \imath\left(\frac{(E')^2}{2m} + \frac{w^2}{2m}\right)\tau_{2X} + \imath\frac{(E'+w)^2}{2m}\tau_X} \tag{368}$$

The left and right halves are again strongly correlated – now in energy/time as well as in three-momentum/space – even though with increasing clock time they are separated by greater and greater distances. Again, they are like Siamese twins separated at birth but still connected, now across time as well as space.

6.7.4. Discussion of particle emission With TQM, to the correlations in three-momentum complained of in the initial EPR paper we add correlations in energy. These provide raw material for a Bell's theorem "in time".

Presumably Einstein would still be unhappy about the "spooky action at a distance", but perhaps he would be partly consoled by the inclusion of time on the same basis as space.

6.8. Absorption of a particle
What does the absorption of a particle look like in TQM?

6.8.1. Overview We look at the case where an A absorbs a B. The initial particle expectations and dispersions are given; we wish to compute the outgoing particle's expectation and dispersion.

As with emission, we expect that this absorption diagram will be part of a larger calculation. We will assume for convenience here that the momentum space expectations of the incoming particle are on-shell, but that need not be true in general.

We start with two Gaussian test functions A, B. These are centered on momenta p_a, k_b with initial expectations for position x_a, x_b. We define $p'_a \equiv p_a + k_b$.

For simplicity we take the A particle as coming in from the left and the B as coming in from the right. Without loss of generality we can assume both are coming in along the x-axis with relative offset b along the y axis. With a slight loss of generality we will assume $b \to 0$.

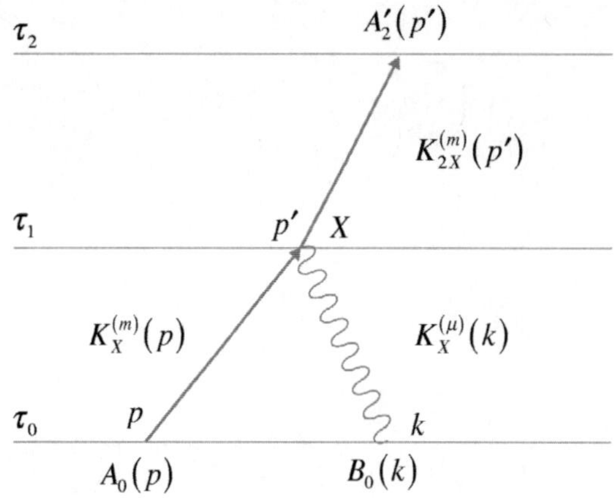

Figure 14. An A particle absorbs a B particle

We have starting velocities, $v > 0, u > 0$:

$$\vec{p}_a = mv\hat{x}, \vec{k}_b = -\mu u\hat{x} \tag{369}$$

and starting points on left and right:

$$\vec{x}_a = -l\hat{x}, \vec{x}_b = d\hat{x} \tag{370}$$

In first order perturbation theory we would compute the final amplitude by integrating over all intermediate clock times τ_1. But as with emission we focus on the interaction at a specific clock time τ_X. Since we have a natural clock time to work with – the time defined by the intersection of the classical paths of A, B – we will use that.

This is defined by:

$$x = -l + v\tau_X = d - u\tau_X \tag{371}$$

giving crossing time τ_X:

$$\tau_X = \frac{d+l}{v+u} \tag{372}$$

and crossing position x_X:

$$x_X = \frac{vd - ul}{v+u} \tag{373}$$

Trivially $y_X = z_X = 0$.

We are left with one integral to do, a convolution of the initial momenta:

$$\hat{A}_2(p') = -\imath\lambda \int dk \hat{K}_{2X}^{(m)}(p') \hat{A}_X(p'-k) \hat{\tilde{B}}_X(k) \tag{374}$$

We will first treat the SQM case, then TQM.

6.8.2. *Absorption of a particle in SQM*

Initial wave functions The initial particles are given by:

$$\begin{aligned} \hat{\tilde{A}}_0(\vec{p}) &= \sqrt[4]{\frac{1}{\pi^3 det\left(\hat{\tilde{\Sigma}}\right)}} e^{-\imath \vec{p}\cdot\vec{x}_a - \frac{1}{2}\vec{p}\cdot\hat{\tilde{\Sigma}}^{-1}\cdot\vec{p}} \\ \hat{\tilde{B}}_0\left(\vec{k}\right) &= \sqrt[4]{\frac{1}{\pi^3 \det\left(\hat{\tilde{S}}\right)}} e^{-\imath \vec{k}\cdot\vec{x}_b - \frac{1}{2}\vec{k}\cdot\hat{\tilde{S}}^{-1}\cdot\vec{k}} \end{aligned} \tag{375}$$

with expectations and dispersions:

$$\begin{aligned} &\vec{p} \equiv \vec{p} - \vec{p}_a, \vec{k} \equiv \vec{k} - \vec{k}_b \\ &\hat{\tilde{\Sigma}} \equiv \begin{pmatrix} \hat{\sigma}_x^2 & 0 & 0 \\ 0 & \hat{\sigma}_x^2 & 0 \\ 0 & 0 & \hat{\sigma}_x^2 \end{pmatrix}, \hat{\tilde{S}} \equiv \begin{pmatrix} \hat{s}_x^2 & 0 & 0 \\ 0 & \hat{s}_y^2 & 0 \\ 0 & 0 & \hat{s}_z^2 \end{pmatrix} \end{aligned} \tag{376}$$

In momentum space the kernels from start to X are:

$$\hat{\tilde{K}}_X^{(m)}(\vec{p}) = \exp\left(-\imath\frac{\vec{p}^2}{2m}\tau_X\right), \hat{\tilde{K}}_X^{(\mu)} = \exp\left(-\imath\frac{\vec{k}^2}{2\mu}\tau_X\right) \tag{377}$$

The wave functions at X are therefore:

$$\begin{aligned} \hat{\tilde{A}}_X(\vec{p}) &= \sqrt[4]{\frac{1}{\pi^3 det\left(\hat{\tilde{\Sigma}}\right)}} e^{-\imath \vec{p}\cdot\vec{x}_a - \frac{1}{2}\vec{p}\cdot\hat{\tilde{\Sigma}}^{-1}\cdot\vec{p} - \imath\frac{\vec{p}^2}{2m}\tau_X} \\ \hat{\tilde{B}}_X\left(\vec{k}\right) &= \sqrt[4]{\frac{1}{\pi^3 \det\left(\hat{\tilde{S}}\right)}} e^{-\imath \vec{k}\cdot\vec{x}_b - \frac{1}{2}\vec{k}\cdot\hat{\tilde{S}}^{-1}\cdot\vec{k} - \imath\frac{\vec{k}^2}{2\mu}\tau_X} \end{aligned} \tag{378}$$

Interaction The final wave function at τ_2 will be given by a convolution of all possible incoming momenta:

$$\hat{\tilde{A}}_2(\vec{p}') = -\imath\lambda\hat{K}_{2X}^{(m)}(\vec{p}') \int d\vec{k}\hat{A}_X\left(\vec{p}' - \vec{k}\right)\hat{B}_X\left(\vec{k}\right) \tag{379}$$

We focus on the convolution integral over x:

$$\hat{I}_X^{(x)}(p'_x) \equiv \int dk_x \hat{A}_X(p'_x - k_x)\hat{B}_X(k_x) \tag{380}$$

Since this is the integral of a Gaussian it can be solved exactly. However we will get more insight by shifting to the position basis. The convolution in momentum space becomes a multiplication in position space:

$$I_X^{(x)}(x) = \sqrt{2\pi}A_X(x)B_X(x) \tag{381}$$

The Gaussian test functions for A and B are centered on their corresponding classical paths.

The close relationship of classical and quantum trajectories is an attractive feature of the approach here; we can think of a particle as traveling along a classical line with quantum fuzz around it. In SQM the fuzz extends out in the three space dimensions; in TQM in the time dimension as well.

At the crossing time, the coordinate forms for A and B at X are therefore:

$$\begin{aligned} A_X(x) &= F_X^{(x)} e^{\imath p_x^{(a)}(x-x_a) - \frac{1}{2\sigma_x^2 f_X^{(x)}}(x-x_X)^2 - \imath\frac{p_x^{(a)2}}{2m}\tau_X} \\ B_X(x) &= G_X^{(x)} e^{\imath k_x^{(b)}(x-x_b) - \frac{1}{2s_x^2 g_X^{(x)}}(x-x_X)^2 - \imath\frac{k_x^{(b)2}}{2\mu}\tau_X} \end{aligned} \tag{382}$$

The dispersion factors f, g and normalization factors F, G are spelled out in Appendix A.4. We work along the same lines as for the single slit above (subsection 5.1.2).
We define effective crossing times τ^* and dispersions σ_x^{*2} via:

$$\frac{1}{\sigma_x^2 + \imath\frac{\tau_X}{m}} + \frac{1}{s_x^2 + \imath\frac{\tau_X}{\mu}} = \frac{1}{\sigma_x^{*2} + \imath\frac{\tau^*}{m}} \tag{383}$$

The effect of the interaction is to change the shape of the A wave function so that it looks as if it had started at time $\tau_X - \tau^*$ with dispersion σ_x^{*2}.
We clear the denominators:

$$\left(\sigma_x^2 + s_x^2\right)\sigma_x^{*2} + \imath\left(\frac{\tau_X}{m} + \frac{\tau_X}{\mu}\right)\sigma_x^{*2} - \left(\frac{\tau_X}{m} + \frac{\tau_X}{\mu}\right)\frac{\tau^*}{m} + \imath\left(\sigma_x^2 + s_x^2\right)\frac{\tau^*}{m} = \left(\sigma_x^2 s_x^2 - \frac{\tau_X^2}{m\mu}\right) + \imath\left(\sigma_x^2\frac{\tau_X}{\mu} + s_x^2\frac{\tau_X}{m}\right) \tag{384}$$

We equate the real and imaginary parts:

$$\begin{aligned} \left(s_x^2 + \sigma_x^2\right)\sigma_t^{*2} - \left(\frac{\tau_X}{m} + \frac{\tau_X}{\mu}\right)\frac{\tau^*}{m} &= \sigma_x^2 s_x^2 - \frac{\tau_X^2}{m\mu} \\ \left(\frac{\tau_X}{m} + \frac{\tau_X}{\mu}\right)\sigma_x^{*2} + \left(\sigma_x^2 + s_x^2\right)\frac{\tau^*}{m} &= \sigma_x^2\frac{\tau_X}{\mu} + s_x^2\frac{\tau_X}{m} \end{aligned} \tag{385}$$

We have a two by two matrix equation for σ_x^{*2}, τ^*, which we invert and apply to the right hand side:

$$\begin{pmatrix} \sigma_x^{*2} \\ \frac{\tau^*}{m} \end{pmatrix} = \frac{1}{D}\begin{pmatrix} s_x^2 + \sigma_x^2 & \frac{\tau_X}{m} + \frac{\tau_X}{\mu} \\ -\left(\frac{\tau_X}{m} + \frac{\tau_X}{\mu}\right) & s_x^2 + \sigma_x^2 \end{pmatrix}\begin{pmatrix} \sigma_x^2 s_x^2 - \frac{\tau_X^2}{m\mu} \\ \sigma_x^2\frac{\tau_X}{\mu} + s_x^2\frac{\tau_X}{m} \end{pmatrix} \tag{386}$$

with determinant D:

$$D = \left(\sigma_x^2 + s_x^2\right)^2 + \left(\frac{\tau_X}{m} + \frac{\tau_X}{\mu}\right)^2 \tag{387}$$

The resulting expressions for the rescaling constants are relatively complex; the shape of A' is a marriage of the shapes of A, B:

$$\begin{aligned} \sigma_x^{*2} &= \frac{1}{D}\left(\left(s_x^2 + \sigma_x^2\right)\left(\sigma_x^2 s_x^2 - \frac{\tau_X^2}{m\mu}\right) + \left(\frac{\tau_X}{m} + \frac{\tau_X}{\mu}\right)\left(\sigma_x^2\frac{\tau_X}{\mu} + s_x^2\frac{\tau_X}{m}\right)\right) \\ \frac{\tau^*}{m} &= \frac{1}{D}\left(-\left(\frac{\tau_X}{m} + \frac{\tau_X}{\mu}\right)\left(\sigma_x^2 s_x^2 - \frac{\tau_X^2}{m\mu}\right) + \left(s_x^2 + \sigma_x^2\right)\left(\sigma_x^2\frac{\tau_X}{\mu} + s_x^2\frac{\tau_X}{m}\right)\right) \end{aligned} \tag{388}$$

The resulting wave function is:

$$I_X^{(x)}(x) = N_X^{(x)}\varphi_X^{(*)}(x) \tag{389}$$

with:

$$\varphi_X^{(*)}(x) = F_\tau^{(x)*} e^{\imath p_x^{(a')}(x - x_X) - \frac{1}{2\sigma_x^{*2} f_X^{*(x)}}(x - x_X)^2 - \imath\frac{p_x^{(a')2}}{2m}\tau_X} \tag{390}$$

and an overall constant $N_X^{(x)}$ independent of x. The overall constant will drop out when we calculate the final expectation and dispersion, so the effective dispersion and crossing time carry all the physically significant information.

Final wave function Since all x dependence is carried by $\varphi_X^{(*)}(x)$, we have the Fourier transform by inspection:

$$\hat{I}_X^{(x)}(p'_x) = N_X^{(x)} \hat{\varphi}_X^{(*)}(p'_x) \tag{391}$$

with the momentum space form of the starred wave function:

$$\hat{\varphi}^*_X(p'_x) = \sqrt[4]{\frac{1}{\pi\hat{\sigma}_X^{*2}}} e^{-\imath p'_x x_X - \frac{\left(p'_x - p_x^{(a')}\right)^2}{2\hat{\sigma}_X^{*2}} - \imath \frac{p'^2}{2m}\tau_X} \tag{392}$$

The other two dimensions work in parallel. Now that we are back in momentum space we have the resulting wave function at τ_2:

$$\hat{A}_2(\vec{p}') = -\imath\lambda \exp\left(-\imath\frac{\vec{p}'^2}{2m}\tau_{2X}\right) N_X^{(x)} N_X^{(y)} N_X^{(z)} \hat{\varphi}^*_X(\vec{p}') \tag{393}$$

For our purposes the most interesting aspect is the associated uncertainty in momentum. The overall normalization drops out:

$$\left\langle \left(\vec{p}' - \vec{p}^{(a')}\right)^2 \right\rangle = \frac{\int d\vec{p}' \left(\vec{p}' - \vec{p}^{(a')}\right)^2 \left|\hat{\varphi}_X(\vec{p}')\right|^2}{\int d\vec{p}' \left|\hat{\varphi}_X(\vec{p}')\right|^2} \tag{394}$$

The uncertainty in p is defined by the post-interaction wave function. For the x direction this is:

$$\hat{\sigma}_X^{*2} = \frac{1}{\sigma_X^{*2}} \tag{395}$$

We can read off the physically important part of the resulting wave function from the starred dispersion.

The simplest case is when B is both narrow in space $s_x \ll \sigma_x$ and heavy $\mu \gg m$. In this case the effective dispersions and start time take the simple form:

$$\begin{aligned} \sigma_x^{*2} &\to s_x^2 \\ \frac{\tau^*}{m} &\to 0 \end{aligned} \tag{396}$$

Operationally, if we detect A' at all, we know to within $s_x \ll \sigma_x$ where A was at τ_X. But this A' has momentum p'_x with a large effective dispersion $\hat{\sigma}_x^* \sim 1/s_x$. The momentum of A' has become much more uncertain.

So the B functions as a classical gate, with the familiar reciprocal relationship between the position and momentum.

6.8.3. Absorption of a particle in TQM We work along the same lines to extend the analysis to TQM. The classical trajectories now include a time component:

$$t = t_0 + \gamma\tau \tag{397}$$

We will treat the time parts of the wave functions as non-relativistic, $\gamma \approx 1$. This is consistent with our use of the non-relativistic approximation for the space parts.

We will assume our initial wave functions are centered on τ_0:

$$t_0 = \tau_0 \Rightarrow t_0 = 0 \tag{398}$$

As a result we have the same intersection point in coordinate time that we have in clock time:

$$t_X = \tau_X \tag{399}$$

Initial wave functions For the initial wave functions we have A and B as products of their time and space parts:

$$\begin{aligned} \hat{A}_0(p) &= \sqrt[4]{\frac{1}{\pi det(\hat{\Sigma})}} e^{-\imath p x_a - \frac{1}{2} p \hat{\Sigma} p} = \hat{\tilde{A}}_0(E)\, \hat{\tilde{A}}_0(\vec{p}) \\ \hat{B}_0(k) &= \sqrt[4]{\frac{1}{\pi^4 \det(\hat{S})}} e^{\imath k x_b - \frac{1}{2} k \hat{S}^{-1} k} = \hat{\tilde{B}}_0(w)\, \hat{\tilde{B}}_0(\vec{k}) \end{aligned} \tag{400}$$

with time parts:

$$\begin{aligned} \hat{\tilde{A}}_0(E) &= \sqrt[4]{\frac{1}{\pi \hat{\sigma}_t^2}} e^{\imath E t_a - \frac{(E-E_a)^2}{2\hat{\sigma}_t^2}} \\ \hat{\tilde{B}}_0(w) &= \sqrt[4]{\frac{1}{\pi \hat{s}_t^2}} e^{\imath w t_b - \frac{1}{2\hat{s}_t^2}(w-w_b)^2} \end{aligned} \tag{401}$$

and:

$$\begin{aligned} & p \equiv p - p_a, k \equiv k - k_b \\ & \hat{\Sigma} \equiv \begin{pmatrix} \hat{\sigma}_t^2 & 0 & 0 & 0 \\ 0 & \hat{\sigma}_x^2 & 0 & 0 \\ 0 & 0 & \hat{\sigma}_y^2 & 0 \\ 0 & 0 & 0 & \hat{\sigma}_z^2 \end{pmatrix}, \hat{S} \equiv \begin{pmatrix} \hat{s}_t^2 & 0 & 0 & 0 \\ 0 & \hat{s}_x^2 & 0 & 0 \\ 0 & 0 & \hat{s}_y^2 & 0 \\ 0 & 0 & 0 & \hat{s}_z^2 \end{pmatrix} \end{aligned} \tag{402}$$

and expectations and dispersions in time/energy:

$$\begin{aligned} & t_a \approx \tau_0 = 0, t_b \approx \tau_0 = 0 \\ & E_a \approx \sqrt{m^2 + \vec{p}_a^2}, w_b \approx \sqrt{\mu^2 + \vec{k}_b^2} \\ & \hat{\sigma}_E^2 \approx \hat{\sigma}_x^2 + \hat{\sigma}_y^2 + \hat{\sigma}_z^2, \hat{s}_t^2 \approx \hat{s}_x^2 + \hat{s}_y^2 + \hat{s}_z^2 \end{aligned} \tag{403}$$

The wave functions at X are:

$$\begin{aligned} \hat{A}_X(p) &= \sqrt[4]{\frac{1}{\pi^4 det(\hat{\Sigma})}} e^{\imath p x_a - \frac{1}{2} p \hat{\Sigma}^{-1} p} e^{-\imath f_p \tau_X} \\ \hat{B}_X(k) &= \sqrt[4]{\frac{1}{\pi^4 \det(\hat{S})}} e^{\imath k x_b - \frac{1}{2} k \hat{S}^{-1} k - \imath f_k \tau_X} \end{aligned} \tag{404}$$

Interaction To compute the wave function at τ_2 we again convolute the incoming wave functions:

$$\hat{A}_2(p') = -\imath\lambda \hat{K}_{2X}^{(m)}(p') \int dk \hat{A}_X(p'-k)\, \hat{B}_X(k) \tag{405}$$

The coordinate energy part of the integral in momentum space is:

$$\hat{\tilde{I}}_X(E') \equiv \int dw \hat{\tilde{A}}(E'-w)\, \hat{\tilde{B}}(w) \tag{406}$$

In coordinate time:

$$\tilde{I}_X(t) = \sqrt{2\pi} \tilde{A}_X(t)\, \tilde{B}_X(t) \tag{407}$$

The wave functions in time are:

$$\begin{aligned} \tilde{A}_X(t) &= F_X^{(t)} e^{-\imath E_a t - \frac{1}{2\sigma_t^2 f_X^{(t)}}\left(t - t_a - \frac{E_a}{m}\tau_X\right)^2 + \imath \frac{E_a^2}{2m}\tau_X} \\ \tilde{B}_X(t) &= G_X^{(t)} e^{-\imath w_b t - \frac{1}{2 s_t^2 g_X^{(t)}}\left(t - t_b - \frac{w_b}{\mu}\tau_X\right)^2 + \imath \frac{w_b^2}{2\mu}\tau_X} \end{aligned} \tag{408}$$

The quadratic arguments of the Gaussians both reduce to $(t - \tau_X)^2$ in our non-relativistic approximation.

IOP Conf. Series: Journal of Physics: Conf. Series **1239** (2019) 012015 doi:10.1088/1742-6596/1239/1/012015

As before, we define effective crossing times τ^* and dispersions σ_t^{*2} via:

$$\frac{1}{\sigma_t^2 - \imath\frac{\tau_X}{m}} + \frac{1}{s_t^2 - \imath\frac{\tau_X}{\mu}} = \frac{1}{\sigma_t^{*2} - \imath\frac{\tau^*}{m}} \tag{409}$$

Again, effect of the interaction is to change the shape of the A wave function so that it looks as if it had started at time $\tau_X - \tau^*$ with dispersion σ_t^{*2}.

As before, we clear the denominators, equate the real and imaginary parts, and invert the two by two matrix equation for σ_t^{*2}, τ^* to get:

$$\begin{aligned} \sigma_x^{*2} &= \frac{1}{D}\left(\left(\sigma_t^2 + s_t^2\right)\left(\sigma_t^2 s_t^2 - \frac{\tau_X^2}{m\mu}\right) + \left(\frac{\tau_X}{m} + \frac{\tau_X}{\mu}\right)\left(\sigma_t^2\frac{\tau_X}{\mu} + s_t^2\frac{\tau_X}{m}\right)\right) \\ \frac{\tau^*}{m} &= \frac{1}{D}\left(-\left(\frac{\tau_X}{m} + \frac{\tau_X}{\mu}\right)\left(\sigma_t^2 s_t^2 - \frac{\tau_X^2}{m\mu}\right) + \left(\sigma_t^2 + s_t^2\right)\left(\sigma_t^2\frac{\tau_X}{\mu} + s_t^2\frac{\tau_X}{m}\right)\right) \end{aligned} \tag{410}$$

with determinant D:

$$D = \left(\sigma_t^2 + s_t^2\right)^2 + \left(\frac{\tau_X}{m} + \frac{\tau_X}{\mu}\right)^2 \tag{411}$$

As before we have:

$$\tilde{I}_X(t) = \tilde{N}_X \tilde{\varphi}_X^{(*)}(t) \tag{412}$$

with starred wave function:

$$\tilde{\varphi}_X^{(*)}(t) \equiv F_X^{(t)*} e^{\imath E'_a (t - t_X) - \frac{1}{2\sigma_t^{*2} f_X^{*(t)}}(t - t_X)^2 - \imath\frac{{E'_a}^2}{2m}\tau_X} \tag{413}$$

Since all t dependence is carried by $\tilde{\varphi}_X^{(*)}(t)$, we have the Fourier transform by inspection:

$$\hat{I}_X^{(x)}(p'_x) = \tilde{N}_X \hat{\tilde{\varphi}}_X^{(*)}(E') \tag{414}$$

with the momentum space form of the starred wave function:

$$\hat{\tilde{\varphi}}_X^{(*)}(E') = \sqrt[4]{\frac{1}{\pi\hat{\sigma}_X^{(t)*2}}}\, e^{\imath E' x_X - \frac{(E' - E'_a)^2}{2\hat{\sigma}_X^{(t)*2}} + \imath\frac{(E')^2}{2m}\tau_X} \tag{415}$$

and therefore the final wave function:

$$\begin{aligned} \hat{A}_2(p') &= \hat{\tilde{A}}_2(E')\,\hat{\vec{A}}_2(\vec{p}') \\ \hat{\tilde{A}}_2(E') &= \exp\left(\imath\frac{E'^2}{2m}\tau_{2X}\right) N_X^{(t)} \hat{\tilde{\varphi}}_X(E') \end{aligned} \tag{416}$$

Uncertainty in time For our purposes the most interesting aspect is the associated uncertainty in energy. The overall normalization drops out:

$$\left\langle\left(E' - E^{(a')}\right)^2\right\rangle = \frac{\int dE'\left(E - E^{(a')}\right)^2\left|\hat{\tilde{\varphi}}_X(E)\right|^2}{\int dE'\left|\hat{\tilde{\varphi}}_X(E)\right|^2} \tag{417}$$

The post-interaction dispersion in E is given by the post-interaction dispersion in t:

$$\sigma_E^{*2} \equiv \hat{\sigma}_t^{*2} = \frac{1}{\sigma_t^{*2}} \tag{418}$$

and the uncertainty in E by the usual formula:

$$\Delta E = \sqrt{\frac{\sigma_E^{*2}}{2}} \tag{419}$$

With the dispersion in coordinate time we can calculate the dispersion in clock time at a detector using the formulas above in subsection 4.3.

The increase in complexity from using a quantum particle as the gate, rather than a classical camera shutter, is significant.

The simplest case is when B is both narrow in time $s_x \ll \sigma_x$ and heavy $\mu \gg m$. In this case the effective dispersions and start time take the form:

$$\begin{aligned} \sigma_t^{*2} &\to s_t^2 \\ \frac{\tau^*}{m} &\to 0 \end{aligned} \tag{420}$$

In this case we can apply the same formulae as with the single slit in time; getting the same arbitrarily large increase in the dispersion in time-of-arrival.

In the more general case, a certain amount of ingenuity may be required to tease out the effects of the increased dispersion in time.

6.8.4. Distinguishing between uncertainty in space and in time And this point takes us to a general problem in looking for uncertainty in time with time-of-arrival measurements. If we use time-of-arrival measurements we are measuring dispersion in time, yes, but we are also measuring dispersion in momentum along the axis of flight. Since we expect the initial dispersions in time to be of order of those in space, it may be difficult to *prove* there was no dispersion in time. Perhaps it was lost in the error bars?

The basic problem is the time-of-arrival measurement is being used to measure two different things: one measurement cannot serve two masters.

One way to separate the two measurements would be to run the post-interaction particle through a magnetic field. Let's say the particle will be bent to the right by the magnetic field. The y position will serve as usual as a measurement of velocity. But if at each y position we also record the time-of-arrival, the time-of-arrival should now serve as a measurement of dispersion in coordinate time.

If we graph the clicks on a y, τ grid, the faster particles will hit earlier in time and more to the left. In SQM we would expect to see a relatively narrow trace from small y, τ to large y, τ. In TQM we would expect to see the same trace on average, but significantly broader in τ at each y.

The SQM trace will look more like the thin scar left by a rapier; the TQM trace more like the thicker scar left by a saber.

6.9. Exchange of a particle

What does the exchange of a particle look like in TQM?

6.9.1. Overview We look at the case where an A and a C exchange a B. The initial particle expectations and dispersions are given; we wish to compute the outgoing particle expectation and dispersions. We have two cases: the A emits a B which is then absorbed by C and the C emits a B which is then absorbed by the A. We will call these the left and right cases.

As with emission and absorption, we expect that this exchange diagram will be part of a larger calculation. We will assume for convenience here that the momentum space expectations

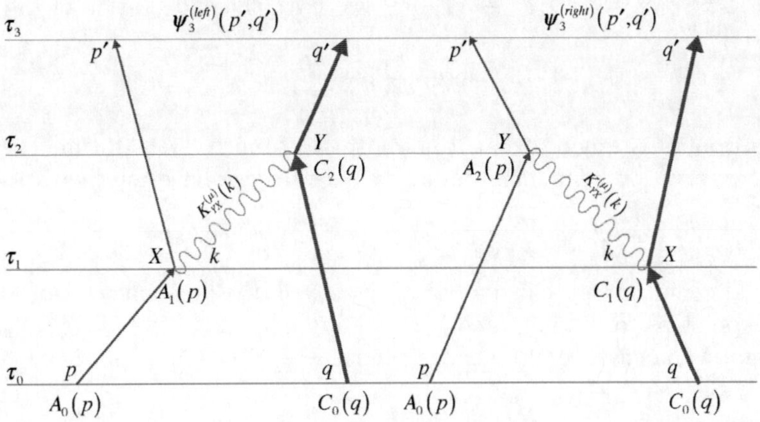

Figure 15. An A particle exchanges a B particle with a C particle

of the incoming particles are on-shell, but that need not be true in general. Note that this is the first time it is possible for both incoming and outgoing particles to have their momentum space expectations on-shell; the exchanged particle will still remain in general off-shell.

To lowest order in perturbation expansion we have convolutions over the two intermediate clock times τ_1, τ_2:

$$\begin{aligned} \hat{\psi}_3^{(left)}(p',q') &= -\lambda\Lambda \int_0^{\tau_3} d\tau_2 \hat{K}_{32}^{(M)}(q') \int d^4q d^4k \hat{C}_2(q) \int_0^{\tau_2} d\tau_1 \int d^4p \hat{K}_{21}^{(\mu)}(k) \hat{A}_1(p) \\ \hat{\psi}_3^{(right)}(p',q') &= -\lambda\Lambda \int_0^{\tau_3} d\tau_2 \hat{K}_{32}^{(m)}(p') \int d^4p d^4k \hat{A}_2(p) \int_0^{\tau_2} d\tau_1 \int d^4q \hat{K}_{21}^{(\mu)}(k) \hat{C}_1(q) \end{aligned} \tag{421}$$

In TQM, the two initial wave functions are each defined by four expectations in position, four in momentum, and four dispersions in either momentum or position space – twenty-four variables total. We have twelve integrals in momentum (or position), over the two initial particles A, C and the exchange particle B. And the two convolutions in clock time. With appropriately programmed mathematics software this is hardly a problem. But it is easy to lose sight of the physics in the course of doing the calculations.

To keep focus we will do as we have in the two previous subsections and fix the clock times of the vertexes as τ_X, τ_Y. Different values of τ_X, τ_Y will let us look at specific cases. The properties of the exchanged particle are key; we will focus on these.

Classical trajectories We will take the same starting wave functions as with absorption, but with $B \to C$. A has expectations $x_a, p^{(a)}$, dispersions $\hat{\sigma}^{(a)}$; C expectations $x_c, q^{(c)}$, dispersions $\hat{s}^{(a)}$.

We assume we have A coming in from the left; C from the right; both along the x axis. Again we assume the collision is head-on with no offset along the $\hat{y}$ or $\hat{z}$ axes.

We start by fixing τ_X, τ_Y. Taking the left case first, we have for A and C:

$$\begin{aligned} x_X^{(left)} &= x_a + \frac{p_x^{(a)}}{m}\tau_X \\ x_Y^{(left)} &= x_c + \frac{q_x^{(c)}}{M}\tau_Y \end{aligned} \tag{422}$$

And for the exchange particle:

$$x_Y^{(left)} = x_X^{(left)} + \frac{k_x^{(b)}}{\mu}(\tau_Y - \tau_X) \tag{423}$$

IOP Conf. Series: Journal of Physics: Conf. Series **1239** (2019) 012015 doi:10.1088/1742-6596/1239/1/012015

With $\delta x^{(left)} \equiv x_Y^{(left)} - x_X^{(left)}, \delta\tau \equiv \tau_Y - \tau_X$ we have the expectation of the momentum of the exchange particle:

$$k_x^{(left)} = \mu \frac{\delta x^{(left)}}{\delta\tau} \tag{424}$$

The momentum of B is exactly what it needs to get from $X \to Y$ in time for its rendezvous with C. From conservation of momentum at each vertex we have for the expectations of the final momenta:

$$\langle p'_x \rangle = p_x^{(a)} - k_x^{(left)}, \langle q'_x \rangle = q_x^{(c)} + k_x^{(left)} \tag{425}$$

With the initial conditions specified and τ_X, τ_Y as well, the final expectations are immediate, with the final dispersions to be computed.

To get the right hand case we interchange the roles $A \leftrightarrow C$:

$$\begin{aligned} x_X^{(right)} &= x_c + \frac{q_x^{(c)}}{M}\tau_Y \\ x_Y^{(right)} &= x_a + \frac{p_x^{(a)}}{m}\tau_X \end{aligned} \tag{426}$$

and:

$$k_x^{(right)} = \mu \frac{\delta x^{(right)}}{\delta\tau} \tag{427}$$

giving final expectations:

$$\langle p'_x \rangle = p_x^{(a)} + k_x^{(right)}, \langle q'_x \rangle = q_x^{(c)} - k_x^{(right)} \tag{428}$$

6.9.2. Exchange particle We can write the wave function for the exchange particle in SQM as:

$$\hat{\hat{B}}_{YX}\left(\vec{k}\right) = \sqrt[4]{\frac{1}{\pi\hat{\hat{\Sigma}}^{(X)}}} e^{-\imath\vec{k}\cdot\vec{x}_X - \frac{1}{2}(\vec{k}-\vec{k}_X)\cdot\frac{1}{\hat{\hat{\Sigma}}^{(X)}}\cdot(\vec{k}-\vec{k}_X) - \imath\frac{\vec{k}^2}{2\mu}\delta\tau} \tag{429}$$

and in TQM in covariant notation:

$$\begin{aligned} \hat{B}_{YX}(k) &= \sqrt[4]{\frac{1}{\pi\hat{\Sigma}^{(X)}}} e^{\imath k x_X - \frac{1}{2}(k-k_X)\frac{1}{\Sigma^{(X)}}(k-k_X) - \imath f_k \delta\tau} \\ f_k &\equiv -\frac{k^2-\mu^2}{2\mu} \end{aligned} \tag{430}$$

and as a product of time and space parts:

$$\hat{B}_{YX}(k) = \hat{\hat{B}}_{YX}(w)\,\hat{\hat{B}}_{YX}\left(\vec{k}\right) \exp\left(\imath\frac{\mu}{2}\delta\tau\right) \tag{431}$$

$$\hat{\hat{B}}_{YX}(w) = \sqrt[4]{\frac{1}{\pi\hat{\hat{\sigma}}^{(X)}}} e^{\imath w t_X - \frac{1}{2}\frac{(w-w_X)^2}{2\hat{\hat{\sigma}}^{(X)}} + \imath\frac{w^2}{2\mu}\delta\tau} \tag{432}$$

Taking the left side for definiteness, the wave function of the intermediate state is properly the direct product of the highly correlated wave functions of A, B with the (as yet) uncorrelated wave function of the C particle. After B encounters C at Y, B is gone and now A and C are highly correlated.

We have proceeded a bit formally, specifying τ_X, τ_Y and then deriving the properties of the exchange particle from these. A more physical approach might be to fix the momentum of the exchange particle and then integrate over all values of τ_X and τ_Y consistent with that.

IOP Conf. Series: Journal of Physics: Conf. Series **1239** (2019) 012015 doi:10.1088/1742-6596/1239/1/012015

6.9.3. Discussion How we develop the exchange depends on what we are interested in:

(i) If we are interested in the use of one particle as a measurement of the other, then we already have what we need: B inherits its dispersions from its parent. If its parent was narrow in time/space, then it will be as well and act as a de facto gate with respect to the other particle.

(ii) If we are interested in subtler effects of dispersion in time, we can look for forces of anticipation and regret: if wave functions are extended in time as they are in space, then in a collision they will start to interact earlier, cease interacting later than would otherwise be the case.

(iii) If we are interested in doing a Bell's theorem correlation in time, then we need to track the correlations through. Because momentum is conserved at each vertex, the outgoing particles will be highly correlated in momentum space.

(iv) If we are interested in symmetry properties in the time direction, then we can:
   (a) Take our C as an A,
   (b) Start with wave functions that are symmetric under particle exchange but which nevertheless have a component which is anti-symmetric in time (as the wide and narrow wave function in section 6.3),
   (c) Look for unexpected anti-symmetries in space.

(v) If we are interested in the bound case, we can look at the exchange particle as creating a Yukawa force in time. The $\exp(-\imath f_k \tau)$ factor will tend to keep the exchange particle on-shell for larger values of $\delta\tau$. In coordinate space this will tend to make the effective potential look like a Liénard–Wiechert potential.

6.10. Loop correction to the mass
How do we calculate loop diagrams in TQM?

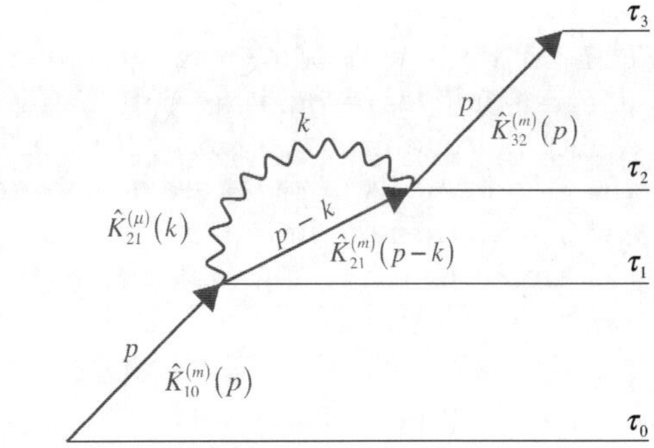

Figure 16. Loop correction to the mass

6.10.1. Loop correction in SQM In the ABC model there is an amplitude λ for an A to emit a B then absorb it. As is well known, in quantum field theory this can be made to look like a correction to the mass, with the effect of taking a bare mass to a corrected mass: $m_0^2 \to m^2 = m_0^2 + \delta m^2$. Unfortunately the integral for this δm^2 correction is divergent. If the A particle has four-momentum p and emits a B with four-momentum k, to compute the amplitude

IOP Conf. Series: Journal of Physics: Conf. Series **1239** (2019) 012015 doi:10.1088/1742-6596/1239/1/012015

associated with the loop we will need to integrate over all possible values of the intermediate k:

$$\delta m^2 \sim \int d^4k \frac{\imath}{(p-k)^2 - m^2} \frac{\imath}{k^2 - \mu^2} \tag{433}$$

This is logarithmically divergent at large k:

$$\delta m^2 \sim \int \frac{d^4k}{k^4} \tag{434}$$

As a result all such integrals have to be regularized: a convergence factor has to be inserted which acts as an effective cutoff, throwing out the high energy part of the loop in a way that does not distort results at lower energies.

To do this we take advantage of the general principle that all physical measurements involve an implicit comparison between two measurements. If we are going to use the mass of a particle as a value in one calculation we must first have found that mass in another. An absolute, standalone measurement is not possible even in principle. All measurements have to be renormalized– normalized by comparison to another – to get physically meaningful numbers.

In quantum field theory renormalization – needed physically – is also used to regularize, to contain and control the infinities. If we make the necessary comparison in the right way, we can use it to subtract off the infinities.

It's like weighing a mouse by first weighing an aircraft carrier without the mouse, weighing the aircraft carrier with the mouse on board, and then subtracting out the weight of the aircraft carrier to get the weight of the mouse.

There is no guarantee that this will work. What if the cutoff function/procedure being used on both sides of the comparison has unintended side-effects at lower energies? But in spite of the obvious risk the procedure works — and brilliantly – producing some of the most accurate predictions in the whole of physics.

6.10.2. Loop correction in TQM With an extra dimension to integrate over we might expect that the corresponding loop integrals in TQM would not only be divergent, but perhaps even be divergent in a way which cannot be contained by renormalization.

We will show that in fact the loop integrals in TQM are not divergent. The combination of finite initial dispersion in time with entanglement in time keep the integrals convergent. We will work through a simple example, computing the familiar mass correction loop, first as a function of clock time, then as a function of clock frequency.

Loop integral for fixed clock time We apply the loop integral to a initial Gaussian test function. In momentum space:

$$L_\tau(p) = \int d^4k \hat{K}_\tau^{(m)}(p-k)\, \hat{K}_\tau^{(\mu)}(k)\, \hat{\varphi}_0(p) \tag{435}$$

with initial Gaussian test function:

$$\hat{\varphi}_0(p) = \sqrt[4]{\frac{1}{\pi^4 \det\left(\hat{\Sigma}\right)}} \exp\left(-\frac{(p-p_0)^2}{2\Sigma}\right) \tag{436}$$

The corresponding loop integral in coordinate space is:

$$L_\tau(x_1) = 4\pi^2 \int d^4x_0 K_\tau^{(m)}(x_1; x_0)\, K_\tau^{(\mu)}(x_1; x_0)\, \varphi_0(x_0) \tag{437}$$

IOP Conf. Series: Journal of Physics: Conf. Series **1239** (2019) 012015 doi:10.1088/1742-6596/1239/1/012015

The kernels in coordinate space are:

$$\begin{aligned} K_\tau^{(m)}(x_1;x_0) &= -\imath\tfrac{m^2}{4\pi^2\tau^2}\exp\left(-\tfrac{\imath m}{2\tau}(x_1-x_0)^2-\imath\tfrac{m}{2}\tau\right)\\ K_\tau^{(\mu)}(x_1;x_0) &= -\imath\tfrac{\mu^2}{4\pi^2\tau^2}\exp\left(-\tfrac{\imath\mu}{2\tau}(x_1-x_0)^2-\imath\tfrac{\mu}{2}\tau\right)\end{aligned} \tag{438}$$

The product equals a single coordinate space kernel:

$$K_\tau^{(M)}(x_1;x_0) = -\imath\frac{M^2}{4\pi^2\tau^2}\exp\left(-\imath\frac{M}{2\tau}(x_1-x_0)^2-\imath\frac{M}{2}\tau\right) \tag{439}$$

with a modified mass $M\equiv m+\mu$ and a prefactor.

So the loop integral in coordinate space is now:

$$L_\tau(x_1) = -\imath\frac{m^2\mu^2}{\tau^2M^2}\int d^4x_0K_\tau^{(M)}(x_1;x_0)\varphi_0(x_0) \tag{440}$$

The presence of the Gaussian test function on the right means we get by inspection:

$$L\tau(x_1) = -\imath\frac{m^2\mu^2}{\tau^2M^2}\varphi_\tau^{(M)}(x_1) \tag{441}$$

We have a correction that shows a spread in time, but at the slightly slower rate associated with the slightly larger mass M. Further the correction is much greater at shorter clock times:

$$\begin{aligned} L_\tau(p_1) &= -\imath\tfrac{m^2\mu^2}{\tau^2M^2}\int d^4p_0\hat{K}_\tau^{(M)}(p_1;p_0)\hat{\varphi}_0(p_0)\\ \hat{K}_\tau^{(M)}(p_1;p_0) &= \exp\left(\imath\tfrac{p_0^2-m^2}{2m}\tau\right)\delta^4(p_1-p_0)\end{aligned} \tag{442}$$

So the loop correction for fixed clock time is:

$$L_\tau(p) = -\imath\frac{m^2\mu^2}{M^2\tau^2}\exp\left(\imath\frac{p^2-M^2}{2M}\tau\right)\hat{\varphi}_0(p) \tag{443}$$

At this point the value of the loop correction at a particular value of p is independent of the specific shape of the incoming wave function. We are therefore free to drop the initial wave function from the analysis:

$$L_\tau(p) = -\imath\frac{m^2\mu^2}{M^2\tau^2}\exp\left(\imath\frac{p^2-M^2}{2M}\tau\right) \tag{444}$$

Fourier transform of the loop integral over clock time Now that we have the loop integral for a specific value of the clock time we can take the Fourier transform with respect to τ:

$$\hat{L}_\omega(p) = -\imath\frac{1}{\sqrt{2\pi}}\int\limits_{-\infty}^{\infty}d\tau\frac{m^2\mu^2}{M^2\tau^2}e^{\imath(\omega-F_p)\tau} \tag{445}$$

We define $F_p\equiv-\frac{p^2-M^2}{2M}$. For small μ, $F_p\approx f_p$. We have the value of the Fourier transform (with conventions per Mathematica):

$$\mathcal{FT}\left[\frac{\exp(-\imath F_p\tau)}{\tau^2}\right] = -\sqrt{\frac{\pi}{2}}|\omega-F_p| \tag{446}$$

or in our case:

$$\hat{L}_{\omega}(p) = \imath \frac{m^2 \mu^2}{M^2} \sqrt{\frac{\pi}{2}} |\omega - F_p| \tag{447}$$

The value of the loop integral is therefore finite, without need for regularization or the introduction of cutoffs or other artificial approaches.

We still need to renormalize. But we have separated the problems associated with renormalization from those associated with regularization.

As we have seen throughout, TQM implies the initial wave functions:

(i) have finite dispersion in time (as well as space)

(ii) are entangled with the loop integration variables

This in turn means that each loop integration picks up a Gaussian factor that guarantees its convergence – a Gaussian easily dominates merely polynomial divergences.

The combination of finite initial dispersion and entanglement therefore forces convergence.

Note this approach will not work in SQM. In SQM there is by assumption no finite initial dispersion in time. And even if there were, each step is cut off from the previous since there is no entanglement in time.

We have only looked at a toy case. But the principles established here apply generally. As noted, the combination of Morlet wavelet analysis and entanglement in time mean that the integrals encountered in a diagram are self-regularized: the Gaussian functions which are passed through a series of integrals easily dominate any polynomial divergence.

An implication is that it is the assumption that quantum mechanics does not apply along the time dimension that is responsible for the ultraviolet divergences. The familiar divergences are a side-effect of not pushing the ideas of quantum mechanics and special relativity hard enough, of our failure to treat time and space symmetrically in quantum mechanics.

6.11. Discussion of the multiple particle case

We have established that we can extend TQM to include the multiple particle case.

TQM is conceptually simpler than SQM: time and space are treated on an equal footing and there are no ultraviolet divergences. But it is calculationally more complex: we have coordinate time to consider and we are required to use Gaussian test functions rather than plane waves as the fundamental unit of analysis.

We have examined the basic parts of a Feynman diagram: free propagators; the emission, absorption, and exchange of a particle; and simple loop diagrams. We are therefore able to work out – in principle at least – the results for any diagram in a perturbation expansion. And therefore to compare TQM to SQM in any experiment which can be described by such expansions.

To falsify TQM we need experiments that work at short times and with individual wave packets. Long times and averages over wave packets kill the effects associated with dispersion in time.

The most decisive such experiments are likely to be ones that emphasize the effects of the time/energy uncertainty principle.

Additional effects include:

(i) anti-symmetry in time,

(ii) forces of anticipation and regret,

(iii) interference and entanglement in time,

(iv) adjustments to the usual loop corrections.

IOP Conf. Series: Journal of Physics: Conf. Series **1239** (2019) 012015 doi:10.1088/1742-6596/1239/1/012015

To be sure, these additional effects likely to be both subtle and small. Therefore they may not be that useful for falsifiability, our primary target in this work. But they are potentially interesting in their own right.

We observe that the results here should be in common across the relativistic dynamics program; these are all consequences of dispersion in time, which is found across the relativistic dynamics program.

7. Discussion

"It is difficult to see what one does not expect to see." – William Feller [97]

We have explored the possibility that the quantum wave function should be extended in time:

$$\psi(\vec{x}) \rightarrow \psi(t, \vec{x}) \tag{448}$$

We have argued that this idea can be developed in a way that is self-consistent, consistent with existing experimental and observational work, and – most importantly – is falsifiable.

7.1. Falsifiability

With the single slit in time we have a decisive test of temporal quantum mechanics. In SQM, the narrower the slit, the *less* the dispersion in subsequent time-of-arrival measurements. In TQM, the narrower the slit, the *greater* the subsequent dispersion in subsequent time-of-arrival measurements. *In principle, the difference may be made arbitrarily great.*

To get to this point we had to develop the rules for calculation in a way that is unambiguously falsifiable. To do this:

(i) We took path integrals as the defining representation. This made the extension from three to four dimensions unambiguous.
(ii) We used Morlet wavelet analysis rather than Fourier analysis to define the initial wave functions. This let us avoid the use of unphysical wave functions and made achieving convergence and normalization of the path integrals possible.
(iii) We distinguished carefully between "clock time" and "coordinate time".

As a result, once we applied the requirement that TQM match SQM in the appropriate limit, we were left with no free parameters.

There are other ways to extend quantum mechanics to include time. We have started with path integrals; one could start with a Hamiltonian approach, see for instance Yau [98]. We have focused on the extension of the wave function along coordinate time; at longer clock times the effects of differences in the handling of the clock time/evolution parameter may become significant.

We have therefore been careful to focus on dimensional and symmetry arguments, which give first order predictions which are independent of the specifics of whatever method we might use:

- The initial dispersion in time is fixed by symmetry between time and space and the principle of maximum entropy.
- The evolution of the wave functions is fixed by the long, slow approximation. This allows for give, but only over times of picoseconds, glacial by the standards of TQM.
- We could choose Alice's frame or Bob's to do the analysis, but the corrections due to this are of second order. Further, the corrections can be eliminated entirely by selecting "the rest frame of the vacuum" as the defining frame for TQM.

The predictions of TQM are therefore falsifiable to first order.

7.2. Experimental effects

We have discussed two primary effects:

(i) generally increased dispersion in time

(ii) and the time/energy uncertainty principle.

These effects should be present in any experimental setup in which the sources vary in time and the detectors are time-sensitive.

Additionally we can look for:

- Shadowing in time – self-interference by detectors and sources.
- Interference, correlations, and entanglement in time. Consider for instance a Bell's theorem in time: particles that have interacted in the past, as in EPR experiments, will be entangled in time as well as space.
- Forces of anticipation and regret. As the paths in TQM advance into the future they will encounter potentials earlier (anticipation) than in SQM. And as they dive back into the past they will continue to interact with potentials later (regret) than is the case in SQM.
- Anti-symmetry in time. Wave functions are free to satisfy their symmetry requirements using the time dimension as well as the three space dimensions.
- Small – probably quite small – corrections to the usual loop integral corrections.

In general, any quantum effect seen in space is likely have an "in time" variation. TQM is to SQM with respect to time as SQM is to classical mechanics with respect to space.

Reviews of foundational experiments in quantum mechanics (for example Lamoreaux [99], Ghose [100], and Auletta [101]) provide a rich source of candidate experiments: the single and double slit as well as many other foundational experiments have an "in time" variant, typically with time and a space dimension flipped.

The experiments are likely to be difficult. The attosecond times here are at the edge of the detectible. The investigation here was partly inspired by Lindner's "Attosecond Double-Slit Experiment" [80]. But the times there, $500as$, are far too long for us. More recent work has reached shorter and shorter times: 12 attoseconds in Koke, Sebastian, and Grebing [102] and as noted the extraordinary sub-attosecond times in Ossiander [41]. Further while the fundamental scale is defined by the time taken by light to cross an atom, in the case of Rydberg atom the width of an atom in space may be made almost arbitrarily big. (We are indebted to Matt Riesen for this suggestion.) Therefore the effects of dispersion in time should now be (barely) within experimental range.

7.3. Further extensions

We have provided only a basic toolkit for TQM.

As TQM falls within the relativistic dynamics program (per subsection 3.6), the existing literature in this area is in general applicable. We get TQM from approaches to relativistic dynamics by assuming that the dependence on clock time is negligible. When this is reasonable, we can use the existing relativistic dynamics literature to get TQM specific calculations by taking the sub-picosecond limit (per subsection 3.4). And we can go the other way: use TQM to get short clock time approximations to other work in relativistic dynamics.

Areas for further investigation include:

(i) Generalizing the treatment of spinless massive bosons to include photons and fermions. Extension to the Standard Model.

(ii) Derivation of the bound state wave functions [103, 104, 36].

(iii) More detailed treatment of scattering experiments [105, 37].

(iv) Exact treatment of the single slit in time, including paths that wander back and forth through the slit [78].
(v) More detailed treatment of the double slit in time and of diffraction in time [106, 38].
(vi) Entanglement in time [38].
(vii) Careful treatment of measurements, including paths that overshoot, undershoot, and loop around the detector.
(viii) Decoherence in time.
(ix) Infrared divergences. From the point of view of TQM, these may be the flip side of the ultraviolet divergences, suppressed in a similar way.
(x) More detailed treatment of the ultraviolet divergences.
(xi) Examination of the spin-statistics connection: the T in the CPT theorem must come under grave suspicion.
(xii) Statistical mechanics. Any statistical ensemble should include fluctuations in time [36, 40, 107].
(xiii) More detailed treatment of the choice of frame.
(xiv) Quantum gravity. Since TQM is by construction highly symmetric between time and space and free of the ultraviolet divergences, it may be a useful starting point for attacks on the problem of quantum gravity [36, 108].

7.4. Five requirements

In their delightfully titled *How to Think about Weird Things* [109] the philosophers Schick and Vaughn lay out five requirements that a hypothesis such as TQM should satisfy:

(i) **Testability** – *are there experimental tests? ideally: is the hypothesis falsifiable?* TQM has no free parameters; it can therefore be falsified by any experiment at appropriate scale looking at time varying quantum phenomena.
(ii) **Fruitfulness** – *does the hypothesis suggest new lines of research, new phenomena to explore?* All time-varying quantum phenomena offers targets for investigation. The list of experimental effects given above is doubtless far from exhaustive.
(iii) **Scope** – *how widespread are the phenomena?* TQM applies to all time-varying quantum phenomena.
(iv) **Simplicity** – *does it make the fewest possible assumptions?* TQM eliminates the assumption that time and space should be treated differently in quantum mechanics. It also eliminates the ultraviolet divergences and the consequent need to regularize the loop integrals in field theories (as QED).
(v) **Conservatism** – *is it consistent with what is known?* TQM matches SQM in the long time (picosecond) limit.

7.5. No null experiments

Meeting the Schick and Vaughn requirements is not enough to establish that TQM is true, only that it may be worth investigating experimentally.

Now suppose that one or more of the proposed experiments is done and conclusively demonstrates that TQM is false. That would in turn raise some interesting questions:

(i) Is there a frame in which TQM is maximally (or minimally) falsified? That would be a preferred frame, anathema to relativity.
(ii) Is TQM equally false in all frames? if it is *equally* false in all frames, how do we reconcile the disparate wave functions of Alice and Bob?

As TQM is a straight-forward extrapolation of quantum mechanics and special relativity, experiments that falsify TQM are likely to require modification of our understanding of either quantum mechanics or special relativity or both.

Therefore it would appear that there are no null experiments.

This concludes the argument for making an experimental investigation of TQM.

Acknowledgments
I thank my long time friend Jonathan Smith for invaluable encouragement, guidance, and practical assistance.

I thank Ferne Cohen Welch for extraordinary moral and practical support.

I thank Martin Land, Larry Horwitz, and the organizers of the IARD 2018 Conference for encouragement, useful discussions, and hosting a talk on this paper at IARD 2018.

I thank Y. S. Kim for organizing the invaluable Feynman Festivals, for several conversations, and for general encouragement and good advice.

I thank the anonymous reviewers at the online journal Quanta, especially for their comments on the need for an emphasis on testable and on clarity.

I thank Catherine Asaro, Julian Barbour, Gary Bowson, Howard Brandt, Daniel Brown, Ron Bushyager, John G. Cramer, J. Ferret, Robert Forward, N. Gisin, Fred Herz, J. Peřina, Linda Kalb, A. Khrennikov, David Kratz, Steve Libby, Andy Love, Walt Mankowski, O. Maroney, John Myers, Paul Nahin, Marilyn Noz, R. Penrose, Stewart Personick, V. Petkov, H. Price, Matt Riesen, Terry Roberts, J. H. Samson, Lee Smolin, L. Skála, Arthur Tansky, R. Tumulka, Joan Vaccaro, L. Vaidman, A. Vourdas, H. Yadsan-Appleby, S. Weinstein, Asher Yahalom, and Anton Zeilinger for helpful conversations.

I thank the organizers of several QUIST, DARPA, Perimeter Institute conferences I've attended and the very much on topic conferences *Quantum Time* in Pittsburgh in 2014 and *Time and Quantum Gravity* in San Diego in 2015.

I thank a host of friends, casual acquaintances, and passers-by for many excellent conversations on the subjects here – ranging from a group of graduate students over good Czech beer at the Cafe Destiny in Olomouc to a pair of Canadian customs agents encountered seriatim on the way to the Perimeter institute.

And I would particularly like to thank my Olomouc experimentalist – I'm sorry I did not get your name – who defined the right objective for this work. I hope I have given you something you can prove wrong!

And none of the above are in any way responsible for any errors of commission or omission in this work.

Appendix A. Conventions

We use natural units $\hbar = c = 1$.

When summing over three dimensions we use i, j, k. We use $\imath$ sans dot for the square root of -1; i with dot for the index variable. When summing over four dimensions we use h for the coordinate time index so such sums run over h, i, j, k.

We define the relativistic time dilation factor $\gamma \equiv \frac{E}{m}$.

We use m as the rest mass. We make no use of a relativistic mass, i.e. γm. Here the mass m is always the rest mass.

Appendix A.1. Clock time

We use τ for clock time. The use of the Greek letter τ for clock time is meant to suggest that this is a "classical" time. We use f for its complementary variable, clock frequency:

$$f^{(op)} \equiv \imath \frac{\partial}{\partial \tau} \tag{A.1}$$

IOP Conf. Series: Journal of Physics: Conf. Series **1239** (2019) 012015 doi:10.1088/1742-6596/1239/1/012015

The clock time τ will usually be found at the bottom right of any symbol it is indexing:

$$\varphi_\tau, K_\tau \tag{A.2}$$

The clock time is in its turn frequently indexed: $\tau_0, \tau_1, \tau_2, \ldots$

As a result, deeply nested subscripts are an occasional hazard of this analysis. To reduce the nesting level we use obvious shortenings, i.e.:

$$\varphi_{\tau_1}(x_1) \to \varphi_1(x_1) \to \varphi_1 \tag{A.3}$$

$$K_{\tau_1\tau_0}(x_1; x_0) \to K_{10}(x_1; x_0) \to K_1(x_1; x_0) \to K_1 \tag{A.4}$$

And we represent differences in clock time by combining indexes:

$$\tau_{21} \equiv \tau_2 - \tau_1 \tag{A.5}$$

Appendix A.2. Coordinate time and space

We use $E, \vec{p}$ for the momentum variables complementary to coordinate time t and space $\vec{x}$:

$$\begin{array}{c} E^{(op)} \equiv \imath\frac{\partial}{\partial t} \\ \vec{p}^{(op)} \equiv -\imath\nabla \end{array} \tag{A.6}$$

When there is a natural split into coordinate time and space parts we use a tilde to mark the time part, an overbar to mark the space part. For example:

$$\psi(t, \vec{x}) = \tilde{\psi}_\tau(t)\bar{\psi}_\tau(\vec{x}) \tag{A.7}$$

This is to reinforce the idea that in this analysis the three dimensional part is the average (hence overbar), while the coordinate time part contributes a bit of quantum fuzziness (hence tilde) on top of that. With that said, we will sometimes omit the overbar and the tilde when they are obvious from context:

$$\tilde{\varphi}_\tau(t)\bar{\varphi}_\tau(\vec{x}) \to \varphi_\tau(t)\varphi_\tau(\vec{x}) \tag{A.8}$$

We use an overdot to indicate the partial derivative with respect to laboratory time:

$$\dot{g}_\tau(t, \vec{x}) \equiv \frac{\partial g_\tau(t, \vec{x})}{\partial \tau} \tag{A.9}$$

Appendix A.3. Fourier transforms

We use a caret to indicate that a function or variable is being taken in momentum space. To keep the Fourier transform itself covariant we use opposite signs for the coordinate time and space parts:

$$\begin{array}{c} \hat{g}(E, \vec{p}) = \frac{1}{\sqrt{2\pi}^4}\int\limits_{-\infty}^{\infty} dt d\vec{x} e^{\imath Et - \imath\vec{p}\cdot\vec{x}} g(t, \vec{x}) \\ g(t, \vec{x}) = \frac{1}{\sqrt{2\pi}^4}\int\limits_{-\infty}^{\infty} dE d\vec{p} e^{-\imath Et + \imath\vec{p}\cdot\vec{x}} \hat{g}(E, \vec{p}) \end{array} \tag{A.10}$$

For plane waves:

$$\begin{array}{c} \phi_p(x) = \tilde{\phi}(t)\bar{\phi}(\vec{x}) = \frac{1}{\sqrt{2\pi}}\exp(-\imath Et)\frac{1}{\sqrt{2\pi}^3}\exp(\imath\vec{p}\cdot\vec{x}) \\ \hat{\phi}_x(p) = \hat{\tilde{\phi}}(E)\hat{\bar{\phi}}(\vec{p}) = \frac{1}{\sqrt{2\pi}}\exp(\imath Et)\frac{1}{\sqrt{2\pi}^3}\exp(-\imath\vec{p}\cdot\vec{x}) \end{array} \tag{A.11}$$

IOP Conf. Series: Journal of Physics: Conf. Series **1239** (2019) 012015 doi:10.1088/1742-6596/1239/1/012015

To shorten the expressions we use:

$$x \equiv (t, \vec{x}) = (t, x, y, z) \tag{A.12}$$

The difference between x the four vector and x the first space coordinate should be clear from context. In momentum space we use:

$$p \equiv (E, \vec{p}) = (E, p_x, p_y, p_z) \tag{A.13}$$

We have similar rules for clock time τ and its complementary energy f:

$$\begin{aligned} \hat{g}_f &= \frac{1}{\sqrt{2\pi}} \int_{-\infty}^{\infty} d\tau e^{\imath f \tau} g_\tau \\ g_\tau &= \frac{1}{\sqrt{2\pi}} \int_{-\infty}^{\infty} df e^{-\imath f \tau} \hat{g}_f \end{aligned} \tag{A.14}$$

With these conventions when a Fourier transform is given by a convolution:

$$\hat{h}(p) = \int dk \hat{f}(p-k)\, \hat{g}(k) \tag{A.15}$$

the function in coordinate space is given by:

$$h(x) = \sqrt{2\pi} f(x)\, g(x) \tag{A.16}$$

When it is obvious that a symbol represents a Fourier transform we may drop the caret:

$$\hat{\varphi}(p) \to \varphi(p) \tag{A.17}$$

Appendix A.4. Gaussian test functions

As Gaussian test functions play a critical role in this investigation it is useful to have a consistent notation with which to describe them.

In general we define a Gaussian test function as a normalized Gaussian function. It may be in position or momentum space. It is defined by its expectations for position and momentum and by its dispersion in either position or momentum. The most important single example here is the Gaussian test function that describes the time part of the wave function of a free particle:

$$\tilde{\varphi}_\tau(t) = F_\tau^{(t)} e^{-\imath E_0 t - \imath \frac{E_0^2}{2m}\tau - \frac{1}{2\sigma_t^2 f_\tau^{(t)}}\left(t - t_0 - \frac{E_0}{m}\tau\right)^2} \tag{A.18}$$

with dispersion:

$$\sigma_t \tag{A.19}$$

dispersion factor:

$$f_\tau^{(t)} \equiv 1 - \imath \frac{\tau}{m\sigma_t^2} \tag{A.20}$$

and normalization factor:

$$F_\tau^{(t)} \equiv \sqrt[4]{\frac{1}{\pi\sigma_t^2}} \sqrt{\frac{1}{f_\tau^{(t)}}} \tag{A.21}$$

The sign of the complex part of a dispersion factor is negative for time; positive for space:

$$f_\tau^{(x)} \equiv 1 + \imath \frac{\tau}{m\sigma_x^2} \tag{A.22}$$

IOP Conf. Series: Journal of Physics: Conf. Series **1239** (2019) 012015 doi:10.1088/1742-6596/1239/1/012015

If we are dealing with multiple Gaussian test functions we may make name changes $f \to g, F \to G$, $f \to h, F \to H$ to the dispersion and normalization factors.

In general we can switch between time/energy and space/momentum forms by taking the complex conjugate and interchanging variables $t \leftrightarrow x$. This is our own small version of the CPT transformations.

As expressions like $\sigma_{p_x}^2$ are cumbersome we sometimes replace them with $\hat{\sigma}_x^2 \equiv \sigma_{p_x}^2$ taking implicit advantage of the fact that with our conventions $\sigma_x^2 = \frac{1}{\sigma_{p_x}^2}$.

If we need to tag various Gaussian wave functions we may assign each specific letter a, b, c as:

$$\tilde{\varphi}_\tau^{(a)}(t) = F_\tau^{(a)} e^{-\imath E_a t - \imath \frac{E_a^2}{2m}\tau - \frac{1}{2\sigma_t^{(a)2} f_\tau^{(a)}}\left(t - t_a - \frac{E_a}{m}\tau\right)^2} \tag{A.23}$$

We usually use φ for Gaussian test functions but may use a capital letter to reduce notational clutter, as $A \equiv \varphi^{(a)}$.

Appendix A.5. Acronyms

CM Classical Mechanics: all four dimensions treated as parameters.

SQM Standard Quantum Mechanics: quantum mechanics with the three space dimensions treated as observables, time as a parameter.

TQM Temporal Quantum Mechanics: SQM but with time treated as an observable on the same basis as the three space dimensions.

Appendix B. Classical equations of motion

We verify that we get the classical equations of motion from the Lagrangian.

Broken out into time and space parts the Lagrangian is:

$$L\left(t, \vec{x}, \dot{t}, \dot{\vec{x}}\right) = -\frac{1}{2}m\dot{t}^2 + \frac{1}{2}m\dot{\vec{x}} \cdot \dot{\vec{x}} - q\dot{t}\Phi(t, \vec{x}) + q\dot{x}_j A_j(t, \vec{x}) - \frac{1}{2}m \tag{B.1}$$

The Euler-Lagrange equations are:

$$\frac{d}{d\tau}\frac{\delta L}{\delta \dot{x}^\mu} - \frac{\delta L}{\delta x^\mu} = 0 \tag{B.2}$$

From the Euler-Lagrange equations we have:

$$m\ddot{t} = -q\dot{\Phi} + q\dot{t}\Phi_{,0} - q\dot{x}_j A_{j,0} = -q\dot{x}_j\left(\Phi_{,j} + A_{j,0}\right) \tag{B.3}$$

$$m\ddot{x}_i = -q\dot{A}_i - q\dot{t}\Phi_{,i} + q\dot{x}_j A_{j,i} = -q\dot{t}A_{i,0} - q\dot{x}_j A_{i,j} - q\Phi_{,i}\dot{t} + q\dot{x}_j A_{j,i} \tag{B.4}$$

Here the Roman indexes, i and j, go from 1 to 3 and if present in pairs are summed over. We use an overdot to indicate differentiation by the laboratory time τ.

By using:

$$\vec{E} = -\nabla\Phi - \frac{\partial \vec{A}}{\partial t} \tag{B.5}$$

$$\vec{B} = \nabla \times \vec{A} \tag{B.6}$$

we get:

$$m\ddot{t} = q\vec{E} \cdot \dot{\vec{x}} \tag{B.7}$$

and:

$$m\ddot{\vec{x}} = q\dot{t}\vec{E} + q\dot{\vec{x}} \times \vec{B} \tag{B.8}$$

which are the familiar equations of motion of a classical particle in an electromagnetic field if we take τ as the proper time of the particle.

In the text we take τ as the laboratory time rather than the proper time of the particle. In the non-relativistic case, they are nearly the same.

IOP Conf. Series: Journal of Physics: Conf. Series **1239** (2019) 012015 doi:10.1088/1742-6596/1239/1/012015

Appendix C. Unitarity

Since there is a chance of the wave function "sneaking past" the plane of the present, we have to be particularly careful to confirm unitarity.

To establish that the path integral kernel is unitary we need to establish that it preserves the normalization of the wave function. The analysis in the subsection 2.7 only established this for the free case. We therefore need to confirm that the normalization of the wave function is preserved in the general case. We use a proof from Merzbacher [110] but in four rather than three dimensions.

We form the probability:

$$P \equiv \int d^4x \psi^* (x) \psi (x) \tag{C.1}$$

We therefore have for the rate of change of probability in time:

$$\frac{dP}{d\tau} = \int d^4x \left(\psi^* (x) \frac{\partial \psi (x)}{\partial \tau} + \frac{\partial \psi^* (x)}{\partial \tau} \psi (x) \right) \tag{C.2}$$

The Schrödinger equations for the wave function and its complex conjugate are:

$$\frac{\partial \psi}{\partial \tau} = -\frac{\imath}{2m} \partial^\mu \partial_\mu \psi + \frac{q}{m} (A^\mu \partial_\mu) \psi + \frac{q}{2m} (\partial^\mu A_\mu) \psi + \imath \frac{q^2}{2m} A^\mu A_\mu \psi - \imath \frac{m}{2} \psi \tag{C.3}$$

$$\frac{\partial \psi^*}{\partial \tau} = \frac{\imath}{2m} \partial^\mu \partial_\mu \psi^* + \frac{q}{m} (A^\mu \partial_\mu) \psi^* + \frac{q}{2m} (\partial^\mu A_\mu) \psi^* - \imath \frac{q^2}{2m} A^\mu A_\mu \psi^* + \imath \frac{m}{2} \psi^* \tag{C.4}$$

We rewrite $\frac{\partial \psi}{\partial \tau}$ and $\frac{\partial \psi^*}{\partial \tau}$ using these and throw out canceling terms. Since the probability density is gauge independent, we choose the Lorentz gauge $\partial A = 0$ to get:

$$\frac{dP}{d\tau} = \int d^4x \left(\psi^* \left(-\frac{\imath}{2m} \partial^\mu \partial_\mu \psi + \frac{q}{m} (A\partial) \psi \right) + \left(\frac{\imath}{2m} \partial^\mu \partial_\mu \psi^* + \frac{q}{m} (A\partial) \psi^* \right) \psi \right) \tag{C.5}$$

We integrate by parts; we are left with zero on the right:

$$\frac{dP}{d\tau} = 0 \tag{C.6}$$

Therefore the rate of change of probability is zero, as was to be shown. And therefore the normalization is correct in the general case.

Appendix D. Gauge transformations

As noted in the text, the ambiguities in the phase of the normalization of the wave function may be seen as representing a kind of gauge transformation. We have all the usual possibilities for gauge transformations. And we have in addition the possibility of gauge transformations which are a function of the laboratory time.

To explore this, we write the wave function as a product of a gauge function in coordinate time, space, and laboratory time and a gauged wave function:

$$\psi'_\tau (t, \vec{x}) = e^{\imath q \Lambda_\tau (t, \vec{x})} \psi_\tau (t, \vec{x}) \tag{D.1}$$

If the original wave function satisfies a gauged Schrödinger equation:

$$\left(\imath \frac{\partial}{\partial \tau} - q\mathcal{A}_\tau (x) \right) \psi_\tau (x) = -\frac{1}{2m} \left((p - qA)^2 - m^2 \right) \psi_\tau (x) \tag{D.2}$$

the gauged wave function also satisfies a gauged Schrödinger equation:

$$\left(\imath\frac{\partial}{\partial\tau} - q\mathcal{A}'_\tau(x)\right)\psi'_\tau(x) = -\frac{1}{2m}\left((p-qA')^2 - m^2\right)\psi'_\tau(x) \tag{D.3}$$

provided we have:

$$\mathcal{A}'_\tau(x) = \mathcal{A}_\tau(x) - \frac{\partial\Lambda_\tau(x)}{\partial\tau} \tag{D.4}$$

and the usual gauge transformations:

$$A'^\mu = A^\mu - \partial^\mu\Lambda_\tau(x) \tag{D.5}$$

or:

$$\begin{array}{c}\Phi' = \Phi - \frac{\partial\Lambda}{\partial t}\\ \vec{A}' = \vec{A} + \nabla\Lambda\end{array} \tag{D.6}$$

If the gauge function Λ is not a function of the laboratory time ($\Lambda = \Lambda(t,\vec{x})$) then we recover the usual gauge transformations for Φ and $\vec{A}$. On the other hand, we could let the gauge depend on the laboratory time, perhaps using different gauges for different parts of the problem in hand.

Appendix E. Free wave functions and kernels

We here assemble the solutions of the free Schrödinger's equation for reference.

In general the free wave functions and kernels can be written as a coordinate time part times a familiar non-relativistic part. The division into coordinate time, space, and – occasionally – clock time parts is to some extent arbitrary.

Appendix E.1. Plane waves

Plane wave in coordinate time:

$$\tilde{\phi}_\tau(t) = \frac{1}{\sqrt{2\pi}}\exp\left(-\imath E_0 t + \imath\frac{E_0^2}{2m}\tau\right) \tag{E.1}$$

Plane wave in space:

$$\bar{\phi}_\tau(\vec{x}) = \frac{1}{\sqrt{2\pi}^3}\exp\left(\imath\vec{p}_0\cdot\vec{x} - \imath\frac{\vec{p}_0^{\,2}}{2m}\tau\right) \tag{E.2}$$

The full plane wave is the product of coordinate time and space plane waves:

$$\phi_\tau(x) = \tilde{\phi}_\tau(t)\,\bar{\phi}_\tau(\vec{x})\exp\left(\imath\frac{m^2}{2m}\tau\right) = \frac{1}{4\pi^2}\exp\left(-\imath p_0 x - \imath f_0\tau\right) \tag{E.3}$$

with definition of clock frequency:

$$f_0 \equiv -\frac{E_0^2 - \vec{p}_0^{\,2} - m^2}{2m} \tag{E.4}$$

The equivalents in momentum space are δ functions with a clock time dependent phase:

$$\begin{array}{l}\hat{\tilde{\phi}}_\tau(E) = \delta(E-E_0)\exp\left(\imath\frac{E_0^2}{2m}\tau\right)\\ \hat{\bar{\phi}}_\tau(\vec{p}) = \delta^{(3)}(\vec{p}-\vec{p}_0)\exp\left(-\imath\frac{\vec{p}_0^2}{2m}\tau\right)\\ \hat{\phi}_\tau(p) = \hat{\tilde{\phi}}_0(E)\,\hat{\bar{\phi}}_0(\vec{p})\exp\left(-\imath\frac{m^2}{2m}\tau\right) = \delta^{(4)}(p-p_0)\exp\left(-\imath f_0\tau\right)\end{array} \tag{E.5}$$

IOP Conf. Series: Journal of Physics: Conf. Series **1239** (2019) 012015 doi:10.1088/1742-6596/1239/1/012015

Appendix E.2. Gaussian test functions

By Morlet wavelet decomposition any normalizable wave function may be written as a sum over Gaussian test functions. We have specified the conventions we are using for Gaussian test functions above; here we look specifically at Gaussian test functions as solutions of the free Schrödinger equation.

Appendix E.2.1. Time and energy Gaussian test function in coordinate time at clock time zero:

$$\tilde{\varphi}_0(t) \equiv \sqrt[4]{\frac{1}{\pi\sigma_t^2}} e^{-\imath E_0(t-t_0)-\frac{(t-t_0)^2}{2\sigma_t^2}} \tag{E.6}$$

$$\hat{\tilde{\varphi}}_0(E) \equiv \sqrt[4]{\frac{1}{\pi\sigma_E^2}} e^{\imath E t_0-\frac{(E-E_0)^2}{2\sigma_E^2}} \tag{E.7}$$

With these conventions, the energy and coordinate time dispersions are reciprocals:

$$\sigma_E = \frac{1}{\sigma_t} \tag{E.8}$$

As noted, it is often convenient to thread a letter through the wave function to label the constants, e.g.:

$$\tilde{\varphi}_a(t) \equiv \sqrt[4]{\frac{1}{\pi\sigma_t^{(a)2}}} e^{-\imath E_a(t-t_a)-\frac{(t-t_a)^2}{2\sigma_t^{(a)2}}} \tag{E.9}$$

$$\hat{\tilde{\varphi}}_a(E) \equiv \sqrt[4]{\frac{1}{\pi\sigma_E^{(a)2}}} e^{\imath E t_a-\frac{(E-E_a)^2}{2\sigma_E^{(a)2}}} \tag{E.10}$$

A typical case is where $a \to 0$, to represent the function at $\tau = 0$.

Gaussian test function for coordinate time as a function of clock time:

$$\tilde{\varphi}_\tau(t) = \sqrt[4]{\frac{1}{\pi\sigma_t^2}} \sqrt{\frac{1}{f_\tau^{(t)}}} e^{-\imath E_0 t+\imath\frac{E_0^2}{2m}\tau-\frac{1}{2\sigma_t^2 f_\tau^{(t)}}\left(t-t_0-\frac{E_0}{m}\tau\right)^2} \tag{E.11}$$

with dispersion factor $f_\tau^{(t)} \equiv 1-\imath\frac{\tau}{m\sigma_t^2}$ and with expectation, probability density, and uncertainty:

$$\begin{aligned} \langle t_\tau\rangle &= t_0 + \tfrac{E}{m}\tau = t_0 + \gamma\tau \\ \tilde{\rho}_\tau(t) &= \sqrt{\tfrac{1}{\pi\sigma_t^2}} \exp\left(-\frac{(t-\langle t_\tau\rangle)^2}{\sigma_t^2\left(1+\frac{\tau^2}{m^2\sigma_t^2}\right)}\right) \\ (\Delta t)^2 &\equiv \langle t^2\rangle - \langle t\rangle^2 = \tfrac{\sigma_t^2}{2}\left|1+\frac{\tau^2}{m^2\sigma_t^4}\right| \end{aligned} \tag{E.12}$$

In the non-relativistic case, if we start with $t_0 = \tau_0$, then we have $\langle t\rangle \approx \tau$ throughout.

For longer clock times the uncertainty in coordinate time is proportional to the dispersion in the energy:

$$\Delta t \sim \frac{\tau}{m\sigma_t} = \frac{\tau}{m}\sigma_E \tag{E.13}$$

Gaussian test function in energy:

$$\hat{\tilde{\varphi}}_\tau(E) \equiv \sqrt[4]{\frac{1}{\pi\sigma_E^2}} e^{\imath E t_0-\frac{(E-E_0)^2}{2\sigma_E^2}+\imath\frac{E^2}{2m}\tau} \tag{E.14}$$

IOP Conf. Series: Journal of Physics: Conf. Series **1239** (2019) 012015 doi:10.1088/1742-6596/1239/1/012015

with expectation, probability density, and uncertainty:

$$\begin{aligned} &\langle E\rangle = E_0 \\ &\hat{\bar{\rho}}_\tau(E) = \hat{\bar{\rho}}_0(E) = \sqrt{\tfrac{1}{\pi\sigma_E^2}}\exp\left(-\tfrac{(E-E_0)^2}{\sigma_E^2}\right) \\ &(\Delta E)^2 = \tfrac{\sigma_E^2}{2} \end{aligned} \tag{E.15}$$

Appendix E.2.2. Single space/momentum dimension Gaussian test function in one space dimension at clock time zero:

$$\bar{\varphi}_0(x) = \sqrt[4]{\frac{1}{\pi\sigma_x^2}}e^{\imath p_0(x-x_0)-\frac{(x-x_0)^2}{2\sigma_x^2}} \tag{E.16}$$

and in momentum:

$$\hat{\bar{\varphi}}_0(p) = \sqrt[4]{\frac{1}{\pi\sigma_p^2}}e^{-\imath p x_0-\frac{(p-p_0)^2}{2\sigma_p^2}} \tag{E.17}$$

As with time and energy, the space and momentum dispersions are reciprocal:

$$\sigma_p \equiv \frac{1}{\sigma_x} \tag{E.18}$$

When we have to consider the dispersion for all three space momentum we may reduce the level of nesting by writing:

$$\hat{\sigma}_x \equiv \sigma_{p_x}, \hat{\sigma}_y \equiv \sigma_{p_y}, \hat{\sigma}_z \equiv \sigma_{p_z} \tag{E.19}$$

Gaussian test function in one space dimension as a function of clock time:

$$\bar{\varphi}_\tau(x) = \sqrt[4]{\frac{1}{\pi\sigma_x^2}}\sqrt{\frac{1}{f_\tau^{(x)}}}e^{\imath p_0 x-\frac{1}{2\sigma_x^2 f_\tau^{(x)}}\left(x-x_0-\frac{p_0}{m}\tau\right)^2-\imath\frac{p_0^2}{2m}\tau} \tag{E.20}$$

The definition of the dispersion factor $f_\tau^{(x)} = 1+\imath\frac{\tau}{m\sigma_x^2}{}_\tau$ is parallel to that for coordinate time (but with the opposite sign for the imaginary part).

Expectation, probability density, and uncertainty for x:

$$\begin{aligned} &\langle x_\tau\rangle = x_0 + \tfrac{p_x}{m}\tau = x_0 + \gamma v_x\tau \\ &\bar{\rho}_\tau(x) = \sqrt{\tfrac{1}{\pi\sigma_x^2}}\exp\left(-\frac{(x-\langle x_\tau\rangle)^2}{\sigma_x^2\left(1+\frac{\tau^2}{m^2\sigma_x^2}\right)}\right) \\ &(\Delta x)^2 \equiv \langle x_\tau^2\rangle - \langle x_\tau\rangle^2 = \tfrac{\sigma_x^2}{2}\left|1+\tfrac{\tau^2}{m^2\sigma_x^4}\right| \end{aligned} \tag{E.21}$$

and similarly for y and z.

As clock time goes to infinity, the uncertainty in space scales as:

$$\Delta x \sim \frac{\tau}{m\sigma_x} = \frac{\tau}{m}\hat{\sigma}_x \tag{E.22}$$

Negative x-momentum is movement to the left, positive to the right. As we require the most complete parallelism between time and space, we therefore have that positive energy corresponds to movement into the future, negative into the past. As most of our wave functions have an energy of order:

$$E \sim m + \frac{\vec{p}^2}{2m} \gg 0 \tag{E.23}$$

IOP Conf. Series: Journal of Physics: Conf. Series **1239** (2019) 012015 doi:10.1088/1742-6596/1239/1/012015

they are usually going into the future. As expected.

Gaussian test function for momentum in one dimension as a function of clock time:

$$\hat{\tilde{\varphi}}_\tau(p) = \sqrt[4]{\frac{1}{\pi\sigma_p^2}} e^{-\imath p x_0 - \frac{(p-p_0)^2}{2\sigma_p^2} - \imath\frac{p^2}{2m}\tau} \tag{E.24}$$

The expectation, probability density, and uncertainty for p are constant:

$$\begin{array}{l} \langle p\rangle_\tau = \langle p\rangle_0 = p_0 \\ \hat{\tilde{\rho}}_\tau(p) = \hat{\tilde{\rho}}_0(p) = \sqrt{\frac{1}{\pi\sigma_p^2}} \exp\left(-\frac{(p-p_0)^2}{\sigma_p^2}\right) \\ (\Delta p)^2 = \frac{\sigma_p^2}{2} \end{array} \tag{E.25}$$

Appendix E.2.3. Covariant forms Usually we treat four dimensions wave functions as simple products of one dimensional wave functions. But it is more appropriate in general to treat them as a single covariant object.

We define the four dimensional dispersion Σ in position space at clock time zero:

$$\Sigma_0^{\mu\nu} \equiv \begin{pmatrix} \sigma_t^2 & 0 & 0 & 0 \\ 0 & \sigma_x^2 & 0 & 0 \\ 0 & 0 & \sigma_y^2 & 0 \\ 0 & 0 & 0 & \sigma_z^2 \end{pmatrix} \tag{E.26}$$

The determinant takes the simple form:

$$\det\Sigma_0 = \sigma_t^2\sigma_x^2\sigma_y^2\sigma_z^2 \tag{E.27}$$

With this we have the wave function at clock time zero:

$$\varphi_0(x) = \sqrt[4]{\frac{1}{\pi^4 \det(\Sigma_0)}} e^{-\imath p_0^\mu (x-x_0)_\mu - \frac{1}{2}(x-x_0)_\mu \Sigma_0^{-1\mu\nu}(x-x_0)_\nu} \tag{E.28}$$

Four dimensional dispersion as a function of clock time τ:

$$\Sigma_\tau^{\mu\nu} \equiv \begin{pmatrix} \sigma_t^2 f_\tau^{(t)} = \sigma_t^2 - \imath\frac{\tau}{m} & 0 & 0 & 0 \\ 0 & \sigma_x^2 f_\tau^{(x)} = \sigma_x^2 + \imath\frac{\tau}{m} & 0 & 0 \\ 0 & 0 & \sigma_y^2 f_\tau^{(y)} = \sigma_y^2 + \imath\frac{\tau}{m} & 0 \\ 0 & 0 & 0 & \sigma_z^2 f_\tau^{(z)} = \sigma_z^2 + \imath\frac{\tau}{m} \end{pmatrix} \tag{E.29}$$

Four dimensional wave function as a function of clock time:

$$\varphi_\tau(x) = \sqrt[4]{\frac{\det(\Sigma_0)}{\pi^4 \det^2(\Sigma_\tau)}} e^{-\imath p_0^\mu (x-x_0-v\tau)_\mu - \frac{1}{2}(x-x_0-v\tau)_\mu \Sigma_\tau^{-1\mu\nu}(x-x_0-v\tau)_\nu - \imath f_0\tau} \tag{E.30}$$

with the obvious definition of the four velocity:

$$v_\mu \equiv \frac{p_\mu}{m} \tag{E.31}$$

In the general case, the wave function does not split into coordinate time and space factors. But it may still be represented as a sum over basis wave functions that do, via Morlet wavelet decomposition.

IOP Conf. Series: Journal of Physics: Conf. Series **1239** (2019) 012015 doi:10.1088/1742-6596/1239/1/012015

We have similar but simpler formulas in momentum space. In momentum space at clock time zero:

$$\hat{\varphi}_0(p) = \sqrt[4]{\frac{1}{\pi^4 \det\left(\hat{\Sigma}\right)}} e^{-\imath p x_0 - \frac{1}{2}(p-p_0)_\mu \left(\Sigma^{-1}\right)^{\mu\nu}(p-p_0)_\nu} \tag{E.32}$$

and as a function of clock time:

$$\hat{\varphi}_\tau(p) = \sqrt[4]{\frac{1}{\pi^4 \det\left(\hat{\Sigma}\right)}} e^{-\imath p x_0 - \frac{1}{2}(p-p_0)_\mu \left(\Sigma^{-1}\right)^{\mu\nu}(p-p_0)_\nu - \imath f_0 \tau} = \hat{\varphi}_0(p)\exp\left(-\imath f_0 \tau\right) \tag{E.33}$$

We define the momentum dispersion matrix $\hat{\Sigma}$ as the reciprocal of the coordinate space dispersion matrix Σ_0:

$$\hat{\Sigma} \equiv \begin{pmatrix} \hat{\sigma}_t^2 & 0 & 0 & 0 \\ 0 & \hat{\sigma}_x^2 & 0 & 0 \\ 0 & 0 & \hat{\sigma}_y^2 & 0 \\ 0 & 0 & 0 & \hat{\sigma}_z^2 \end{pmatrix} \tag{E.34}$$

Appendix E.3. Free kernels

We list the kernels corresponding to the free Schrödinger equation in time. These are retarded kernels going from clock time zero to clock time τ, so include an implicit $\theta(\tau)$.

Appendix E.3.1. Coordinate time and space Kernel in coordinate time:

$$\tilde{K}_\tau\left(t''; t'\right) = \sqrt{\frac{\imath m}{2\pi\tau}} \exp\left(-\imath m \frac{\left(t'' - t'\right)^2}{2\tau}\right) \tag{E.35}$$

In three space we have the familiar non-relativistic kernel (e.g. Merzbacher [110]):

$$\bar{K}_\tau\left(\vec{x}''; \vec{x}'\right) = \sqrt{-\frac{\imath}{2\pi\tau}}^{\,3} \exp\left(\imath m \frac{\left(\vec{x}'' - \vec{x}'\right)^2}{2\tau}\right) \tag{E.36}$$

The full kernel is:

$$K_\tau\left(x''; x'\right) = \tilde{K}_\tau\left(t''; t'\right) \bar{K}_\tau\left(\vec{x}''; \vec{x}'\right) \exp\left(-\imath \frac{m}{2}\tau\right) \tag{E.37}$$

Explicitly:

$$K_\tau\left(x''; x'\right) = -\imath \frac{m^2}{4\pi^2\tau^2} e^{-\imath m \frac{\left(t''-t'\right)^2}{2\tau} + \imath m \frac{\left(\vec{x}''-\vec{x}'\right)^2}{2\tau} - \imath \frac{m}{2}\tau} \tag{E.38}$$

Appendix E.3.2. Momentum space Energy part:

$$\hat{\tilde{K}}_\tau\left(E''; E'\right) = \delta\left(E'' - E'\right) \exp\left(\imath \frac{E'^2 - m^2}{2m}\tau\right) \tag{E.39}$$

Three momentum part:

$$\hat{\bar{K}}_\tau\left(\vec{p}''; \vec{p}'\right) = \delta^{(3)}\left(\vec{p}'' - \vec{p}'\right) \exp\left(-\imath \frac{\left(\vec{p}'\right)^2}{2m}\tau\right) \tag{E.40}$$

IOP Conf. Series: Journal of Physics: Conf. Series **1239** (2019) 012015 doi:10.1088/1742-6596/1239/1/012015

Again, the same as the usual non-relativistic kernel.

The full kernel is:

$$\hat{K}_\tau\left(p'' - p'\right) = \hat{\hat{K}}_\tau\left(E''; E'\right)\hat{\vec{K}}_\tau\left(\vec{p}''; \vec{p}'\right) \tag{E.41}$$

Spelled out:

$$\hat{K}_\tau\left(p''; p'\right) = \delta^{(4)}\left(p'' - p'\right)\exp\left(\imath\frac{E'^2 - (\vec{p}')^2 - m^2}{2m}\tau\right) \tag{E.42}$$

or:

$$\hat{K}_\tau\left(p''; p'\right) = \delta^{(4)}\left(p'' - p'\right)\exp\left(-\imath f_{p'}\tau\right) \tag{E.43}$$

With definition of clock frequency as above:

$$f_p \equiv -\frac{E^2 - \vec{p}^2 - m^2}{2m} \tag{E.44}$$

Appendix F. Checkpoint Copenhagen

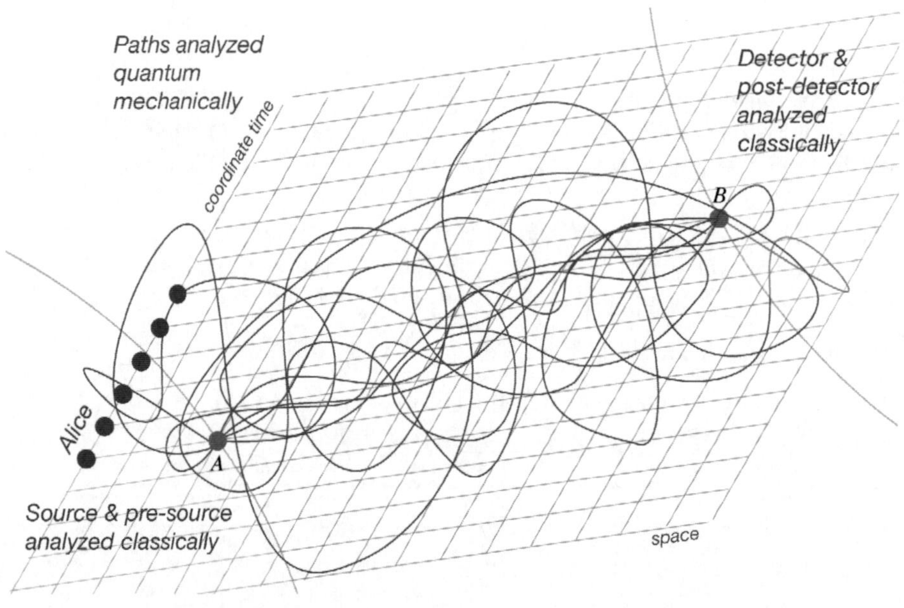

Figure F1. Clock time and coordinate time

"Atoms are completely impossible from the classical point of view, since the electrons would spiral into the nucleus." – Richard P. Feynman [79]

Appendix F.1. Four dimensions and an approximation scheme

Naively we might appear to have a five dimensional coordinate system here: clock time, coordinate time, and the three space dimensions. When this work was presented at the 2018 International Association for Relativistic Dynamics (IARD) conference Dr. Asher Yahalom observed that this is potentially a bit cumbersome.

Per subsequent discussion with Dr. Yahalom, it is more accurate to describe it as four dimensions plus an approximation scheme.

Let's start with a sheet of four dimensional graph paper representing discretized space time. Alice is off to one side, drawing paths on it. She is interested in the amplitude for a particle to get from A to B. And to compute this she draws all possible paths from A to B, planning to sum them using the rules in the text.

But Alice herself is a part of the universe she is observing.

Therefore we draw Alice on the left as a series of blue dots representing her at successive clock ticks. At each clock tick she has a corresponding three dimensional hyper-surface representing her rest frame. These are the horizontal lines in the diagram. She is off to one side, because she is after all observing the particle going from A to B, but she is on the same piece of four dimensional graph paper because she is part of the same universe.

If she is using SQM to do the calculation then at each clock tick each of her paths will slide side to side on the corresponding hyper-surface. But if she is using TQM then the paths will also go forwards and backwards in time, going off the current hyper-surface, often in quite elaborate ways.

If Bob is also present, we can represent him by a series of green dots on the right with his own coordinate system. (We have left this off the illustration because it would make it too busy. Please imagine Bob's green dots and hyper-surfaces are present.) He will be looking at the same set of paths, but slicing them up differently because he has in general a different set of three dimensional hyper-surfaces.

And if we need to resolve the slight differences between Alice and Bob's descriptions – per subsection 3.5 – we can call in Vera in the V frame to provide the definitive story.

So far so good, we have only one four dimensional coordinate system with different observers. Each observer has his or her clock time, but these clock times are present, tick by tick, on the same piece of graph paper. And we have well-defined rules for going from one observer's frame to another's.

The problem comes in when we try to reconcile the quantum descriptions Alice uses for particles with the classical descriptions she uses for detectors, emitters, or herself. We have an impedance mismatch between quantum and classical descriptions. This is the problem of measurement.

Appendix F.2. Quantum descriptions and classical approximations

The critical observation here is Feynman's: there are no classical atoms. Since the emitters, detectors, and observers are all made of atoms all are quantum objects.

This means that there is no transition from a quantum to a classical realm. Everything on both sides of the act of detection is a quantum system.

So what is going on here?

Most of the systems we deal with can treated for all practical purposes (FAPP) as if they were classical systems. It is only when we look at certain parts of the system that we need to get down to the quantum level. When we describe a particle being ejected from an atom we need to use a quantum description. When we describe a particle in flight we need to use a quantum description. When we describe its encounter with a detector we need to use a quantum description. But once the detector has registered a click, we can use classical approaches to describe the counting and processing of those clicks.

The rules for detection and emission are ways to navigate the associated approximations. In a detector, the wave function is not collapsing, instead we are passing from a quantum to a classical description. And in emission, the particle is not originally classical, it is just that up to the starting gun, it can be treated as if it were.

In human terms, picture Alice going thru passport control. Bob, now a customs official (he gets around), stamps her passport with a visa stamp. She then heads to her ultimate destination. Both Alice and Bob are – per Feynman – quantum systems. Their previous and subsequent paths are unknown to official customs. But we have that visa stamp and associated computer records.

They are the measurement. They certify that at time T position X with uncertainties ΔT and ΔX Alice and Bob encountered each other.

Since Bob's location is highly localized – he is in a customs booth, they are not big – we treat this as a measurement of Alice's position. But the situation is in reality completely symmetrical. We can think of it as Alice measuring Bob or Bob measuring Alice. But for customs purposes only the visa stamp matters. The visa stamp is the measurement.

Ultimately we must always be prepared to go down to the quantum level. The quantum rules are decisive; the classical rules a mere useful approximation. Even the visa stamp itself is made of atoms, of quantum mechanical objects. But in some cases we can get away with a classical analysis.

The division between quantum and classical is a division of analysis. It is not part of the physical universe, it is part of how we describe that universe – allowing for the fact that we are a part of what we are describing.

And we get our definition of clock time from the rest of the universe:

$$\tau = \langle \text{rest of universe} | \, t \, | \text{rest of universe} \rangle \tag{F.1}$$

This may be seen as a variation on Fanchi's "historical time" [36]. He used an outside system, $S2$, to define a reference clock. We extend his $S2$ to be the entire rest of the universe (fortunately there are no budget constraints on thought experiments).

Appendix F.3. Where, when, and to what extent?

But to take this point from the realm of philosophy to the realm of science we need to throw some numbers into the mix, we need to answer the questions: "when, where, and to what extent does the classical approximation break down?"

The rules in the Copenhagen interpretation are not specific. "Somewhere" between where the particle is being described by quantum mechanics and where it is detected, the particle goes through a "checkpoint Copenhagen", where its description collapses from the fuzziness of quantum mechanics to the determinism and specificity of classical mechanics.

This is both a strength and a weakness of the Copenhagen interpretation.

It is a strength because it let physicists get on with physics. And because it does not try to "explain" what is going on.

But it is a weakness because the terms of the transition are not specified. Schrödinger's cat experiment is the most striking illustration of this point [111].

Decoherence provides part of the answer [112, 113, 114, 115, 116]. But decoherence is not yet as quantitative as one might like (although see Venugopalan, Qureshi, and Mishra [117]).

It is possible that new physics plays a role in the transition. The Ghirardi, Rimini, and Weber (GRW) [118] approach and more generally the continuous spontaneous localization (CSL) approaches (see Dickson [119]) hypothesize additional physics. These alternatives are helpful in parameterizing the transition, but have not had any experimental confirmation.

Meanwhile, the quantum cats continue to get bigger and bigger, less and less microscopic [120, 121]. At some point we will be able to see the transition itself, or show that there is none.

TQM does *not* specifically address this question. We have taken the existing rules as given. However the temporal fluctuations in TQM (see especially the entropic estimate of the initial wave function subsubsection 4.1.2) provide a source of "internal decoherence" so would affect estimates of the size and rate of decoherence. (We owe the phrase "internal decoherence" to Dr. Daniel Braun at the 2007 Feynman Festival.)

References

[1] Thorne K S 2009 *Science* **320** 1603

[2] Feynman R P 1965 *The Character of Physical Law* (Modern Library) ISBN 0 679 60127 9
[3] Adler R, Bazin M and Schiffer M 1965 *Introduction to General Relativity* (New York: McGraw-Hill)
[4] Gödel K 1949 *Rev. Mod Phys.* **21** 447–450
[5] Thorne K S 1994 *Black Holes & Time Warps: Einstein's Outrageous Legacy* (New York: W. W. Norton) ISBN 0 395 05505 0
[6] Hilgevoord J 1996 *American Journal of Physics* **64** 1451–1456
[7] Hilgevoord J 1998 *American Journal of Physics* **66** 396–402
[8] Schilpp P A and Bohr N 1949 *Discussion with Einstein on Epistemological Problems in Atomic Physics* (Evanston: The Library of Living Philosophers) pp 200–241
[9] Heisenberg W 1930 *The Physical Principles of the Quantum Theory* (Chicago: University of Chicago Press)
[10] Busch P 2001 *The Time Energy Uncertainty Relation* (Berlin: Springer-Verlag) pp 69–98 Lecture Notes in Physics
[11] Pauli W 1980 *General principles of quantum mechanics* (Springer-Verlag) URL `http://www.springer.com/us/book/9783540098423`
[12] Dirac P A M 1958 *General Principles of Quantum Mechanics* 4th ed International series of monographs on physic (Oxford, Clarendon Press)
[13] Muga J G, Sala Mayato R and Egusquiza I L 2002 *Time in quantum mechanics* (Berlin ; New York: Springer) ISBN 3540432949
[14] Muga J G, Sala Mayato R and Egusquiza I L 2008 *Time in quantum mechanics - Vol 2* (Berlin ; New York: Springer-Verlag) ISBN 3540734724 9783540734727 0075-8450 ;
[15] Stueckelberg 1941 *Helv. Phys. Acta.* **14** 588
[16] Stueckelberg 1941 *Helv. Phys. Acta.* **14** 322–323
[17] Feynman R P 1948 *Rev. Mod. Phys.* **20**(2) 367–387
[18] Feynman R P 1949 *Phys Rev* **76** 749–759
[19] Feynman R P 1949 *Phys Rev* **76** 769–789
[20] Feynman R P 1950 *Physical Review* **80** 440–457
[21] Schulman L S 1997 *Time's arrows and quantum measurement* (New York: Cambridge University Press) ISBN 0 521 56775 0
[22] Zeh H D 2001 *The Physical Basis of the Direction of Time* (Berlin: Springer-Verlag)
[23] Callender C 2017 *What Makes Time Special?* (Oxford University Press)
[24] Kirk G S, Raven J E and Schofield M 1983 *The presocratic philosophers: a critical history with a selection of texts* 2nd ed (Cambridge: Cambridge University Press) ISBN 0521254442 URL `http://www.loc.gov/catdir/description/cam022/82023505.html`
[25] Price H 1996 *Time's Arrow and Archimedes' Point* (New York: Oxford University Press) ISBN 0 19 510095 6
[26] Barbour J 2000 *The End of Time: the Next Revolution in Physics* (Oxford: Oxford University Press)
[27] Cramer J G 1986 *Rev. Mod. Phys.* **58** 647–687
[28] Cramer J G 1988 *Intl. J. Theoretical Phys.* **27** 227–236
[29] Kastner R E 2013 *The new transactional interpretation of quantum mechanics: the reality of possibility* (New York: Cambridge University Press) ISBN 9780521764155 (hardback) URL `http://assets.cambridge.org/97805217/64155/cover/9780521764155.jpg`
[30] Cramer J G 2016 *The Quantum Handshake: Entanglement, Nonlocality and Transactions* (Springer International Publishing)
[31] Aharonov Y and Rohrlich D 2005 *Quantum Paradoxes: Quantum Theory for the Perplexed* (Wiley-VCH)
[32] Reznik B and Aharonov Y 1995 *Phys. Rev. A* **52** 2538–2550
[33] Horwitz L P and Piron C 1973 *Helvetica Physica Acta* **46**
[34] Fanchi J R and Collins R E 1978 *Found Phys* **8** 851–877
[35] Fanchi J R 1993 *Found. Phys.* **23**
[36] Fanchi J R 1993 *Parameterized Relativistic Quantum Theory* (*Fundamental Theories of Physics* vol 56) (Kluwer Academic Publishers)
[37] Land M C and Horwitz L P 1996 *ArXiv e-prints* URL `http://lanl.arxiv.org/abs/hep-th/9601021v1`
[38] Horwitz L P 2005 *ArXiv e-prints* URL `http://lanl.arxiv.org/abs/quant-ph/0507044`
[39] Fanchi J R 2011 *Found Phys* **41** 4–32
[40] Horwitz L P 2015 *Relativistic Quantum Mechanics* Fundamental Theories of Physics (Springer Dordrecht Heidelberg New York London)
[41] Ossiander M, Siegrist F, Shirvanyan V, Pazourek R, Sommer A, Latka T, Guggenmos A, Nagele S, Feist J, Burgdörfer J, Kienberger R and Schultze M 2016 *Nature Physics* **13** 280 EP – URL `https://doi.org/10.1038/nphys3941`
[42] Abbott E A 1884 *Flatland: A Romance of Many Dimensions* (Seeley and Co. of London)

[43] Feynman R P, Hibbs A R and Styer D F 2010 *Quantum mechanics and path integrals* (Mineola, N.Y.: Dover Publications) ISBN 9780486477220 0486477223 URL http://www.loc.gov/catdir/enhancements/fy1006/2010004550-d.html
[44] Schulman L S 1981 *Techniques and Applications of Path Integrals* (New York: John Wiley and Sons, Inc.) ISBN 0 471 16610 3
[45] Rivers R J 1987 *Path integral methods in quantum field theory* (Cambridge [Cambridgeshire] ; New York: Cambridge University Press) ISBN 0521259797 URL http://www.loc.gov/catdir/description/cam032/86026861.html http://www.loc.gov/catdir/toc/cam032/86026861.html
[46] Swanson M 1992 *Path Integrals and Quantum Processes* (Academic Press, Inc.) ISBN 0 12 678945 2
[47] Khandekar D C, Lawande S V and Bhagwat K V 1993 *Path-integral Methods and their Applications* (Singapore: World Scientific Publishing Co. Pte. Ltd.)
[48] Kashiwa T, Ohnuki Y and Suzuki M 1997 *Path integral methods* (Oxford Oxford ; New York: Clarendon Press ; Oxford University Press) ISBN 0198517718
[49] Grosche C and Steiner F 1998 *Handbook of Feynman path integrals* (New York: Springer Verlag) ISBN 3540571353 (alk. paper)
[50] Zinn-Justin J 2005 *Path Integrals in Quantum Mechanics* (Oxford: Oxford University Press)
[51] Kleinert H 2009 *Path integrals in quantum mechanics, statistics, polymer physics, and financial markets* (New Jersey: World Scientific) ISBN 9789814273558 9814273554
[52] Zee A 2010 *Quantum field theory in a nutshell* (Princeton, N.J.: Princeton University Press) ISBN 9780691140346 (hardcover alk. paper)
[53] Morlet J, Arens G, Fourgeau I and Giard D 1982 *Geophysics* **47** 203–236
[54] Chui C K 1992 *An introduction to wavelets* (Boston: Academic Press) ISBN 0121745848 (acid-free paper)
[55] Meyer Y and Salinger D H 1992 *Wavelets and operators* (Cambridge ; New York, N.Y.: Cambridge University Press) ISBN 0521420008
[56] Kaiser G 1994 *A friendly guide to wavelets* (Boston: Birkhäuser)
[57] Berg J C v d 1999 *Wavelets in physics* (Cambridge ; New York: Cambridge University Press) ISBN 0521593115 (hb)
[58] Bratteli O and Jørgensen P E T 2002 *Wavelets through a looking glass: the world of the spectrum* (Boston: Birkhäuser) ISBN 0817642803 (acid-free paper)
[59] Addison P S 2002 *The illustrated wavelet transform handbook : introductory theory and applications in science, engineering, medicine and finance* (Bristol: Institute of Physics Publishing) ISBN 0750306920
[60] Visser M 2003 *Phys. Lett. A* **315** 219–224 URL http://xxx.lanl.gov/abs/hep-th/0304081v1
[61] Antoine J P, Murenzi R, Vandergheynst P and Ali S T 2004 *Two-dimensional wavelets and their relatives* (Cambridge, UK; New York: Cambridge University Press) ISBN 0521624061 (hb) URL http://www.loc.gov/catdir/description/cam032/2003061134.html
[62] Ashmead J 2012 *Quanta* **1** ISSN 1314-7374 URL http://quanta.ws/ojs/index.php/quanta/article/view/5
[63] Goldstein H 1980 *Classical Mechanics Second Edition* (Reading, MA: Addison-Wesley)
[64] Feynman R P and Hibbs A R 1965 *Quantum Mechanics and Path Integrals* (McGraw-Hill)
[65] Tannor D J 2007 *Introduction to quantum mechanics: a time-dependent perspective* (Sausalito, Calif.: University Science Books) ISBN 1891389238 (alk. paper) 9781891389238 (alk. paper) URL http://www.loc.gov/catdir/toc/fy0708/2006043933.html
[66] Weinberg S 1972 *Gravitation and Cosmology: Principles and Applications of the General Theory of Relativity* (New York: John Wiley and Sons, Inc.) ISBN 0 471 92567 5
[67] Fleischhauer M, Imamoglu A and Marangos J P 2005 *Rev. Mod. Phys.* **77**(2) 633–673 URL https://link.aps.org/doi/10.1103/RevModPhys.77.633
[68] Muga J G and Leavens C R 2000 *Physics Reports* **338** 353–438
[69] Kijowski J 1974 *Reports on Mathematical Physics* **6** 361–386
[70] Moshinsky M 1951 *Phys. Rev.* **84** 525–532
[71] Moshinsky M 1952 *Phys. Rev.* **88** 625–631
[72] Gerhard G Paulus D B 2009 *Time in quantum mechanics - vol 2* (*Lecture Notes in Physics* vol 789) ed Gonzalo Muga Andreas Ruschhaupt A d C (Springer) chap 11, pp 303–339
[73] Umul Y Z 2009 *Turkish Journal of Physics* **33** 1–9
[74] Marchewka A and Schuss Z 1998 *Phys. Lett. A* **240** 177–184 URL http://xxx.lanl.gov/abs/quant-ph/9708034
[75] Marchewka A and Schuss Z 1999 *ArXiv e-prints* URL http://xxx.lanl.gov/abs/quant-ph/9906003
[76] Yabuki H 1986 *Int J Theor Ph* **25** 159–174
[77] De Raedt H, Michielsen K and Hess K 2012 *Phys. Rev. A* **85**(1) 012101 URL http://link.aps.org/doi/10.1103/PhysRevA.85.012101

[78] Sawant R, Samuel J, Sinha A, Sinha S and Sinha U 2014 *Phys. Rev. Lett* **113** 120406 (*Preprint* 1308.2022v2) URL http://arxiv.org/abs/1308.2022
[79] Feynman R P, Leighton R B and Sands M 1965 *The Feynman Lectures on Physics* vol Volumes I, II, and III (Addison-Wesley)
[80] Lindner F, Schätzel M G, Walther H, Baltuška A, Goulielmakis E, Krausz F, Milošević D B, Bauer D, Becker W and Paulus G G 2005 *Physical Review Letters* **95** 040401 URL http://link.aps.org/doi/10.1103/PhysRevLett.95.040401 http://lanl.arxiv.org/abs/quant-ph/0503165
[81] Sakurai J J 1967 *Advanced Quantum Mechanics* (Addison-Wesley) ISBN 0 201 06710 2
[82] Bjorken J D and Drell S D 1965 *Relativistic Quantum Mechanics* (New York: McGraw Hill)
[83] Bjorken J D and Drell S D 1965 *Relativistic Quantum Fields* (New York: McGraw Hill)
[84] Roman P 1969 *Introduction to quantum field theory* (New York,: Wiley) ISBN 0471731986
[85] Ramond P 1990 *Field theory* (Addison Wesley) ISBN 0 201 54611 6
[86] Kaku M 1993 *Quantum field theory : a modern introduction* (New York: Oxford University Press) ISBN 0195076524 (acid-free paper) URL http://www.loc.gov/catdir/enhancements/fy0604/92027704-d.html http://www.loc.gov/catdir/enhancements/fy0604/92027704-t.html
[87] Peskin M E and Schroeder D V 1995 *An introduction to quantum field theory* (Reading, Mass.: Addison-Wesley Pub. Co.) ISBN 0201503972 URL http://www.loc.gov/catdir/enhancements/fy0831/89027170-b.html http://www.loc.gov/catdir/enhancements/fy0831/89027170-d.html
[88] Weinberg S 1995 *Quantum Theory of Fields: Foundations* vol I (Cambridge University Press) ISBN 0 521 55001 7
[89] Huang K 1998 *Quantum field theory : from operators to path integrals* (New York: Wiley) ISBN 0471141208 (cloth alk. paper) URL http://www.loc.gov/catdir/bios/wiley047/98013437.html
[90] Weinberg S 1995 *Quantum Theory of Fields: Modern Applications* vol II (Cambridge University Press) ISBN 0 521 55002 5
[91] Itzykson C and Zuber J B 2005 *Quantum field theory* (Mineola, N.Y.: Dover Publications) ISBN 0486445682 (pbk.) URL http://www.loc.gov/catdir/enhancements/fy0625/2005053026-d.html
[92] Maggiore M 2005 *A modern introduction to quantum field theory* vol 12 (Oxford: Oxford University Press) ISBN 0198520735 (hbk.) URL http://www.loc.gov/catdir/enhancements/fy0627/2005295989-d.html
[93] Nazarov Y V and Danon J 2013 *Advanced Quantum Mechanics: A Practical Guide* (Cambridge University Press) ISBN 978-0-521-76150-5
[94] Das A 2014 *Field Theory : A Path Integral Approach* (Singapore, UNITED STATES: World Scientific Publishing Co Pte Ltd) ISBN 9789812773265 URL http://ebookcentral.proquest.com/lib/upenn-ebooks/detail.action?docID=1209926
[95] Lancaster T and Blundell S 2014 *Quantum field theory for the gifted amateur* first edition ed (Oxford University Press) ISBN 9780199699339 (pbk.)
[96] Einstein A B P N R 1935 *Phys. Rev.* **47** 777–780
[97] Feller W 1968 *An Introduction to Probability Theory and Its Applications* vol I (John Wiley and Sons, Inc.)
[98] Yau H 2015 *University of Pittsburgh Eprints* URL http://philsci-archive.pitt.edu/id/eprint/11622
[99] Lamoreaux S K 1992 *Intl. J. of Mod. Physics* **7** 6691–6762
[100] Ghose P 1999 *Testing quantum mechanics on new ground* (Cambridge: Cambridge University Press)
[101] Auletta G 2000 *Foundations and Interpretation of Quantum Mechanics: In the Light of a Critical-Historical Analysis of the Problems and of a Synthesis of the Results* (Singapore: World Scientific Publishing Co. Pte. Ltd.)
[102] Koke S, Grebing C, Frei H, Anderson A, Assion A and Steinmeyer G 2010 *Nat Photon* **4** 462–465 URL http://dx.doi.org/10.1038/nphoton.2010.91
[103] Arshansky R and Horwitz L P 1989 *Journal of Mathematical Physics* **30** 66–80 (*Preprint* https://doi.org/10.1063/1.528591) URL file:///.file/id=6571367.1323100154
[104] Arshansky R and Horwitz L P 1989 *Journal of Mathematical Physics* **30** 380–392 (*Preprint* https://doi.org/10.1063/1.528456) URL https://doi.org/10.1063/1.528456
[105] Horwitz L P and Lavie Y 1982 *Phys. Rev. D* **26**(4) 819–838 URL https://link.aps.org/doi/10.1103/PhysRevD.26.819
[106] Horwitz L P and Y R 1976 *Lettere Al Nuovo Cimento* **17** 501–507
[107] Horwitz L, Schieve W and Piron C 1981 *Annals of Physics* **137** 306 – 340 ISSN 0003-4916 URL http://www.sciencedirect.com/science/article/pii/0003491681901998
[108] Horwitz L 2018 *ArXiv e-prints* URL https://arxiv.org/abs/1810.09248
[109] Schick Jr T and Vaughan L 1995 *How to Think About Weird Things: Critical Thinking for a New Age*

(Mayfield Publishing Company)
[110] Merzbacher E 1998 *Quantum Mechanics* (New York: John Wiley and Sons, Inc.)
[111] Schrödinger E 1980 *Proceedings of the American Philosophical Society* **124** 323–38
[112] Omnès R 1994 *The Interpretation of Quantum Mechanics* (Princeton: Princeton University Press) ISBN 0 691 03669 1
[113] Giulini D, Joos E, Kiefer C, Kupsch J, Stamatescu I O and Zeh H D 1996 *Decoherence and the Appearance of a Classical World in Quantum Theory* (Berlin: Springer-Verlag)
[114] Heiss W D 2002 *Fundamentals of quantum information: quantum computation, communication, decoherence, and all that* vol 587 (Berlin: Springer) ISBN 3540433678 (alk. paper) URL `http://www.loc.gov/catdir/enhancements/fy0817/2002021847-d.html`
[115] Joos E 2003 *Decoherence and the appearance of a classical world in quantum theory* 2nd ed (Berlin: Springer) ISBN 3540003908 (acid-free paper) URL `http://www.loc.gov/catdir/toc/fy038/2003042813.html`
[116] Schlosshauer M A 2007 *Decoherence and the quantum-to-classical transition* (Berlin: Springer) ISBN 9783540357735 (hbk.) URL `http://www.loc.gov/catdir/enhancements/fy0814/2007930038-b.html`
[117] Venugopalan A, Mishra S and Qureshi T 2019 *Physica A: Statistical Mechanics and its Applications* **516** 308–316 (*Preprint* 1807.05021) URL `https://arxiv.org/pdf/1807.05021`
[118] Ghirardi G C, Rimini A and Weber T 1986 *Phys. Rev. D* **34** 470
[119] Dickson W M 1998 *Quantum chance and non-locality: probability and non-locality in the interpretations of quantum mechanics* (Cambridge: Cambridge University Press) ISBN 0 521 58127 3
[120] Norte R A, Moura J P and Gröblacher S 2016 *Physical Review Letters* **116** 147202 (*Preprint* 1511.06235) URL `https://arxiv.org/pdf/1511.06235`
[121] Vinante A, Mezzena R, Falferi P, Carlesso M and Bassi A 2017 *Physical Review Letters* **119** 110401 (*Preprint* 1611.09776) URL `https://arxiv.org/pdf/1611.09776`

IOP Conf. Series: Journal of Physics: Conf. Series **1239** (2019) 012016 doi:10.1088/1742-6596/1239/1/012016

How vacuum fluctuations determine the properties of the vacuum

G. B. Mainland[1] and Bernard Mulligan[2]

[1] Department of Physics, The Ohio State University at Newark, 1179 University Dr., Newark, OH 43055, USA, [2] Department of Physics, The Ohio State University, Columbus, OH 43210, USA

E-mail: mainland.1@osu.edu, mulligan.3@osu.edu

Abstract. Particle-antiparticle pairs are predicted by quantum field theory to appear as vacuum fluctuations. The model of the vacuum used here is postulated to have the following properties: To minimize the violation of conservation energy allowed by the Heisenberg uncertainty principle and to avoid violating conservation of angular momentum, vacuum fluctuations of charged particle-antiparticle pairs appear as bound states in the lowest energy level that has zero angular momentum. These transient atoms are polarized by electric fields somewhat similarly to the way that ordinary matter is polarized. As a consequence, the permittivity ϵ_0 of the vacuum can be calculated. Once the permittivity of the vacuum has been calculated, formulas for the speed of light c in the vacuum and the fine-structure constant α immediately follow. The values for ϵ_0, c, and α calculated here agree with the accepted values to within a few percent. Only the leading terms in the formulas have been retained in the calculations. The absence of dispersion in the vacuum is discussed and explained.

1. Introduction: vacuum fluctuations – the vacuum as a dielectric

Values of the permittivity ϵ_0 of the vacuum and the speed c of light in the vacuum are properties of the vacuum. Initially it is not obvious that the value of the fine-structure constant α also is a property of the vacuum. However, as will be shown later in the Introduction, if one of the three quantities ϵ_0, c, or α has been calculated, values for the other two immediately follow. Consequently, the value of α must also be a property of the vacuum.

Among the three quantities ϵ_0, c, and α, physicists have had little idea how to calculate either c or α. But for almost a century they have known how to calculate the permittivity of a dielectric[1]. As a consequence, most of this article is devoted to calculating ϵ_0, guided along the way by the well-established procedures for calculating the permittivity of dielectrics. Once ϵ_0 has been calculated, values for c and α immediately follow.

Vacuum fluctuations (VFs) are particle-antiparticle pairs that appear spontaneously in the vacuum as predicted by relativistic quantum field theory[2, 3, 4], violating conservation of energy to the extent allowed by the Heisenberg uncertainty principle. VFs are created on-mass-shell and are not associated with a Green's function or perturbation theory.

At this point a word of caution is required: Vacuum bubbles[3, 5] are a class of perturbation effects in quantum field theories that are sometimes confused with VFs because their Feynman diagrams look like bubbles that originate from and terminate in the vacuum. Vacuum bubbles are not VFs and, in fact, do not make a contribution to physical processes[3, 5].

IOP Conf. Series: Journal of Physics: Conf. Series **1239** (2019) 012016 doi:10.1088/1742-6596/1239/1/012016

The idea that VFs play a role in determining the permittivity of the vacuum is very old: in 1934 Furry and Oppenheimer[6] wrote that VFs of charged particle-antiparticle pairs would affect the value of the dielectric constant of the vacuum; "Because of the polarizability of the nascent pairs, the dielectric constant of space into which no matter has been introduced differs from that of truly empty space." In 1936 the idea of treating the vacuum as a medium with electric and magnetic polarizability was discussed by Weisskopf and Pauli[7, 8].

For VFs to be responsible for the permittivity of the vacuum, it is crucial that they possess elasticity (an effective spring constant). The possibility that a charged lepton-antilepton pair can form an atomic bound state, which possesses precisely this property, was discussed by Ruark[9]. At almost the same time, the formation of a charged lepton-antilepton bound state was elaborated on by Wheeler[10] when he calculated the decay rate of parapositronium, the spin-0, ground state of a bound electron and positron, into two photons. The experimental proof that such atoms exist was provided by Deutsch and Brown[11].

In 1957 Dicke[12] wrote about the possibility that the vacuum could be considered as a dielectric medium. More recently the possibility that the properties of the quantum vacuum determine, in the vacuum, the speed of light and the permittivity has been explored by a number of authors[13, 14, 15].

As will be shown, the VFs that primarily contribute to the permittivity of the vacuum are charged lepton-antilepton pairs. To conserve angular momentum and minimize the violation of conservation of energy, charged lepton-antilepton VFs must appear in the vacuum as atoms. The term "dielectric" can then be used in the usual sense: a photon passing through the vacuum is slowed by its interactions with transient atoms, a concept familiar from discussions of a physical dielectric[1, 16, 17].

In the discussion that follows, a quantum calculation of the the permittivity of the vacuum is presented that is similar both to a calculation of the permittivity of a physical dielectric[18] and to a previous quantum calculation of the the permittivity[19] of the vacuum by the authors. The result of the calculation is [1]

$$\epsilon_0 \cong \frac{6\mu_0}{\pi}\left(\frac{8e^2}{\hbar}\right)^2 = 9.10 \times 10^{-12}\frac{\mathrm{C}}{\mathrm{Vm}}\,. \tag{1}$$

In the above equation e is the magnitude of the charge on an electron, $\hbar$ is Planck's constant divided by 2π, and μ_0 is the permeability of the vacuum with an assigned value $4\pi \times 10^{-7}$H/m. Thus in Equation (1), ϵ_0 is expressed in terms of two experimentally-determined quantities, e and $\hbar$. The experimental value for ϵ_0 is 2.8 % less than the calculated value. The calculation of ϵ_0 is simplified, and the numerical accuracy is reduced, by including (a) only contributions to lowest order in what turns out to be an expansion in powers of the fine-structure constant α and (b) only the interactions of photons with bound states of charged lepton-antilepton VFs.

Once ϵ_0 has been calculated, formulas for c and α immediately follow. Using $c = 1/\sqrt{\mu_0\epsilon_0}$ and the formula for ϵ_0 in Equation (1),

$$c \cong \sqrt{\frac{\pi}{6}\frac{\hbar}{8e^2\mu_0}} = 2.96 \times 10^8 \mathrm{m/s}\,. \tag{2}$$

The accepted value is $c = 3.00 \times 10^8$m/s, which is 1.3% more than the calculated value. The progress of a photon traveling through the vacuum is slowed when it interacts with and has a polarizing effect on a VF consisting of a charged lepton and antilepton bound into an atom, implying that the interaction of photons with vacuum fluctuations determines the speed of light in the vacuum.

[1] SI units are used throughout.

IOP Conf. Series: Journal of Physics: Conf. Series **1239** (2019) 012016 doi:10.1088/1742-6596/1239/1/012016

Any observer in an inertial frame of reference cannot detect relative motion of the vacuum and the inertial frame. Thus the observer would conclude that the vacuum is at rest with respect to the inertial frame. Leonhardt et al.[20] state, "In free space the quantum vacuum is Lorentz invariant, so a uniformly moving observer would not see any effect due to motion, but an accelerated observer would. This is known as the Unruh effect[21]." Since the vacuum determines the speed of light in the vacuum and the vacuum is at rest with respect to any inertial frame, it then follows that the speed of light in the vacuum is the same in any inertial frame. The previous statement is one of the two postulates in Einstein's 1905 paper "On the Electrodynamics of Moving Bodies"[22] in which he introduced special relativity.

In the early universe when the temperature was sufficiently high that it was difficult for charged lepton-antilepton VFs to bind into charged lepton-antilepton atoms, the number density of charged lepton-antilepton atoms that are VFs would have been less than today. With fewer bound, charged lepton-antilepton VFs per unit volume to slow the progress of photons, the speed of light would have been greater. Moffat[23] has pointed out that inflation[24] with the speed of light much greater in the early universe than it is now allows "all regions in the universe to be causally connected", solving the horizon problem. A sufficiently high superluminary speed also, "leads to a mechanism of monopole suppression in cosmology and can resolve the flatness problem."

The fine-structure constant α, which is defined by

$$\alpha \equiv \frac{e^2}{4\pi\epsilon_0\hbar c}, \tag{3}$$

has historically attracted theoretical interest because it is dimensionless. The fine-structure constant was introduced by Sommerfeld in 1916[25] to quantify relativistic corrections in the Bohr theory of the hydrogen atom and is a measure of the strength of the electromagnetic interaction. However, Sommerfeld was not the first to recognize the importance of the combined symbols used to define α: in the period from 1905-1910[26] both Planck[27] and Einstein[28] discussed the fact that Planck's constant h and $e^2/(4\pi\epsilon_0 c)$ had the same dimensions. α grows logarithmically with energy scale, and equals $1/137.036\ldots$ at zero energy scale. Using the formula for ϵ_0 in Equation (1) and the formula for c in Equation (2), Equation (3) becomes

$$\frac{1}{\alpha} \cong 8^2\sqrt{3\pi/2} = 138.93\ldots. \tag{4}$$

The experimental value for $1/\alpha$ is 1.4% less than the value calculated in Equation (4).

This article is organized as follows: In §2 examples of attempts to calculate the fine-structure constant are briefly discussed. In §3 properties of charged lepton-antilepton VFs are discussed, including the energy source for their creation and a proof of their existence. In §4 the role that the Heisenberg uncertainty principle plays in describing VFs is discussed, yielding an ansatz for the number density of VFs. §5 discusses how VFs act as harmonic oscillators. In §6 dielectric properties of the vacuum are calculated and discussed in five subsections: In §6.1 the interaction Hamiltonian describing the interaction of photons with VFs is constructed, in §6.2 a quantum calculation of the polarization of VFs is performed, and in §6.3 a general formula for the permittivity ϵ_0 of the vacuum is derived. In §6.4 and §6.5 the contributions to ϵ_0 from charged lepton-antilepton vacuum fluctuations and quark-antiquark VFs are, respectively, calculated and discussed. Finally, in §6.6 the permittivity ϵ_0 of the vacuum is calculated. The process that primarily determines the value of ϵ_0, photon capture by atoms consisting of charged lepton-antilepton vacuum fluctuations and the subsequent decay of these atoms, is discussed.

2. Attempts to calculate the fine-structure constant

In 1988 Richard Feynman wrote as follows about the fine-structure constant: "It has been a mystery ever since it was discovered more than fifty years ago, and all good theoretical physicists

IOP Conf. Series: Journal of Physics: Conf. Series **1239** (2019) 012016 doi:10.1088/1742-6596/1239/1/012016

put this number up on their wall and worry about it.... It's one of the greatest damn mysteries of physics: a magic number that comes to us with no understanding by man. You might say the 'hand of God' wrote that number, and 'we don't know how He pushed his pencil.' "[29]

Kragh [30] writes that there are three reasons that the fine-structure constant is not thought to be as important today as it was in the early decades after its discovery: (a) α is only one of four coupling constants, the other three being the gravitational, weak, and strong. (b) As mentioned previously, α grows logarithmically with energy scale, and equals 1/137.036... only at zero energy scale. (c) There is evidence that the fine-structure constant is not constant over time[31]. On the other hand, in Kinoshita's review article[32] written 80 years after Sommerfeld's introduction of α, he points out that α "characterizes the whole range of physics including elementary particle, atomic, macroscopic, and microscopic systems."

In the past there have been two general theoretical methods for attempting to calculate the fine-structure constant. (a) physics nonsense: a physics-based discussion that makes no sense, but nevertheless yields a numerical expression for α. (b) numerical coincidence (numerology): a number is found that closely approximates the experimental value for α.

In the first category, the most famous physics nonsense determination of α was by Eddington in about 1930. At that time there were only two known "elementary" particles, the electron and the proton. Dirac had incorrectly assumed that a proton is a hole in the, generally filled, sea of negative energy states of the electron. Eddington started with Dirac's theory that uses a 4×4 matrix to describe an electron. Eddington also used a 4×4 matrix to describe the proton. To describe both particles requires a 16×16 matrix, which, if symmetric, would have 136 independent elements. Later, when $1/\alpha$ was found to be closer to 137, Eddington stated that the orbital motion of the electron about the proton added one more degree of freedom, bringing the total to 137. The physics itself is incorrect, and even if it were correct, there is no reason that the number of degrees of freedom should equal the reciprocal of α. A summary of Eddington's approach was given by Birge[33] in 1934 in which the value Eddington calculates for α is characterized as simply being a result of numerology.

Hans Bethe wrote,"Beck, Riesler and I were at Cambridge on fellowships and had listened to Eddington's unbelievable talk about the number 137". To make fun of Eddington's "derivation", they submitted the article *On the Quantum theory of the Temperature of Absolute Zero*[34]. They had noticed the numerical coincidence that $-273 = -(2 \times 137 - 1)$ so they proposed the following equation relating the temperature T_0 of absolute zero to α:

$$T_0 = -(\frac{2}{\alpha} - 1) \text{ degrees Celsius}. \tag{5}$$

They then wrote,"Putting $T_0 = -273°$, we obtain for $1/\alpha$ the value 137 in perfect agreement within the limits of accuracy with the value obtained by totally independent methods." Of course, the value of absolute zero depends on the temperature scale while α is independent of units. The editor of Die Naturwissenschaften approved the article, and it was published in January, 1931. In March, 1931, a "Correction" was published in Die Naturwissenschaften stating that the article "was intended to characterize a certain class of papers in theoretical physics of recent years which are purely speculative and based on spurious numerical agreements."

Apparently neither the criticism by Bethe et al. nor Birge had a significant affect on Eddington's self-assurance that he had theoretically calculated the value of α. Eddington incorporated the above argument into his 1936 book [35] and continued to "clarify" his method for calculating his value for α in his final book[36] published after his death in 1944. In 1994 Kilmister[37] published an analysis of Eddington's work, noting that Eddington always used the term *fine-structure constant* for $1/\alpha$, rather than for α.

The second group of calculations rely on a numerical coincidence. For example, Allen[38] noted that the mass m_e of the electron divided by 1 atomic mass unit (u) satisfies

$$\frac{m_e}{u} \cong 10\alpha^2 . \quad (6)$$

Other examples consist of formulas involving integers, π, square roots, cube roots, etc. that yield a number close to the experimental value for α. Such an example by Wyler[39], which is associated with the symmetric space of the group of Maxwell's equations, is very close to the accepted value, $1/\alpha = 137.03599\ldots$.

$$\frac{1}{\alpha} \cong \frac{16\pi^3}{9}\sqrt[4]{\frac{5!}{\pi}} \cong 137.03608 . \quad (7)$$

So many examples similar to Equation (7) are now known that it is likely that no physical insight into the value of α is to be gained from them.

3. Vacuum fluctuations of charged lepton-antilepton pairs

The appearance of VFs is a stochastic process: as such, either VFs appear on mass shell or they don't appear at all. As discussed previously, a charged lepton-antilepton VF that results from a fluctuation of the Dirac field will appear in the vacuum as a transient atom. During the time while such an atom exists, it can interact with a photon. The fundamental interaction of a photon with a VF is the capture of the photon by the charged lepton-antilepton VF.

Field theory provides a simple explanation for the source of the energy available for the creation of a VF and a proof that VFs must exist. The structure of an atom that is a VF does not play a significant role in the discussion of the energy required to produce the VF. (Of course, the structure of the VF, which results from the electromagnetic interaction of the charged lepton-antilepton pair, is important for the calculation of the decay rate of the atom that is a VF.) Thus an atom consisting of a charged lepton and antilepton in its ground state with zero angular momentum can, as far as its creation is concerned, be approximately represented by a free, neutral, spin-0, Klein-Gordon field $\phi(x)$, as first suggested by Pauli and Weisskopf[8] and elaborated on by Wentzel[40]. Using a field to describe a particle with internal degrees of freedom is discussed in Ref.[2]; representing a particle by a quantum field is essential to understanding the mathematical structure of a VF of a quantum field. Indeed, it is the quanta of a free field that behave as free particles[4].

The energy available for the creation of VFs is the energy in the vacuum, called the zero-point energy, and is given by the vacuum expectation value $\langle 0|H|0\rangle$ of the Hamiltonian H for a free, neutral Klein-Gordon field[5]. If there is no cutoff in momentum, the vacuum expectation $\langle 0|H|0\rangle$ is infinite; if there is a cutoff, the vacuum energy is finite, but still large. As Peskin and Schroeder point out, since experiments measure only energy differences, the energy associated with VFs cannot be measured directly in elementary particle experiments[5].

The *presence* of Klein-Gordon VFs, however, is easily established. To show that vacuum fluctuations of charged lepton-antilepton pairs must exist, first note that for a free field $\phi(x)$, the vacuum expectation value $\langle 0|\phi(x)|0\rangle = 0$. In contrast, the expectation value of the product of the free field at two different locations x and x' is[2, 3] $\langle 0|\phi(x)\phi(x')|0\rangle \neq 0$. Accordingly, the vacuum expectation of the square of the field $\phi(x)$ deviates from the square of the vacuum expectation value of the field $\phi(x)$. This demonstrates that the free field $\phi(x)$ in the vacuum is nonzero and means that field theory predicts that fluctuations of a free (noninteracting) field occur in the vacuum.

The purpose of the present paper is to demonstrate a measurable effect of the presence of charged lepton-antilepton VFs, namely, the importance of these VFs in establishing the value of ϵ_0. This calculation does not depend on an absolute value for the zero-point energy. As Peskin and Schroeder point out, a potential method for establishing (measuring) an absolute value for the zero-point energy in the vacuum lies in understanding the coupling of that energy to gravity via the cosmological constant in Einsteins equations. Such a direct comparison seems at present unlikely, since the calculation and measurement appear to differ by a factor of around 10^{120}[5].

4. Heisenberg uncertainty principle and charged lepton-antilepton vacuum fluctuations

There are three types of charged lepton-antilepton vacuum fluctuations: (a) electron-positron (b) muon-antimuon (c) tau-antitau. The electron-positron VF in the lowest energy level that has zero angular momentum is parapositronium, which is a singlet spin state[11, 41, 42]. Initially attention will be restricted to parapositronium since the corresponding calculations for muon-antimuon and tau-antitau VFs can be obtained by replacing the mass of the electron with the mass of the muon or tau, respectively.

For parapositronium that is a VF, the Heisenberg uncertainty principle is

$$\Delta E_{\mathrm{p-Ps}}\,\Delta t_{\mathrm{p-Ps}} \geq \frac{\hbar}{2}\,. \tag{8}$$

Denoting the mass of an electron (or positron) by m_e, $\Delta E_{\mathrm{p-Ps}}$ is the energy $2m_ec^2$ for the production of parapositronium that is a VF [2]. The minimum uncertainty in time is the average time $\Delta t_{\mathrm{p-Ps}}$ that a parapositronium VF exists. Then Equation (8) yields

$$\Delta t_{\mathrm{p-Ps}} = \frac{\hbar}{4m_ec^2}\,. \tag{9}$$

During the time $\Delta t_{\mathrm{p-Ps}}$, a beam of light travels a distance $L_{\mathrm{p-Ps}}$ given by

$$L_{\mathrm{p-Ps}} = c\,\Delta t_{\mathrm{p-Ps}} = \frac{\hbar}{4m_ec}\,. \tag{10}$$

Since a parapositronium VF appears from the vacuum at essentially a single location and since nothing can travel faster than the speed of light, while they exist the maximum displacement between the electron and positron in parapositronium is $L_{\mathrm{p-Ps}}$: for a time $\Delta t_{\mathrm{p-Ps}}/2$ the electron and positron can move apart, and for a time $\Delta t_{\mathrm{p-Ps}}/2$ they must move back toward each other in order to annihilate on average at the time $\Delta t_{\mathrm{p-Ps}}$. Since an energy $2m_ec^2$ has already been borrowed from the volume $L^3_{\mathrm{p-Ps}}$ of the vacuum for the creation of a parapositronium atom, it is reasonable to assume that another parapositronium VF is unlikely to form in the same volume. Thus for the model being discussed,

$$\text{number of parapositronium atoms/volume} = \frac{1}{L^3_{\mathrm{p-Ps}}}\,. \tag{11}$$

The result Equation (11) can immediately be generalized to other charged lepton-antilepton VFs and quark-antiquark VFs. The number density of charged lepton-antilepton VFs ranges from $1.12 \times 10^{39}/\mathrm{m}^3$ for electron-positron VFs to $4.70 \times 10^{49}/\mathrm{m}^3$ for tau-anti-tau VFs. In a 6,000 W, CO_2 cutting laser with a beam diameter of 0.32 mm, the number density of photons is on the order of 10^{22} photons/m^3. Even in such an intense laser beam, the number density of charged lepton-antilepton VFs is much greater than the number density of photons. Thus if a charged lepton-antilepton VF interacts with a photon at all, it essentially always interacts with only one photon.

[2] The binding energy of parapositronium, which is small in comparison with $2m_ec^2$, is being neglected.

The number density of atoms or molecules of an ideal gas at STP is $2.68 \times 10^{25}/\mathrm{m}^3$. It is possible for the number density of VFs to be many orders of magnitude greater than the number density of atoms or molecules in a gas because a VF cannot exert a force. Consider first the electromagnetic force: if a VF has not already absorbed radiation, it cannot spontaneously emit radiation. If it did, the radiated photons would exist after the VF has disappeared back into the vacuum, permanently violating conservation of energy. If a VF has interacted with a photon, when the VF vanishes back into the vacuum the VF must emit a photon identical to the incident photon in order to conserve energy, momentum, and angular momentum. Since a VF cannot "permanently" exchange a photon with either a VF or a physical quantum, it cannot exert a force on either. Similar arguments establish that VFs cannot exert a force of any type.

In addition to absorbing and emitting photons, VFs can also interact through the annihilation of a physical particle and a VF or a physical antiparticle and a VF. For example, a physical electron can annihilate with a positron that is part of a VF. The electron that was also part of the VF then becomes a physical electron with a location different from that of the original, physical electron, giving rise to zitterbewegung[2].

5. Charged lepton-antilepton vacuum fluctuations as harmonic oscillators

Here attention is again restricted to parapositronium. In the center-of-mass rest frame, the relative position of the positron and electron are given by $\vec{r} = \vec{r}_+ - \vec{r}_-$, where $\vec{r}_+$ and $\vec{r}_-$ are, respectively, the positions of the positron and electron. Note that $\vec{r}$ points in the direction of the electric dipole moment of parapositronium. For the hydrogen atom $\vec{r}$ usually points from the positive nucleus to the electron. However, the Schrödinger equation for parapositronium, which in the center-of-mass reference frame has a spherically symmetric wave function, is identical to the Schrödinger equation describing states of the hydrogen atom with spherically symmetric wave functions except that the reduced mass μ is different: for hydrogen the reduced mass is approximately m_e, where m_e is the mass of an electron. For parapositronium the reduced mass is $m_e/2$. The non-relativistic binding energy for parapositronium, E_{p-Ps}, is obtained from the $n = 1$ binding energy of hydrogen[43] just by changing the reduced mass:

$$E_{p-Ps} = -\frac{(m_e/2)e^4}{2(4\pi\epsilon_0)^2\hbar^2} = -\frac{m_e\alpha^2c^2}{4}\,. \tag{12}$$

From both the classical[1, 16, 17] and quantum[18] calculations of the permittivity of physical matter consisting of atoms (or molecules), it follows that for the matter to possess permittivity, the atoms must be able to oscillate when interacting with an electric field (or photons). Taking the electric field to point in the x-direction, since the atom oscillates along the direction of the electric field, only the oscillatory properties of the atom in the x-direction are of significance in the interaction. Thus, when calculating the permittivity, the interaction of the electromagnetic field with the atom can be described by a one-dimensional potential.

Similarly to many other systems in physics that can be described by a one-dimensional potential $U(x)$, the parapositronium atom can be expected to oscillate if the potential has a minimum at x_e and can be expanded in a Taylor series about that minimum:

$$U(x) = U(x_e) + \left.\frac{\mathrm{d}U(x)}{\mathrm{d}x}\right|_{x=x_e}(x - x_e) + \frac{1}{2!}\left.\frac{\mathrm{d}^2U(x)}{\mathrm{d}x^2}\right|_{x=x_e}(x - x_e)^2 + \ldots. \tag{13a}$$

In the above formula $x = x_+ - x_-$. At x_e there is a relative minimum of the potential so $\frac{\mathrm{d}U(x)}{\mathrm{d}x}$ is zero at x_e. Choosing the origin of the x-axis at the equilibrium position x_e so that $x_e = 0$, Equation (13a) can be rewritten as

$$U(x) \cong U(x_e = 0) + \frac{1}{2}Kx^2\,. \tag{13b}$$

Equation (13b) is the equation for a one-dimensional harmonic oscillator potential with a spring constant $K = \frac{d^2U(x)}{dx^2}\big|_{x=x_e}$. The constant term $U(x_e = 0)$ just shifts all energy levels by the same amount.

As pointed out by Feynman[44], when interacting with an electric field, an atom in its ground state interacts with the electric field as if it were a harmonic oscillator with the first two energy levels separated by the binding energy of the atom. Since adjacent harmonic oscillator energy levels are separated by an energy $\hbar\omega^0$, from the expression Equation (12) for the binding energy E_{p-Ps} of parapositronium,

$$\omega^0 = \frac{|E_{p-Ps}|}{\hbar} = \frac{m_e\alpha^2c^2}{4\hbar} . \tag{14}$$

The spring constant K of the harmonic oscillator corresponding to parapositronium is

$$K = \mu(\omega^0)^2 = \frac{m_e}{2}\left(\frac{m_e\alpha^2c^2}{4\hbar}\right)^2 . \tag{15}$$

Using the formula for the energy[43] of a harmonic oscillator in one dimension, $E = \hbar\omega^0(n + 1/2)$, $n = 0, 1, 2 \ldots$, the energy eigenvalues E_n^0 of parapositronium are

$$E_n^0 = U(x_e = 0) + \hbar\omega^0(n + \frac{1}{2}) . \tag{16}$$

The harmonic oscillator matrix element that appears in the calculation of the permittivity does not depend on the constant term $U(x_e = 0)$ in the potential, so there is no need to determine a specific value.

In the center-of-mass rest frame of the parapositronium VF, the parapositronium VF is described by the harmonic oscillator Hamiltonian H^0,

$$H^0 = \frac{1}{2}\mu\left(\frac{dx}{dt}\right)^2 + U(x_e = 0) + \frac{1}{2}\mu(\omega^0)^2x^2 , \tag{17}$$

where ω^0 is defined in Equation (14).

6. Calculation of the permittivity of the vacuum

6.1. Interaction Hamiltonian for photons interacting with vacuum fluctuations

The one-dimensional harmonic oscillator described by the Hamiltonian Equation (17) can undergo an electric dipole transition between its ground state and first excited state because the initial and final states have opposite parity. The interaction Hamiltonian describing the dipole interaction of a photon with a VF follows:

The dipole moment (operator) p_x of an atom in the presence an electromagnetic wave with its electric field in the x-direction is $p_x = e(x_+ - x_-) = ex$, where e is the magnitude of the charge on an an electron. The Hamiltonian H^1 describing the interaction of the electric dipole of the atom with the electromagnetic wave is

$$H^1 = -\mathbf{p} \cdot \mathbf{E(t)} = -exE_0 \cos\omega t . \tag{18}$$

The two Hamiltonian equations for the Hamiltonian $H = H^0 + H^1$ can be combined to yield

$$-Kx + eE_0 \cos\omega t = \mu\frac{d^2x(t)}{dt^2} , \tag{19}$$

which is just Newton's second law. The classical calculation of the permittivity of ordinary matter is based on Equation (19) with one exception: for ordinary matter there is also a phenomenological term proportional to velocity that describes damping that results from radiation and collisions. For gases damping is often very small and can be neglected. While neglecting damping is an approximation for physical particles in a dielectric, it is exactly true for VFs. VFs can neither radiate energy nor lose energy in collisions with other quanta. If they did, after they vanished they would permanently leave behind energy, violating the principle of conservation of energy.

To describe the interaction of VFs with an electric field, one change has to be made to H^1. If an atom is continually interacting with photons from an electric field $E_0 \cos \omega t\, \hat{\mathbf{x}}$, then the interaction H^1 of the atom with the field is given by Equation (18). As discussed in §4, even in an intense laser beam the number density of photons is much less than the number density of VFs. Thus the probability that at a given time a VF interacts with more than one photon is very small. That is, if a VF interacts with an electromagnetic wave, it almost always does so with only one photon.

If a VF absorbs a photon at time t_i, the photon vanishes at that instant. The electric field at the moment of interaction is $E(t_i) = E_0 \cos \omega t_i \equiv \mathbb{E}_0$, implying that the VF interacts with the electric field $\mathbb{E}_0\, \hat{\mathbf{x}}$. Thus, for VFs Equation (18) becomes

$$H^{1^{VF}} = -exE_0 \cos \omega t_i \equiv -ex\mathbb{E}_0\,. \tag{20}$$

As a part of the capture process, a parapositronium atom will, to some extent, be excited by $\mathbb{E}_0$ to resonate at its characteristic frequency ω^0, as discussed in the following section.

6.2. Quantum calculation of the polarization of vacuum fluctuations

Because photons interact with atoms that are VFs somewhat similarly to the the way that they interact with ordinary atoms, it is possible to calculate the permittivity ϵ_0 of the vacuum using techniques similar to those employed for calculating the permittivity ϵ of a physical dielectric. In the vacuum the interactions that primarily contribute to the value of ϵ_0 are photon capture by charged lepton-antilepton VFs bound into the lowest energy state that has zero angular momentum. The charged lepton-antilepton pair quickly annihilate, emitting a photon identical to the incident, captured photon.

Let $\psi_n^0(x,t)$ be solutions to the unperturbed Schrödinger equation,

$$i\hbar \frac{\partial \psi_n^0(x,t)}{\partial t} = H^0 \psi_n^0(x,t) = E_n^0 \psi_n^0(x,t)\,, \tag{21}$$

where E_n^0 and H^0 are given in Equation (16) and Equation (17), respectively. From Equation (21) it follows that the energy dependence of $\psi_n^0(x,t)$ can be factored,

$$\psi_n^0(x,t) = e^{-E_n^0 t/\hbar} \psi_n^0(x)\,. \tag{22}$$

The "exact" wave function $\psi(x,t)$ satisfies the Schrödinger equation,

$$i\hbar \frac{\partial \psi(x,t)}{\partial t} = (H^0 + H^{1^{VF}})\, \psi(x,t)\,. \tag{23}$$

Perturbation theory is now used to calculate $\psi(x,t)$ to first order in the perturbation $H^{1^{VF}}$: the "exact" wave function $\psi(x,t)$ is written as

$$\begin{aligned} \psi(x,t) &= a_0(t)\psi_0^0(x,t) + a_1(t)\psi_1^0(x,t)\,, \\ &= a_0(t)e^{-iE_0^0 t/\hbar}\psi_0^0(x) + a_1(t)e^{-iE_1^0 t/\hbar}\psi_1^0(x)\,. \end{aligned} \tag{24}$$

Substituting Equation (24) into Equation (23),

$$i\hbar \frac{\mathrm{d}a_0(\mathrm{t})}{\mathrm{dt}} e^{-iE_0^0 t/\hbar} \psi_0^0(x) + i\hbar \frac{\mathrm{d}a_1(\mathrm{t})}{\mathrm{dt}} e^{-iE_1^0 t/\hbar} \psi_1^0(x) \\ = H^{1^{VF}} a_0(t) e^{-iE_0^0 t/\hbar} \psi_0^0(x) + H^{1^{VF}} a_1(t) e^{-iE_1^0 t/\hbar} \psi_1^0(x) \,. \tag{25}$$

Using the orthogonality relations,

$$\int_{-\infty}^{\infty} \mathrm{d}x \, \psi_{n'}^{0*}(x) \, \psi_n^0(x) = \delta_{n',n} \,, \tag{26}$$

and noting that $\psi_0^0(x)$ and $\psi_1^0(x)$ have opposite parity while $H^{1^{VF}} \equiv -ex\mathbb{E}_0$ has negative parity, equations for $\frac{\mathrm{d}a_0(\mathrm{t})}{\mathrm{dt}}$ and $\frac{\mathrm{d}a_1(\mathrm{t})}{\mathrm{dt}}$ are immediately obtained from Equation (25):

$$\frac{\mathrm{d}a_0(\mathrm{t})}{\mathrm{dt}} = a_1(t) \frac{ie\mathbb{E}_0}{\hbar} \langle x \rangle_{0,1} e^{-i\omega^0 t} \,, \tag{27a}$$

$$\frac{\mathrm{d}a_1(\mathrm{t})}{\mathrm{dt}} = a_0(t) \frac{ie\mathbb{E}_0}{\hbar} \langle x \rangle_{1,0} e^{i\omega^0 t} \,. \tag{27b}$$

In the above equation

$$\langle x \rangle_{n',n} \equiv \int_{-\infty}^{\infty} \mathrm{d}x \, \psi_{n'}^{0*}(x) \, x \, \psi_n^0(x) \tag{28}$$

and, as discussed in §5,

$$\omega^0 \equiv \frac{1}{\hbar}(E_1^0 - E_0^0) \,. \tag{29}$$

Letting the time t_i that the photon interacts with the parapositronium VF be infinitesimally greater than $t = 0$ implies that

$$a_0(t=0) = 1 \text{ and } a_1(t=0) = 0 \,, \tag{30}$$

because the parapositronium VF is initially in its ground state, which, when interacting with a photon, is represented by the ground state $\psi_0^0(x,t)$ of the harmonic oscillator. Using Equation (30), Equation (27b) becomes

$$\frac{\mathrm{d}a_1(\mathrm{t})}{\mathrm{dt}} = \frac{ie\mathbb{E}_0}{\hbar} \langle x \rangle_{1,0} e^{i\omega^0 t} \,, \tag{31}$$

which can immediately be integrated to yield

$$a_1(t) = \frac{e\mathbb{E}_0}{\hbar\omega^0} \langle x \rangle_{1,0} e^{i\omega^0 t} \,. \tag{32}$$

Note that Equation (32) satisfies Equation (30) because the interaction has not yet occurred at time $t_i = 0$. If the formula for $a_1(t)$ in Equation (32) is substituted into Equation (27a), a correction to $a_0(t)$ is obtained that is on the order of $(e\mathbb{E}_0)^2$. Thus from Equation (30) and Equation (32), to order $e\mathbb{E}_0$ the "exact" wave function $\psi(x,t)$ is given by

$$\psi(x,t) = e^{-iE_0^0 t/\hbar} \psi_0^0(x) + \frac{e\mathbb{E}_0}{\hbar\omega^0} \langle x \rangle_{1,0} e^{i\omega^0 t} e^{-iE_1^0 t/\hbar} \psi_1^0(x) \,, \\ = e^{-iE_0^0 t/\hbar} [\psi_0^0(x) + \frac{e\mathbb{E}_0}{\hbar\omega^0} \langle x \rangle_{1,0} \psi_1^0(x)] \,. \tag{33}$$

From Equation (18) and Equation (20) the expectation value $\langle p^{VF}\rangle$ of the electric dipole moment in the state characterized by $\psi(x,t)$ is

$$\begin{aligned}
\langle p^{VF}\rangle &= \int_{-\infty}^{\infty} dx\, \psi^*(x,t)\,(ex)\,\psi(x,t)\,, \\
&= \int_{-\infty}^{\infty} dx\, e^{iE_0^0 t/\hbar}[\psi_0^{0*}(x) + \frac{e\mathbb{E}_0}{\hbar\omega^0}\langle x\rangle_{1,0}\psi_1^{0*}(x)](ex)e^{-iE_0^0 t/\hbar}[\psi_0^0(x) + \frac{e\mathbb{E}_0}{\hbar\omega^0}\langle x\rangle_{1,0}\psi_1^0(x)]\,, \\
&\cong \int_{-\infty}^{\infty} dx\,[\psi_0^{0*}(x)(ex)\frac{e\mathbb{E}_0}{\hbar\omega^0}\langle x\rangle_{1,0}\psi_1^0(x) + \frac{e\mathbb{E}_0}{\hbar\omega^0}\langle x\rangle_{1,0}\psi_1^{0*}(x)(ex)\psi_0^0(x)]\,, \\
&= \frac{2e^2\mathbb{E}_0}{\hbar\omega^0}\langle x\rangle_{1,0}^2\,.
\end{aligned} \tag{34}$$

To obtain the final equality, phases were chosen so that $\psi_0^0(x)$ and $\psi_1^0(x)$ are both real. Using the one-dimensional harmonic oscillator result,

$$\langle x\rangle_{1,0} = \sqrt{\frac{\hbar}{2\mu\omega^0}}\,, \tag{35}$$

it immediately follows from Equation (34) that

$$\langle p^{VF}\rangle_{0,0} = \frac{(e^2/\mu)\mathbb{E}_0}{(\omega^0)^2}\,. \tag{36}$$

Describing an oscillator as a harmonic oscillator, the quantum formula for the expectation value of the electric dipole of a VF agrees with the classical formula for the electric dipole of an atom of ordinary matter with two exceptions that have already been discussed: (1) Since VFs can neither radiate energy nor loose energy in a collision, there is no damping term. (2) Because VFs essentially always only interact with a single photon, only the value of the electric field at the instant of interaction is relevant so there is no dependence on the angular frequency ω of the incident photon.

To allow for the possibility that there is more than one type of atom made from particle-antiparticle VFs, an index j is added: accordingly, the charge $e \to q_j$, the reduced mass $\mu \to \mu_j$, and the resonant frequency $\omega^0 \to \omega_j^0$.

$$\langle p_j^{VF}\rangle_{0,0} = \frac{(q_j^2/\mu_j)\mathbb{E}_0}{(\omega_j^0)^2}\,. \tag{37}$$

6.3. General formula for the permittivity ϵ_0 of the vacuum

In a dielectric[45] the electric displacement $D(t)$ satisfies

$$D(t) = \epsilon E(t) = \epsilon_0 E(t) + P(t)\,, \tag{38}$$

where ϵ is the permittivity of the dielectric, $E(t)$ is the electric field, and $P(t)$ is the electric polarization (dipole moment per unit volume). $P(t)$ can be expressed in terms of the individual dipole moments $p_j(t)$:

$$P(t) = \sum_j N_j p_j(t)\,. \tag{39}$$

In Equation (39) N_j is the number of oscillators per unit volume of the jth variety that are available to interact. In a uniform, classical electric field, the field is everywhere $E_0 \cos\omega t$.

IOP Conf. Series: Journal of Physics: Conf. Series **1239** (2019) 012016 doi:10.1088/1742-6596/1239/1/012016

Since the electric polarization is proportional to the electric field, the electric field cancels out of Equation (39).

The electric polarization $P(t)$ is responsible for the increase from $\epsilon_0 E(t)$ to $\epsilon E(t)$ because of photons interacting with oscillators in the material dielectric and results entirely from polarization of the atoms, molecules or both in the dielectric. It then follows that in the vacuum $\epsilon_0 E(t)$ must result entirely from the electric polarization $P^{VF}(t)$ of atoms, molecules, or both that are VFs. Thus,

$$\epsilon_0 E(t) = P^{VF}(t)\,, \tag{40a}$$

or

$$\epsilon_0 = \frac{P^{VF}(t)}{E(t)}\,. \tag{40b}$$

The measurement of the permittivity ϵ_0 of the vacuum occurs over a time interval Δt. For any time t_i in the interval Δt for which a photon-VF interaction occurs, the electric field at that instant is $E(t_i) = E_0 \cos \omega t_i \equiv \mathbb{E}_0$. As shown in Equation (37), the polarization density is proportional to $\mathbb{E}_0$, implying that the instantaneous value of the electric field will cancel out in Equation (40). From Equation (40b), Equation (39) and Equation (37)

$$\epsilon_0 = \sum_j N_j^{VF} \frac{p_j^{VF}(t_i)}{\mathbb{E}_0} = \sum_j N_j^{VF} \frac{\langle p_j^{VF} \rangle_{0,0}}{\mathbb{E}_0} = \sum_j N_j^{VF} \frac{(q_j^2/\mu_j)}{(\omega_j^0)^2}\,. \tag{41}$$

Since the value of the electric field has canceled out of Equation (41), ϵ_0 is independent of the polarizing electric field, including its frequency. As a result, there is no dispersion in the vacuum.

Contributions from various types of VFs must now be summed over in Equation (41) to obtain the formula for ϵ_0. The three types of VFs first considered are atomic, bound states of a charged lepton and antilepton, namely, parapositronium, muon-antimuon bound states, and tau-antitau bound states. Quark-antiquark states will be discussed later. Initially attention is restricted to a parapositronium VF.

6.4. Contribution to ϵ_0 from charged lepton-antilepton vacuum fluctuations

The progress of a photon traveling through the vacuum is slowed when it interacts with and has a polarizing effect on a VF consisting of a charged lepton and antilepton bound into an atom in its ground state. The fundamental interaction of a photon with a VF is the capture of the photon by the charged lepton-antilepton VF. Labeling the initial (incident) and final (emitted) photons, respectively, by γ_i and γ_f, to lowest order the two Feynman diagrams that contribute to the process γ_i+charged lepton-antilepton VF $\rightarrow \gamma_f$ are shown in Fig. 1. In the diagrams p_-, p_+, k_i, and k_f are, respectively, the four-momenta of the lepton, antilepton, initial photon, and final photon.

A quantum electrodynamics calculation[19] determines the decay rate Γ of this quasi-stationary atom. Using Γ, the effect of this interaction on the permittivity of the vacuum can be calculated. Coherence between the initial and final state is maintained. As required by conservation of energy and momentum, when the photon-excited, quasi-stationary state annihilates, the energy and momentum that were originally borrowed from the vacuum are returned to the vacuum. An isolated, ordinary atom consisting of a charged lepton-antilepton pair is kinematically forbidden from capturing a photon, annihilating, and then emitting a single photon; however, because the kinematics is different for the capture and release of a photon by a VF, the process is kinematically allowed for charged lepton-antilepton VFs.

The electromagnetic decay rate $\Gamma_{\mathrm{p-Ps}}$ for a parapositronium VF after it has interacted with the incident photon to form a quasi-stationary state is[19]

$$\Gamma_{\mathrm{p-Ps}} = \frac{\alpha^5 m_e c^2}{\hbar}\,. \tag{42}$$

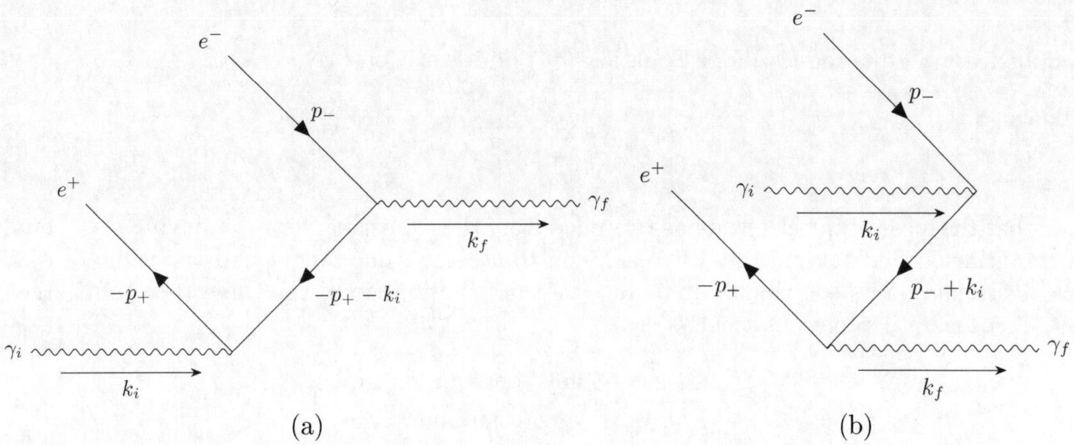

Figure 1. (a) Photon γ_i interacts with an antilepton that then annihilates with an lepton, emitting photon γ_f. (b) Photon γ_i interacts with a lepton that then annihilates with an antilepton, emitting photon γ_f.

The above rate is twice the decay rate of ordinary parapositronium into two photons[10, 46].

The probability that an excited parapositronium VF has not decayed during a time t is $e^{-\Gamma_{\text{p-Ps}}t}$, and the probability that it has decayed electromagnetically is $1 - e^{-\Gamma_{\text{p-Ps}}t}$. The quantity N_j^{VF} in Equation (41) for a parapositronium VF is the number of parapositronium VFs per unit volume with which a photon actually interacts. At equilibrium the average rate at which a parapositronium VF absorbs a photon equals the average rate for a parapositronium VF to annihilate and emit a photon. As a consequence, the average probability that parapositronium absorbs a photon during a time t is $1 - e^{-\Gamma_{\text{p-Ps}}t}$.

For a parapositronium VF the quantity N_j^{VF}, denoted by $N_{\text{p-Ps}}$, is the number density of parapositronium VFs multiplied by the probability that a parapositronium VF will absorb an incoming photon during the lifetime $\Delta t_{\text{p-Ps}}$ of the parapositronium VF:

$$N_{\text{p-Ps}}^{VF} \cong \frac{1}{L_{\text{p-Ps}}^3} \times (1 - e^{-\Gamma_{\text{p-Ps}}\Delta t_{\text{p-Ps}}}) \,. \tag{43}$$

Since $\Gamma_{\text{p-Ps}}\Delta t_{\text{p-Ps}} \ll 1$, the term $1 - e^{-\Gamma_{\text{p-Ps}}\Delta t_{\text{p-Ps}}}$ is very nearly equal to $\Gamma_{\text{p-Ps}}\Delta t_{\text{p-Ps}}$. Thus

$$N_{\text{p-Ps}}^{VF} \cong \frac{1}{L_{\text{p-Ps}}^3} \times \Gamma_{\text{p-Ps}}\Delta t_{\text{p-Ps}} = \frac{\alpha^5}{4}\left(\frac{4m_e c}{\hbar}\right)^3 . \tag{44}$$

To better understand Equation (44), for a particular atom (or molecule) let N be the number of atoms per unit volume, let Γ be the decay rate of a photon-excited atom into the atom in its ground state plus a photon, and let t be the average time that the atom exists. For ordinary matter N_j in Equation (39) is N, but for a vacuum fluctuation it is $N\Gamma t$ as given in Equation (44). How does this difference arise? Making no assumptions about the magnitude of Γt and using the logic that led to Equation (43), $N_j = N(1 - e^{-\Gamma t})$. If the atom is stable, the lifetime t is infinite so Γt is infinite, and $N_j = N$ as expected. On the other hand, if the lifetime is sufficiently small that $\Gamma t \ll 1$, as is the case for a parapostronium VF, $N_j^{VF} \cong N\Gamma t$ as given in Equation (44) [3].

[3] As pointed out by A. Bohm[47], a physical state prepared in a scattering experiment contains background information about the reaction in which it was created. The above description of a single photon interacting with

IOP Conf. Series: Journal of Physics: Conf. Series **1239** (2019) 012016 doi:10.1088/1742-6596/1239/1/012016

Substituting Equation (14) and Equation (44) into Equation (41),

$$\epsilon_0 \cong \sum_j \frac{8^3 \alpha e^2}{\hbar c}\,. \tag{45}$$

Note that the mass of the electron has cancelled from the expression for ϵ_0, implying that bound muon-antimuon and tau-antitau VFs each contribute the same amount to the value of ϵ_0 as parapositronium VFs contribute. Including the contributions from the three types of charged, bound lepton-antilepton VFs yields

$$\epsilon_0 \cong 3\,\frac{8^3 \alpha e^2}{\hbar c} + \begin{pmatrix}\text{any contribution from}\\ \text{quark-antiquark VFs.}\end{pmatrix} \tag{46}$$

6.5. Contribution to ϵ_0 from quark-antiquark vacuum fluctuations
The only other charged particle-antiparticle VFs that might exhibit elastic behavior are quark-antiquark VFs; however, as will now be shown, contributions to ϵ_0 from quark-antiquark VFs is substantially reduced in comparison with those from charged lepton-antilepton VFs.

First consider the heavy quarks $Q = c, b$, or t, where it is appropriate to think in terms of static quark potentials for $Q\bar{Q}$ bound states. Let $m_{Q\bar{Q}}$ be the mass of the least massive $Q\bar{Q}$ bound state that has zero angular momentum. From the Heisenberg uncertainty principle, the $Q\bar{Q}$ bound state will have an average lifetime $\Delta t_{Q\bar{Q}} = \hbar/(2M_{Q\bar{Q}}c^2)$. During the time $\Delta t_{Q\bar{Q}}$ light will travel a distance $L_{Q\bar{Q}} = c\Delta t_{Q\bar{Q}}$. As previously noted, the decay rate of a photon-excited parapositronium VF into a photon is twice the decay rate of ordinary parapositronium into two photons. Since the decay rate of a photon-excited $Q\bar{Q}$ VF into a photon is not known, it is approximated by twice the decay rate $\Gamma_{Q\bar{Q}\to\gamma+\gamma}$ of an ordinary $Q\bar{Q}$ bound state into two photons. From Equation (44) it then follows that

$$N^{VF}_{Q\bar{Q}} \sim \frac{1}{L^3_{Q\bar{Q}}} \times 2\Gamma_{Q\bar{Q}\to\gamma+\gamma}\Delta t_{Q\bar{Q}} \sim 8c\left(\frac{M_{Q\bar{Q}}}{\hbar}\right)^2 \Gamma_{Q\bar{Q}\to\gamma+\gamma}\,. \tag{47}$$

Letting m_Q and q_Q denote, respectively, the mass and charge of the heavy quark Q and letting $\omega^0_{Q\bar{Q}}$ denote the resonant angular frequency of the $Q\bar{Q}$ VF, from Equation (41) it the follows that the contribution of the $Q\bar{Q}$ VF to ϵ_0 is

$$\epsilon_{0\begin{pmatrix}\text{contribution}\\ \text{from } Q\bar{Q}\text{ VF}\end{pmatrix}} \sim 8c\left(\frac{M_{Q\bar{Q}}}{\hbar}\right)^2 \Gamma_{Q\bar{Q}\to\gamma+\gamma}\,\frac{q_Q^2/(m_Q/2)}{(\omega^0_{Q\bar{Q}})^2}\,. \tag{48}$$

The least massive $c\bar{c}$ bound state, $\eta_c(1S)$, has $J = 0$ and has positive charge conjugation parity[48], implying that a photon excited $\eta_c(1S)$ VF can decay into a single photon. To obtain an order-of-magnitude estimate of the maximum contribution that an $\eta_c(1S)$ VF could make to ϵ_0, the experimental decay rate $\Gamma_{\eta_c(1S)\to\gamma+\gamma}$ of $\eta_c(1S)$ into two photons is 7.69×10^{18}/s[48]. It is not obvious which energy should be used to calculate the angular frequency $\omega^0_{Q\bar{Q}}$, but the minimum possible energy $E_{\min}$ yields a minimum possible value for $\omega^0_{Q\bar{Q}}$ and a maximum possible value for the contribution of an $\eta_c(1S)$ VFs to ϵ_0. $E_{\min}$ is the difference between the mass $m_{\eta_c(1S)}$ of $\eta_c(1S)$ and the masses m_c of the charm and anti-charm quarks when they are

a VF clearly depicts such a circumstance despite the fact that the details of the discussion involve a decay rate Γ. Usually, a decay rate such as Γ_{p-Ps} is associated only with a stochastic process. In each individual interaction of a photon with a VF, the coherence of the final photon with the initial photon is assured, as promised at the beginning of §6.4.

IOP Conf. Series: Journal of Physics: Conf. Series **1239** (2019) 012016 doi:10.1088/1742-6596/1239/1/012016

weakly bound: $E_{\rm min} = m_{\eta_c(1S)} - 2m_{\rm c} \sim 2.98\,{\rm GeV} - 2\times 1.27\,{\rm GeV} = 0.44$ GeV[48]. From Equation (48) the maximum possible contribution to ϵ_0 from $\eta_c(1S)$ VFs is then calculated to be

$$\epsilon_0 \begin{pmatrix} \text{maximum possible} \\ \text{contribution from } \eta_c(1S) \text{ VFs} \end{pmatrix} \sim 1.3 \times 10^{-3} \frac{e^2}{\hbar c}. \tag{49}$$

Comparing Equation (49) with Equation (46), the maximum contribution to ϵ_0 from $\eta_c(1S)$ VFs is about 10^{-4} times smaller than the combined contribution from the three charged lepton-antilepton VFs.

The the least massive $b\bar{b}$ bound state, $\eta_b(1S)$, has $J = 0$ and has positive charge conjugation parity[48]. Its mass is known experimentally, but the decay rate into two photons is not[48]. There are, however, theoretical calculations of the decay rate that range from 0.22 keV to 0.45 keV[49, 50, 51, 52]. To determine the maximum contribution to ϵ_0 from oscillations of $\eta_b(1S)$ VFs, the maximum value for the decay rate and the minimum value of energy associated with the state are used: $E_{\rm min}$ is the difference between the mass $m_{\eta_b(1S)}$ of $\eta_b(1S)$ and the masses $m_{\rm b}$ of the bottom and anti-bottom quarks when they are weakly bound: $E_{\rm min} = m_{\eta_b(1S)} - 2m_{\rm b} \sim 9.40\,{\rm GeV} - 2 \times 4.3\,{\rm GeV} = 0.8$ GeV[48]. From Equation (48) The maximum contribution to ϵ_0 from oscillations of $\eta_b(1S)$ VFs is then calculated to be

$$\epsilon_0 \begin{pmatrix} \text{maximum possible} \\ \text{contribution from } \eta_b(1S) \text{ VFs} \end{pmatrix} \sim 2.6 \times 10^{-5} \frac{e^2}{\hbar c}, \tag{50}$$

which is about 10^{-6} times smaller than the contribution from the three charged lepton-antilepton VFs.

For the heavy quarks c and b, as the mass increases from m_c to m_b, the minimum possible angular frequency increases, and the decay rate of $Q\bar{Q}$ into two photons decreases. Both effects decrease the maximum possible contribution to ϵ_0 and suggest the contribution to ϵ_0 from the top quark t would be smaller than that from either c or b. On the other hand, the square of the charge for t is four times larger than that for b. Although there is no experimental information about $\eta_t(1S)$[48], from the above discussion the contribution of $\eta_t(1S)$ VFs to ϵ_0 is expected to be small compared with the contribution from the three charged lepton-antilepton VFs.

For the light quarks $q = u, d$, or s, the π^0, η, and η' are the least massive $J = 0$ combinations of $q\bar{q}$ bound states that decay into two photons. Here, however, the comparison with an oscillator is less appropriate; a Bethe-Saltpeter approach to $q\bar{q}$ bound states fails completely[53]. As discussed in Ref. [54], a completely relativistic approach is required, an approach that shows no indication that a $q\bar{q}$ pair can be characterized by an oscillator potential energy. Moreover, since the strong interactions are primarily responsible for the binding of these relativistic states, $q\bar{q}$ bound state VFs would have much higher natural frequencies than the electromagnetically bound, charged lepton-antilepton VFs that are much more weakly bound. Accordingly $q\bar{q}$ bound state VFs contribute little to ϵ_0 and to lowest order need not be considered.

6.6. Calculation of ϵ_0

Ignoring the small contributions to ϵ_0 from quark-antiquark VFs, an approximate formula for ϵ_0 is immediately obtained from Equation (46) using the defining formula $\alpha = e^2/(4\pi\epsilon_0\hbar c)$ to eliminate α and then using $c = 1/\sqrt{\mu_0\epsilon_0}$:

$$\epsilon_0 \cong \frac{6\mu_0}{\pi}\left(\frac{8e^2}{\hbar}\right)^2 = 9.10 \times 10^{-12} \frac{\rm C}{\rm Vm}. \tag{51}$$

IOP Conf. Series: Journal of Physics: Conf. Series **1239** (2019) 012016 doi:10.1088/1742-6596/1239/1/012016

The calculation of ϵ_0 in Equation (51) is strictly a quantum calculation: (1) The existence of on-mass-shell, charged lepton-antilepton VFs are predicted by quantum field theory. (2) The appearance of charged lepton-antilepton VFs as bound states in the lowest energy level with $J = 0$ is predicted by the Heisenberg uncertainty principle and conservation of angular momentum. (3) The energy levels of the bound state VFs are calculated using non-relativistic quantum mechanics. (4) The decay rates of the charged lepton-antilepton bound state VFs are calculated using quantum electrodynamics. (5) The calculation of ϵ_0 from (1)-(4) is a quantum derivation. Once (1)-(4) are known, however, it is possible to obtain the formula Equation (51) using a classical derivation similar to that used to obtain the permittivity of a physical dielectric[55].

From Equation (51) equations for c and α immediately follow as shown in the Introduction. Only the lowest-order terms in α have been retained in calculating ϵ_0. The binding energy of parapositronium was neglected when calculating Δt_{p-Ps} in Equation (9), and only the leading term was retained when calculating ω_j^0 and $\Gamma_{\text{p-Ps}}$. The formulas for ϵ_0, c, and α exhibit the point of view that e and $\hbar$ are fundamental constants. (In SI units μ_0 is a parameter with assigned a value.) The calculation of ϵ_0 also describes how the properties of the vacuum determine the speed of light in the vacuum and the fine-structure constant.

References

[1] Kramers H A 1924 *Nature* **113** 673–674
[2] Thirring W E 1958 *Principles of Quantum Electrodynamics* (New York: Academic Press) p. 77; p. 70
[3] Bjorken J D and Drell S D 1965 *Relativistic Quantum Fields* (New York: McGraw-Hill) pp. 35-37; pp. 187-189
[4] Roman P 1969 *Introduction to Quantum Field Theory* (New York: Wiley)
[5] Peskin M E and Schroeder D V 1995 *An Introduction To Quantum Field Theory (Frontiers in Physics)* (Reading Mass.: Perseus Books) pp. 112-113; pp. 19-24; pp. 21-22 and pp. 790-791; pp. 790-791
[6] Furry W H and Oppenheimer J R 1934 *Phys. Rev.* **45** 245 p. 261
[7] Weisskopf V 1994 The electrodynamics of the vacuum based on the quantum theory of the electron *Early quantum electrodynamics: a source book* ed Miller A I (Cambridge: Cambridge University Press) pp 206–226. English translation of V. Weisskopf, Kgl. Danske Videnskab. Selskabs Math.-Fys. Medd. **14**, 3 (1936).
[8] Pauli W and Weisskopf V 1994 The quantization of the scalar relativistic wave equation *Early quantum electrodynamics: a source book* ed Miller A I (Cambridge: Cambridge University Press) pp 188–205. English translation of W. Pauli and V. Weisskopf, Helvetica Physica Acta **7**, 709 (1934).
[9] Ruark A E 1945 *Phys. Rev.* **68** 278
[10] Wheeler J A 1946 *Ann. N. Y. Acad. Sci.* **48** 219–238
[11] Deutsch M and Brown C 1952 *Phys. Rev.* **85** 1047
[12] Dicke R H 1957 *Rev. Mod. Phys.* **29** 363–376
[13] Leuchs G, Villar A S and Sánchez-Soto L L 2010 *Appl. Phys. B* **100** 9–13
[14] Leuchs G and Sánchez-Soto L L 2013 *Eur. Phys. J. D* **67** 57
[15] Urban M, Couchot F, Sarazin X and Djannati-Atai A 2013 *Eur. Phys. J. D* **67** 58
[16] Rossi B 1957 *Optics* (Reading, Massachusetts: Addison-Wesley) pp. 358-365
[17] Feynman R P, Leighton R B and Sands M 1963 *The Feynman Lectures on Physics, Vol. I* (Reading, Massachusetts: Addison-Wesley) pp. 31-1–5
[18] Sokolov A A, Ternov I M and Zhukovskii V C 1984 *Quantum Mechanics* (Moscow: MIR) pp. 157-164
[19] Mainland G B and Mulligan B 2018 *ArXiv e-prints* (*Preprint* 1810.04341)
[20] Leonhardt U, Griniasty I, Wildeman S, Fort E and Fink M 2018 *Phys. Rev. A* **98** 022118–1–022118–11 p. 022118-1
[21] Unruh W G 1976 *Phys. Rev. D* **14** 870
[22] Einstein A 1905 *Annalen der Physik* **17** 891
[23] Moffat J W 1993 *Int. J. Mod. Phys. D.* **2** 351–365
[24] Guth A H 1981 *Phys. Rev. D* **23** 347–356
[25] Sommerfeld A 1916 *Ann. Physik* **51** 1–94 p. 51
[26] Kragh H 2003 *Eur. Phys. J. D* **24** 169–173
[27] Hermann A 1971 *The Genesis of Quantum Theory (1899-1913)* (Cambridge, Mass.: MIT Press) English

translation of Hermann A 1969 *Frühgeschichte de Quantentheorie (1899-1931)* (Mosbach in Baden, Deutschland: Physik Verlag)
[28] Einstein A 1909 *Phys. Z.* **10** 185–193
[29] Feynman R P 1988 *QED. The Strange Theory of Light and Matter* (Princeton: Princeton University Press) p. 129
[30] Kragh H 2003 *Arch. Hist. Exact Sci.* **57** 395–431
[31] Murphy M T et al 2001 *Monthly Notices of the Royal Astronomical Society* **327** 1208–1222
[32] Kinoshita T 1996 *Rep. Prog. Phys.* **59** 1459–1492
[33] Birge R T 1934 *Nature* **133** 648
[34] Beck G, Bethe H A and Riezler W 1931 *Die Naturwissenschaften* **19** 39
[35] Eddington A S 1936 *Relativity Theory of Protons and Electrons* (New York: MacMillan)
[36] Eddington A S 1948 *Fundamental Theory* (Cambridge: Cambridge University Press) pp. 64-67
[37] Kilmister C W 1994 *Eddington's search for a fundamental theory* (Cambridge: Cambridge University Press) p. 115
[38] Allen H S 1915 *Proceedings of the Physical Society (London)* **27** 425–431
[39] Wyler C R 1969 *C. R. Acad. Sci.* **269** A743–A745
[40] Wentzel G 2003 *Quantum Theory of Fields* (New York: Dover) English translation of Wentzel G 1942 *Einführung in die Quantentheorie der Wellenfelder* (Vienna: Franz Deuticke) § 8.
[41] DeBenedetti S and Corben H C 1954 *Annu. Rev. Nucl. Sci.* **4** 191–218
[42] Hughes V W, Marder S and Wu C S 1957 *Phys. Rev.* **106** 934–947
[43] Schiff L I 1955 *Quantum Mechanics* (New York: McGraw-Hill) eq. (16.15); eq. (13.8); eq. (11.13)
[44] Feynman R P, Leighton R B and Sands M 1964 *The Feynman Lectures on Physics, Vol. II* (Reading, Massachusetts: Addison-Wesley) pp. 11-1–3
[45] Jackson J D 1999 *Classical Electrodynamics, Third Ed,* (New Jersey: Wiley) pp. 152-153
[46] Jauch J M and Rohrlich F 1955 *The Theory of Photons and Electrons* (Cambridge, Massachusetts: Addison-Wesley) pp. 274-286
[47] Bohm A 1981 *J. Math. Phys.* **22** 2813–2823
[48] Patrignani C et al P D G 2016 *Chin. Phys. C* **40** 100001
[49] Munz C R 1996 *Nucl. Phys. A* **609** 364–376
[50] Ebert D, Faustov R N and Galkin V O 2003 *Mod. Phys. Lett. A* **18** 601–607
[51] Kim C S, Lee T and Wang G L 2005 *Phys. Lett. B* **606** 323–328
[52] Huang H W, Liu J H, Tang J and Chao K T 1997 *Phys. Rev. D* **56** 368–376
[53] Gara A, Durand B and Durand L 1990 *Phys. Rev. D* **42** 1651–1660
[54] Biernat E P, Gross F, Peña M T and Stadler A 2014 *Phys. Rev. D* **89** 016005
[55] Mainland G B and Mulligan B 2017 *ArXiv e-prints* (*Preprint* 1705.11068)

IOP Conf. Series: Journal of Physics: Conf. Series **1239** (2019) 012017 doi:10.1088/1742-6596/1239/1/012017

Critical Scaling and a Dynamical Higgs Boson

Philip D. Mannheim
Department of Physics, University of Connecticut, Storrs, CT 06269, USA
E-mail: philip.mannheim@uconn.edu

Abstract. In a quantum electrodynamics theory that is realized by critical scaling and anomalous dimensions, the action is not chiral invariant and there are no dynamical Goldstone or Higgs boson bound states. In the mean-field approximation to a chiral invariant four-fermion theory the associated mean-field sector action is not chiral invariant either and it also possesses no dynamical bound states, with Goldstone and Higgs bosons instead being generated by an accompanying four-fermion residual interaction. In this paper we show that if a critical scaling electrodynamics in which the dimension d_θ of $\bar{\psi}\psi$ is reduced from three to two is augmented with a four-fermion interaction, precisely because it possesses no dynamical bound states the electrodynamic sector can be reinterpreted as a mean-field approximation to a larger theory that is chiral symmetric. And with $d_\theta = 2$ we show in this larger theory there is a residual interaction that then does generate dynamical Goldstone and Higgs bosons in scattering amplitudes that are completely finite. While the dynamically generated Goldstone boson is massless, the dynamically induced Higgs boson is found to be a narrow resonance just above threshold, with its width being a diagnostic that could potentially enable one to distinguish between a dynamical Higgs boson and an elementary one.

1. Introduction

With the discovery [1, 2] of the Higgs boson associated with the Higgs mechanism [3, 4, 5, 6], it has become imperative to determine whether the Higgs boson is an elementary field that appears in the fundamental Lagrangian of nature, or whether it is a fermion-antifermion composite bound state. If the Higgs boson is to be elementary, one has to explain why its divergent self-energy is not regularized at a very high grandunified or Planck mass scale (the hierarchy problem), and one has to explain why the cosmological constant term that the Higgs field induces when it acquires a non-vanishing vacuum expectation value is not at the same mass scale as the 125 GeV mass that the Higgs boson is now known to have. If the Higgs boson is to be composite, one needs to develop a suitable dynamical mass generation scheme in a renormalizable quantum field theory that would not only not lead to an unacceptably high mass for the Higgs boson or an unacceptably large cosmological constant term, but would recover the successful standard phenomenology associated with an elementary Higgs boson. Here we present a dynamical scheme that does this, a scheme that does not differ from an elementary Higgs boson theory when the Higgs boson is off shell (off-shell Higgs exchange and loop diagrams), but does differ from it on the mass shell. On shell the dynamical scheme requires that the Higgs boson be a narrow resonance just above threshold, with its width being a diagnostic that could potentially distinguish a dynamical Higgs boson from an elementary one.

IOP Conf. Series: Journal of Physics: Conf. Series **1239** (2019) 012017 doi:10.1088/1742-6596/1239/1/012017

The dynamical scheme that we study is a quantum electrodynamics (QED) theory as realized by critical scaling with anomalous dimensions in the ultraviolet, and by the non-vanishing of the vacuum expectation value of the composite operator $\bar{\psi}\psi$ due infrared dynamics, with the dynamical dimension of $\bar{\psi}\psi$ being reduced from three to two. It has long been known that because of the Baker-Johnson mass renormalization anomaly [7], if such critical scaling is to occur no dynamical pseudoscalar Goldstone or scalar Higgs boson is generated, with the chiral symmetry being broken in the Lagrangian, and with the theory actually corresponding to a theory with an intrinsic mass with Lagrangian $\mathcal{L}^m_{\rm QED} = \mathcal{L}^0_{\rm QED} - m\bar{\psi}\psi$ where $\mathcal{L}^0_{\rm QED} = -(1/4)F_{\mu\nu}F^{\mu\nu} + \bar{\psi}\gamma^\mu(i\partial_\mu - eA_\mu)\psi$. The fact that there would be no composite bound states in a critical scaling QED has a parallel in the chiral-invariant Nambu-Jona-Lasinio (NJL) four-fermion model [8], where the Lagrangian is broken into two pieces, a mean-field piece and a residual interaction, with the mean-field sector piece containing a mass term that is not present in the original NJL Lagrangian. Neither of these two pieces is separately chirally symmetric, with the composite bound states being found not in the mean-field sector at all but in the residual interaction sector. Thus by augmenting a massless fermion $\mathcal{L}^0_{\rm QED}$ with a four-fermion interaction, we can recognize an $\mathcal{L}^m_{\rm QED}$ with its fermion mass term as a mean-field theory, and as such this sector should not have any composite bound states since a mean-field sector never does. However, dynamical Goldstone and Higgs boson bound states are generated by the four-fermion residual interaction. And with $(\bar{\psi}\psi)^2$ acting as an operator whose dynamical dimension has been reduced from six to four, the four-fermion sector is power-counting renormalizable, and the mass of the ensuing Higgs boson is found to be both finite and at the same mass scale as that of the dynamically generated fermion rather than at a high regulator mass scale. By coupling the theory to an equally conformal invariant gravity theory (conformal gravity), the interplay between the gravity and four-fermion sector is found to take care of the cosmological constant that is induced by the dynamical symmetry breaking procedure.

2. The NJL Chiral Four-Fermion Model

2.1. NJL Model as a Mean-Field Theory

Since our interest in this paper is in studying critical scaling in a four-fermion theory, we will need to generalize the standard analysis of the NJL model. In fact our approach will be seen to lead us to a renormalizable version of the NJL model. The NJL action

$$I_{\rm NJL} = \int d^4x \left[i\bar{\psi}\gamma^\mu\partial_\mu\psi - \frac{g}{2}[\bar{\psi}\psi]^2 - \frac{g}{2}[\bar{\psi}i\gamma^5\psi]^2 \right] \tag{1}$$

is invariant under chiral transformations of the form $\psi(x) \to \exp(i\alpha\gamma^5)\psi(x)$, with the chiral symmetry preventing the presence of any fermion mass term in the action. Nonetheless, a fermion mass can be induced dynamically in the model. A direct way to show this is to introduce a mass term, and break the action into mean-field and residual interaction pieces according to $I_{\rm NJL} = I_{\rm MF} + I_{\rm RI}$, where

$$I_{\rm MF} = \int d^4x \left[i\bar{\psi}\gamma^\mu\partial_\mu\psi - m\bar{\psi}\psi + \frac{m^2}{2g} \right], \quad I_{\rm RI} = \int d^4x \left[-\frac{g}{2}\left(\bar{\psi}\psi - \frac{m}{g}\right)^2 - \frac{g}{2}\left(\bar{\psi}i\gamma^5\psi\right)^2 \right]. \tag{2}$$

In this decomposition neither $I_{\rm MF}$ nor $I_{\rm RI}$ is separately chiral invariant, only their sum is. One then treats m as a trial parameter and looks to show that a state $|\Omega_m\rangle$ with non-zero m has lower energy than the state $|\Omega_0\rangle$ with $m = 0$, to thus be energetically favored. In the Hartree-Fock approximation one sets

$$\langle\Omega_{\rm m}| \left[\bar{\psi}\psi - \frac{m}{g}\right]^2 |\Omega_{\rm m}\rangle = \langle\Omega_{\rm m}| \left[\bar{\psi}\psi - \frac{m}{g}\right] |\Omega_{\rm m}\rangle^2 = 0, \qquad \langle\Omega_{\rm m}|\bar{\psi}i\gamma^5\psi|\Omega_{\rm m}\rangle = 0, \tag{3}$$

$$\langle \Omega_{\rm m}|\bar{\psi}\psi|\Omega_{\rm m}\rangle = -i\int \frac{d^4k}{(2\pi)^4}{\rm Tr}\left[\frac{1}{\not{p}-m+i\epsilon}\right] = \frac{m}{g}, \quad (4)$$

and defines the physical mass m to be that value of m that satisfies $\langle \Omega_{\rm M}|\bar{\psi}\psi|\Omega_{\rm M}\rangle = M/g$, i.e. which satisfies the gap equation

$$-\frac{M\Lambda^2}{4\pi^2} + \frac{M^3}{4\pi^2}{\rm ln}\left(\frac{\Lambda^2}{M^2}\right) = \frac{M}{g}, \quad (5)$$

where Λ is a cutoff that is required since the NJL model is not renormalizable. The energy-density difference in a volume V, viz. $\epsilon(m) = (\langle \Omega_m|H|\Omega_m\rangle - \langle \Omega_0|H|\Omega_0\rangle)/V$, associated with $I_{\rm MF}$ is given (see e.g. [9]) by the infinite sum $\sum(1/n!)G_0^{(n)}(q_\mu = 0, m = 0)m^n$ of massless fermion graphs with point (i.e. not dressed) $m\bar{\psi}\psi$ insertions as exhibited in figure (1), to yield an $\tilde{\epsilon}(m) = \epsilon(m) - m^2/2g$ of the form

$$\tilde{\epsilon}(m) = i\int \frac{d^4p}{(2\pi)^4}{\rm Tr}\left[{\rm ln}\left(\frac{\not{p}-m+i\epsilon}{\not{p}+i\epsilon}\right)\right] - \frac{m^2}{2g} = \frac{m^4}{16\pi^2}{\rm ln}\left(\frac{\Lambda^2}{m^2}\right) - \frac{m^2M^2}{8\pi^2}{\rm ln}\left(\frac{\Lambda^2}{M^2}\right) + \frac{m^4}{32\pi^2}, \quad (6)$$

with $\tilde{\epsilon}(M)$ indeed being negative, just as required.

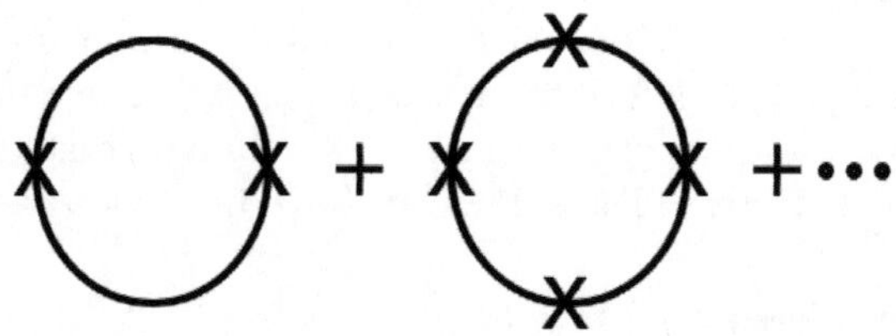

Figure 1. Vacuum energy density $\epsilon(m)$ via an infinite summation of massless graphs with zero-momentum point $m\bar{\psi}\psi$ insertions.

2.2. Higgs-Like Lagrangian

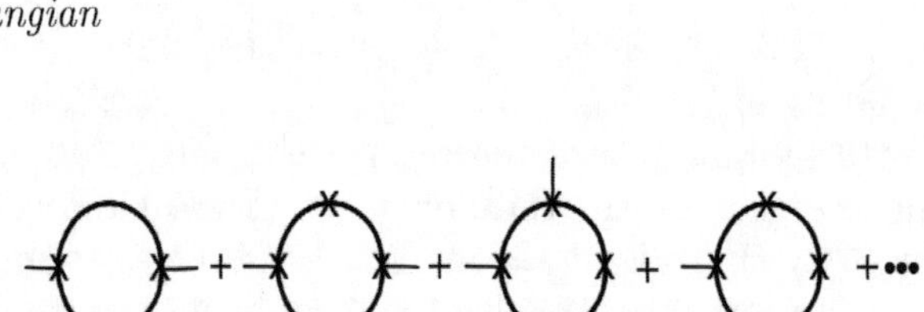

Figure 2. $\Pi_{\rm S}(q^2, m(x))$ developed as an infinite summation of massless graphs, each with two point $m\bar{\psi}\psi$ insertions carrying momentum q_μ (shown as external lines), with all other point $m\bar{\psi}\psi$ insertions carrying zero momentum.

To generate a spacetime dependence to $m(x)$, we introduce a spacetime-dependent coherent state $|C\rangle$ and consider spacetime-dependent matrix elements of the form $\langle C|\bar{\psi}(x)\psi(x)|C\rangle$. For $\tilde{\epsilon}(m)$ we can replace m by $m(x)$, since even though $m(x)$ is not spacetime independent, the set of graphs given in figure (1) is a set of graphs in each one of which the fermion is massless, so that $m(x)$ does not appear inside the individual momentum integrals, but only multiplies them by $m^n(x)$. (We thus do the momentum integrals with constant m first and then set $m = m(x)$ afterwards.) For a kinetic energy term for $m(x)$ we introduce $\Pi_{\rm S}(q^2, m)$ as defined in (9) below, and evaluate $\Pi_{\rm S}(q^2, m(x))$ as [10] the series of massless graphs given in figure (2), where again $m(x)$ does not appear inside the momentum integrals. We can then identify $-d\Pi_{\rm S}(q^2, m(x))/dq^2|_{q^2=0} = [\ln(\Lambda^2/m^2(x)) - 5/3]/8\pi^2$ as the coefficient of $(1/2)\partial_\mu m(x)\partial^\mu m(x)$, to obtain a leading cutoff dependent effective action of the form [11, 10]

$$I_{\rm EFF} = \int \frac{d^4x}{8\pi^2}\ln\left(\frac{\Lambda^2}{M^2}\right)\left[\frac{1}{2}\partial_\mu m(x)\partial^\mu m(x) + m^2(x)M^2 - \frac{1}{2}m^4(x)\right]. \tag{7}$$

If we make the chiral symmetry local and introduce a coupling $g_{\rm A}\bar{\psi}\gamma^\mu\gamma^5 A_{\mu5}\psi$ to an axial gauge field $A_{\mu5}(x)$, on setting $\phi(x)/g = \langle C|\bar{\psi}(x)(1+\gamma^5)\psi(x)|C\rangle$ the effective action becomes the local [11, 10]

$$I_{\rm LOC} = \int \frac{d^4x}{8\pi^2}\ln\left(\frac{\Lambda^2}{M^2}\right)\left[\frac{1}{2}|(\partial_\mu - 2ig_{\rm A}A_{\mu5})\phi(x)|^2 + |\phi(x)|^2M^2 - \frac{1}{2}|\phi(x)|^4 - \frac{g_{\rm A}^2}{6}F_{\mu\nu5}F_5^{\mu\nu}\right]. \tag{8}$$

We recognize this action as a double-well Ginzburg-Landau type Higgs Lagrangian with order parameter $\phi(x)$, only now generated dynamically.

In $I_{\rm EFF}$ of (7) there is a double-well Higgs potential, but since $m(x)/g = \langle C|\bar{\psi}(x)\psi(x)|C\rangle$ is a c-number, $m(x)$ does not itself represent a q-number scalar field. Rather, as we now show, the q-number fields are to be found as collective bound state modes generated by the residual interaction, with no fundamental scalar fields being needed at all.

2.3. *The Collective Goldstone and Higgs Modes*

To identify dynamical bound states we introduce the scalar and pseudoscalar vacuum polarizations associated with $I_{\rm MF}$, viz. (cf. figure (2) and its pseudoscalar analog)

$$\begin{aligned}
\Pi_{\rm S}(q^2, M) &= \int d^4x e^{iq\cdot x}\langle\Omega_M|\bar{\psi}(x)\psi(x)\bar{\psi}(0)\psi(0)|\Omega_M\rangle \\
&= -i\int\frac{d^4p}{(2\pi)^4}{\rm Tr}\left[\frac{1}{\not{p} - M + i\epsilon}\frac{1}{\not{p} + \not{q} - M + i\epsilon}\right] \\
&= -\frac{\Lambda^2}{4\pi^2} + \frac{M^2}{4\pi^2}\ln\left(\frac{\Lambda^2}{M^2}\right) + \frac{(4M^2 - q^2)}{8\pi^2} + \frac{(4M^2 - q^2)}{8\pi^2}\ln\left(\frac{\Lambda^2}{M^2}\right) \\
&\quad -\frac{1}{8\pi^2}\frac{(4M^2 - q^2)^{3/2}}{(-q^2)^{1/2}}\ln\left(\frac{(4M^2 - q^2)^{1/2} + (-q^2)^{1/2}}{(4M^2 - q^2)^{1/2} - (-q^2)^{1/2}}\right),
\end{aligned} \tag{9}$$

$$\begin{aligned}
\Pi_{\rm P}(q^2, M) &= \int d^4x e^{iq\cdot x}\langle\Omega_M|\bar{\psi}(x)i\gamma^5\psi(x)\bar{\psi}(0)i\gamma^5\psi(0)|\Omega_M\rangle \\
&= -i\int\frac{d^4p}{(2\pi)^4}{\rm Tr}\left[i\gamma^5\frac{1}{\not{p} - M + i\epsilon}i\gamma^5\frac{1}{\not{p} + \not{q} - M + i\epsilon}\right] \\
&= -\frac{\Lambda^2}{4\pi^2} + \frac{M^2}{4\pi^2}\ln\left(\frac{\Lambda^2}{M^2}\right) - \frac{q^2}{8\pi^2}\ln\left(\frac{\Lambda^2}{M^2}\right) + \frac{(4M^2 - q^2)}{8\pi^2} \\
&\quad +\frac{(8M^4 - 8M^2q^2 + q^4)}{8\pi^2(-q^2)^{1/2}(4M^2 - q^2)^{1/2}}\ln\left(\frac{(4M^2 - q^2)^{1/2} + (-q^2)^{1/2}}{(4M^2 - q^2)^{1/2} - (-q^2)^{1/2}}\right).
\end{aligned} \tag{10}$$

The scattering matrices in the two channels are given by using $I_{\rm RI}$ to iterate the vacuum polarizations according to $T = g + g\Pi g + g\Pi g\Pi g + ...$, to yield

$$T_{\rm S}(q^2, M) = \frac{1}{g^{-1} - \Pi_{\rm S}(q^2, M)}, \qquad T_{\rm P}(q^2, M) = \frac{1}{g^{-1} - \Pi_{\rm P}(q^2, M)}. \tag{11}$$

With g^{-1} being given by $\langle \Omega_{\rm M}|\bar{\psi}\psi|\Omega_{\rm M}\rangle = M/g$, we find poles in each of the two channels, and near them the scattering matrices behave as

$$T_{\rm S}(q^2, M) = \frac{R_{\rm S}^{-1}}{(q^2 - 4M^2)}, \quad T_{\rm P}(q^2, M) = \frac{R_{\rm P}^{-1}}{q^2}, \tag{12}$$

where $R_{\rm S} = R_{\rm P} = \ln(\Lambda^2/M^2)/8\pi^2$. Dynamical pseudoscalar Goldstone and scalar Higgs boson states are thus induced by the residual interaction even though neither is contained in the mean-field sector, with the mass of the scalar Higgs boson naturally being of order the mass of the fermion and not of order the cutoff Λ. The NJL model thus contains the key ingredients needed for dynamical symmetry breaking. However the model is not renormalizable. To make it renormalizable we turn now to critical scaling and anomalous dimensions, as this will serve to sufficiently soften the point four-fermion couplings.

3. Critical Scaling in Quantum Electrodynamics

3.1. Vanishing of the Bare Fermion Mass

In a study of quantum electrodynamics, Johnson, Baker, and Willey [12, 13, 7] found that the standard Z_3 and δm renormalization constants of the theory would be finite in the event of critical scaling and anomalous dimensions. Their results can be summarized by noting that in the generalized Landau gauge the asymptotic renormalized fermion propagator $\tilde{S}^{-1}(p)$ obeys [14] the Callan-Symanzik equation

$$\left[m\frac{\partial}{m} + \beta(\alpha)\frac{\partial}{\partial\alpha}\right]\tilde{S}^{-1}(p) = m[\gamma_\theta(\alpha) - 1]\tilde{\Gamma}_{\rm S}(p, p, 0), \tag{13}$$

where $\tilde{\Gamma}_{\rm S}(p, p, 0)$ is the renormalized Green's function associated with the insertion of a zero-momentum composite operator $\theta = \bar{\psi}\psi$ into the inverse fermion propagator. Critical scaling and the finiteness of Z_3 is achievable if $\beta(\alpha) = 0$, with one then asymptotically having

$$\tilde{S}^{-1}(p) = \not{p} - m\left(\frac{-p^2 - i\epsilon}{m^2}\right)^{\gamma_\theta(\alpha)/2}, \qquad \tilde{\Gamma}_{\rm S}(p, p, 0) = \left(\frac{-p^2 - i\epsilon}{m^2}\right)^{\gamma_\theta(\alpha)/2}. \tag{14}$$

The vanishing of $\beta(\alpha)$ can be achieved in various ways. The function $\beta(\alpha)$ could have a non-trivial zero for some specific value of the coupling constant α. For an Abelian gauge theory this could only occur non-perturbatively, while for a non-Abelian one, with an appropriate fermion content the second-order term in α could have the opposite sign to that of the first-order term, to then permit a perturbative cancellation. However, for our purposes here, one does not need to actually require that $\beta(\alpha)$ have such a zero, since one could instead work in the much studied quenched approximation in which one keeps the photon canonical (i.e. $\beta(\alpha)$ is zero for all α and there is no charge renormalization). In this approximation, one obtains the asymptotic scaling form for $\tilde{S}^{-1}(p)$ if one sums all photon exchange diagrams both planar and non-planar combined [12], doing so for any value of the coupling constant. Or one could keep quenched planar graphs alone and then [15, 16, 17] get scaling if $\alpha \leq \pi/3$, with the dynamical dimension

$d_\theta(\alpha) = 3 + \gamma_\theta(\alpha)$ of $\bar{\psi}\psi$ being given by $d_\theta(\alpha) = 2 + (1 - 3\alpha/\pi)^{1/2}$. In all of these cases the dressing of the $\bar{\psi}\psi$ vertex with quenched photons converts the point $\tilde{\Gamma}_S(p,p,0) = 1$ Green's function into the dressed $\tilde{\Gamma}_S(p,p,0) = (-p^2/m^2)^{\gamma_\theta(\alpha)/2}$, and that is the only requirement that we will need for this work as it will soften the point vertices if $\gamma_\theta(\alpha) < 0$, and lead to completely finite scalar and pseudoscalar channel fermion-antifermion scattering amplitudes if $\gamma_\theta(\alpha) = -1$. Since the summation of all planar plus non-planar quenched photon graphs leads to scaling for any value of α, and since $\gamma_\theta(\alpha)$ is a continuous function of α, there will (presumably) be some value of α for which $\gamma_\theta(\alpha) = -1$, and in fact this is already seen in the quenched planar graph approximation if $\alpha = \pi/3$. In the following we thus explore the implications of $\gamma_\theta(\alpha) = -1$.

Once there is critical scaling no matter what the cause, the bare mass behaves as

$$m_0 = m\left(\frac{\Lambda^2}{m^2}\right)^{\gamma_\theta(\alpha)/2}, \tag{15}$$

and thus vanishes if $\gamma_\theta(\alpha) < 0$. In consequence, $\delta(m) = m - m_0$ is finite. With a zero m_0 and a non-zero m this looks like dynamical symmetry breaking. However, with $Z_\theta^{-1/2} = (\Lambda^2/m^2)^{\gamma_\theta(\alpha)/2}$ effecting $Z_\theta^{-1/2}(\bar{\psi}\psi)_0 = \bar{\psi}\psi$, $m_0(\bar{\psi}\psi)_0 = m(\bar{\psi}\psi)$ is not zero. Because of this, chiral symmetry is broken in the Lagrangian, and there is no associated Goldstone boson [7].

In the quenched planar graph approximation it was found [15, 16, 17] that if $\alpha > \pi/3$ the fermion propagator does not scale asymptotically and there then is a Goldstone boson. Thus the prevailing wisdom from that time on has been that the generation of dynamical Goldstone bosons is associated with strong coupling only (cf. $\alpha > \pi/3$). In this paper and the more detailed [18, 9] we revisit this wisdom and show that there can be a Goldstone boson even for weak coupling and critical scaling. However, since there is none in the fermion gauge boson theory itself, we will need to embed the theory in a larger chiral invariant one, namely an NJL type theory, one that would be power-counting renormalizable if $d_\theta(\alpha) = 2$, since then $(\bar{\psi}\psi)^2$ would act as an operator whose dynamical dimension is reduced from six to four. (In [18, 9] we discuss dynamical Goldstone boson studies by other authors, some of which also involve $d_\theta(\alpha) = 2$.)

3.2. Non-vanishing of Physical Fermion Mass

With the exact photon-fermion-antifermion vertex $\Gamma^\mu(p+q,p)$ obeying the unrenormalized $\Gamma^\mu(p+q,p) = \gamma^\mu - \int d^4k/(2\pi)^4 S(p+q+k)\Gamma^\mu(p+q+k,p+k)S(p+k)K(p+k,p+q+k,k)$ where $S(p) = 1/(\not{p} - m_0 - \Sigma(p))$ and $K(p+k,p+q+k,k)$ is the Bethe-Salpeter kernel, from the Ward identity $q_\mu\Gamma^\mu(p+q,p) = S^{-1}(p+q) - S^{-1}(p)$, we obtain

$$\begin{aligned}
&\Sigma(p) - \Sigma(p+q)\\
&= -\int \frac{d^4k}{(2\pi)^4} S(p+q+k)(\Sigma(p+k) - \Sigma(p+q+k))S(p+k)K(p+k,p+q+k,k)\\
&- \int \frac{d^4k}{(2\pi)^4} \frac{[\not{p}+\not{q}+\not{k}+m_0+\Sigma(p+q+k)]}{[(p+q+k)^2 - (m_0+\Sigma(p+q+k))^2]} \not{q} \frac{[(\not{p}+\not{k}+m_0+\Sigma(p+k))]}{[(p+k)^2 - (m_0+\Sigma(p+k))^2]}\\
&\times K(p+k,p+q+k,k).
\end{aligned} \tag{16}$$

Now one can find a gauge [12], the generalized Landau gauge, in which $\Sigma(p)$ has no Dirac gamma matrix dependence. Thus in the second term in (16) the only terms that will survive will be those that are linear in $m_0 + \Sigma$. With the kernel being no more divergent than $1/k^2$ if the photon is canonical, then given (14) and (15), the $m_0 + \Sigma$ terms will appear in an integral that is finite. One can thus drop the m_0 dependence in the second term in (16). In consequence, (16)

IOP Conf. Series: Journal of Physics: Conf. Series **1239** (2019) 012017 doi:10.1088/1742-6596/1239/1/012017

becomes a self-consistent homogeneous equation for $\Sigma(p)$, and since (16) involves a difference of two propagators it has better asymptotic convergence properties than the unrenormalized Schwinger-Dyson equation $\Sigma(p) = ie_0^2 \int d^4k/(2\pi)^4 D_{\mu\nu}(k)\Gamma^\mu(p, p-k)S(p-k)\gamma^\nu$. As such, (16) can have the asymptotic solution $\Sigma(p) = m(-p^2/m^2)^{\gamma_\theta(\alpha)/2}$ as given in (14) above. However, since (16) is homogeneous, it can also have the completely trivial solution $\Sigma(p) = 0$ (i.e. both m_0 and m identically zero). The trivial solution to (16) is thus associated with the Lagrangian $\mathcal{L}^0_{\rm QED}$, while the non-trivial solution is associated with $\mathcal{L}^m_{\rm QED} = \mathcal{L}^0_{\rm QED} - m\bar{\psi}\psi$.

We thus need a criterion that would oblige us to select the non-trivial solution, and thus need to show that it has lower energy than the trivial one, i.e. just as in the NJL case, we need to show that the $\epsilon(m) = (\langle\Omega_m|H|\Omega_m\rangle - \langle\Omega_0|H|\Omega_0\rangle)/V$ associated with $I^m_{\rm QED}$ is negative when $m \neq 0$. Now $\epsilon(m)$ is given by $\sum(1/n!)G_0^{(n)}(q_\mu = 0, m = 0)m^n$. In this sum each Green's function is associated not with a massive fermion but a massless one. Now if the massless fermion theory has critical scaling, the relations

$$\tilde{S}^{-1}(p, m=0) = \not{p}, \qquad \tilde{\Gamma}_{\rm S}(p,p,0) = \left(\frac{-p^2 - i\epsilon}{\mu^2}\right)^{\gamma_\theta(\alpha)/2} \tag{17}$$

will not only hold for asymptotic momenta, they will hold for all momenta in the massless theory since the massless theory has no intrinsic mass scale. (As is typical of massless theories, we renormalize Green's functions off shell at some spacelike point with $p^2 = -\mu^2$, and while we can set μ equal to the physical fermion mass M, it is more instructive to keep μ as is and only set it equal to M at the end.) To determine $\epsilon(m)$ we thus replace the $\tilde{\Gamma}_{\rm S}(p,p,0) = 1$ point vertices of figure (1) by the dressed $\tilde{\Gamma}_{\rm S}(p,p,0)$ vertices [19, 20, 21] to obtain figure (3).

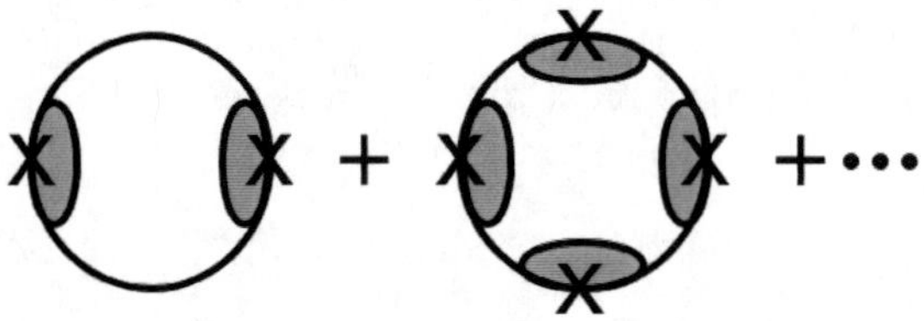

Figure 3. Vacuum energy density $\epsilon(m)$ via an infinite summation of massless graphs with zero-momentum dressed $m\bar{\psi}\psi$ insertions.

The graphs in figure (3) can be summed analytically, and in terms of the propagator

$$\tilde{S}_\mu^{-1}(p) \;=\; \not{p} - m\left(\frac{-p^2 - i\epsilon}{\mu^2}\right)^{\gamma_\theta(\alpha)/2} + i\epsilon, \tag{18}$$

yield [19, 20, 21]

$$\epsilon(m) = i\int \frac{d^4p}{(2\pi)^4} {\rm Tr}\left[\ln(\tilde{S}_\mu^{-1}(p)) - \ln(\not{p} + i\epsilon)\right]. \tag{19}$$

As a function of $\gamma_\theta(\alpha)$, $\epsilon(m)$ is found [19, 20, 21] to have the shape of a single well if $-1 < \gamma_\theta(\alpha) < 0$, and to have the shape of an unbounded, upside down single well if $\gamma_\theta(\alpha) < -1$. However $\epsilon(m)$ is found to have the shape of a double-well potential if $\gamma_\theta(\alpha) = -1$, i.e. if $d_\theta(\alpha)$ is

IOP Conf. Series: Journal of Physics: Conf. Series **1239** (2019) 012017 doi:10.1088/1742-6596/1239/1/012017

precisely at the value that makes $(\bar{\psi}\psi)^2$ act as a renormalizable operator [20], with $\epsilon(m)$ taking the form

$$\epsilon(m) = -\frac{m^2\mu^2}{16\pi^2}\left[\ln\left(\frac{\Lambda^4}{m^2\mu^2}\right)+1\right]. \tag{20}$$

While taking $\gamma_\theta(\alpha)$ to be negative softens the short-distance behavior of the theory (for any negative $\gamma_\theta(\alpha)$ the bare mass vanishes), as $\gamma_\theta(\alpha)$ is taken to be more and more negative, the theory becomes more and more divergent in the infrared, and at $\gamma_\theta(\alpha) = -1$ the infrared divergences become so severe that the theory is forced into a new, dynamically broken, vacuum. (In the quenched planar graph approximation of [15, 16, 17] the condition $\gamma_\theta(\alpha) = -1$ is realized right at the critical $\alpha = \pi/3$ value.) As constructed, $\epsilon(m)$ is logarithmically divergent (the drop in dimension of $\bar{\psi}\psi$ from three to two converts the original quadratic divergence in the point-coupled NJL model to logarithmic). As with the mean-field sector of the NJL model, we introduce (and actually generate in $\mathcal{L}_{\rm QED-MF}$ below) a $-m^2/2g$ counterterm and set $\tilde{\epsilon}(m) = \epsilon(m) - m^2/2g$. We then identify the physical mass as the one that obeys $\tilde{\epsilon}'(M) = \epsilon'(M) - M/g = 0$ (this relation serves to express g in terms of M). Since $\epsilon'(m) = \langle\Omega_m|\bar{\psi}\psi|\Omega_m\rangle$, the physical mass and $\tilde{\epsilon}(m)$ are given by [19, 20, 21]

$$\begin{aligned}\frac{M}{g} &= \langle\Omega_M|\bar{\psi}\psi|\Omega_M\rangle = -i\int\frac{d^4p}{(2\pi)^4}{\rm Tr}[\tilde{\Gamma}_{\rm S}(p,p,0)\tilde{S}_\mu(p)]\\ &= i\int\frac{d^4p}{4\pi^4}\frac{M\mu^2}{(p^2+i\epsilon)^2+M^2\mu^2} = -\frac{M\mu^2}{4\pi^2}\ln\left(\frac{\Lambda^2}{M\mu}\right),\end{aligned} \tag{21}$$

$$\tilde{\epsilon}(m) = \frac{m^2\mu^2}{16\pi^2}\left[\ln\left(\frac{m^2}{M^2}\right)-1\right], \tag{22}$$

with (21) having a non-trivial solution no matter how small a negative, viz. attractive, g might be. We recognize $\tilde{\epsilon}(m)$ as being completely finite and having the shape of the double-well potential exhibited in figure (4). (In contrast, the point-coupled NJL $\tilde{\epsilon}(m)$ is log divergent.)

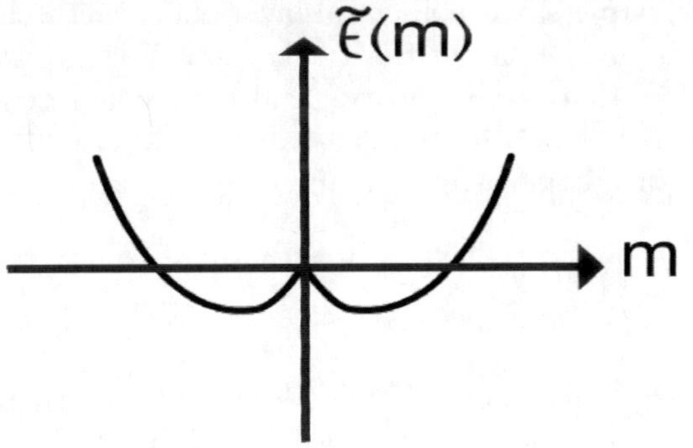

Figure 4. Dynamically generated double-well potential for the renormalized vacuum energy density when $\gamma_\theta(\alpha) = -1$.

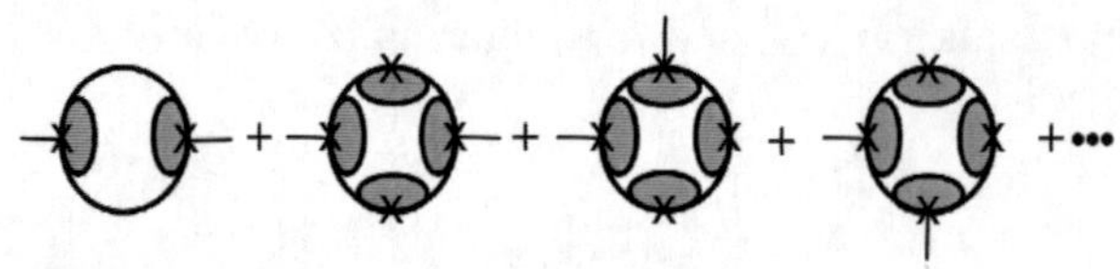

Figure 5. $\Pi_S(q^2, m(x))$ developed as an infinite summation of massless graphs, each with two dressed $m\bar{\psi}\psi$ insertions carrying momentum q_μ (shown as external lines), with all other dressed $m\bar{\psi}\psi$ insertions carrying zero momentum.

We can also parallel the NJL discussion of coherent states by taking a spacetime-dependent $m(x)$. We replace figure (2) by its dressed version given in figure (5). On summing the graphs in figure (5) we obtain [21]

$$\Pi_S(q^2, m) = -i\int \frac{d^4p}{(2\pi)^4}\mathrm{Tr}\left[\tilde{\Gamma}_S(p+q,p,-q)\tilde{S}_\mu(p)\tilde{\Gamma}_S(p,p+q,q)\tilde{S}_\mu(p+q)\right], \tag{23}$$

where $\tilde{\Gamma}_S(p,p+q,q) = \tilde{\Gamma}_S(p+q,p,-q) = [(-p^2/\mu^2)(-(p+q)^2/\mu^2)]^{\gamma_\theta(\alpha)/4}$, and thus obtain the effective Higgs Lagrangian [21]

$$\mathcal{L}_{\mathrm{EFF}} = -\frac{m^2(x)\mu^2}{16\pi^2}\left[\ln\left(\frac{m^2(x)}{M^2}\right) - 1\right] + \frac{3\mu}{256\pi m(x)}\partial_\mu m(x)\partial^\mu m(x), \tag{24}$$

together with higher-order derivative terms. (Since $m(x)$ is a c-number rather than a q-number, higher-order derivative terms do not affect renormalizability.)

As we see, at critical scaling the theory based on the Lagrangian $\mathcal{L}^0_{QED} - m\bar{\psi}\psi + m^2/2g = -(1/4)F_{\mu\nu}F^{\mu\nu} + \bar{\psi}\gamma^\mu(i\partial_\mu - eA_\mu)\psi - m\bar{\psi}\psi + m^2/2g$ has all the trappings of dynamical symmetry breaking seen in the NJL model save only that it has no pseudoscalar or scalar bound states, with the presence of the $m\bar{\psi}\psi$ term (as expressly required by the non-vanishing of $m_0(\bar{\psi}\psi)_0$ even as m_0 vanishes) indicating that the chiral symmetry is broken at the level of the Lagrangian. Now in our study above of the NJL model we encountered an analogous situation, with its mean-field sector having precisely this same non-chiral-invariant structure. Thus if we could have $\mathcal{L}^0_{QED} - m\bar{\psi}\psi + m^2/2g$ emerge in a larger theory that is chiral invariant, we could reinterpret it as the mean-field sector of that larger theory. And if we could do so, we would not expect a critical scaling $\mathcal{L}^0_{QED} - m\bar{\psi}\psi + m^2/2g$ to possess dynamical bound states, since a mean-field sector never does. Rather, such bound states should be generated by a residual interaction, just as we now show.

3.3. The Collective Goldstone and Higgs Modes

To generate bound states we augment massless QED with a four-fermion interaction to give the chiral invariant Lagrangian $\mathcal{L}_{\mathrm{QED-FF}}$, and then decompose it as

$$\begin{aligned}\mathcal{L}_{\mathrm{QED-FF}} &= -\frac{1}{4}F_{\mu\nu}F^{\mu\nu} + \bar{\psi}\gamma^\mu(i\partial_\mu - eA_\mu)\psi - \frac{g}{2}[\bar{\psi}\psi]^2 - \frac{g}{2}[\bar{\psi}i\gamma^5\psi]^2 \\ &= -\frac{1}{4}F_{\mu\nu}F^{\mu\nu} + \bar{\psi}\gamma^\mu(i\partial_\mu - eA_\mu)\psi - m\bar{\psi}\psi + \frac{m^2}{2g} - \frac{g}{2}\left(\bar{\psi}\psi - \frac{m}{g}\right)^2 - \frac{g}{2}\left(\bar{\psi}i\gamma^5\psi\right)^2 \\ &= \mathcal{L}_{\mathrm{QED-MF}} + \mathcal{L}_{\mathrm{QED-RI}}.\end{aligned} \tag{25}$$

With $\Pi_{\rm S}(q^2, m)$ of (23) being generated via figure (5), analogously one can evaluate $\Pi_{\rm P}(q^2, m)$, to obtain

$$\Pi_{\rm P}(q^2,m) = -i\int \frac{d^4p}{(2\pi)^4}{\rm Tr}\left[\tilde{\Gamma}_{\rm S}(p+q,p,-q)i\gamma^5\tilde{S}_\mu(p)\tilde{\Gamma}_{\rm S}(p,p+q,q)i\gamma^5\tilde{S}_\mu(p+q)\right]. \quad (26)$$

On translating p_μ to $p_\mu - q_\mu/2$, we can thus set

$$\Pi_{\rm S}(q^2,m) = -4i\mu^2\int \frac{d^4p}{(2\pi)^4}\frac{N(q,p)+m^2\mu^2}{D(q,p,m)}, \quad \Pi_{\rm P}(q^2,m) = -4i\mu^2\int \frac{d^4p}{(2\pi)^4}\frac{N(q,p)-m^2\mu^2}{D(q,p,m)}, \quad (27)$$

where

$$\begin{aligned} N(q,p) &= (p^2+i\epsilon-q^2/4)(-(p-q/2)^2-i\epsilon)^{1/2}(-(p+q/2)^2-i\epsilon)^{1/2}, \\ D(q,p,m) &= (((p-q/2)^2+i\epsilon)^2+m^2\mu^2)(((p+q/2)^2+i\epsilon)^2+m^2\mu^2). \end{aligned} \quad (28)$$

Evaluating $\Pi_{\rm P}(q^2, m)$ at $q^2 = 0$ yields

$$\Pi_{\rm P}(q^2=0,m) = -4i\mu^2\int \frac{d^4p}{(2\pi)^4}\frac{(p^2)(-p^2)-m^2\mu^2}{((p^2+i\epsilon)^2+m^2\mu^2)^2} = 4i\mu^2\int \frac{d^4p}{(2\pi)^4}\frac{1}{(p^2+i\epsilon)^2+m^2\mu^2}. \quad (29)$$

With $m = M$ we recognize this expression as precisely being equal to g^{-1}. We thus find a pole at $q^2 = 0$ in the pseudoscalar $T_{\rm P}(q^2, M) = [g^{-1} - \Pi_{\rm P}(q^2, M)]^{-1}$, with the scattering amplitude behaving as

$$T_{\rm P}(q^2,M) = \frac{128\pi M}{7\mu q^2} \quad (30)$$

near the pole, with (unlike NJL) the residue at the pole being completely finite. Thus just as required, with dynamical symmetry breaking the residual interaction generates a massless pseudoscalar Goldstone boson that is not present in the critical scaling mean-field sector.

The evaluation of $\Pi_{\rm S}(q^2, m)$ is not nearly as straightforward since we are interested in finding a pole in $T_{\rm S}(q^2, M)$ at some non-zero q^2. Moreover, the analytic structure of $\Pi_{\rm S}(q^2, m)$ is much more complicated than in the point-coupled case [18]. Specifically, if for timelike q^2 we set $q_\mu = (q_0, 0, 0, 0)$ we find that for $(p_1^2 + p_2^2 + p_3^2)^{1/2} = p < q_0/2$, as well as branch points in the lower right- and upper left-hand quadrants in the complex p_0 plane, the function $N(q, p)$ also has branch points in the upper right- and lower left-hand quadrants. The Wick rotation for timelike q^2 thus has to follow the contour given in figure (6).

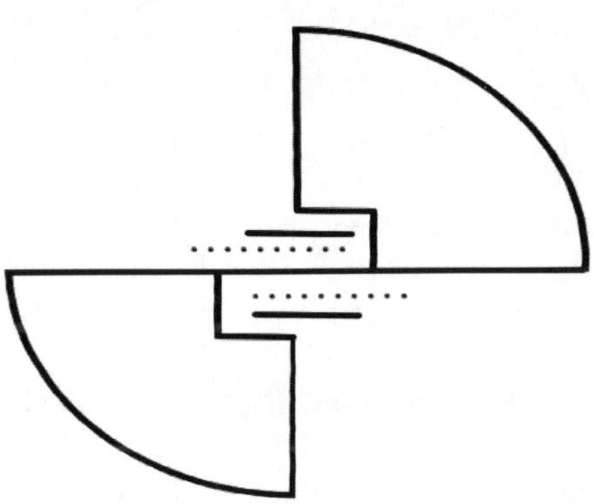

Figure 6. The Wick contour in the complex p_0 plane. The branch cuts are shown as lines and the poles as dots.

Because of the location of the branch points in $N(q,p)$, the function $\Pi_{\rm S}(q^2,m)$ is complex for all $q^2>0$, and thus poles in $T_{\rm S}(q^2,M)=[g^{-1}-\Pi_{\rm S}(q^2,M)]^{-1}$ cannot have real q^2. To cancel this complex piece we will need some other complex contribution. Thus is provided by $D(q,p,m)$ as it has a branch point of its own at $q^2=2m\mu$. Solutions to $g^{-1}-\Pi_{\rm S}(q^2,M)=0$ must thus lie above the $q^2=2M\mu$ threshold. Thus once we dress the point vertices of the original point-coupled NJL model the dynamical scalar bound state pole in $T_{\rm S}(q^2,M)$ must move into the complex q^2 plane and become an above-threshold resonance. This provides a quite sharp contrast with an elementary Higgs boson since the mass of an elementary Higgs boson is given by the magnitude of the second derivative of the Higgs potential at its minimum, a quantity that must be real if the potential itself is.

On numerically evaluating $T_{\rm S}(q^2,M)$ [18, 9] a dynamical Higgs boson pole is found at $q_0=(1.48-0.02i)(M\mu)^{1/2}$, $q^2=(2.19-0.05i)M\mu$, i.e. just above $q^2=2M^2$ (on setting $\mu=M$). Near the pole the scattering amplitude is found to have the Breit-Wigner form

$$T_{\rm S}(q^2,M)=\frac{46.14+1.03i}{q^2-2.22M\mu+0.05iM\mu}, \tag{31}$$

with an expressly negative imaginary part just as required for decay, with the associated decay width being fairly narrow. Such a width could potentially serve to distinguish a dynamical Higgs boson from an elementary one. With the dynamically generated Higgs mass naturally being at the fermion mass scale rather than at some large regulator mass scale, unlike in the elementary Higgs case, there automatically is no Higgs boson hierarchy problem.

Inspection of (31) shows that in the dynamical Higgs case the magnitude of the residue at the pole is completely determined, with the theory thus determining the Yukawa coupling of the Higgs boson to a fermion-antifermion pair (and likewise the Goldstone boson Yukawa coupling as per (30)). This stands in sharp contrast to the standard electroweak theory with its elementary Higgs boson, where the magnitudes of Yukawa couplings have to be introduced by hand.

In the above discussion of the collective modes we have iterated the T matrix as $T=g+g\Pi g+g\Pi g\Pi g+...=1/(g^{-1}-\Pi)$, by taking the Bethe-Salpeter scattering amplitude kernel to be given by $\Pi_{\rm S}(q^2,m)$ in the scalar channel and $\Pi_{\rm P}(q^2,m)$ in the pseudoscalar channel, and have found that $T_{\rm S}(q^2,M)$ and $T_{\rm P}(q^2,M)$ are both completely finite. To understand this in detail, we note in the expansion for $\Pi_{\rm S}(q^2,m)$ given in figure (5), that when $\gamma_\theta(\alpha)=-1$ the only term that is ultraviolet divergent is the very first term, with this term diverging as a single logarithm. Similarly, in the expansion of $\epsilon(m)$ given in figure (3) the only term that is ultraviolet divergent is the very first term, with this term not only also diverging as a single logarithm, but doing so with the very same coefficient as the first term in figure (5) since the first terms in figures (3) and (5) differ only by the value of the external momentum q_μ. With $\epsilon(m)=\sum(1/n!)G_0^{(n)}(q_\mu=0,m=0)m^n$, it follows that in $\epsilon'(m)=\sum(1/(n-1)!)G_0^{(n)}(q_\mu=0,m=0)m^{n-1}$ the only divergence is associated with the $n=2$ term, i.e. the $G_0^{(2)}(q_\mu=0,m=0)m$ term. However, in the Hartree-Fock approximation given in (21) we have precisely identified $\epsilon'(M)$ with M/g. In consequence, the quantity $g^{-1}-\Pi_{\rm S}(q^2,M)$ is completely finite. Since the ultraviolet behavior of $\Pi_{\rm P}(q^2,M)$ is the same as that of $\Pi_{\rm S}(q^2,M)$ (the short-distance behavior of the theory being chirally symmetric as the chiral symmetry is only broken by mass generation in the infrared), the quantity $g^{-1}-\Pi_{\rm P}(q^2,M)$ is completely finite as well.

Thus to lowest order in g in the T matrix kernel, the four-fermion interaction gives cutoff-independent results, due to the softening of the $\tilde{\Gamma}_{\rm S}(p,p+q,q)$ and $\tilde{\Gamma}_{\rm P}(p,p+q,q)$ vertices from their point-coupled 1 and $i\gamma^5$ values in the NJL model to the dressed $[(-p^2/\mu^2)(-(p+q)^2/\mu^2)]^{-1/4}$ and $[(-p^2/\mu^2)(-(p+q)^2/\mu^2)]^{-1/4}i\gamma^5$ forms that they take when $\gamma_\theta(\alpha)=-1$. To establish the renormalizability of a four-fermion interaction with these dressed vertices, we need to dress the kernel to all orders in the four-fermion coupling constant g. We thus need to dress $\Pi_{\rm S}(q^2,m)$,

IOP Conf. Series: Journal of Physics: Conf. Series **1239** (2019) 012017 doi:10.1088/1742-6596/1239/1/012017

$\Pi_{\rm P}(q^2,m)$, $\epsilon(m)$ and $\epsilon'(m)$ with internal fermion loop graphs. When this is done, it is found [22] that no new divergences beyond the lowest order single logarithm are generated, with their coefficients still being equal. The all-order in g renormalizability of the four-fermion theory with dressed $\tilde{\Gamma}_{\rm S}(p,p+q,q)$ and $\tilde{\Gamma}_{\rm P}(p,p+q,q)$ vertices with $\gamma_\theta(\alpha)=-1$ is thus established.

In the literature there have been many studies of an Abelian gluon model coupled to a four-fermion interaction. One of the first times the $\mathcal{L}_{\rm QED-FF}$ theory was explored in the literature was in [21], though in that paper only its mean-field aspects were studied and not the residual interaction aspects that we have presented here. One of the first studies of the residual interaction dynamics associated with $\mathcal{L}_{\rm QED-FF}$ was presented in [23], with this and other studies (as recently detailed in [18, 9]) finding that dynamical symmetry breaking would occur if α and g were related according to $-g\Lambda^2=\pi^2(1+(1-3\alpha/\pi)^{1/2})^2$. And as the presence of Λ indicates, these studies all involved a cutoff. The studies that lead to this relation differ from the study presented here in a crucial way. While these studies were based on the quenched planar graph approximation so that they did involve a fermion propagator $\tilde{S}(p)$ that scaled with an anomalous dimension just as given as per (14) in our study, these studies did not implement the Callan-Symanzik equation (13) and did not use the dressed $\tilde{\Gamma}_{\rm S}(p,p,0)$ given in (14), but instead took it to be given by the point-coupled $\tilde{\Gamma}_{\rm S}(p,p,0)=1$. Thus for the tadpole $\langle\Omega_M|\bar{\psi}\psi|\Omega_M\rangle$, instead of setting it equal to$-i\int d^4p/(2\pi)^4{\rm Tr}[\tilde{\Gamma}_{\rm S}(p,p,0)\tilde{S}_\mu(p)]$ as we did in (21), in these studies it was set equal to $-i\int d^4p/(2\pi)^4{\rm Tr}[\tilde{S}_\mu(p)]$. Without the dressed $\tilde{\Gamma}_{\rm S}(p,p,0)$ the tadpole was not softened enough (even with $\gamma_\theta(\alpha)=-1$), to hence require the use of a cutoff. However, in our case the tadpole is softened sufficiently to make it renormalizable. With our study also being an all-order planar plus non-planar graph quenched photon study, we are not constrained by the planar graph $-g\Lambda^2=\pi^2(1+(1-3\alpha/\pi)^{1/2})^2$ relation, a relation that imposes a lower bound on g. Rather, in our case (21) imposes no lower bound on g, with it only being required to be negative (viz. attractive). Moreover, if $\beta(\alpha)$ does have a zero, and if it is given by the fine-structure constant, α and g could then both be weak and still lead to dynamical symmetry breaking. One is thus able to associate dynamical symmetry breaking with weak coupling, with strong coupling not being required at all.

3.4. Distinguishing Dynamical and Elementary Higgs Bosons

The path integral associated with the massless fermion $\mathcal{L}_{\rm QED-FF}$ is of the form

$$Z(\bar{\eta},\eta)=\int D[\bar{\psi},\psi]\exp\left[i\int d^4x\left(\bar{\psi}i\gamma^\mu\partial_\mu\psi-e\bar{\psi}\gamma^\mu A_\mu\psi-\frac{g}{2}(\bar{\psi}\psi)^2+\bar{\eta}\psi+\bar{\psi}\eta\right)\right]. \tag{32}$$

(Not displayed are the A_μ integration measure, a source term for A_μ, or $-(g/2)(\bar{\psi}i\gamma^5\psi)^2$.) On introducing a dummy field σ we can rewrite the path integral as

$$\begin{aligned}&Z(\bar{\eta},\eta)=\\&\int D[\bar{\psi},\psi,\sigma]\exp\left[i\int d^4x\left(\bar{\psi}i\gamma^\mu\partial_\mu\psi-e\bar{\psi}\gamma^\mu A_\mu\psi-\frac{g}{2}(\bar{\psi}\psi)^2+\frac{g}{2}\left(\frac{\sigma}{g}-\bar{\psi}\psi\right)^2+\bar{\eta}\psi+\bar{\psi}\eta\right)\right]\\&=\int D[\bar{\psi},\psi,\sigma]\exp\left[i\int d^4x\left(\bar{\psi}\gamma^\mu i\partial_\mu\psi-e\bar{\psi}\gamma^\mu A_\mu\psi-\sigma\bar{\psi}\psi+\frac{\sigma^2}{2g}+\bar{\eta}\psi+\bar{\psi}\eta\right)\right].\end{aligned} \tag{33}$$

We recognize the path integral Lagrangian that we obtain as being of precisely the same form as the mean-field $\mathcal{L}_{\rm QED-MF}$ given above, with $\sigma(x)$ replacing $m(x)$. On introducing $\tilde{\Gamma}_{\rm S}(x)$ as the Fourier transform of $\tilde{\Gamma}_{\rm S}(p,p,0)$, on doing a path integration over the fermions we obtain an effective action in the $\sigma(x)$ sector of the form $\int D[\sigma]\exp[i\int d^4x\mathcal{L}_{\rm EFF}]$ where $\mathcal{L}_{\rm EFF}={\rm Trln}[i\partial\!\!\!/_x-\int d^4x'\sigma(x')\tilde{\Gamma}_{\rm S}(x-x')]$. On expanding in derivatives of $\sigma(x)$, we recognize

$\mathcal{L}_{\rm EFF}$ as being the effective Higgs Lagrangian given in (24), to thus generate a kinetic energy term for $\sigma(x)$.

However, while the path integral looks like that of an elementary scalar field, there is one key difference: there is no $J(x)\sigma(x)$ source term for $\sigma(x)$. The function $Z(\bar{\eta},\eta)$ only depends on the fermion sources, while if the scalar field were to be elementary the path integral would be associated with $Z(\bar{\eta},\eta,J)$ instead. With one and the same Lagrangian, the off-shell scalar field (internal exchange and loop diagrams) contributions to Green's functions with external fermion legs as generated by either $Z(\bar{\eta},\eta)$ or $Z(\bar{\eta},\eta,J)$ would be identical, with it being the all-order iteration of internal σ exchange diagrams that would generate the dynamical Goldstone and Higgs bosons that are not present in the σ field action itself. However, $Z(\bar{\eta},\eta,J)$ would also allow for Green's functions with external boson legs as well. Thus elementary and dynamical Higgs bosons only differ when the Higgs field goes on shell, while not differing off shell at all. With the Higgs width being an on-shell property of the Higgs field, again we see that it is in the width of the Higgs boson that one could potentially distinguish between a dynamical Higgs boson and an elementary one.

3.5. Conformal Gravity Treatment of the Vacuum Energy Density

If we evaluate the zero-point vacuum energy density of a free massive fermion with equation of motion $[i\gamma^{\mu}\partial_{\mu}-m]\psi=0$ and energy-momentum tensor $T^{\mu\nu}=i\hbar\bar{\psi}\gamma^{\mu}\partial^{\nu}\psi$, from the filled negative energy sea fermion modes we obtain

$$\langle\Omega_m|T^{00}|\Omega_m\rangle=-\frac{\Lambda^4}{16\pi^2}\ln(\Lambda^2)+\frac{\Lambda^4}{32\pi^2}-\frac{m^2\Lambda^2}{8\pi^2}+\frac{m^4}{16\pi^2}\ln\left(\frac{\Lambda^2}{m^2}\right)+\frac{m^4}{32\pi^2},\tag{34}$$

to thus generate quartic, quadratic, and logarithmic divergences. In flat spacetime one is only interested in energy-density differences, and so one can subtract off the zero-point vacuum energy density $\langle\Omega_0|T^{00}|\Omega_0\rangle=-\Lambda^4\ln(\Lambda^2)16\pi^2+\Lambda^4/32\pi^2$ associated with a free massless fermion, to yield the energy-density difference $\epsilon(m)$. If one then makes a further subtraction via the mean-field $\tilde{\epsilon}(m)=\epsilon(m)-m^2/2g$, one is left with a logarithmic divergence. If, on the other hand, one is in a critical scaling theory with $\gamma_{\theta}(\alpha)=-1$, the mass independent quartic term in $\epsilon(m)$ is not affected, while the quadratic divergence term becomes logarithmically divergent and the logarithmic divergence becomes finite. Then, the mean-field $\tilde{\epsilon}(m)$ is completely finite.

Now if one couples to gravity, one is not actually free to simply subtract away any infinite energy density contribution since gravity couples to energy density itself and not to energy-density difference. Thus to take care of the critical scaling logarithmically divergent term we need a dynamics that would induce an appropriate counterterm. As we have seen, such a counterterm can be supplied by a four-fermion interaction, an interaction that would even be renormalizable if $\gamma_{\theta}(\alpha)=-1$. It is thus gravity that forces the four-fermion interaction upon us, with it then serving to both renormalize the mean-field energy density and generate dynamical bound states. To cancel off the quartically divergent term we would need some additional field contribution, one that would have to be due to a boson since a bosonic zero-point energy density has the opposite sign to that of a fermion. Such a contribution could be provided by a supersymmetric partner of the fermion, but when the superpartner gets a mass it generates a new quadratic divergence, to thus give an unacceptably large vacuum energy density. An alternative bosonic field would be the gravitational field itself since it is also bosonic. However, no cancellation can be achieved via standard Einstein gravity, since as a quantum theory it is not renormalizable. However, conformal gravity [24, 25], viz. gravity based on the conformal invariant action $I_{\rm W}=-\alpha_g\int d^4x(-g)^{1/2}C_{\lambda\mu\nu\kappa}C^{\lambda\mu\nu\kappa}$ where $C^{\lambda\mu\nu\kappa}$ is the Weyl conformal tensor, is renormalizable, and ghost free [26, 27]. So now the cancellation can consistently be effected. Specifically, if one defines $(-g)^{-1/2}\delta I_{\rm W}/\delta g_{\mu\nu}=-2\alpha_g W^{\mu\nu}$, the conformal gravity equations of

IOP Conf. Series: Journal of Physics: Conf. Series **1239** (2019) 012017 doi:10.1088/1742-6596/1239/1/012017

motion take the form $-4\alpha_g W^{\mu\nu} + T^{\mu\nu} = 0$, with the gravitational and matter vacuum zero-point energy densities then canceling each other, with all the vacuum contributions, including those that arise due to the change in vacuum from $|\Omega_0\rangle$ to $|\Omega_m\rangle$ taking care of each other identically. (This differs from the elementary Higgs field case because there the shift in energy density is not cancelled.) With the vacuum contributions all taking care of each other when dynamical symmetry breaking is coupled to conformal gravity, what is observed in cosmology is only the much smaller contribution of the positive energy one-particle states (one-particle matrix elements of $4\alpha_g W^{\mu\nu} = T^{\mu\nu}$) that lie above the vacuum, and not the contribution due the entire filled vacuum itself [18]. In this way the cosmological constant is under control.

References

[1] ATLAS Collaboration, Aad G et. al. 2012 *Phys. Lett. B* **716** 1
[2] CMS Collaboration, Chatrchyan S et. al. 2012 *Phys. Lett. B* **716** 30
[3] Englert F and Brout R 1964 *Phys. Rev. Lett.* **13** 321
[4] Higgs P W 1964 *Phys. Lett.* **12** 132
[5] Higgs P W 1964 *Phys. Rev. Lett.* **13** 508
[6] Guralnik G S, Hagen C R and Kibble T W B 1964 *Phys. Rev. Lett.* **13** 585
[7] Baker M and Johnson K 1971 *Phys. Rev. D* **3** 2516
[8] Nambu Y and Jona-Lasinio G 1961 *Phys. Rev.* **122** 345
[9] Mannheim P D 2017 *Prog. Part. Nucl. Phys.* **94** 125
[10] Mannheim P D 1976 *Phys. Rev. D* **14** 2072
[11] Eguchi T and Sugawara H 1974 *Phys. Rev. D* **10** 4257
[12] Johnson K, Baker M and Willey R 1964 *Phys. Rev.* **136** B1111
[13] Johnson K, Willey R and Baker M 1967 *Phys. Rev.* **163** 1699
[14] Adler S L and Bardeen W A 1971 *Phys. Rev. D* **4**, 3045; 1972 **6** 734E
[15] Maskawa T and Nakajima H 1974 *Prog. Theor. Phys.* **52** 1326
[16] Maskawa T and Nakajima H 1975 *Prog. Theor. Phys.* **54** 860
[17] Miransky V A 1985 *Il Nuovo. Cim. A* **90** 149
[18] Mannheim P D 2017 *J. Phys. G* **44** 115003
[19] Mannheim P D 1974 *Phys. Rev. D* **10** 3311
[20] Mannheim P D 1975 *Phys. Rev. D* **12** 1772
[21] Mannheim P D 1978 *Nucl. Phys. B* **143** 285
[22] Mannheim P D 2017 *Phys. Lett. B* **773** 604
[23] Leung C N, Love S T and Bardeen W A 1986 *Nucl. Phys. B* **273** 649
[24] Mannheim P D 2006 *Prog. Part. Nucl. Phys.* **56** 340
[25] Mannheim P D 2012 *Found. Phys.* **42** 388
[26] Bender C M and Mannheim P D 2008 *Phys. Rev. Lett.* **100** 110402
[27] Bender C M and Mannheim P D 2008 *Phys. Rev. D* **78** 025022

IOP Conf. Series: Journal of Physics: Conf. Series **1239** (2019) 012018 doi:10.1088/1742-6596/1239/1/012018

Quantum equations for massless particles of any spin in curved spacetime

Amnon Moalem and Alexander Gersten
Department of Physics, Ben Gurion University of the Negev, Beer-Sheva, Israel
E-mail: amnon.moalem@gmail.com

Abstract. Quantum equations for massless particles of any spin in curved spacetime are presented. These equations can be extended to a global spacetime from the ones derived in a flat spacetime. Here we demonstrate that these equations can be derived also from a scalar Lagrangian, and they form a basis for quantum theory of spins 1 and 2 in in the vicinity of massive objects. The conserved current is calculated by applying Noether theorem and a complete set of commutative observables is determined. The solutions of these equations are considered in a flat Minkowski spacetime and in axially symmetric spacetimes.

1. Introduction

In previous contributions [1]-[4], it was demonstrated that free massless particles of any spin in flat Minkowski spacetime satisfy the following quantum wave equations,

$$\left(\frac{\hat{E}}{c}\gamma^0 - \gamma^k \hat{p}_k\right)\Phi^{(4s)} = 0, \tag{1}$$

where s is the spin, the gamma's $(\gamma^0, \gamma^1, \gamma^2, \gamma^3)$ denote $4s \times 4s$ Hermitian matrices, $\hat{E} = i\hbar\partial_t$ and $\hat{p}_k = -i\hbar\partial_k$ are the energy and momentum operators. Properties of these matrices are given in the Appendix. Here we note that the gamma matrices form representations of the Pauli matrices and factorize the d'Almbertian operator. For spin 1 and 2 they are chosen to reproduce the main equations and the subsidiary conditions [1]-[3] and have eigenvalues of ± 1 only, corresponding to forward and backward helicities. These are the only possible helicity values for massless particles. It is to be noted that the gamma matrices satisfy the Pauli algebra with γ^0 being the identity matrix.The wave functions $\Phi^{(4s)}$ in the expression above form bases for $D^{(s-1/2,1/2)}$representations of the Lorentz group, which we write as a $4s$ column matrix with $(2s-1)$ components of spin $(s-1)$ and $(2s+1)$ components of spin s, i.e.,

$$\Phi^{(4s)} = \begin{pmatrix} \psi^{s-1}_{s-1} \\ \cdot \\ \cdot \\ \psi^{s-1}_{-s+1} \\ \psi^{s}_{s} \\ \cdot \\ \cdot \\ \psi^{s}_{-s} \end{pmatrix}. \tag{2}$$

Using natural units with light velocity $c = 1$ and Planck constant $\hbar = 1$ Eq.1 can be rewritten as,

$$\gamma^b \eta_{ba} \partial_a \Phi^{(4s)} = 0, \tag{3}$$

where $\eta_{\mu\nu} = diag(1, -1, -1, -1)$ stands for the Minkowski metric.

Our main purpose in the present work is to discuss the properties of the solutions of a covariant form of the equations above. Such a covariant form was reported already in [5]. We recall that the derivation reported is based on the Principle of Equivalence which we may phrase as follows: A curved spacetime is locally a Minkowski spacetime in a sufficiently small region around any point of a curved spacetime . Accordingly, Eqs.(1) should be valid in within sufficiently small regions around any point of a curved spacetime and can be used as our starting point. In order to reformulate the above equations in a global curved spacetime some modifications are in order. First the Minkowski metric must be replaced by a curved spacetime tensor metric $g_{\mu\nu}$. Secondly, the partial local derivatives must be replaced by the covariant derivatives $(\partial_\mu + \Omega_\mu)$, where Ω_μ is the connection coefficient for the wave function $\Phi^{(4s)}$. Thirdly, the gamma matrices in Eq.(3) carry tangent space indices so that they maintain a flat spacetime form and must be transformed to global gamma matrices by means of vierbein fields *i.e.*, $\gamma^\mu = E^\mu_a \gamma^a$ [6]. Here $E^\mu_a(x)$ denotes an inverse vierbein field. A vierbein field is denoted by $e^a_\mu(x)$. As shown in [5], the connection coefficient of the wave function is related to the spin connection,

$$\Omega_\mu = \frac{1}{4}\omega_{\mu bc}\gamma^b\gamma^c \;\; ; \;\; \omega^a_{\mu b} = \frac{1}{2}\gamma^b \left[E^\sigma_b \partial_\sigma \ln e + \partial_\sigma E^\sigma_b\right], \tag{4}$$

where $e = \sqrt{-\det g}$ is the square root of the determinant of the global metric tensor $g_{\mu\nu}$. Inserting these modifications in Eq.(3) gives the covariant form of Eq.(1),

$$E^\mu_a \gamma^a \partial_\mu \left(\sqrt{e}\Phi^{4s}\right) = -\frac{1}{2}\gamma^b \partial_\sigma E^\sigma_b \Phi^{(4s)}, \tag{5}$$

as reported in [5]. We note that the Hermitian conjugate of $\Phi^{(4s)}$ is defined as $\left[\Phi^{(4s)}\right]^H = \left[\Phi^{*(4s)}\right]^T$. Clearly, $\Phi^H\Phi$ is positive definite which we interpret as probability density.

It is straightforward to guess that the Lagrangian density is given by,

$$\mathcal{L} = \Phi^H (E^\mu_a \gamma^a \partial_\mu \left(\sqrt{e}\Phi\right) + \frac{1}{2}\gamma^b \partial_\sigma E^\sigma_b \Phi). \tag{6}$$

The variation w.r.t. Φ^H obviously yields Eq.(5). Variation w.r.t. Φ gives the Hermitian conjugate equation,

$$\left(\partial_\mu \left(\sqrt{e}\Phi^H\right)\gamma^a E^\mu_a + \frac{1}{2}\partial_\sigma E^\sigma_b \Phi^H \gamma^b\right) = \left(E^\mu_a \gamma^a \partial_\mu \left(\sqrt{e}\Phi^*\right) + \frac{1}{2}\gamma^b \partial_\sigma E^\sigma_b \Phi^*\right) = 0. \tag{7}$$

The solutions Φ and Φ^* correspond to forward and backward helicity, each with two eigenvalues of energy one positive and one negative. In what follows we disregard negative energy solutions and use positive energy only but with opposite helicities. Our main interest in the present note is to demonstrate that Eqs.(5) and (7) form a basis for a quantum theory of massless particles of spin 1 (the photon) and spin 2 (the graviton). The conserved current density of Eq.(5) above can be obtained from the Lagrangian, Eq.(6), using Noether's Theorem. Let Φ be a solution of Eq.(5), and let $\delta\Phi^b$ be the variation of the b component of Φ of some conserved quantity, $\delta\Phi^b = i\Phi^b\delta\alpha$ and $\delta(\Phi^b)^* = -i(\Phi^b)^*\delta\alpha$, then,

$$j^\mu = \frac{\delta\mathcal{L}}{\delta\left(\partial_\mu\Phi_b\right)}\frac{\delta\Phi^b}{\delta\alpha} - J_0^\mu = \Phi_b^H E_a^\mu \gamma^a \Phi^b = \Phi^H \gamma^\mu \Phi. \tag{8a}$$

With $j^0 = \Phi^H\gamma^0\Phi = \Phi_b^H\Phi^b$ being the probability density, the current above satisfies the continuity equation, *i.e.*,

$$\partial_\mu j^\mu = \partial_\mu[\Phi^H\gamma^\mu\Phi]. \tag{9}$$

The conserved current above is independent of the spin connection. A more intuitive derivation of this expression given in the Appendix, explains why the conserved current does not depend on the spin connection. In what follows we consider Eq.(5) in Minkowski flat spacetime, and in axially symmetric spacetimes, namely, the Friedman-Walker-Robertson(FRW) and Schwarzschild spacetimes.

2. Free Particle Equations In Minkowski Spacetime

In Minkowski flat spacetime, $g_{\mu\nu} = diag(1,-1,-1,-1)$, $\sqrt{e} = 1$ so that Eqs.(5) and (7) reduce to,

$$i\partial_t\Phi = -i\boldsymbol{\gamma}\cdot\boldsymbol{\nabla}\Phi, \tag{10}$$

$$i\partial_t\Phi^* = i\boldsymbol{\gamma}\cdot\boldsymbol{\nabla}\Phi^*, \tag{11}$$

where $\boldsymbol{\nabla}$ is the del operator, and $\hat{E} = i\partial_t$ and $\hat{\boldsymbol{p}} = -i\boldsymbol{\nabla}$ are the energy and momentum operators, respectively. We identify $-i\,\boldsymbol{\gamma}\,\cdot\boldsymbol{\nabla}$ with the Hamiltonian H_0 of the unperturbed free space. In spherical coordinates,

$$\boldsymbol{\nabla} = \hat{\boldsymbol{r}}\,\gamma^r\partial_r + \hat{\boldsymbol{\theta}}\,\gamma^\theta\partial\theta + \hat{\boldsymbol{\phi}}\,\gamma^\phi\partial_\phi, \tag{12}$$

and the angular momentum operator is defined as,

$$\boldsymbol{L} = -i\hat{\boldsymbol{r}}\times\boldsymbol{\nabla}. \tag{13}$$

By multiplying this expression on the left side by $\hat{\boldsymbol{r}}$ we can write the $\boldsymbol{\nabla}$ as,

$$\boldsymbol{\nabla} = -\hat{\boldsymbol{r}}\boldsymbol{\partial}_r - \frac{1}{r}\hat{\boldsymbol{r}}\times\boldsymbol{L}. \tag{14}$$

It is simple to show, using the gamma matrices properties listed in the Appendix, that for any three vectors $\boldsymbol{A}$ and $\boldsymbol{B}$,

$$(\boldsymbol{\gamma}\,\cdot\boldsymbol{A})(\boldsymbol{\gamma}\,\cdot\boldsymbol{B}) = \gamma^0\boldsymbol{A}\cdot\boldsymbol{B} + \frac{i}{2}\boldsymbol{\gamma}\cdot(\boldsymbol{A}\times\boldsymbol{B})\,. \tag{15}$$

Then taking $\boldsymbol{A} = \hat{\boldsymbol{r}}$ and $\boldsymbol{B} = \boldsymbol{L}$, one obtains,

$$\boldsymbol{\gamma}\,\cdot\hat{\boldsymbol{r}}\times\boldsymbol{L} = -i\,(\boldsymbol{\gamma}\cdot\hat{\boldsymbol{r}})\,(2\boldsymbol{\gamma}\,\cdot\boldsymbol{L}). \tag{16}$$

IOP Conf. Series: Journal of Physics: Conf. Series **1239** (2019) 012018 doi:10.1088/1742-6596/1239/1/012018

With the above relation we can rewrite the Hamiltonian in the form,

$$H_0 = -i\gamma \cdot \boldsymbol{\nabla} = -i\gamma \cdot \hat{\boldsymbol{r}}\, \partial_r - \frac{1}{r}\gamma \cdot \hat{\boldsymbol{r}}\, (2\gamma \cdot \boldsymbol{L}). \tag{17}$$

Finally we identify the spin $\boldsymbol{S}$ with γ and define an angular operator,

$$K = 2\,\gamma \cdot \boldsymbol{L} + 1 = 2\boldsymbol{S} \cdot \boldsymbol{L} + 1 = J^2 - L^2 - S^2 + 1, \tag{18}$$

where $\boldsymbol{J} = \boldsymbol{L} + \boldsymbol{S}$ is the total angular momentum. Substituting this in Eq.(17) yields,

$$H_0 = -i\gamma \cdot \hat{\boldsymbol{r}}\, [\partial_r + \frac{1}{r} - \frac{1}{r}K]. \tag{19}$$

This expression of the Hamiltonian suggests that we may construct the solution Φ by separation of variables. In the Appendix we show that H_0, J^2, J_3, K and parity $\mathcal{P}$ form a complete set of commuting operators. Then the angular part of the spinor Φ is a spinor of spherical harmonics which we write as,

$$\mathcal{Y}^j_{lm_j}(\Omega) = \sum_{m_s} C\,(l, s, j; m_j - m_s, m_s, m_j)\, Y^{m_l}_l(\theta, \varphi)\, \chi^{m_s}_s, \tag{20}$$

where $C\,(l, s, j; m_j - m_s, m_s, m_j)$ is a Clebsch-Gordan coefficient for combining angular momenta l and s to total angular momentum j with magnetic quantum numbers $m_j - m_s, m_s, m_j$, respectively, $Y^{m_l}_l(\theta, \varphi)$ are the usual spherical harmonics which are eigenfunctions of L^2, L_3 and $\chi^{m_s}_s$ are eigenfunctions of S^2 and S_3. The spinor spherical harmonics are orthonormal, *i.e.*,

$$\int d\Omega \left(\mathcal{Y}^{j'}_{l'm_{j'}}(\Omega)\right)^\dagger \mathcal{Y}^j_{lm_j}(\Omega) = \delta_{j'j}\delta_{ll'}\delta_{m_{j'}m_j}. \tag{21}$$

Further more, the function Φ is an eigenfunction of the operators J^2 , J_3 angular operator K and parity $(\mathcal{P})$,

$$J^2\Phi^j_{m_j,\kappa} = j\,(j+1)\,\Phi^j_{m_j,\kappa_j}, \tag{22}$$

$$J_3\Phi^j_{m_j,\kappa} = m_j\Phi^j_{m_j,\kappa_j}, \tag{23}$$

$$K\Phi^j_{m_j,\kappa_j} = \kappa_j\Phi^j_{m_j,\kappa_j}. \tag{24}$$

Let us evaluate the eigenvalues of the operator $\kappa = (1 + 2\boldsymbol{L} \cdot \boldsymbol{S})^2$. From addition of angular momenta $j = l \pm s$.Then,

$$\kappa = (2\boldsymbol{S} \cdot \boldsymbol{L} + 1)^2 = \left(J^2 - L^2 - S^2 + 1\right)^2$$

$$= [j(j+1) - l(l+1) - s(s+1) + 1] = (2ls + 1)^2\,, \tag{25}$$

so that for for $j = l + s$, the eigenvalues are,

$$\kappa_j = (2ls + 1)\,, \tag{26}$$

and similarly, for $j = l - s$,

$$\kappa_j = -2s\,(l+1)\,. \tag{27}$$

We can use κ_j to label Φ since once it is set so are l and j.

IOP Conf. Series: Journal of Physics: Conf. Series **1239** (2019) 012018 doi:10.1088/1742-6596/1239/1/012018

2.1. Quantization of the Function Φ

The free particle solutions of Eqs.(10-11) correspond respectively to forward and backward helicity. Let $u(\boldsymbol{p}, s)\exp(-ipx)$ denote the positive energy and positive helicity solution and let $v(\boldsymbol{p}, s)\exp(ipx)$ denote the positive energy and negative helicity solution. Inserting these into Eqs.(10-11) gives,

$$(\omega_{\boldsymbol{p}} - \boldsymbol{\gamma}\cdot\boldsymbol{p})u(\boldsymbol{p}) = 0, \quad \text{and} \quad (\omega_{\boldsymbol{p}} + \boldsymbol{\gamma}\cdot\boldsymbol{p})v(\boldsymbol{p}) = 0. \tag{28}$$

Here $\omega_{\boldsymbol{p}} = p_0 = \pm\sqrt{\boldsymbol{p}\cdot\boldsymbol{p}}$. The spinors $u(\boldsymbol{p})$ and $v(\boldsymbol{p})$ are orthogonal so that we add the conditions,

$$u(\boldsymbol{p})^H v(\boldsymbol{p}) = v^H(\boldsymbol{p})\, u(\boldsymbol{p}) = 0, \tag{29}$$

and normalize the spinors according to,

$$u^a(\boldsymbol{p})\, u_a^\dagger(\boldsymbol{p}') = \omega_{\boldsymbol{p}} \ ; \quad v^a(\boldsymbol{p})\, v_a^\dagger(\boldsymbol{p}') = \omega_{\boldsymbol{p}}. \tag{30}$$

We then write,

$$\Phi(x) = \int \frac{d^3\boldsymbol{p}}{(2\pi)^{3/2}}\sqrt{\frac{1}{\omega_{\boldsymbol{p}}}}[b(\boldsymbol{p})\, u(\boldsymbol{p})\exp(-ipx) + d^\dagger(\boldsymbol{p})\, v(\boldsymbol{p})\exp(ipx)], \tag{31}$$

$$\Phi^H(x) = \int \frac{d^3\boldsymbol{p}}{(2\pi)^{3/2}}\sqrt{\frac{1}{\omega_{\boldsymbol{p}}}}\left[b^\dagger(\boldsymbol{p})\, u^\dagger(\boldsymbol{p})\exp(ipx) + d(\boldsymbol{p})\, v^\dagger(\boldsymbol{p})\exp(-ipx)\right], \tag{32}$$

where $b(\boldsymbol{p})$ and $b^\dagger(\boldsymbol{p})$ are annihilation and creation operators for a positive energy and forward helicity, and $d(\boldsymbol{p})$, $d^\dagger(\boldsymbol{p})$ are annihilation and creation operators for a positive energy and backward helicity. Next in order to quantize Φ we impose the following commutators,

$$\left[\Phi(x), \Phi(x')^H\right] = i\delta^3(\boldsymbol{x} - \boldsymbol{x}'),$$

$$[\Phi(x), \Phi(x')] = 0,$$

$$[\Phi^H(x), \Phi^H(x')] = 0. \tag{33}$$

These are equivalent to the following commutators for creation and annihilation operators,

$$\left[b(\boldsymbol{p}), b^\dagger(\boldsymbol{p}')\right] = \left[d(\boldsymbol{p}), d^\dagger(\boldsymbol{p}')\right] = \delta_{pp'}\delta_{\sigma\sigma'}, \tag{34}$$

$$\left[b^\dagger(\boldsymbol{p}), b^\dagger(\boldsymbol{p}')\right] = \left[d^\dagger(\boldsymbol{p}), d^\dagger(\boldsymbol{p}')\right] = [b(\boldsymbol{p}), b(\boldsymbol{p}')] = [d(\boldsymbol{p}), d(\boldsymbol{p}')] = 0, \tag{35}$$

$$\left[b(\boldsymbol{p}), b^\dagger(\boldsymbol{p}')\right] = \left[d(\boldsymbol{p}), b^\dagger(\boldsymbol{p}')\right] = \left[d^\dagger(\boldsymbol{p}), b(\boldsymbol{p}')\right] = 0, \tag{36}$$

where σ stands for the helicity. This can be shown rather straightforward by inserting Eq.(31) and Eq.(32) in Eqs.(33). Then substituting in the double integral obtained the normalization conditions Eq.(30), the orthogonality conditions Eq.(29), the commutator Eqs.(34-36), and by integrating Eqs.(33) are recovered.

3. Particle Equations In Axially Symmetric Spacetime

The equations (5) assume relatively simpler form in axially symmetric spacetime. Following refs.[9],[10] we assume that the wave function Φ factorizes as,

$$\Phi\left(x^0, x^1, x^2, x^3\right) = \psi\left(x^0, x^1, x^2, x^3\right) f\left(x^1, x^2\right), \tag{37}$$

and require that the function ψ satisfies an equation independent of the spin connection, i.e.,

$$E^{\mu}_{a}\gamma^{a}\partial_{\mu}\psi = 0. \tag{38}$$

Then by inserting Eq.(37) in Eq.(5) and using Eq.(38) one finds that the function $f\left(x^1, x^2\right)$ satisfies the equation,

$$\partial_{\mu}\ln\left(f\left(x^1, x^2\right)\sqrt{e}\right) = -\frac{1}{2}e^{b}_{\mu}\partial_{\sigma}E^{\sigma}_{b}. \tag{39}$$

The above expression depends on the metric tensor and contains the dependence on the spin connection. Obviously, axial symmetry restricts the form of the metric tensor and consequently the corresponding vierbein fields. In what follows we show that in FRW and Schwarzschild spacetimes Eq.(39) is integrable and its solution assumes a simple analytic form. We note that black holes are stationary axis-symmetric solutions of the Einstein equation[7],[8], so that the factorization procedure proposed above may serve to study the behavior of free massless particles such as photons and gravitons in the vicinity of black holes and massive objects. The Schwarzschild and the FRW metric tensors are diagonal and Eq.(39) assumes simple forms. Explicitly,

$$\partial_{\mu}\ln\left(f\left(x^1, x^2\right)\sqrt{e}\right) = -\frac{1}{2}[e^{t}_{t}\partial_{t}E^{t}_{t} + e^{r}_{r}\partial_{r}E^{r}_{r} + e^{\theta}_{\theta}\partial_{\theta}E^{\theta}_{\theta} + e^{\varphi}_{\varphi}\partial_{\varphi}E^{\varphi}_{\varphi}. \tag{40}$$

The last term on the r.h.s. vanishes because of symmetry. In what follows we consider the equations above in static FRW and Schwarzschild spacetimes.

3.1. Particle Equations In Static FRW Spacetime

In FRW spacetime the metric is $g_{\mu\nu} = diag\left(1, -a^2/F^2, -a^2r^2, -a^2r^2\sin^2\theta\right)$, where a representing the size of the universe with dimension of length, r is a dimensionless parameter and $F = \sqrt{1-kr^2}$. Thus the vierbein and inverse vierbein fields are respectively,

$$e^{a}_{\mu} = diag(1, a/F, ar, ar\sin\theta), \tag{41}$$

and,

$$E^{\mu}_{a} = diag\left(1, F/a, 1/\left(ar\right), 1/(ar\sin\theta)\right). \tag{42}$$

Inserting these into Eq.(40) and integrating gives,

$$\partial_{\mu}\ln\left(f\left(x^1, x^2\right)\sqrt{e}\right) = \partial_{\mu}\ln\left(1-kr^2\right)^{-1/4}, \tag{43}$$

and,

$$f\left(r, \theta\right) = a^{3/2}r^{1/2}\sin^{1/2}\theta. \tag{44}$$

Consider now Eq.(38). The FRW spacetime is spherically symmetric so that we can apply separation of variables as in the Minkowski case. In fact by inserting the vierbeins fields Eq.(38)

can be written in the form,

$$i\gamma^{t}\partial_{t}\psi = H_0\psi = -i\boldsymbol{\gamma}\cdot\hat{\boldsymbol{r}}\ [\frac{F}{a}\boldsymbol{\partial}_r + \frac{1}{ar} - \frac{1}{ar}K]\psi, \tag{45}$$

so that the angular part is identical up to a factor a to that found for the Minkowski case. The radial function is different because of the factor F/a. It is to be indicated that in the expression above the vierbein axes are chosen in the directions of t, x, y, z, so that the gamma matrices which were determined in [1]-[3], must be transformed to,

$$\begin{pmatrix} \gamma^t \\ \gamma^r \\ \gamma^\theta \\ \gamma^\varphi \end{pmatrix} = \begin{pmatrix} 1 & 0 & 0 & 0 \\ 0 & F & 0 & 0 \\ 0 & 0 & \frac{1}{r} & 0 \\ 0 & 0 & 0 & \frac{1}{r\sin\theta} \end{pmatrix} \begin{pmatrix} \gamma^0 \\ \gamma^1 \\ \gamma^2 \\ \gamma^3 \end{pmatrix}. \tag{46}$$

Here $(\gamma^0, \gamma^1, \gamma^2, \gamma^3)$ are the gamma matrices worked out in refs.[1]-[3] to reproduce the subsidiary conditions along with the main equation.

3.2. Particle Equations In Static Schwarzschild Spacetime

In Schwarzschild spacetime the metric is $g_{\mu\nu} = diag\left(F, -1/F, -r^2, -r^2\sin^2\theta\right)$, where $F = \left(1 - \frac{2GM}{r}\right)$. Here G is the gravitation constant and M is a constant with dimension of mass. This metric approaches the Minkowski metric as M approaches zero. Similarly, when r goes to infinity, the Schwarzschild metric approaches the Minkowski metric. The corresponding vierbein fields and their inverse are

$$e^a_\mu = diag(\sqrt{F}, 1/\sqrt{F}, r, r\sin\theta)$$

and

$$E^\mu_a = diag(1/\sqrt{F}, \sqrt{F}, 1/r, 1/\left(r\sin\theta\right)).$$

Inserting these into Eq.(40) and integrating gives,

$$f\left(r,\theta\right) = \left(1 - \frac{2MG}{r}\right)^{1/4} r\sin^{1/2}\theta. \tag{47}$$

The equation for ψ will be,

$$H_0\psi = -i\gamma\cdot\hat{\boldsymbol{r}}\ [\left(1 - \frac{2MG}{r}\right)\partial_r + \frac{1}{r} - \frac{1}{r}K]\psi. \tag{48}$$

Here as well the angular part is given by spherical harmonics as in the Minkowski case. However the radial function is different because of the factor $\left(1 - \frac{2MG}{r}\right)$.We may thus conclude that as in the Minkowski spacetime the solutions of Eq. (5) in the axially symmetric FRW and Schwarzschild spacetimes are simple modes which are specified by the orbital angular momentum and its z component $(l = 0, 1, 2, \cdots)$, $(m_l = -l, -l+1\cdots, l-1, l)$, the spin and its z component $(s\ ,\ m_s = -s, s)$,the total angular momentum $(j = l + s)$ and its z component $(m_j = -j, \cdots, j)$, and $\kappa_j = (2sl + 1)$ and $-2s\,(l+1)$ for $J = L \pm S$, respectively. To completely specify the solution Φ we have to add of course the radial quantum number $(n_r = 0, 1, 2, \cdots)$.

3.3. Quantization of the Function ψ

The quantization procedure which we have applied to the function Φ in Minkowski spacetime is applicable also in axially symmetric spacetime. In fact Eqs. (38) for the function ψ and its complex conjugate in axially symmetric spacetime are similar to Eqs.(10-11),

$$i\partial_t\psi = -i\gamma\cdot\nabla\psi, \tag{49}$$

$$i\partial_t\psi^* = i\gamma \cdot \nabla\psi^*. \tag{50}$$

Then we may quantize ψ following exactly the same procedure of quantizing Φ in the Minkowski case. We first expand ψ as,

$$\psi(x) = \int \frac{d^3\boldsymbol{p}}{(2\pi)^{3/2}}\sqrt{\frac{1}{\omega_{\boldsymbol{p}}}}[b(\boldsymbol{p})\,u(\boldsymbol{p})\exp(-ipx) + d^\dagger(\boldsymbol{p})\,v(\boldsymbol{p})\exp(ipx)], \tag{51}$$

$$\psi^H(x) = \int \frac{d^3\boldsymbol{p}}{(2\pi)^{3/2}}\sqrt{\frac{1}{\omega_{\boldsymbol{p}}}}\left[b^\dagger(\boldsymbol{p})\,u^\dagger(\boldsymbol{p})\exp(ipx) + d(\boldsymbol{p})\,v^\dagger(\boldsymbol{p})\exp(-ipx)\right], \tag{52}$$

and require that ψ satisfies the following commutators,

$$\left[\psi(x), \psi(x')^H\right] = i\delta^3(\boldsymbol{x} - \boldsymbol{x}'),$$

$$[\psi(x), \psi(x')] = 0,$$
$$[\psi^H(x), \psi^H(x')] = 0. \tag{53}$$

By substituting Eqs.(51-52) in the above expressions and imposing the orthogonality and normalization conditions Eqs.(29-30) we obtain exactly the same set of the commutators Eqs.(34-36).

4. Concluding Remarks

Quantum equations for free massless particles of any spin were derived in curved spacetime. These equations can be obtained by extending the ones presented in Refs.[1]-[3], to a global curved spacetime as demonstrated in Ref. [5]. Here we have shown that the same equations can be derived from a scalar Lagrangian and a complete set of commuting variables is determined. By applying the Noether theorem to this Lagrangian a conserved current which satisfies the continuity equation was found. The free particle equations assume a simple form in Minkowski spacetime and can be solved by variable separation. The solution of the equations in curved spacetime is more involved. We have shown though that in stationary and axially symmetric spacetime a solution can be obtained by factorizing the solutions into factors $\psi\left(x^t, x^r, x^\theta, x^\varphi\right)$ and $f\left(x^r, x^\theta\right)$ each satisfying a manageable equation. In particular the factor $f\left(x^r, x^\theta\right)$ which contains the spin connection can be calculated analytically for both the FRW and Schwarzschild space time. The angular part of $\psi\left(x^t, x^r, x^\theta, x^\varphi\right)$ is practically equal to that of Φ in the Minkowski case but the radial part needs to be calculated numerically. We believe that the scheme presented above may serve to study the eigenvalues and degeneracies of massless particles on the 3-sphere.

Appendix A. Properties of the Gamma Matrices

We recall that the gamma matrices satisfy the Pauli algebra and that γ^0 is the identity matrix.We recall that the gamma matrices satisfy the Pauli algebra and that γ^0 is the identity matrix. The gamma matrices satisfy the following relations,

$$(\gamma^a)^\dagger = (\gamma^a); \quad (\gamma^a)^2 = \gamma^0, \tag{A.1}$$

$$\left\{\gamma^a, \gamma^b\right\} = 2\gamma^0\delta^{ab}; \quad a, b = 0, 1, 2, 3, \tag{A.2}$$

$$\left[\gamma^a, \gamma^b\right] = i\epsilon^{abc}\gamma_c; \quad a, b, c = 1, 2, 3, \tag{A.3}$$

IOP Conf. Series: Journal of Physics: Conf. Series **1239** (2019) 012018 doi:10.1088/1742-6596/1239/1/012018

$$trace(\gamma^a) = 0; \quad a = 1, 2, 3. \tag{A.4}$$

We stress that although the anticommutators Eqs.(A.2) are common features of the Clifford algebra. Our gamma matrices do not form a Clifford algebra and are different from the Dirac's gamma matrices.

The gamma matrices satisfy the following identity: For any two three vectors $\boldsymbol{A}$ and $\boldsymbol{B}$,

$$(\gamma \cdot \boldsymbol{A})(\gamma \cdot \boldsymbol{B}) = \boldsymbol{A} \cdot \boldsymbol{B} + \frac{i}{2}\gamma \cdot \boldsymbol{A} \times \boldsymbol{B}. \tag{A.5}$$

Proof:

$$(\gamma^i A_i)(\gamma^j B_j) = \gamma^i \gamma^j A_i B_j$$

$$= \delta_{ij}(\gamma^i)^2 A_i B_j + \frac{i}{2}\epsilon^{ijk}\gamma_k A_i B_j = \gamma^0 A_i B_i + \frac{i}{2}\gamma_k(\boldsymbol{A} \times \boldsymbol{B})^k, \tag{A.6}$$

Where we have used Eqs.(A.1),(A.2) and (A.3).

Appendix B. Conserved Current

By factorizing the d'Alembertian operator we obtain two equations (see ref. [5] for more details),

$$\gamma^\nu(x)(\partial_\nu + \Omega_\nu)\Phi(x) = 0, \tag{B.1}$$

$$\gamma^\mu(x)\left(\partial_\mu + \Omega^*_\mu\right)\Phi^*(x) = 0, \tag{B.2}$$

where Φ^* is the complex conjugate of Φ, $\Omega_\mu(x)$ and $\Omega^*_\mu(x)$ are the connection coefficients for the functions $\Phi(x)$ and $\Phi^*(x)$. Here we note that $\Omega(x) = -\Omega^*(x)$. We use the equations above to prove our claim for the conserved current. We multiply Eq.(B.1) by the Hermitian conjugate Φ^H and Eq.(B.2) by the transposed Φ^T of Φ, and sum both equations to obtain,

$$E^\nu_a\psi^H_1\gamma^a\partial_\nu\psi_1 + \cdots + E^\nu_a\psi^H_{(4s)}\gamma^a\partial_\nu\psi_{(4s)} + E^\nu_a\psi_1\gamma^a\partial_\nu\psi^H_1 + \cdots + E^\nu_a\psi_{(4s)}\gamma^a\partial_\nu\psi^H_{(4s)} = 0. \tag{B.3}$$

Now we can sum pairs of terms like,

$$E^\mu_a\psi^H_n\gamma^a\partial_\mu\psi_n + E^\mu_a\psi_n\gamma^a\partial_\mu\psi^H_n = \partial_a(\psi^H_n\gamma^a\psi_n). \tag{B.4}$$

Inserting Eq.(B.4) in Eq.(B.3) yields,

$$\partial_0(\Phi^H\Phi) + \partial_k\left(\Phi^H\gamma^k\Phi\right) = 0; k = 1, 2, 3. \tag{B.5}$$

This explains intuitively why the conserved current does not depend on the spin connection. Note that $\Phi^H\Phi$ is positive definite and is interpreted as probability density.

IOP Conf. Series: Journal of Physics: Conf. Series **1239** (2019) 012018 doi:10.1088/1742-6596/1239/1/012018

Appendix C. Commuting Variables

In what follows we show that the total angular momentum $\boldsymbol{J} = \boldsymbol{L} + \boldsymbol{S}$, J^2, J_3, $\kappa = 2\boldsymbol{\gamma} \cdot \boldsymbol{L} - 1$ and parity $\mathcal{P}$ form complete set of commuting observables. To show that we recall that the Hamiltonian of the free field and the ith component of the orbital and spin angular momentum are given by,

$$H_0 = -i\gamma^l \partial_l, \tag{C.1}$$

$$L_i = -i\epsilon^{ijk} x_j \partial_k, \tag{C.2}$$

and,

$$S_i = \epsilon^{ijk} \gamma_j \partial_k. \tag{C.3}$$

Then by simple algebraic manipulations one finds that,

$$[L_i, H_0] = +\epsilon^{ijk} \gamma_j \partial_k, \tag{C.4}$$

and,

$$[S_i, H_0] = -\epsilon^{ijk} \gamma_j \partial_k. \tag{C.5}$$

Then from the expressions Eq.(C.4) and Eq.(C.5), The ith component of the total angular momentum commute with the Hamiltonian so that,

$$[J_3, H_0] = 0, \tag{C.6}$$

$$\left[J^2, H_0\right] = 0. \tag{C.7}$$

The operator $\kappa = 2\boldsymbol{\gamma} \cdot \boldsymbol{L} - 1$ also commutes with J_i since,

$$[L_i, K] = [L_i, 2\gamma \cdot L] = 2\left(L_i \gamma^j L_j - \gamma^j L_j L_i\right) = 2i\gamma^j \epsilon^{ijk} L_k, \tag{C.8}$$

$$[S_i, K] = [\gamma_i, 2\gamma \cdot L] = 2\left(\gamma_i \gamma^j L_j - \gamma^j L_j \gamma_i\right) = 2\left[\gamma^i, \gamma^j\right] L_j = 2i\epsilon^{ikj} \gamma_j L_k. \tag{C.9}$$

These two expressions above sum to zero and hence,

$$[J_i, K] = 0, \tag{C.10}$$

Now we show that the parity operator commutes with the Hamiltonian. By definition,

$$\mathcal{P} = \beta P; \quad \beta = \begin{pmatrix} I_{2s} & 0 \\ 0 & -I_{2s} \end{pmatrix}, \tag{C.11}$$

where $\mathcal{P}$ and P are the parity operators acting on the spinors and on coordinates, respectively. Indeed,

$$[\mathcal{P}, H_0] = [\beta P, -i\gamma^k \partial_k] = -i\beta P \gamma^k \partial_k + \gamma^k \partial_k i\beta P = -i\beta\gamma^k P \partial_k + i\gamma^k \beta \partial_k P$$

$$= -i\beta P \gamma^k (-\partial_k P) + i\gamma^k \beta \partial_k P = i\left\{\beta, \gamma^k\right\} \partial_k P = 0. \tag{C.12}$$

We have used the fact that P does not depend on the coordinates and therefore $\partial_k P = 0$. Furthermore the parity operator commutes with J_i since $[\mathcal{P}, J_i] = \beta [P, J_i]$. The coordinate parity operator P commutes with J_i since the angular momentum is a pseudovector. To conclude H_0, J^2, J_3, K and P form a complete set of commuting operators.

IOP Conf. Series: Journal of Physics: Conf. Series **1239** (2019) 012018 doi:10.1088/1742-6596/1239/1/012018

References

[1] A. Gersten and A. Moalem, *J. Phys. Conf. Series*, **330**(2011),012010.
[2] A. Gersten and A. Moalem, *J. Phys. Conf. Series*, **437**(2013),012019.
[3] A. Gersten and A. Moalem, *J. Phys. Conf. Series*, **615**(2015),012011.
[4] A. Gersten and A. Moalem, J. Phys. Conf. Series, **845**(2017), 012011.
[5] A. Moalem and A. Gersten, *J. Phys. Conf. Series*, **845**(2017),012029.
[6] See for example Jeffery Yepez, arXiv:1106.3037,(2011)10.
[7] See for example, Robert M. Wald, *General Relativity* , University of Chicago Press(1984).
[8] C.I.S. Clarke, *Relativity on Curved Manifolds*, Cambridge University Press(1990).
[9] Dag Morten Sjostrom , *Boson and Fermions in mCurved Spacetime*, Thesis, Norwegian University of Science and Technology,2013.
[10] Seyd Alwi B. Ahmad, *Proceedings of the Conference in Honor of the 90th Birthday of Freeman Dyson*, Edited by Phua K K et al. Published by World Scientific Publications Co. Ltd., 2014,p 419-424.

IOP Conf. Series: Journal of Physics: Conf. Series **1239** (2019) 012019 doi:10.1088/1742-6596/1239/1/012019

Solutions in the $(\frac{1}{2},0)\oplus(0,\frac{1}{2})$ Representation of the Lorentz Group

Valeriy V. Dvoeglazov

UAF, Universidad Autónoma de Zacatecas
Apartado Postal 636, Suc. 3, Zacatecas 98061 Zac., México

E-mail: valeri@fisica.uaz.edu.mx

Abstract. I present explicit examples of generalizations in relativistic quantum mechanics. First of all, I discuss the generalized spin-1/2 equations for neutrinos. They have been obtained by means of the Gersten-Sakurai method for derivations of arbitrary-spin relativistic equations. Possible physical consequences are discussed. Next, it is easy to check that both Dirac algebraic equation $Det(\hat{p}-m)=0$ and $Det(\hat{p}+m)=0$ for $u-$ and $v-$ 4-spinors have solutions with $p_0=\pm E_p=\pm\sqrt{\mathbf{p}^2+m^2}$. The same is true for higher-spin equations. Meanwhile, every book considers the equality $p_0=E_p$ for both $u-$ and $v-$ spinors of the $(1/2,0)\oplus(0,1/2))$ representation only, thus applying the Dirac-Feynman-Stueckelberg procedure for elimination of the negative-energy solutions. The recent work by Ziino (and, independently, the articles of several others) show that the Fock space can be doubled. We re-consider this possibility on the quantum field level for $S=1/2$ particles. The third example is: we postulate the non-commutativity of 4-momenta, and we derive the mass splitting in the Dirac equation. Some applications are discussed.

1. Generalized Neutrino Equations.

A. Gersten [1] proposed a method for derivations of massless equations of arbitrary-spin particles. In fact, his method is related to the van der Waerden-Sakurai [2] procedure for the derivation of the massive Dirac equation. I commented on the derivation of the Maxwell equations (the $S=1$ first-quantized equations) in [3]: Then I showed that one can obtain *generalized* $S=1$ equations, which connect the antisymmetric tensor field with additional scalar fields. The problem of physical significance of additional scalar fields should be solved by experiment (see, however, the note on QED [4]). In the present talk I apply the similar procedure to the spin-1/2 fields. As a result one obtains equations which *generalize* the well-known Weyl equations. However, these equations have been known for a long time [5]. Raspini [6, 7] analyzed them again in detail. I add some comments on physical significance of the generalized spin-1/2 equations.

Let me use the equation (4) of the first Gersten paper [1] for the two-component spinor field function:

$$(E^2-c^2\vec{\mathbf{p}}^2)I^{(2)}\psi=\left[EI^{(2)}-c\vec{\mathbf{p}}\cdot\vec{\boldsymbol{\sigma}}\right]\left[EI^{(2)}+c\vec{\mathbf{p}}\cdot\vec{\boldsymbol{\sigma}}\right]\psi=0\,. \quad (1)$$

Actually, this equation is the massless limit of the equation which was presented in the Sakurai book [2]. In the latter case one should substitute m^2c^4 into the right-hand side of Eq. (1):

$$\left[EI^{(2)}-c\vec{\mathbf{p}}\cdot\vec{\boldsymbol{\sigma}}\right]\left[EI^{(2)}+c\vec{\mathbf{p}}\cdot\vec{\boldsymbol{\sigma}}\right]\psi=m_2^2c^4\psi\,. \quad (2)$$

IOP Conf. Series: Journal of Physics: Conf. Series **1239** (2019) 012019 doi:10.1088/1742-6596/1239/1/012019

However, instead of equation (3.25) of [2] one can define the two-component 'right' field function

$$\phi_R = \frac{1}{m_1 c}(i\hbar\frac{\partial}{\partial x_0} - i\hbar\boldsymbol{\sigma}\,\boldsymbol{\nabla})\psi, \quad \phi_L = \psi \tag{3}$$

with the different mass parameter m_1. In such a way we come to the system of the first-order differential equations

$$(i\hbar\frac{\partial}{\partial x_0} + i\hbar\boldsymbol{\sigma}\,\boldsymbol{\nabla})\phi_R = \frac{m_2^2 c}{m_1}\phi_L\,, \tag{4}$$

$$(i\hbar\frac{\partial}{\partial x_0} - i\hbar\boldsymbol{\sigma}\,\boldsymbol{\nabla})\phi_L = m_1 c\phi_R\,. \tag{5}$$

It can be re-written in the 4-component form:

$$\begin{pmatrix} i\hbar(\partial/\partial x_0) & i\hbar\boldsymbol{\sigma}\cdot\boldsymbol{\nabla} \\ -i\hbar\boldsymbol{\sigma}\cdot\boldsymbol{\nabla} & -i\hbar(\partial/\partial x_0)\end{pmatrix}\begin{pmatrix}\psi_A \\ \psi_B\end{pmatrix} = \tag{6}$$

$$= \frac{c}{2}\begin{pmatrix} (m_2^2/m_1 + m_1) & (-m_2^2/m_1 + m_1) \\ (-m_2^2/m_1 + m_1) & (m_2^2/m_1 + m_1)\end{pmatrix}\begin{pmatrix}\psi_A \\ \psi_B\end{pmatrix}$$

for the function $\Psi = column(\psi_A \quad \psi_B) = column(\phi_R + \phi_L \quad \phi_R - \phi_L)$. The equation (6) can be written in the covariant form:

$$\left[i\gamma^\mu\partial_\mu - \frac{m_2^2 c}{m_1\hbar}\frac{(1-\gamma^5)}{2} - \frac{m_1 c}{\hbar}\frac{(1+\gamma^5)}{2}\right]\Psi = 0\,. \tag{7}$$

The standard representation of γ^μ matrices has been used here.

If $m_1 = m_2$ we can recover the standard Dirac equation. As noted in [5b] this procedure can be viewed as the simple change of the representation of γ^μ matrices. However, this is valid only if the mass to not be equal to zero. Otherwise, the entries in the transformation matrix become singular. The unitary matrix does not exist.

Furthermore, one can either repeat a similar procedure (the modified Sakurai procedure) starting from the *massless* equation (4) of [1a] or put $m_2 = 0$ in eq. (7). The *massless equation* is

$$\left[i\gamma^\mu\partial_\mu - \frac{m_1 c}{\hbar}\frac{(1+\gamma^5)}{2}\right]\Psi = 0\,. \tag{8}$$

It is necesary to stress that the term *'massless'* is used in the sense that $p_\mu p^\mu = 0$. The solutions of the equation (8) are

$$u_\sigma^{(1)}(\mathbf{p}) = \begin{pmatrix} [E + \boldsymbol{\sigma}\cdot\mathbf{p}]\phi_\sigma \\ [E + m_1 - \boldsymbol{\sigma}\cdot\mathbf{p}]\chi_\sigma \end{pmatrix}, \quad v_\sigma^{(1)}(\mathbf{p} = \gamma^5 u_\sigma^{(1)}(\mathbf{p})\,. \tag{9}$$

in the spinorial representation of γ-matrices. Then, we may have different physical consequences following from (8) comparing with those which follow from the Weyl equation.[1] The mathematical reason for such a possibility of different massless limits is that the corresponding change of representation of γ^μ matrices involves the mass parameters m_1 and m_2 themselves.

It is interesting to note that we can also repeat this procedure for the definition (or for even more general definitions):

$$\phi_L = \frac{1}{m_3 c}(i\hbar\frac{\partial}{\partial x_0} + i\hbar\boldsymbol{\sigma}\cdot\boldsymbol{\nabla})\psi, \quad \phi_R = \psi\,, \tag{10}$$

[1] Remember that the Weyl equation is obtained as $m \to 0$ limit of the usual Dirac equation.

with the additional arbitrary mass parameter m_3. This is due to the fact that the parity properties of the two-component spinor are undefined in the two-component equation. The resulting equation is

$$\left[i\gamma^\mu\partial_\mu - \frac{m_4^2 c}{m_3 \bar{h}}\frac{(1+\gamma^5)}{2} - \frac{m_3 c}{\bar{h}}\frac{(1-\gamma^5)}{2}\right]\tilde{\Psi} = 0\,, \tag{11}$$

which gives us yet another equation in the massless limit (the physical mass $m_4 \to 0$):

$$\left[i\gamma^\mu\partial_\mu - \frac{m_3 c}{\bar{h}}\frac{(1-\gamma^5)}{2}\right]\tilde{\Psi} = 0\,, \tag{12}$$

differing in the sign at the γ_5 term. Solutions of the equation (12) are

$$u_\sigma^{(2)}(\mathbf{p}) = \begin{pmatrix} [E + m_3 + \boldsymbol{\sigma}\cdot\mathbf{p}]\tilde{\phi}_\sigma \\ [E - \boldsymbol{\sigma}\cdot\mathbf{p}]\tilde{\chi}_\sigma \end{pmatrix}, \quad v_\sigma^{(2)}(\mathbf{p} = \gamma^5 u_\sigma^{(2)}(\mathbf{p})\,. \tag{13}$$

At this moment, neither (9) nor (13) are orthonormalized.

The above procedure can be generalized to *any* Lorentz group representations, *i. e.*, to any spins. In some sense the equations (8),(12) are analogous to the $S = 1$ equations [3, (4-7,10-13)], which also contain additional parameters.

The physical content of the generalized $S = 1/2$ *massless* equations may be different from that of the Weyl equation. The excellent discussion can be found in [5a,b]. First of all, the theory does *not* have chiral invariance. Those authors call the additional parameters the measure of the degree of chirality. Apart of this, Tokuoka introduced the concept of the gauge transformations (not to confuse with phase transformations) for the 4-spinor fields. He also found some strange properties of the anti-commutation relations (see §3 in [5a] and cf. [8] and [9, 10]). And finally, the equations (8,12) describe *four* states, two of which correspond to the positive energy $p_0 = |\mathbf{p}|$, and two others correspond to the negative energy $p_0 = -|\mathbf{p}|$.

I just want to add the following to the discussion. The operator of the *chiral-helicity* $\hat{\eta} = (\boldsymbol{\alpha}\cdot\hat{\mathbf{p}})$ (in the spinorial representation) used in [5b] does *not* commute, *e.g.*, with the Hamiltonian of the equation (8):[2]

$$[\mathcal{H}, \boldsymbol{\alpha}\cdot\hat{\mathbf{p}}]_- = 2\frac{m_1 c}{\bar{h}}\frac{1-\gamma^5}{2}(\boldsymbol{\gamma}\cdot\hat{\mathbf{p}})\,, \tag{14}$$

thus not having the common eigenstates. For the eigenstates of the *chiral-helicity* the system of corresponding equations can be read ($\eta =\uparrow, \downarrow$)

$$i\gamma^\mu\partial_\mu\Psi_\eta - \frac{m_1 c}{\bar{h}}\frac{1+\gamma^5}{2}\Psi_{-\eta} = 0\,. \tag{15}$$

The conjugated eigenstates of the Hamiltonian $|\Psi_\uparrow + \Psi_\downarrow >$ and $|\Psi_\uparrow - \Psi_\downarrow >$ are connected, in fact, by γ^5 transformation $\Psi \to \gamma^5\Psi \sim (\boldsymbol{\alpha}\cdot\hat{\mathbf{p}})\Psi$ (or $m_1 \to -m_1$). However, the γ^5 transformation is related to the PT ($t \to -t$ only) transformation, which, in its turn, can be interpreted as the change of the sign of the energy, if one accepts the Stueckelberg idea about antiparticles. We associate $|\Psi_\uparrow + \Psi_\downarrow >$ with the positive-energy eigenvalue of the Hamiltonian and $|\Psi_\uparrow - \Psi_\downarrow >$, with the negative-energy eigenvalue of the Hamiltonian. Thus, the free chiral-helicity massless eigenstates may oscillate one to another with the frequency $\omega = E/\bar{h}$(as the massive chiral-helicity eigenstates, see [11a] for details). Moreover, a special kind of interaction which is not symmetric with respect to the chiral-helicity states (for instance, if the left chiral-helicity

[2] Do not confuse with the Dirac Hamiltonian.

IOP Conf. Series: Journal of Physics: Conf. Series **1239** (2019) 012019 doi:10.1088/1742-6596/1239/1/012019

eigenstates interact with the matter only) may induce changes in the oscillation frequency, like in the Wolfenstein (MSW) formalism.

The question is: how can these frameworks be connected with the Ryder method of derivation of relativistic wave equations [12], and with the subsequent analysis of the problems of the choice of normalization and that of the choice of phase factors in the papers [11, 9, 13]? However, the conclusion may be similar to that which was achieved before: the dynamical properties of the massless particles (*e. g.*, neutrinos and photons) may differ from those defined by the well-known Weyl and Maxwell equations [13, 14].

2. Negative Energies in the Dirac Equation.

The recent problems of superluminal neutrinos, negative mass-squared neutrinos, various schemes of oscillations including sterile neutrinos, e. g. [15], etc require much attention. Next, the problem of the lepton mass splitting (e, μ, τ) has long history [16]. This suggests that something is missed in the foundations of relativistic quantum theories. Modifications seem to be necessary in the Dirac sea concept, and in the even more sophisticated Stueckelberg concept of the backward propagation in time. The Dirac sea concept is intrinsically related to the Pauli principle. However, the Pauli principle is intrinsically connected with the Fermi statistics and the anticommutation relations of fermions. Recently, the concept of *bi-orthonormality* has been proposed; the (anti) commutation relations and statistics are assumed to be different for *neutral* particles [9] (cf. [23]).

We observe some interisting things related to the negative-energy concept. Usually, everybody uses the following definition of the field operator [17] in the pseudo-Euclidean metrics:

$$\Psi(x) = \frac{1}{(2\pi)^3}\sum_h \int \frac{d^3\mathbf{p}}{2E_p}[u_h(\mathbf{p})a_h(\mathbf{p})e^{-ip\cdot x} + v_h(\mathbf{p})b_h^\dagger(\mathbf{p})]e^{+ip\cdot x}]\,, \tag{16}$$

as given *ab initio.* After actions of the Dirac operator on $\exp(\mp ip\cdot x)$ the 4-spinors ($u-$ and $v-$) satisfy the momentum-space equations: $(\hat{p}-m)u_h(p) = 0$ and $(\hat{p}+m)v_h(p) = 0$, respectively; h is the polarization index. However, it is easy to prove from the characteristic equations $Det(\hat{p} \mp m) = (p_0^2 - \mathbf{p}^2 - m^2)^2 = 0$ that the solutions should satisfy the energy-momentum relation $p_0 = \pm E_p = \pm\sqrt{\mathbf{p}^2 + m^2}$ in both cases.

Let me recall the general scheme of construction of the field operator, which was presented in [18]. In the case of the $(1/2, 0) \oplus (0, 1/2)$ representation we have:

$$\begin{aligned}
\Psi(x) &= \frac{1}{(2\pi)^3}\int d^4p\, \delta(p^2 - m^2)e^{-ip\cdot x}\Psi(p) = \\
&= \frac{1}{(2\pi)^3}\sum_h \int d^4p\, \delta(p_0^2 - E_p^2)e^{-ip\cdot x}u_h(p_0,\mathbf{p})a_h(p_0,\mathbf{p}) = \\
&= \frac{1}{(2\pi)^3}\int \frac{d^4p}{2E_p}[\delta(p_0 - E_p) + \delta(p_0 + E_p)] \\
&\quad [\theta(p_0) + \theta(-p_0)]e^{-ip\cdot x}\sum_h u_h(p)a_h(p) = \\
&= \frac{1}{(2\pi)^3}\sum_h \int \frac{d^4p}{2E_p}[\delta(p_0 - E_p) + \delta(p_0 + E_p)] \\
&\quad \left[\theta(p_0)u_h(p)a_h(p)e^{-ip\cdot x} + \theta(p_0)u_h(-p)a_h(-p)e^{+ip\cdot x}\right] = \\
&= \frac{1}{(2\pi)^3}\sum_h \int \frac{d^3\mathbf{p}}{2E_p}\theta(p_0)\left[u_h(p)a_h(p)|_{p_0=E_p}e^{-i(E_pt-\mathbf{p}\cdot\mathbf{x})} + \right. \\
&+ \left. u_h(-p)a_h(-p)|_{p_0=E_p}e^{+i(E_pt-\mathbf{p}\cdot\mathbf{x})}\right]
\end{aligned} \tag{17}$$

IOP Conf. Series: Journal of Physics: Conf. Series **1239** (2019) 012019 doi:10.1088/1742-6596/1239/1/012019

During the calculations above we had to represent $1=\theta(p_0)+\theta(-p_0)$ in order to get positive- and negative-frequency parts.[3] Moreover, during these calculations we did not yet assume, which equation this field operator (namely, the $u-$ spinor) does satisfy, with negative- or positive- mass (energy). In general we should transform $u_h(-p)$ to the $v(p)$. The procedure is the following one [20]. In the Dirac case we should assume the following relation in the field operator:

$$\sum_{h=\pm 1/2} v_h(p) b_h^\dagger(p) = \sum_{h=\pm 1/2} u_h(-p) a_h(-p)\,. \tag{18}$$

We know that [12][4]

$$\bar{u}_{(\mu)}(p) u_{(\lambda)}(p) = +m\delta_{\mu\lambda}\,, \tag{19}$$
$$\bar{u}_{(\mu)}(p) u_{(\lambda)}(-p) = 0\,, \tag{20}$$
$$\bar{v}_{(\mu)}(p) v_{(\lambda)}(p) = -m\delta_{\mu\lambda}\,, \tag{21}$$
$$\bar{v}_{(\mu)}(p) u_{(\lambda)}(p) = 0\,, \tag{22}$$

but we need $\Lambda_{(\mu)(\lambda)}(p)=\bar{v}_{(\mu)}(p)u_{(\lambda)}(-p)$. By direct calculations, we find

$$-mb_{(\mu)}^\dagger(p) = \sum_\lambda \Lambda_{(\mu)(\lambda)}(p) a_{(\lambda)}(-p)\,. \tag{23}$$

Hence, $\Lambda_{(\mu)(\lambda)}=-im(\boldsymbol{\sigma}\cdot\mathbf{n})_{(\mu)(\lambda)}$, $\mathbf{n}=\mathbf{p}/|\mathbf{p}|$, and

$$b_{(\mu)}^\dagger(p) = +i\sum_\lambda (\boldsymbol{\sigma}\cdot\mathbf{n})_{(\mu)(\lambda)} a_{(\lambda)}(-p)\,. \tag{24}$$

Multiplying (18) by $\bar{u}_{(\mu)}(-p)$ we obtain

$$a_{(\mu)}(-p) = -i\sum_\lambda (\boldsymbol{\sigma}\cdot\mathbf{n})_{(\mu)(\lambda)} b_{(\lambda)}^\dagger(p)\,. \tag{25}$$

The equations are self-consistent.[5]

However, other ways of thinking are possible. First of all to mention, we have, in fact, $u_h(E_p,\mathbf{p})$ and $u_h(-E_p,\mathbf{p})$, and $v_h(E_p,\mathbf{p})$ and $v_h(-E_p,\mathbf{p})$ originally, which satisfy the equations:[6]

$$\left[E_p(\pm\gamma^0) - \boldsymbol{\gamma}\cdot\mathbf{p} - m\right] u_h(\pm E_p,\mathbf{p}) = 0\,. \tag{27}$$

Due to the properties $U^\dagger\gamma^0 U = -\gamma^0$, $U^\dagger\gamma^i U = +\gamma^i$ with the unitary matrix $U=\begin{pmatrix}0 & -1\\ 1 & 0\end{pmatrix} = \gamma^0\gamma^5$ in the Weyl basis,[7] we have

$$\left[E_p\gamma^0 - \boldsymbol{\gamma}\cdot\mathbf{p} - m\right] U^\dagger u_h(-E_p,\mathbf{p}) = 0\,. \tag{28}$$

[3] See Ref. [19] for some discussion.

[4] (μ) and (λ) are the polazrization indices here. According to the referee advice I use parenthesis here to stress this.

[5] In the $(1,0)\oplus(0,1)$ representation the similar procedure leads to somewhat different situation:

$$a_{(\mu)}(p) = [1-2(\mathbf{S}\cdot\mathbf{n})^2]_{(\mu)(\lambda)} a_{(\lambda)}(-p)\,. \tag{26}$$

This signifies that in order to construct the Sankaranarayanan-Good field operator (which was used recently), it satisfies $[\gamma_{\mu\nu}\partial_\mu\partial_\nu - \frac{(i\partial/\partial t)}{E}m^2]\Psi(x)=0$, we need additional postulates. For instance, one can try to construct the left- and the right-hand side of the field operator separately each other [19].

[6] Remember that, as before, we can always make the substitution $\mathbf{p}\to-\mathbf{p}$ in any of the integrands of (17).

[7] The properties of the U-matrix are opposite to those of $P^\dagger\gamma^0P=+\gamma^0$, $P^\dagger\gamma^iP=-\gamma^i$ with the usual $P=\gamma^0$, thus giving $\left[-E_p\gamma^0+\boldsymbol{\gamma}\cdot\mathbf{p}-m\right]Pu_h(-E_p,\mathbf{p}) = -[\hat{p}+m]\tilde{v}_?(E_p,\mathbf{p}) = 0$. While, the relations of the spinors $v_h(E_p,\mathbf{p})=\gamma_5 u_h(E_p,\mathbf{p})$ are well-known, it seems that the relations of the $v-$ spinors of the positive energy to $u-$ spinors of the negative energy are frequently forgotten, $\tilde{v}_?(E_p,\mathbf{p})=\gamma^0 u_h(-E_p,\mathbf{p})$. Bogoliubov and Shirkov [18, p.55-56] used to construct the complete set of solutions of the relativistic equations, fixing the sign of $p_0=+E_p$.

Thus, unless the unitary transformations do not change the physical content, we have that the negative-energy spinors $\gamma^5\gamma^0 u^-$ (see (28)) satisfy the accustomed "positive-energy" Dirac equation. We should then expect the same physical content. Their explicit forms $\gamma^5\gamma^0 u^-$ are different from the textbook "positive-energy" Dirac spinors, while, of course, they should be superpositions of the latter. They are the following ones:

$$\tilde{u}_1(p) = \frac{N}{\sqrt{2m(-E_p+m)}} \begin{pmatrix} -p^+ + m \\ -p_r \\ p^- - m \\ -p_r \end{pmatrix}, \tag{29}$$

$$\tilde{u}_2(p) = \frac{N}{\sqrt{2m(-E_p+m)}} \begin{pmatrix} -p_l \\ -p^- + m \\ -p_l \\ p^+ - m \end{pmatrix}. \tag{30}$$

$E_p = \sqrt{\mathbf{p}^2 + m^2} > 0$, $p_0 = \pm E_p$, $p^\pm = E_p \pm p_z$, $p_{r,l} = p_x \pm i p_y$. Their normalization is to $(-2N^2)$. What about the $\tilde{v}(p) = \gamma^0 u^-$ transformed with the γ_0 matrix? They are not equal to $v_h(p) = \gamma^5 u_h(p)$. Obviously, they also do not have well-known forms of the usual $v-$ spinors in the Weyl basis, differing by phase factors and in the signs at the mass terms. Their transformation properties are different:

$$\tilde{v}_\tau(\mathbf{p}) = -i \frac{(\boldsymbol{\sigma}^* \cdot \mathbf{p})_{\tau\sigma}}{p} v_\sigma(\mathbf{p}). \tag{31}$$

Next, one can prove that the matrix

$$P = e^{i\theta}\gamma^0 = e^{i\theta} \begin{pmatrix} 0 & 1_{2\times 2} \\ 1_{2\times 2} & 0 \end{pmatrix} \tag{32}$$

can be used in the parity operator as well as in the original Weyl basis. The parity-transformed function $\Psi'(t, -\mathbf{x}) = P\Psi(t, \mathbf{x})$ must satisfy

$$[i\gamma^\mu \partial'_\mu - m]\Psi'(t, -\mathbf{x}) = 0, \tag{33}$$

with $\partial'_\mu = (\partial/\partial t, -\nabla_i)$. This is possible when $P^{-1}\gamma^0 P = \gamma^0$ and $P^{-1}\gamma^i P = -\gamma^i$. The matrix (32) satisfies these requirements, as in the textbook case. However, if we would take the phase factor to be zero we obtain that while spinors $u_h(p)$ have the eigenvalues $+1$ of the parity, but $(R = (\mathbf{x} \to -\mathbf{x}, \mathbf{p} \to -\mathbf{p}))$

$$PR\tilde{u}(p) = PR\gamma^5\gamma^0 u(-E_p, \mathbf{p}) = -\tilde{u}(p), \tag{34}$$

Perhaps, one should choose the phase factor $\theta = \pi$. Thus, we again confirm that only the relative (particle-antiparticle) intrinsic parity has physical significance.

Similar formulations have been presented in Refs. [21], and [22]. Namely, the reflection properties are different for some solutions of relativistic equations therein. Two opposite signs at the mass terms have been taken into account. The group-theoretical basis for such doubling has been given in the papers by Gelfand, Tsetlin and Sokolik [23], who first presented the theory of 5-dimensional spinors (or, the one in the 2-dimensional projective representation of the inversion group) in 1956 (later called as "the Bargmann-Wightman-Wigner-type quantum field theory" in 1993). The corresponding connection with the time reversion has been clarified therein. It was one of the first attempts to explain the K-meson decays. M. Markov proposed *two* Dirac equations with the opposite signs at the mass term [21] to be taken into account:

$$[i\gamma^\mu \partial_\mu - m]\,\Psi_1(x) = 0, \tag{35}$$
$$[i\gamma^\mu \partial_\mu + m]\,\Psi_2(x) = 0. \tag{36}$$

IOP Conf. Series: Journal of Physics: Conf. Series **1239** (2019) 012019 doi:10.1088/1742-6596/1239/1/012019

In fact, he studied all properties of this relativistic quantum model (while the quantum field theory has not yet been completed in 1937). Next, he added and subtracted these equations. What did he obtain?

$$i\gamma^\mu \partial_\mu \varphi(x) - m\chi(x) = 0\,, \tag{37}$$
$$i\gamma^\mu \partial_\mu \chi(x) - m\varphi(x) = 0\,. \tag{38}$$

Thus, the corresponding φ and χ solutions can be presented as some superpositions of the Dirac 4-spinors $u-$ and $v-$. These equations, of course, can be identified with the equations for the Majorana-like $\lambda-$ and $\rho-$, which we presented in Ref. [11a].[8]

$$i\gamma^\mu \partial_\mu \lambda^S(x) - m\rho^A(x) = 0\,, \tag{39}$$
$$i\gamma^\mu \partial_\mu \rho^A(x) - m\lambda^S(x) = 0\,, \tag{40}$$
$$i\gamma^\mu \partial_\mu \lambda^A(x) + m\rho^S(x) = 0\,, \tag{41}$$
$$i\gamma^\mu \partial_\mu \rho^S(x) + m\lambda^A(x) = 0\,. \tag{42}$$

Neither of them can be regarded as the Dirac equation. However, they can be written in the 8-component form as follows:

$$[i\Gamma^\mu \partial_\mu - m]\,\Psi_{(+)}(x) = 0\,, \tag{43}$$
$$[i\Gamma^\mu \partial_\mu + m]\,\Psi_{(-)}(x) = 0\,, \tag{44}$$

with

$$\Psi_{(+)}(x) = \begin{pmatrix}\rho^A(x)\\ \lambda^S(x)\end{pmatrix}, \Psi_{(-)}(x) = \begin{pmatrix}\rho^S(x)\\ \lambda^A(x)\end{pmatrix}, \Gamma^\mu = \begin{pmatrix}0 & \gamma^\mu\\ \gamma^\mu & 0\end{pmatrix}. \tag{45}$$

It is possible to find the corresponding Lagrangian, projection operators, and the Feynman-Dyson-Stueckelberg propagator. For example,

$$\begin{aligned}\mathcal{L} &= \frac{i}{2}\,[\overline{\Psi}_{(+)}\Gamma^\mu \partial_\mu \Psi_{(+)} - (\partial_\mu \overline{\Psi}_{(+)})\Gamma^\mu \Psi_{(+)} + \\ &+ \overline{\Psi}_{(-)}\Gamma^\mu \partial_\mu \Psi_{(-)} - (\partial_\mu \overline{\Psi}_{(-)})\Gamma^\mu \Psi_{(-)}\Big] - m[\overline{\Psi}_{(+)}\Psi_{(+)} - \overline{\Psi}_{(-)}\Psi_{(-)}]\,.\end{aligned} \tag{46}$$

The projection operator P_+ can be easily found, as usual,

$$P_+ = \frac{\Gamma_\mu p^\mu + m}{2m}\,. \tag{47}$$

However, due to the fact that P_- satisfies the Dirac equation with the opposite sign, we cannot have $P_+ + P_- = 1$. This is not surprising because the corresponding states $\Psi_\pm$ do not form the complete system of the 8-dimensional space. One should consider the states $\Gamma_5\Psi_\pm(\mathbf{p})$ too. See also [25] for the methods of obtaining the propagators in the non-trivial cases.

In the previous papers I explained: the connection with the Dirac spinors has been found [11, 24] through the unitary matrix. For instance,

$$\begin{pmatrix}\lambda^S_\uparrow(\mathbf{p})\\ \lambda^S_\downarrow(\mathbf{p})\\ \lambda^A_\uparrow(\mathbf{p})\\ \lambda^A_\downarrow(\mathbf{p})\end{pmatrix} = \frac{1}{2}\begin{pmatrix}1 & i & -1 & i\\ -i & 1 & -i & -1\\ 1 & -i & -1 & -i\\ i & 1 & i & -1\end{pmatrix}\begin{pmatrix}u_{+1/2}(\mathbf{p})\\ u_{-1/2}(\mathbf{p})\\ v_{+1/2}(\mathbf{p})\\ v_{-1/2}(\mathbf{p})\end{pmatrix}, \tag{48}$$

[8] Of course, the signs at the mass terms depend on, how do we associate the positive- or negative- frequency solutions with λ and ρ.

provided that the 4-spinors have the same physical dimension.[9] Thus, this represents itself the rotation of the spin-parity basis. However, it is usually assumed that the $\lambda-$ and $\rho-$ spinors describe the neutral particles, meanwhile, the $u-$ and $v-$ spinors describe the charged particles. Kirchbach [24] found the amplitudes for neutrinoless double beta decay ($00\nu\beta$) in this scheme. It is obvious from (48) that there are some additional terms comparing with the standard calculations of those amplitudes. One can also re-write the above equations into the two-component forms. Thus, one obtains the Feynman-Gell-Mann [26] equations.

Barut and Ziino [22] proposed yet another model. They considered γ^5 operator as the operator of the charge conjugation. In their case the self/anti-self charge conjugate states are, at the same time, the eigenstates of the chirality. Thus, the charge-conjugated Dirac equation has a different sign compared with the ordinary formulation:

$$[i\gamma^\mu\partial_\mu + m]\Psi^c_{BZ} = 0\,, \tag{49}$$

and the so-defined charge conjugation applies to the whole system, fermion + electromagnetic field, $e \to -e$ in the covariant derivative. The superpositions of the Ψ_{BZ} and Ψ^c_{BZ} also give us the "doubled Dirac equation", as the equations for $\lambda-$ and $\rho-$ spinors. The concept of the doubling of the Fock space has been developed in the Ziino works (cf. [23, 27]) in the framework of the quantum field theory [28]. Next, it is interesting to note that we have for the Majorana-like field operators ($a_\eta(\mathbf{p}) = b_\eta(\mathbf{p})$):

$$\left[\nu^{^{ML}}(x^\mu) + \mathcal{C}\nu^{^{ML\,\dagger}}(x^\mu)\right]/2 = \int \frac{d^3\mathbf{p}}{(2\pi)^3}\frac{1}{2E_p} \tag{50}$$

$$\sum_\eta \left[\begin{pmatrix} i\Theta\phi^{*\,\eta}_{_L}(p^\mu) \\ 0\end{pmatrix} a_\eta(p^\mu)e^{-ip\cdot x} + \begin{pmatrix} 0 \\ \phi^\eta_{_L}(p^\mu)\end{pmatrix} a^\dagger_\eta(p^\mu)e^{ip\cdot x}\right]\,,$$

$$\left[\nu^{^{ML}}(x^\mu) - \mathcal{C}\nu^{^{ML\,\dagger}}(x^\mu)\right]/2 = \int \frac{d^3\mathbf{p}}{(2\pi)^3}\frac{1}{2E_p} \tag{51}$$

$$\sum_\eta \left[\begin{pmatrix} 0 \\ \phi^\eta_{_L}(p^\mu)\end{pmatrix} a_\eta(p^\mu)e^{-ip\cdot x} + \begin{pmatrix} -i\Theta\phi^{*\,\eta}_{_L}(p^\mu) \\ 0\end{pmatrix} a^\dagger_\eta(p^\mu)e^{ip\cdot x}\right]\,.$$

This naturally leads to the Ziino-Barut scheme of massive chiral fields, Ref. [22]. See, however, the recent paper [29] which deals with the problems of the Majorana field operator.

Finally, I would like to mention that, in general, in the Weyl basis the $\gamma-$ matrices are *not* Hermitian, $\gamma^{\mu^\dagger} = \gamma^0\gamma^\mu\gamma^0$. So, $\gamma^{i^\dagger} = -\gamma^i$, $i = 1, 2, 3$, the pseudo-Hermitian matrix. The energy-momentum operator $i\partial_\mu$ is obviously Hermitian. So, the question, if the eigenvalues of the Dirac operator $i\gamma^\mu\partial_\mu$ (the mass, in fact) would be always real? The question of the complete system of the eigenvectors of the *non*-Hermitian operator deserve careful consideration [30].

The main points of this Section are: there are "negative-energy solutions" in that is previously considered as "positive-energy solutions" of relativistic wave equations, and vice versa. Their explicit forms have been presented in the case of spin-1/2. Next, relations to previous works have been found. For instance, the doubling of the Fock space and the corresponding solutions of the Dirac equation obtained additional mathematical bases. Similar conclusion can be deduced for the higher-spin equations.

3. Non-commutativity in the Dirac equation.

The non-commutativity [31, 32] manifests interesting peculiarities in the Dirac case. We analized Sakurai-van der Waerden method of derivations of the Dirac (and higher-spins too) equation [33]. We can start from

$$(EI^{(2)} - \boldsymbol{\sigma}\cdot\mathbf{p})(EI^{(2)} + \boldsymbol{\sigma}\cdot\mathbf{p})\Psi_{(2)} = m^2\Psi_{(2)}\,, \tag{52}$$

[9] The reasons of the change of the fermion mass dimension are unclear in the recent works on *elko*.

or (in the 4-component case)

$$(EI^{(4)} + \boldsymbol{\alpha}\cdot\mathbf{p} + m\beta)(EI^{(4)} - \boldsymbol{\alpha}\cdot\mathbf{p} - m\beta)\Psi_{(4)} = 0\,. \qquad (53)$$

Obviously, the inverse operators of the Dirac operators of the positive- and negative- masses exist in the non-commutative case too. As in the original Dirac work, we have

$$\beta^2 = 1\,, \quad \alpha^i\beta + \beta\alpha^i = 0\,, \quad \alpha^i\alpha^j + \alpha^j\alpha^i = 2\delta^{ij}\,. \qquad (54)$$

For instance, their explicite forms can be chosen

$$\alpha^i = \begin{pmatrix}\sigma^i & 0\\ 0 & -\sigma^i\end{pmatrix}, \quad \beta = \begin{pmatrix}0 & 1_{2\times 2}\\ 1_{2\times 2} & 0\end{pmatrix}, \qquad (55)$$

where, again, σ^i are the ordinary Pauli 2×2 matrices.

We postulate the non-commutativity relations for the components of 4-momenta:

$$[E, \mathbf{p}^i]_- = \Theta^{0i} = \theta^i\,, \qquad (56)$$

as usual. Therefore the equation (53) will *not* lead to the well-known equation $E^2 - \mathbf{p}^2 = m^2$. Instead, we have

$$\left\{E^2 - E(\boldsymbol{\alpha}\cdot\mathbf{p}) + (\boldsymbol{\alpha}\cdot\mathbf{p})E - \mathbf{p}^2 - m^2 - i(I_{(2)}\otimes\boldsymbol{\sigma})[\mathbf{p}\times\mathbf{p}]\right\}\Psi_{(4)}$$
$$= 0 \qquad (57)$$

For the sake of simplicity, we may assume the last term to be zero. Thus, we come to

$$\left\{E^2 - \mathbf{p}^2 - m^2 - (\boldsymbol{\alpha}\cdot\boldsymbol{\theta})\right\}\Psi_{(4)} = 0\,. \qquad (58)$$

However, let us apply a unitary transformation. It is known, Refs. [34, 11], that one can transform[10]

$$U_1(\boldsymbol{\sigma}\cdot\mathbf{a})U_1^{-1} = \sigma_3|\mathbf{a}|\,. \qquad (59)$$

For $\boldsymbol{\alpha}$ matrices we re-write (59) to

$$\mathcal{U}_1(\boldsymbol{\alpha}\cdot\boldsymbol{\theta})\mathcal{U}_1^{-1} = |\boldsymbol{\theta}|\begin{pmatrix}1 & 0 & 0 & 0\\ 0 & -1 & 0 & 0\\ 0 & 0 & -1 & 0\\ 0 & 0 & 0 & 1\end{pmatrix} = \alpha_3|\boldsymbol{\theta}|\,. \qquad (60)$$

The explicit form of the U_1 matrix is ($a_{r,l} = a_1 \pm ia_2$):

$$\begin{aligned} U_1 &= \frac{1}{\sqrt{2a(a+a_3)}}\begin{pmatrix}a + a_3 & a_l\\ -a_r & a+a_3\end{pmatrix} = \frac{1}{\sqrt{2a(a+a_3)}}\\ &\times\ [a + a_3 + i\sigma_2 a_1 - i\sigma_1 a_2]\,,\\ \mathcal{U}_1 &= \begin{pmatrix}U_1 & 0\\ 0 & U_1\end{pmatrix}. \end{aligned} \qquad (61)$$

[10] Some relations for the components $\mathbf{a}$ should be assumed. Moreover, in our case $\boldsymbol{\theta}$ should not depend on E and $\mathbf{p}$. Otherwise, we must take the non-commutativity $[E, \mathbf{p}^i]_- \neq 0$ into account again.

Let me apply the second unitary transformation:

$$\mathcal{U}_2\alpha_3\mathcal{U}_2^\dagger = \begin{pmatrix}1&0&0&0\\0&0&0&1\\0&0&1&0\\0&1&0&0\end{pmatrix}\alpha_3\begin{pmatrix}1&0&0&0\\0&0&0&1\\0&0&1&0\\0&1&0&0\end{pmatrix} = \begin{pmatrix}1&0&0&0\\0&1&0&0\\0&0&-1&0\\0&0&0&-1\end{pmatrix}. \tag{62}$$

The final equation is

$$[E^2 - \mathbf{p}^2 - m^2 - \gamma^5_{chiral}|\boldsymbol{\theta}|]\Psi'_{(4)} = 0\,. \tag{63}$$

In the physical sense this implies the mass splitting for a Dirac particle over the non-commutative space, $m_{1,2} = \pm\sqrt{m^2 \pm \theta}$. This procedure may be attractive for explanation of the mass creation and the mass splitting for fermions. One can also use the non-commutativity

$$[\mathbf{p}^i, \mathbf{p}^j] = \Xi^{ij} = \epsilon^{ijk}\xi^k \tag{64}$$

with the corresponding substitutions: $\theta^i = 0$, $\mathcal{U}_1(\theta) \to \mathcal{U}_1(\xi)$ and

$$\mathcal{U}'_2 = \begin{pmatrix}1&0&0&0\\0&0&-1&0\\0&1&0&0\\0&0&0&1\end{pmatrix}. \tag{65}$$

In such a way we obtain the same splitting as in (63), $|\boldsymbol{\theta}| \to |\boldsymbol{\xi}|$.

Acknowledgments. I greatly appreciate old discussions with Prof. A. Raspini and useful information from Prof. A. F. Pashkov. I acknowledge discussions with participants of several recent Conferences. This work has been partly supported by the ESDEPED UAZ, México.

[1] A. Gersten, *Found. Phys. Lett.* **12**, 291 (1999); ibid. **13**, 185 (2000).
[2] J. J. Sakurai, *Advanced Quantum Mechanics* (Addison-Wesley, 1967), §3.2.
[3] V. V. Dvoeglazov, *J. Phys A: Math. Gen.* **33**, 5011 (2000).
[4] S. I. Kruglov, *EJTP-Zacatecas Proc. II*, 107 (2011), ArXiV: 1204.6709.
[5] Z. Tokuoka, *Prog. Theor. Phys.* **37**, 603 (1967); N. D. S. Gupta, *Nucl. Phys.* **B4**, 147 (1967); T. S. Santhanam and P. S. Chandrasekaran, *Prog. Theor. Phys.* **41**, 264 (1969); M. T. Simon, *Lett. Nuovo Cim.* **2**, 616 (1971); T. S. Santhanam and A. R. Tekumalla, *Lett. Nuovo Cim.* **3**, 190 (1972).
[6] A. Raspini, *Int. J. Theor. Phys.* **33**, 1503 (1994); *Fizika B* **5**, 159 (1996); ibid. **6**, 123 (1997); ibid. **7**, 83 (1998).
[7] A. Raspini, A Review of Some Alternative Descriptions of Neutrino. In *Photon and Poincaré Group*. Ed. V. V. Dvoeglazov. (Commack, NY: Nova Science, 1999), pp. 181-188.
[8] T. S. Santhanam, *Lett. Nuovo Cim.* **20**, 13 (1977).
[9] D. V. Ahluwalia, *Int. J. Mod. Phys.* A**11**, 1855 (1996).
[10] D. V. Ahluwalia and A. Ch. Nayak, *Int. J. Mod. Phys.* D**23**, 1430026 (2014).
[11] V. V. Dvoeglazov, *Int J. Theor. Phys.* **34**, 2467 (1995); ibid **37**, 1909 (1998); *Nuovo Cim.* B**111**, 483 (1996); ibid. A**108**, 1467 (1995); *Hadronic J.* **20**, 435 (1997); *Fizika* B**6**, 111 (1997); *Adv. Appl. Clifford Algebras* **7**C, 303 (1997); ibid. **9**, 231 (1999); *Acta Physica Polon.* B**29**, 619 (1998); *Found. Phys. Lett.* **13**, 387 (2000).
[12] L. H. Ryder, *Quantum Field Theory.* (Cambridge University Press, Cambridge, 1985).
[13] V. V. Dvoeglazov, *Czech. J. Phys.* **50**, 225 (2000); Longitudinal Nature of Antisymmetric Tensor Field After Quantization and Importance of the Normalization. In *Photon: Old Problems in Light of New Ideas.* Ed. V. V. Dvoeglazov, pp. 319-337. (Huntington, NY: Nova Science Publishers, 2000).
[14] V. V. Dvoeglazov,*Adv. Appl. Clifford Algebras.* **26**, 1173 (2016).
[15] S. M. Bilenky and C. Giunti, *Nucl. Phys. Proc. Suppl.* **48**, 381 (1996); W. Stecker and S. T. Scully, *Phys. Rev. D***90**, 043012 (2014); Yu. G. Kudenko, *Phys. Usp.* **61** (2018).
[16] A. O. Barut, *Phys. Lett.* B**73**, 310 (1978); *Phys. Rev. Lett.* **42**, 1251 (1979).
[17] C. Itzykson and J.-B. Zuber, *Quantum Field Theory.* (McGraw-Hill Book Co., 1980), p. 156.

[18] N. N. Bogoliubov and D. V. Shirkov, *Introduction to the Theory of Quantized Fields.* 2nd Edition. (Nauka, Moscow, 1973).
[19] V. V. Dvoeglazov, *J. Phys. Conf. Ser.* **284**, 012024 (2011), arXiv:1008.2242.
[20] V. V. Dvoeglazov, *Hadronic J. Suppl.* **18**, 239 (2003), physics/0402094; *Int. J. Mod. Phys.* B**20**, 1317 (2006).
[21] M. Markov, *ZhETF* **7**, 579 (1937); ibid. 603; *Nucl. Phys.* **55**, 130 (1964).
[22] A. Barut and G. Ziino, *Mod. Phys. Lett.* A**8**, 1099 (1993); G. Ziino, *Int. J. Mod. Phys.* A **11**, 2081 (1996).
[23] I. M. Gelfand and M. L. Tsetlin, *Sov. Phys. JETP* **4**, 947 (1957); G. A. Sokolik, *Sov. Phys. JETP* **6**, 1170 (1957); ibid. **9**, 781 (1959); *Dokl. Akad. Nauk SSSR* **114**, 1206 (1957).
[24] M. Kirchbach, C. Compean and L. Noriega, *Eur. Phys. J.* A**22**, 149 (2004).
[25] V. V. Dvoeglazov, *Int. J. Theor. Phys.*, accepted.
[26] R. P. Feynman and M. Gell-Mann, *Phys. Rev.* **109**, 193 (1958).
[27] V. V. Dvoeglazov, *Int. J. Theor. Phys.* **37**, 1915 (1998).
[28] G. Ziino, *Int. J. Mod. Phys.* A**11**, 2081 (1996); *Int. J. Theor. Phys.* **39**, 51 (2000); ibid. 2605 (2000); ibid. **45**, 1993 (2006); ibid. **55**, 1775 (2016).
[29] V. V. Dvoeglazov, *Mod. Phys. Lett.*, submitted.
[30] V. A. Ilyin, *Spektralnaya Teoriya Differencialnyh Operatorov.* (Nauka, Moscow, 1991); V. D. Budaev, *Osnovy Teorii Nesamosopryazhennyh Differencialnyh Operatorov.* (SGMA, Smolensk, 1997).
[31] H. Snyder, *Phys. Rev.* **71**, 38 (1947); ibid. **72**, 68 (1947).
[32] A. Kempf, G. Mangano and R. B. Mann, *Phys. Rev.* D**52**, 1108 (1995); G. Amelino-Camelia, *Nature* **408**, 661 (2000); *Int. J. Mod. Phys.* D**11**, 35 (2002), gr-qc/0012051; *Phys. Lett.* B**510**, 255 (2001), hep-th/0012238; *AIP Conf. Proc.* **589**, 137 (2001), gr-qc/0106004; J. Kowalski-Glikman, *Phys. Lett.* A**286**, 391 (2001), hep-th/0102098; G. Amelino-Camelia and M. Arzano, *Phys. Rev.* D**65**, 084044 (2002), hep-th/0105120; N. R. Bruno, G. Amelino-Camelia and J. Kowalski-Glikman, *Phys. Lett.* B**522**, 133 (2001), hep-th/0107039.
[33] V. V. Dvoeglazov, *Rev. Mex. Fis. Supl.* **49**, 99 (2003) (*Proceedings of the DGFM-SMF School, Huatulco, 2000*)
[34] R. A. Berg, *Nuovo Cimento* **42**A, 148 (1966).

IOP Conf. Series: Journal of Physics: Conf. Series **1239** (2019) 012020 doi:10.1088/1742-6596/1239/1/012020

Maxwell's Equations, Fields or Potentials?

Alexander Gersten and Amnon Moalem

Department of Physics, Ben Gurion University of the Negev, Beer-Sheva, Israel

E-mail: alex.gersten@gmail.com

Abstract. Fields and potentials of Maxwell's Equations transform differently under the Lorentz transformation. Electric and magnetic fields are tridimensional objects which transform like massless fields as a subgroup of the Lorentz group. On the other hand, potentials are four-vectors. Therefore interactions of the electromagnetic fields with four-currents of massive particles cannot be achieved with fields, but only with potentials. Thus potentials seemed to be preferable and a correction to Maxwell's equations seemed unavoidable. We have generalized the Maxwell's equations, with 4-vector fields to allow interaction with currents. We suggest a new way to define the potentials and relate them to the new fields.
Key words: Maxwell's Equations, Generalized Maxwell's Equations, massless particles, any spin.

1. Introduction

Maxwell's Equations are one of the world's wonders. Created in 1862 they satisfy today the rules of Spacial Relativity[5], and of Relativistic Quantum Mechanics[1],[11],[9],[13]. Four exceptional scientists contributed to them, Gauss –the king of mathematicians, Ampere, who at 12 years of age mastered the calculus, and later laid the foundations of electrodynamics, Michael Faraday, the paradoxical scientist who was not acquainted with elementary mathematics, who could not write a simple equation, whose books contain no mathematical formulas, and Maxwell, who corrected Ampere's law demanding charge conservation and who unified the electric and magnetic laws into Maxwell's Equations and predicted the existence of electromagnetic waves travelling with the speed of light. Maxwell, a Scottish gentleman, gave most of the credit to Michael Faraday. Michael Faraday not only discovered the Faraday law (without mathematics) but also imagined the existence of fields of energy which led Maxwell to formulate his theory as field theory. Potentials were introduced, which simplified Maxwell's Equations.

Everything seemed to work perfectly (but for the Planck radiation law and second quantization, which we do not treat here), until in 1958, a Ph.D. student of David Bohm, by the name of Yakir Aharonov found an effect which questioned the foundations of Maxwell's Equations. The Aharonov-Bohm[2] effect demonstrated, that different results are obtained using the electromagnetic fields compared to electromagnetic potentials . Namely, outside an electromagnetically isolated object the fields were equal to zero while the potentials were different from zero. With potentials, measurable results were reproduced within the framework of quantum mechanics. The conclusion was that potentials were preferred.

IOP Conf. Series: Journal of Physics: Conf. Series **1239** (2019) 012020 doi:10.1088/1742-6596/1239/1/012020

We will show that Maxwell's Equations without sources can be considered as quantum type equations of the helicity operator of a massless particle, the photon. The fields are 3-dimensional objects, while potentials are 4-vectors. It is difficult to formulate an interaction of 3-dimensional fields with 4-currents in a covariant way, while it is possible with potentials. Electrodynamics with charges and electric currents is presently formulated with potentials and not with fields.

Wigner[3] has shown that wavefunctions of massless particles of any spin may have only two non-zero helicity components corresponding to the forward or backward directions of motion. Dirac [6], [8] in 1936 derived wavefunctions for massless particles of any spin. The angular momentum basis of the wavefunction of the helicity operator has $2s+1$ components of spin s. But for massless particles only two components, for forward and backward directions, are non-zero. The other $2s-1$ components become equal to zero by imposing $2s-1$ subsidiary conditions. We have developed a formalism equivalent to the Dirac equations using the angular momentum basis of the $[s-\frac{1}{2},\frac{1}{2}]$ representation of the Lorentz Group[25],[26] . The angular momentum basis consists of two spins s and $s-1$, and has $(2s+1)+2(s-1)+1=4s$ components. By equating to zero the spin $s-1$ components of the wave function, the 2s-1 subsidiary conditions (needed to eliminate the non-forward and non-backward helicities.), are fulfilled. The zeros will remain under Lorentz transformations and the wave function transforms under a subgroup of the Lorentz group with only two non-zero helicity projections.

We will show that (in Gaussian units system), that Maxwell's Equations without sources can be considered as quantum equations of a massless particle of spin 1. The wavefunction is proportional to

$$\begin{pmatrix} 0 \\ E_x+iB_x \\ E_y+iB_y \\ E_z+iB_z \end{pmatrix}, \tag{1}$$

where $\boldsymbol{E}$ and $\boldsymbol{B}$ are electric and magnetic fields. The complex 3-vector $\boldsymbol{E}+i\boldsymbol{B}$ transforms under the Lorentz transformation like a massless particle. Thus the electric and magnetic fields are 3 dimensional. This is an obstacle for constructing interactions with 4-currents. On the other hand, the introduction of 4-potentials allows the formulation of interactions with 4-currents. All Lagrangian formulations of electrodynamics are done with 4-potentials. Thus we come to the conclusion that Maxwell's equations should be generalized, so that fields should be 4-vectors.

There have been attempts by Majorana in the 1930's[21], to reformulate Maxwell's Equations with 4-fields in order to replace the treatment with potentials. It did not replace, it only gave a new presentation of Maxwell's equations without sources. Nevertheless, the ideas of Majorana are very interesting and may lead to new developments. It allowed Ivezic[20] to use them for developing electromagnetic field equations for moving media.

Our generalization of Maxwell's Equations is straightforward with a wave function proportional to

$$\begin{pmatrix} E_0+iB_0 \\ E_x+iB_x \\ E_y+iB_y \\ E_z+iB_z \end{pmatrix}. \tag{2}$$

where the component E_0+iB_0 is the spin 0 contribution to the wavefunction. Here we suggest a new method of defining potentials for generalized Maxwell's Equations.

In next sections we will show that Maxwell's Equations can be considered as quantum equations, we will discuss our presentation of massless fields with two helicity projections and we will present our generalization of Maxwell's Equations.

IOP Conf. Series: Journal of Physics: Conf. Series **1239** (2019) 012020 doi:10.1088/1742-6596/1239/1/012020

2. Maxwell's Equations as wave equations of the helicity operator

Minkowski[[5]] found the covariant form of Maxwell's Equations [10], which in the metric $[1,-1,-1,-1]$ and Gaussian system units, are

$$\partial_\mu F^{\mu\nu} = \frac{4\pi}{c} j^\nu, \quad \partial_\mu \tilde{F}^{\mu\nu} = 0, \quad \nu = 0,1,2,3, \tag{3}$$

where $\partial_0 = \frac{1}{c}\frac{\partial}{\partial t}$ and the antisymmetric tensor $F^{\mu\nu}$ and its dual $\tilde{F}^{\mu\nu}$ are defined via the electric and magnetic fields $\mathbf{E}$ and $\mathbf{B}$ respectively as,

$$(F^{\mu\nu}) = \begin{pmatrix} 0 & -E_x & -E_y & -E_z \\ E_x & 0 & -B_z & B_y \\ E_y & B_z & 0 & -B_x \\ E_z & -B_y & B_x & 0 \end{pmatrix}, \tag{4}$$

$$\tilde{F}^{\mu\nu} = \frac{1}{2}\epsilon^{\mu\nu\alpha\beta} F_{\alpha\beta} = \begin{pmatrix} 0 & -B_x & -B_y & -B_z \\ B_x & 0 & E_z & -E_y \\ B_y & -E_z & 0 & E_x \\ B_z & E_y & -E_x & 0 \end{pmatrix}, \tag{5}$$

where $\epsilon^{\mu\nu\alpha\beta}$ is the totally antisymmetric tensor ($\epsilon^{0123} = 1, \quad \epsilon^{\mu\nu\alpha\beta} = -\epsilon_{\mu\nu\alpha\beta}$). The sum of the $F^{\mu\nu}$ and $i\tilde{F}^{\mu\nu}$ is the self-dual antisymmetric tensor, it depends only on the combination,

$$\mathbf{\Psi} = (\mathbf{E}+i\mathbf{B}), \tag{6}$$

$$F^{\mu\nu} + i\tilde{F}^{\mu\nu} = \begin{pmatrix} 0 & -E_x - iB_x & -E_y - iB_y & -E_z - iB_z \\ E_x + iB_x & 0 & iE_z - B_z & B_y - iE_y \\ E_y + iB_y & B_z - iE_z & 0 & iE_x - B_x \\ E_z + iB_z & iE_y - B_y & B_x - iE_x & 0 \end{pmatrix}, \tag{7}$$

$$= \begin{pmatrix} 0 & -\Psi_x & -\Psi_y & -\Psi_z \\ \Psi_x & 0 & i\Psi_z & -i\Psi_y \\ \Psi_y & -i\Psi_z & 0 & i\Psi_x \\ \Psi_z & i\Psi_y & -i\Psi_x & 0 \end{pmatrix} \tag{8}$$

The self-dual Maxwell Equations take the form,

$$\begin{aligned} \partial_\mu \left(F^{\mu\nu} + i\tilde{F}^{\mu\nu}\right) &= -\left(\Gamma_\mu^{(4)}\right)^{\nu k} \partial_\mu \Psi_k = \frac{4\pi}{c} j^\nu, \\ \mu &= 0,x,y,z, = 0,1,2,3 \qquad k = x,y,z, = 1,2,3, \end{aligned} \tag{9}$$

where,

$$\Gamma_0^{(4)} = \begin{pmatrix} 1 & 0 & 0 & 0 \\ 0 & 1 & 0 & 0 \\ 0 & 0 & 1 & 0 \\ 0 & 0 & 0 & 1 \end{pmatrix}, \quad \Gamma_1^{(4)} = \begin{pmatrix} 0 & -1 & 0 & 0 \\ -1 & 0 & 0 & 0 \\ 0 & 0 & 0 & -i \\ 0 & 0 & i & 0 \end{pmatrix},$$

,

$$\Gamma_2^{(4)} = \begin{pmatrix} 0 & 0 & -1 & 0 \\ 0 & 0 & 0 & i \\ -1 & 0 & 0 & 0 \\ 0 & -i & 0 & 0 \end{pmatrix}, \quad \Gamma_3^{(4)} = \begin{pmatrix} 0 & 0 & 0 & -1 \\ 0 & 0 & -i & 0 \\ 0 & i & 0 & 0 \\ -1 & 0 & 0 & 0 \end{pmatrix} \tag{10}$$

and the Ψ_k is one of the three components of the wave function,

$$\Phi = \begin{pmatrix} 0 \\ \Psi_x \\ \Psi_y \\ \Psi_z \end{pmatrix} = \begin{pmatrix} 0 \\ E_x + iB_x \\ E_y + iB_y \\ E_z + iB_{zz} \end{pmatrix}. \tag{11}$$

Multiplying Eq. (9) by $i\hbar c$, where $\hbar$ is the reduced Planck constant, we obtain,

$$\left(\Gamma_\mu^{(4)}\right)^{\nu k} (i\hbar c\partial_\mu)\,\Psi_k = -i\hbar c\frac{4\pi}{c} j^\nu. \tag{12}$$

Replacing $i\hbar\partial_\mu$ by their quantum energy-momentum operators, Eq. (12) takes the form,

$$\left(\Gamma_0^{(4)} E - c\mathbf{\Gamma^{(4)}} \cdot \boldsymbol{p}\right) \Psi = -i\hbar 4\pi J, \tag{13}$$

where J is the column of the four vector j^ν. Maxwell's (quantum) Equations without sources (the photon equation) for the helicity 1 are,

$$\left(\Gamma_0^{(4)} E - c\mathbf{\Gamma^{(4)}} \cdot \boldsymbol{p}\right) \Psi = 0. \tag{14}$$

The equation for helicity -1 can be obtained using exactly the same procedure, but starting from the anti-self-dual tensor $F^{\mu\nu} - i\tilde{F}^{\mu\nu}$.

We obtain,

$$\begin{aligned} \partial_\mu \left(F^{\mu\nu} - i\tilde{F}^{\mu\nu}\right) &= -\left(\Gamma_\mu^{(4)}\right)^{\nu k} \partial_\mu \Psi_k^* = \frac{4\pi}{c} j^\nu, \\ \mu &= 0, x, y, z, = 0, 1, 2, 3 \qquad k = x, y, z, = 1, 2, 3, \end{aligned} \tag{15}$$

where Ψ_k^* is the complex conjugate of Ψ_k. The helicity -1 equation is,

$$\left(\Gamma_0^{(4)} E + c\mathbf{\Gamma^{(4)}} \cdot \boldsymbol{p}\right) \Psi^* = -i\hbar 4\pi J. \tag{16}$$

As we have stated in the introduction, it is difficult to couple 3-vctors with 4 currents. Therefore we can write a Lagrangian formalism only for Maxwell's Equations without sources,

$$\left[E\Gamma_0^{(4)} - c\mathbf{p}\cdot\mathbf{\Gamma^{(4)}}\right] \begin{pmatrix} 0 \\ E_x + iB_x \\ E_y + iB_y \\ E_z + iB_{zz} \end{pmatrix} = 0, \ \mathit{Helicity/s} = 1; \tag{17}$$

$$\left[E\Gamma_0^{(4)} + c\mathbf{p}\cdot\mathbf{\Gamma^{(4)}}\right] \begin{pmatrix} 0 \\ E_x - iB_x \\ E_y - iB_y \\ E_z - iB_{zz} \end{pmatrix} = 0, \ \mathit{Helicity/s} = -1. \tag{18}$$

2.1. The structure of the $\Gamma^{(4)}$ matrices

The $\Gamma^{(4)}$ matrices have the following structure,

$$\Gamma_0^{(4)} = \begin{pmatrix} 1 & 0 \; 0 \; 0 \\ 0 & \\ 0 & (I_0) \\ 0 & \end{pmatrix}, \text{ is the unit matrix} \tag{19}$$

and

$$\Gamma_x^{(4)} = \begin{pmatrix} 0 & -1 \; 0 \; 0 \\ -1 & \\ 0 & (S_x) \\ 0 & \end{pmatrix}, \; \Gamma_y^{(4)} = \begin{pmatrix} 0 & 0 \; -1 \; 0 \\ 0 & \\ -1 & (S_y) \\ 0 & \end{pmatrix}, \tag{20}$$

$$\Gamma_z^{(4)} = \begin{pmatrix} 0 & 0 \; 0 \; -1 \\ 0 & \\ 0 & (S_z) \\ -1 & \end{pmatrix}, \tag{21}$$

$$S_x = \begin{pmatrix} 0 & 0 & 0 \\ 0 & 0 & -i \\ 0 & i & 0 \end{pmatrix}, \quad S_y = \begin{pmatrix} 0 & 0 & i \\ 0 & 0 & 0 \\ -i & 0 & 0 \end{pmatrix}, \quad S_z = \begin{pmatrix} 0 & -i & 0 \\ i & 0 & 0 \\ 0 & 0 & 0 \end{pmatrix}, \tag{22}$$

$$I_0 = \begin{pmatrix} 1 & 0 & 0 \\ 0 & 1 & 0 \\ 0 & 0 & 1 \end{pmatrix}, \tag{23}$$

where S_i are the spin 1 matrices with the properties,

$$[S_x, S_y] = iS_z, \quad [S_z, S_x] = iS_y, \quad [S_y, S_z] = iS_x, \quad \mathbf{S}^2 = 2I_0. \tag{24}$$

Substituting Eqs. (19-20) in Eq. (14) we obtain for the spin 1 projection Φ of the wave function

$$-c\boldsymbol{p} \cdot \boldsymbol{\Phi} = \mathbf{0}, \tag{25}$$

$$\left(I_0 E - c\frac{\boldsymbol{S} \cdot \boldsymbol{p}}{s}\right) \boldsymbol{\Phi} = \mathbf{0}, \; Helicity/s = 1, \tag{26}$$

where

$$\boldsymbol{\Phi} = \begin{pmatrix} \Phi_x \\ \Phi_y \\ \Phi_z \end{pmatrix} = \begin{pmatrix} E_x + iB_x \\ E_y + iB_y \\ E_z + iB_z \end{pmatrix}, \tag{27}$$

and

$$c\boldsymbol{p} \cdot \boldsymbol{\Phi}^* = \mathbf{0}, \quad \left(I_0 E + c\frac{\boldsymbol{S} \cdot \boldsymbol{p}}{s}\right) \boldsymbol{\Phi}^* = \mathbf{0}, \quad Helicity/s = -1, \tag{28}$$

Eq. (26) is the unconstrained helicity equation and Eq. (25) is the subsidiary condition (the perpendicularity condition) which eliminates the zero helicity component.

The $\Gamma^{(4)}$ matrices satisfy the same relations as the Pauli matrices, namely

$$\left(\Gamma_x^{(4)}\right)^2 = \left(\Gamma_y^{(4)}\right)^2 = \left(\Gamma_z^{(4)}\right)^2 = \left(\Gamma_0^{(4)}\right)^2 = \Gamma_0^{(4)}, \tag{29}$$

$$\Gamma_x^{(4)}\Gamma_y^{(4)} = i\Gamma_z^{(4)} \quad \text{(and cyclic permutation)}, \tag{30}$$

$$\Gamma_k^{(4)}\Gamma_l^{(4)} = -\Gamma_l^{(4)}\Gamma_k^{(4)}, \quad \text{for } k \neq l, \quad k, l = x, y, z, \tag{31}$$

therefore the $\Gamma^{(4)}$ matrices form a reducible representation of the Pauli matrices. Furthermore

$$\left(\Gamma_0^{(4)}E - c\mathbf{\Gamma}^{(4)} \cdot \boldsymbol{p}\right)\left(\Gamma_0^{(4)}E + c\mathbf{\Gamma}^{(4)} \cdot \boldsymbol{p}\right) = \left(E^2 - c^2\mathbf{p}^2\right)\Gamma_0^{(4)}, \tag{32}$$

As a result we obtain for the photon the two helicity equations:

$$\left[E\Gamma_0^{(4)} - c\mathbf{p}{\cdot}\mathbf{\Gamma}^{(4)}\right]\psi_R = 0, \quad Helicity/s = 1. \tag{33}$$

$$\left[E\Gamma_0^{(4)} + c\mathbf{p}{\cdot}\mathbf{\Gamma}^{(4)}\right]\psi_L = 0, \quad Helicity/s = -1. \tag{34}$$

One should note that both equations are equivalent to the free Maxwell's Equations.

2.2. Conserved current of probability

In our previous paper[23] we have evaluated the conserved current of probability of the free Maxwell's Equations, we obtained,

$$\frac{\hat{E}}{c}\left(\mathbf{\Phi}^H\mathbf{\Phi}\right) - \hat{\mathbf{p}}{\cdot}\left(\mathbf{\Phi}^H\mathbf{S}\mathbf{\Phi}\right) = 0, \tag{35}$$

or

$$\partial_t\left(\mathbf{\Phi}^H\mathbf{\Phi}\right) + c\mathbf{\nabla}{\cdot}\left(\mathbf{\Phi}^H\mathbf{S}\mathbf{\Phi}\right) = 0. \tag{36}$$

Eq.(36) is the conserved current of probability and $\Phi^H\Phi$ is interpreted as the density of probability which should be normalized to unity,

$$\int\!\!\int\!\!\int dxdydz\left(\mathbf{\Phi}^H\mathbf{\Phi}\right) = 1$$

Using Eqs.(24,27)The conserved probability current is

$$\partial_t\left(\mathbf{\Phi}^H\mathbf{\Phi}\right) + c\mathbf{\nabla}{\cdot}\left(\mathbf{\Phi}^H\boldsymbol{S}\mathbf{\Phi}\right) = \mathcal{N}^2\left[\partial_t\left(\boldsymbol{E}^2 + \boldsymbol{B}^2\right) + 2c\mathbf{\nabla}{\cdot}\left(\boldsymbol{E}\times\boldsymbol{B}\right)\right] = 0, \tag{37}$$

which coincides with the Poynting theorem. The normalization coefficient $\mathcal{N}$ should correspond to the normalization factor.

2.3. Lagrangian Formalism

Here we develop a formalism for the photon wavefunction

$$\Phi = \begin{pmatrix} 0 \\ \Psi_x \\ \Psi_y \\ \Psi_z \end{pmatrix}, \tag{38}$$

The Hermitian conjugate of Φ is

$$\Phi^H = \begin{pmatrix} 0 & \Psi_x^* & \Psi_y^* & \Psi_z^* \end{pmatrix} \tag{39}$$

With these new definitions one can define the Lagrangian density L

$$L = \frac{\Phi^H}{2}\left(E\Gamma_0^{(4)} - c\mathbf{p}\cdot\mathbf{\Gamma}^{(4)}\right)\Phi - \frac{\Phi^T}{2}\left(E\Gamma_0^{(4)} + c\mathbf{p}\cdot\mathbf{\Gamma}^{(4)}\right)\Phi^*, \tag{40}$$

where Φ^T is the transpose of Φ and Φ^* its complex conjugate.

$$\Phi^T = \begin{pmatrix} 0 & \Psi_x & \Psi_y & \Psi_z \end{pmatrix}. \tag{41}$$

The probability current is

$$\partial_t\left(\mathbf{\Psi}^H\mathbf{\Psi}\right) + c\mathbf{\nabla}\cdot\left(\mathbf{\Psi}^H\boldsymbol{S}\mathbf{\Psi}\right) = 0. \tag{42}$$

Thus $\mathbf{\Psi}^H\mathbf{\Psi}$ is interpreted as the density of probability which should be normalized to unity (and is a constant of motion),

$$\int\int\int dxdydz\left(\mathbf{\Psi}^H\mathbf{\Psi}\right) = 1 \tag{43}$$

Having a Lagrangian, one can compute the corresponding energy-momentum tensor,

$$T_{\mu\nu} = \sum_j \frac{\partial L}{\partial\left(\frac{\partial u_j}{\partial x_\mu}\right)}\frac{\partial u_j}{\partial x_\nu} - L\delta_{\mu\nu}, \tag{44}$$

where u_j stand for the components of $\mathbf{\Psi}$ and $\mathbf{\Psi}^*$. For T^{00} we obtain,

$$\frac{2T^{00}}{c} = 2\frac{T_{00}}{c} = \mathbf{\Psi}^H\left(\hat{\boldsymbol{p}}\cdot\boldsymbol{S}\right)\mathbf{\Psi} + \mathbf{\Psi}^T\left(\hat{\boldsymbol{p}}\cdot\boldsymbol{S}\right)\mathbf{\Psi}^* = \mathbf{\Psi}^H\frac{\hat{E}}{c}\mathbf{\Psi} - \mathbf{\Psi}^T\frac{\hat{E}}{c}\mathbf{\Psi}^* \tag{45}$$

and

$$2\int\int\int dxdydzT_{00} = \int\int\int dxdydz\left[\mathbf{\Psi}^H\hat{E}\mathbf{\Psi} - \mathbf{\Psi}^T\hat{E}\mathbf{\Psi}^*\right] \tag{46}$$

Applying the energy operator on Eq.(43) we obtain

$$\hat{E}\int\int\int dxdydz\left(\mathbf{\Psi}^H\mathbf{\Psi}\right) = \int\int\int dxdydz\left[\mathbf{\Psi}^H\hat{E}\mathbf{\Psi} + \mathbf{\Psi}^T\hat{E}\mathbf{\Psi}^*\right] = 0 \tag{47}$$

Hence

$$\int\int\int \mathbf{\Psi}^T\hat{E}\mathbf{\Psi}^* dxdydz = -\int\int\int \mathbf{\Psi}^H\hat{E}\mathbf{\Psi}dxdydz. \tag{48}$$

Substituting Eq.(48) into Eq.(46) we finally obtain

$$\int\int\int dxdydzT^{00} = \int\int\int dxdydz\mathbf{\Psi}^H \hat{E}\mathbf{\Psi} =< \hat{E} > \tag{49}$$

which is the expectation value of the energy operator. For the T^{0k} components we have

$$\frac{T^{0k}}{c} = -\frac{T_{0k}}{c} = \mathbf{\Psi}^H \frac{\hat{p}_k}{2}\mathbf{\Psi} - \mathbf{\Psi}^T \frac{\hat{p}_k}{2}\mathbf{\Psi}^* = \mathbf{\Psi}^H \frac{\hat{p}_k}{2}\mathbf{\Psi} + \left(\mathbf{\Psi}^H \frac{\hat{p}_k}{2}\mathbf{\Psi}\right)^* = \mathrm{Re}\left(\mathbf{\Psi}^H \hat{p}_k \mathbf{\Psi}\right), \tag{50}$$

and

$$\int\int\int dxdydzT^{0k} = \int\int\int dxdydz\mathbf{\Psi}^H \left(c\hat{p}_k\right)\mathbf{\Psi} =< c\hat{p}_k >, \quad k = 1,2,3, \tag{51}$$

which are the expectation values of the momenta.

Let us prove that $\hat{E}$ is a self-adjoint operator. Indeed, from Eq.(48) $\hat{E}^* = -\hat{E}$, we obtain,

$$\int\int\int dxdydz\left(\hat{E}\mathbf{\Psi}\right)^H \mathbf{\Psi} = -\int\int\int dxdydz\left(\hat{E}\mathbf{\Psi}^H\right)\mathbf{\Psi} \tag{52}$$

$$= -\int\int\int \mathbf{\Psi}^T \hat{E}\mathbf{\Psi}^* dxdydz = \int\int\int dxdydz\mathbf{\Psi}^H \hat{E}\mathbf{\Psi}. \tag{53}$$

In a similar way one can prove that the momenta $\hat{p}_k$ are self adjoint. Therefore Eq.(14), with the normalization condition Eq.(43), is the first quantized equation of the photon.

3. Generalization to any spin

In previous papers we have generalized the helicity equations to any spin [23],[24],[25]]. We have used the angular momentum basis of the $D^{(s-1/2,1/2)}$ representation of the Lorentz group. This basis is the sum of the bases of spins s and spin $s-1$ with

$$(2s+1) + (2s-1) = 4s \text{ components.}$$

In this basis the wavefunction is,

$$\Phi^{(4s)} = \begin{pmatrix} \psi_{s-1}^{(2s-1)} \\ \vdots \\ \psi_{-s+1}^{(2s-1)} \\ \psi_{s}^{(2s+1)} \\ \vdots \\ \psi_{-s}^{(2s+1)} \end{pmatrix} \Longrightarrow \begin{pmatrix} 0 \\ \vdots \\ 0 \\ \psi_{s}^{(2s+1)} \\ \vdots \\ \psi_{-s}^{(2s+1)} \end{pmatrix}. \tag{54}$$

The advantage of this formalism, as we have shown in Refs. [24],[25] is that by equating to zero the spin $(s-1)$ components of the wavefunction $\psi_\mu^{(2s-1)}$, the $2s-1$ subsidiary conditions (needed to eliminate the non-forward and non-backward helicities.) are automatically satisfied. The equations for any spin have the form

$$\left[\hat{E}\Gamma_0^{(4s)} - c\mathbf{\Gamma}^{(4s)} \cdot \hat{\boldsymbol{p}}\right]\Phi_R^{(4s)} = 0, \tag{55}$$

$$\left[\hat{E}\Gamma_0^{(4s)} + c\mathbf{\Gamma}^{(4s)} \cdot \hat{\boldsymbol{p}}\right] \Phi_L^{(4s)} = 0, \tag{56}$$

where the $(4s \times 4s)$ gamma matrices have the same multiplication table as the Pauli matrices, therefore they are reducible representation of the Pauli matrices. The equations have built in subsidiary conditions, which are needed to eliminate the non-forward and non-backward helicities. By equating to zero the spin s-1 components of the wave function, the 2s-1 subsidiary conditions (needed to eliminate the non-forward and non-backward helicities.), are automatically satisfied. For spins $\frac{1}{2}$ and spin 1 they are the σ and the $\Gamma^{(4)}$ matrices respectively. The spin 3/2 and spin 2 gamma matrices are given in Refs.[14],[23].

3.1. Lagrangian formalism

The continuity equation for any spin s is

$$\partial_t \left(\left(\Phi^{(4s)} \right)^H \Phi^{(4s)} \right) + \frac{c}{s} \mathbf{\nabla} \cdot \left(\left(\Phi^{(4s)} \right)^H \tilde{\mathbf{S}} \Phi^{(4s)} \right) = 0. \tag{57}$$

Thus the probability density (which should be normalized) is,

$$\rho = \left(\Phi^{(4s)} \right)^H \Phi^{(4s)}, \qquad \int\int\int dx dy dz \rho = 1. \tag{58}$$

In the above formula the density is non-negative and the scalar product is positive definite. Having this result we can find a Lagrangian density,

$$\mathcal{L} = i\hbar \left(\Phi^{(4s)} \right)^H \left(\Gamma_0^{(4s)} \frac{\partial}{\partial t} + c\mathbf{\Gamma}^{(4s)} \cdot \mathbf{\nabla} \right) \Phi^{(4s)}, \tag{59}$$

and using the definition of the energy momentum tensor $T^{\mu\nu}$, we find,

$$\int\int\int dx dy dz T^{00} = \int\int\int dx dy dz \left(\Phi^{(4s)} \right)^H \mathcal{H} \Phi^{(4s)} = < \mathcal{H} >, \tag{60}$$

$$\begin{aligned} \int\int\int dx dy dz T^{0k} &= \int\int\int dx dy dz \left(\Phi^{(4s)} \right)^H (c\hat{p}_k) \Phi^{(4s)} = < c\hat{p}_k >, \\ k &= 1, 2, 3, \end{aligned} \tag{61}$$

i.e. consistent with the expectation values of energy and momentum.

4. Generalized Maxwell's Equations

As we have mentioned earlier, Maxwell's Equations have to be generalized in order to have an interaction term proportional to $\Psi_\mu j^\mu$.

The generalization of Maxwell's Equations is straightforward with replacing the zero of the spin zero component with $\Psi_0 = E_0 + iB_0$,

$$\left(\Gamma_0^{(4)} \partial_0 + \mathbf{\Gamma}^{(4)} \cdot \mathbf{\nabla} \right) \Phi^{(4)} = -\alpha J, \quad \Phi^{(4)} = \begin{pmatrix} \Psi_0 \\ \Psi_x \\ \Psi_y \\ \Psi_z \end{pmatrix}, \quad J = \begin{pmatrix} j_0 \\ j_x \\ j_y \\ j_z \end{pmatrix}. \tag{62}$$

from which one can extract generalized Maxwell's Equations in the form

$$\partial_0 E_0 + \boldsymbol{\nabla} \cdot \boldsymbol{E} = \alpha j_0, \tag{63}$$

$$\partial_0\, \boldsymbol{E} + \boldsymbol{\nabla} E_0 - \boldsymbol{\nabla} \times \mathbf{B} = -\alpha \boldsymbol{j}, \tag{64}$$

$$\partial_0 B_0 + \nabla \cdot \boldsymbol{B} = 0, \tag{65}$$

$$\boldsymbol{\nabla} \times \boldsymbol{E} + \boldsymbol{\partial_0 B} + \boldsymbol{\nabla} B_0 = 0. \tag{66}$$

Similar equations were derived by Dvoeglazov [16] as we show below.

4.1. Gersten's method[15],[18],[19] and Dvoeglazov's[16] generalization of Maxwell's Equations
Gersten[15] has derived the photon equation from the following decomposition

$$\left(\frac{E^2}{c^2} - \tilde{\mathbf{p}}^2\right) I^{(3)} =$$

$$= \left(\frac{E}{c} I^{(3)} - \tilde{\mathbf{p}} \cdot \tilde{\mathbf{S}}\right) \left(\frac{E}{c} I^{(3)} + \tilde{\mathbf{p}} \cdot \tilde{\mathbf{S}}\right) - \begin{pmatrix} p_x^2 & p_x p_y & p_x p_z \\ p_y p_x & p_y^2 & p_y p_z \\ p_z p_x & p_z p_y & p_z^2 \end{pmatrix} = 0, \tag{67}$$

where $I^{(3)}$ is a 3×3 unit matrix, and $\tilde{\mathbf{S}}$ is a spin one vector matrix with components

$$S_x = \begin{pmatrix} 0 & 0 & 0 \\ 0 & 0 & -i \\ 0 & i & 0 \end{pmatrix}, \quad S_y = \begin{pmatrix} 0 & 0 & i \\ 0 & 0 & 0 \\ -i & 0 & 0 \end{pmatrix}, \quad S_z = \begin{pmatrix} 0 & -i & 0 \\ i & 0 & 0 \\ 0 & 0 & 0 \end{pmatrix}, \tag{68}$$

and with the properties

$$[S_x, S_y] = iS_z, \quad [S_z, S_x] = iS_y, \quad [S_y, S_z] = iS_x, \quad \tilde{\mathbf{S}}^2 = 2I^{(3)}. \tag{69}$$

The decomposition Eq.(67) can be verified directly by substitution.

The matrix on the right hand side of Eq.(67) can be rewritten as:

$$\begin{pmatrix} p_x^2 & p_x p_y & p_x p_z \\ p_y p_x & p_y^2 & p_y p_z \\ p_z p_x & p_z p_y & p_z^2 \end{pmatrix} = \begin{pmatrix} p_x \\ p_y \\ p_z \end{pmatrix} \begin{pmatrix} p_x & p_y & p_z \end{pmatrix}. \tag{70}$$

From Eqs.(67 − 68) and Eq.(70), the photon equation can be obtained form

$$\left(\frac{E^2}{c^2} - \tilde{\mathbf{p}}^2\right) \tilde{\boldsymbol{\Psi}} = \left(\frac{E}{c} I^{(3)} - \tilde{\mathbf{p}} \cdot \tilde{\mathbf{S}}\right) \left(\frac{E}{c} I^{(3)} + \tilde{\mathbf{p}} \cdot \tilde{\mathbf{S}}\right) \tilde{\boldsymbol{\Psi}} - \begin{pmatrix} p_x \\ p_y \\ p_z \end{pmatrix} \left(\tilde{\mathbf{p}} \cdot \tilde{\boldsymbol{\Psi}}\right) = 0, \tag{71}$$

where $\tilde{\boldsymbol{\Psi}}$ is a 3 component (column) wave function. Eq. (71) will be satisfied if the two equations

$$\left(\frac{E}{c} I^{(3)} + \tilde{\mathbf{p}} \cdot \tilde{\mathbf{S}}\right) \tilde{\boldsymbol{\Psi}} = 0, \tag{72}$$

$$\tilde{\mathbf{p}} \cdot \tilde{\boldsymbol{\Psi}} = 0, \tag{73}$$

IOP Conf. Series: Journal of Physics: Conf. Series **1239** (2019) 012020 doi:10.1088/1742-6596/1239/1/012020

will be simultaneously satisfied.

Dvoeglazov[16] has found other solution to Eqs.(71),

$$\left(\frac{E}{c}I^{(3)} + \tilde{\mathbf{p}} \cdot \tilde{\mathbf{S}}\right) \tilde{\boldsymbol{\Psi}} = \tilde{\mathbf{p}}\chi, \tag{74}$$

$$\tilde{\mathbf{p}} \cdot \tilde{\boldsymbol{\Psi}} = \frac{E}{c}\chi, \tag{75}$$

where χ is an arbitrary scalar function. It can be shown[16] that $\chi = \Psi_0$ in Eq.(62), i.e. the solution of Dvoeglazov is equivalent to the generalized Maxwell's Equations as in Eq.(62).

5. New way of defining potentials

Let us introduce the 4-potentials,

$$A^{(4)} = \begin{pmatrix} A_0 \\ A_x \\ A_y \\ A_z \end{pmatrix}, \tag{76}$$

where the scalar potential is $V = cA_0$. Their relation to the wavefunction $\Phi^{(4)}$, generalizing previous definitions, can be obtained by requiring that they satisfy the wave equation,

$$\partial_\mu \partial^\mu A^{(4)} = \alpha J. \tag{77}$$

The Γ matrices, which are a representation of the Pauli matrices, will factorize the d'Alembertian the same way as the Pauli matrices factorize it, i.e.,

$$\left(\Gamma_0^{(4)}\partial_0 - \boldsymbol{\Gamma}^{(4)} \cdot \boldsymbol{\nabla}\right)\left(\Gamma_0^{(4)}\partial_0 + \boldsymbol{\Gamma}^{(4)} \cdot \boldsymbol{\nabla}\right) = \Gamma_0^{(4)}\partial_\mu\partial^\mu \tag{78}$$

In order to get Eq. (77), using Eqs. (78) and (62) the relation between the $\Phi^{(4)}$ and $A^{(4)}$ has to be

$$\Phi^{(4)} = \begin{pmatrix} E_0 + iB_0 \\ E_x + iB_x \\ E_y + iB_y \\ E_z + iB_z \end{pmatrix} = -\left(\Gamma_0^{(4)}\partial_0 - \boldsymbol{\Gamma}^{(4)} \cdot \boldsymbol{\nabla}\right) \begin{pmatrix} A_0 \\ A_x \\ A_y \\ A_z \end{pmatrix} \tag{79}$$

$$= -\begin{pmatrix} \partial_0 A_0 + \partial_1 A_x + \partial_2 A_y + \partial_3 A_z \\ \partial_0 A_x + \partial_1 A_0 + i\left(\partial_2 A_z - \partial_3 A_y\right) \\ \partial_0 A_y + \partial_2 A_0 - i\left(\partial_1 A_z - \partial_3 A_x\right) \\ \partial_0 A_z + \partial_3 A_0 + i\left(\partial_1 A_y - \partial_2 A_x\right) \end{pmatrix}, \tag{80}$$

from which, assuming that the potentials are real, the relation between the fields and the potentials is,

$$E_0 = -\partial_0 A_0 - \nabla \cdot \mathbf{A}, \tag{81}$$

$$B_0 = 0, \tag{82}$$

$$\mathbf{E} = -\boldsymbol{\nabla} A_0 - \partial_0 \mathbf{A}, \tag{83}$$

$$\mathbf{B} = \boldsymbol{\nabla} \times \mathbf{A}. \tag{84}$$

These are new relations which can now link two four vectors! With this new formalism we have generalized the fields from 3-vectors to 4-vectors as should be expected from a reliable relativistic theory. The case $E_0 + iB_0 = 0$ brings us back to the usual Maxwell's equations in the Lorenz gauge[17],

$$\partial_0 A_0 + \nabla \cdot \mathbf{A} = 0, \tag{85}$$

In the generalized equations the Lorentz condition does not have to be satisfied. If it is satisfied it enforces the spin 0 component E_0 to be zero. Thus only when the Lorenz condition is not satisfied (while Eq. (81) is valid) the E_0 can be different from zero.

When the Lorenz condition is satisfied the Generalized Maxwell's Equations reduces to the regular spin 1 Maxwell's Equations.

5.1. Wave equations for the Generalized Maxwell's Equations

Let us multiply Eq. (62) by $\left(\Gamma_0^{(4)}\partial_0 - \mathbf{\Gamma}^{(4)} \cdot \boldsymbol{\nabla}\right)$,

$$\left(\Gamma_0^{(4)}\partial_0 - \mathbf{\Gamma}^{(4)} \cdot \boldsymbol{\nabla}\right)\left(\Gamma_0^{(4)}\partial_0 + \mathbf{\Gamma}^{(4)} \cdot \boldsymbol{\nabla}\right)\Phi^{(4)} = \Gamma_0^{(4)}\partial_\mu\partial^\mu\Phi^{(4)} \tag{86}$$

$$= \Box\Phi^{(4)} = -\alpha\left(\Gamma_0^{(4)}\partial_0 - \mathbf{\Gamma}^{(4)} \cdot \boldsymbol{\nabla}\right)J, \tag{87}$$

where $\Box = \partial_\mu\partial^\mu$, from which we obtain,

$$\Box E_0 = \alpha\left(\partial_0 j_0 + \boldsymbol{\nabla} \cdot \mathbf{j}\right), \tag{88}$$

$$\Box B_0 = 0, \tag{89}$$

$$\Box\mathbf{E} = \alpha\left(\partial_0\mathbf{j} + \boldsymbol{\nabla} j_0\right), \tag{90}$$

$$\Box\mathbf{B} = -\alpha\boldsymbol{\nabla} \times \mathbf{j}, \tag{91}$$

Current conservation requires,

$$\partial_0 j_0 + \boldsymbol{\nabla} \cdot \mathbf{j} = \mathbf{0}, \tag{92}$$

therefore when the current is conserved

$$\Box E_0 = 0, \tag{93}$$

and only if the current is not conserved the possibility of $\Box E_0 \neq 0$ will exist. As indicated before, a spin changing interaction can induce the E_0 component.

We will now look for new features of these equations.

IOP Conf. Series: Journal of Physics: Conf. Series **1239** (2019) 012020 doi:10.1088/1742-6596/1239/1/012020

5.2. Energy density

According to Eq.60) the energy density is proportional to,

$$\begin{pmatrix} \Psi_0^* & \Psi_x^* & \Psi_y^* & \Psi_z^* \end{pmatrix} (i\hbar c\partial_0) \begin{pmatrix} \Psi_0 \\ \Psi_x \\ \Psi_y \\ \Psi_z \end{pmatrix} = i\hbar c \left(\Psi_0^* \partial_0 \Psi_0 + \mathbf{\Psi}^* \cdot (\partial_0 \mathbf{\Psi})\right) \tag{94}$$

$$= i\hbar c \left[(E_0 - iB_0)\, \partial_0 \, (E_0 + iB_0) + (\mathbf{E} - i\mathbf{B}) \cdot (\partial_0 \, (\mathbf{E} + i\mathbf{B}))\right], \tag{95}$$

which is different from the conventional form proportional to

$$\mathbf{\Psi}^* \cdot \mathbf{\Psi} = (\mathbf{E} - i\mathbf{B}) \cdot (\mathbf{E} + i\mathbf{B}) = \mathbf{E}^2 + \mathbf{B}^2 \tag{96}$$

6. Summary

We have shown that Maxwell's Equations without sources, can be considered as wave equations of the helicity operator of a massless particle of spin 1, the photon. The fields are 3-vectors, while potentials are 4-vectors. It is almost impossible to formulate an interaction of 3-dimensional fields with 4-currents, while it was possible with potentials. Thus electrodynamics with charges and electric currents can be formulated with potentials, but not with fields. Therefore there are contradictions in Maxwell's Equations that should be corrected.

Wigner[3] has shown that wavefunctions of massless particles of any spin may have only two non-zero helicity components in the forward or backward directions of motion. Dirac[6], [8] in 1936 has derived wavefunctions for massless particles of any spin. The angular momentum basis of the wavefunction of the helicity operator had $2s+1$ components of spin s. But for massless particles only two components, for forward and backward directions, are non-zero. The other $2s-1$ components become equal to zero by imposing $2s-1$ subsidiary conditions.

We have developed a formalism equivalent to Dirac equations using the angular momentum basis of the $[s - \frac{1}{2}, \frac{1}{2}]$ representation of the Lorentz Group[25],[26]. The angular momentum basis consists of two spins s and $s-1$, and has $(2s+1) + 2(s-1) + 1 = 4s$ base vectors. we found that all these equations, are equations of the helicity operator and are similar in form to the spin one-half equation. In these equations the helicity operators are reducible representations of the helicity operator of spin one-half. Moreover, for any spin, there are only two helicity equations similar to the spin one-half equations.

By equating to zero the spin $s-1$ components of the wave function, the 2s-1subsidiary conditions (needed to eliminate the non-forward and non-backward helicities.), are created. The zeros will remain under Lorentz transformations and the wave function will transform under a subgroup of the Lorentz group with only two non-zero helicity projections.

We have generalized the Maxwell's Equations with fields being 4-vectors, allowing thus to have an interaction term of the fields with the electromagnetic currents. We found a new way to define the potentials, applicable to the generalized Maxwell's Equations, and to massless field equations of any spin.

We did not resolve all difficulties in the covariant form of Maxwell's Equations with our generalization. For example we found (see Eq.(90))

$$\Box E_0 = \alpha \left(\partial_0 j_0 + \mathbf{\nabla} \cdot \mathbf{j}\right) = 0,$$

when the electromagnetic current is conserved. Although the fields are 4-vectors, but $\Psi_0 = 0$ brings us back to previous problems of Maxwell's Equations.

More work is needed in order to have equivalent results with fields and potentials.

IOP Conf. Series: Journal of Physics: Conf. Series **1239** (2019) 012020 doi:10.1088/1742-6596/1239/1/012020

References

[1] J. R. Oppenheimer, Note on Light Quanta and the Electromagnetic Field, *Phys. Rev.* **38,** 725-746 (1931)

[2] Y. Aharonov and D. Bohm, Significance of Electromagnetic Potentials in the Quantum Theory, *Phys. Rev.* **115**, 485-491 (1959)

[3] E. P. Wigner, On Unitary Representations of the Inhomogeneous Lorentz Group, Ann. Math. **40**, 149-204 (1939).

[4] LIU Chang-Li and GE Feng-Jun, Proof of Massless Particles Only Having Two Helicity States, e-Print: arXiv:1403.2698 [physics.gen-ph] (2014)

[5] H. Minkowski, "Die Grundgleichungen fur die elektromagnetischen Vorgiinge in bewegten Körpern", *Nachrichten von der Gesellschaft der Wissenschaften zu Gottingen, Mathematisch-Physikalische Klasse*, pp. 53-111, (1908). Presented in the session of December 21, 1907. Published in 1908.
English translation "The Fundamental Equations for Electromagnetic Processes in Moving Bodies" in M. Saha, *The Principle of Relativity*, Calcutta: University Press, pp. 1-69, (1920), and by Wikisource: Translation:The Fundamental Equations for Electromagnetic Processes in Moving Bodies http://en.wikisource.org/?curid=674267

[6] P.A.M. Dirac, Relativistic wave equations, *Proc. Roy. Soc.* **A155**, 447-59 (1936).

[7] P.A.M. Dirac, The Theory of Magnetic Poles, *Phys. Rev.* **74** 67-79 {1948)

[8] H. Bacry, A set of Wave Equations for Massless Fields which Generalize Weyl and Maxwell Equations, *Nuovo Cimento*, **A32**, 448-60 (1976).

[9] I. Bialynicki-Birula, On the Wave Function of the Photon, *Acta Physica Polonica* **86**, 97-116 (1994)

[10] J.D. Jackson, *Classical Electrodynamics, Third Edition,* John Wiley & sons, New York 1999, p.781

[11] S. Weinberg, Feynman Rules for Any Spin. II. Massless Particles, *Phys. Rev.* **134**, B882-96 (1964)

[12] A. Gersten, Conserved Currents of the Maxwell Equations with Electric and Magnetic Sources *Annales de la Fondation Louis de Broglie,* **21,** 67-79 (1996)

[13] A. Gersten, "Maxwell's Equations as the one-photon quantum equation", *Found. Phys. Lett.* **12**, 291-7 (1999).

[14] A. Gersten, Quantum equations for massless particles of any spin, *Found. Phys. Lett.* **13**, 185-192 (2000).

[15] A. Gersten, Maxwell Equations– The One-Photon Quantum Equation, *Found. Phys.*, **31**, No. 8, 1211-1231 (2001)

[16] Valeri V. Dvoeglazov, Generalized Maxwell Equations from the Einstein postulate, *J. Phys. A: Math. Gen.* **33**, 5011-5016 (2000).

[17] L. Lorenz, On the Identity of the Vibrations of Light with Electrical Currents, *Philosophical Magazine. Series 4*. **34** (230): 287–301. (1867)

[18] F. Smarandache, *Neutrosophic Physics: More Problems, More Solutions,* p 69

[19] V. Christianto and F. Smarandache, A derivation of Maxwell Equations in Quaternion Space, *Progress in Physics* **2,** 23-47 (2010)

[20] T. Ivezic, The electromagnetic field equations for moving media, *J. of Phys. Conf. Series* **845** (2017) 012013

[21] S. Esposito, Covariant Majorana Formulation of Electrodynamics, *Found. Phys.* **28**, 231-44, (1998)

[22] J.S. Lomont, Dirac-Like Wave Equations for Particles of Zero Rest Mass, and Their Quantization, *Phys. Rev.* **111**, 1710-1716 (1958)

[23] A. Gersten and A. Moalem, Maxwell's Equations, quantum physics and the quantum graviton, *Journal of Physics: Conference Series* **330** (2011) 012010.

[24] A. Gersten and A. Moalem, Consistent quantization of massless fields of any spin, *Journal of Physics: Conference Series* **437** (2013), 012019.

[25] A. Gersten and A. Moalem, Consistent quantization of massless fields of any spin and the generalized Maxwell's Equations, *Journal of Physics: Conference Series* 615 (2015) 012011

[26] A. Gersten and A. Moalem, Spin 1/2 properties of massless particles of any spin, : *Journal of Physics: Conf. Series* **845** (2017) 012011

IOP Conf. Series: Journal of Physics: Conf. Series **1239** (2019) 012021 doi:10.1088/1742-6596/1239/1/012021

The Einstein-Podolsky-Rosen paradox and SU(2) relativity

Paul O'Hara

Istituto Universitario Sophia,Via San Vito, 28 - Loppiano, 50064 Figline e Incisa Valdarno (FI), Italy

E-mail: paul.ohara@sophiauniversity.org

Abstract. The EPR paradox dates back to 1935 when Einstein et al., through the use of non commuting operators, proposed that quantum mechanics was not complete[1] in that it suggested a "spooky action at a distance." Later in 1964 John Bell was able to express the dilemma in a simple inequality[2] involving spin singlet states. If the inequality were satisfied then Einstein was correct and if it were violated then it favored the quantum mechanics point of view. In what follows, we present two approaches to Bell's inequality, and then offer an interpretation from the viewpoint of quantum mechanics based on the principle that *the whole is more than the sum of its parts.* This is then combined with the properties of the $SU(2)$ group to give a more intuitive understanding of non-locality and spin entanglement from the perspective of the relativistic characteristics of rotations.
KEY Words: Bell's inequality, whole more that sum of parts, semigroup invariance, $SU(2)$ relativity.

1. Understanding the EPR Paradox

The original version of the EPR paradox dates back to 1935, when Einstein, Podoloski and Rosen published a paper entitled *Can Quantum-Mechanical Description of Physical Reality Be Considered Complete?* The authors of the paper noted that the success of a physical theory rests on two epistemological questions: (1) Is the theory correct? and (2) Is the description given by the theory complete? They then analyzed conjugate variables (position and momentum) from the perspective of the second question and came to the conclusion that "the wave function does not provide a complete description of the physical reality" because its existence depended upon conjugacy and the measuring instrument. This contrasted with their belief that physical realities existed in themselves and can be measured (in principle) with certainty "*without disturbing the system.*" By way of response, Niels Bohr pointed out an ambiquity in Einstein's use of the expression 'without in any way disturbing a system,' and that quantum mechanics characterizes '... an entirely new situation as regards the description of physical phenomena."[3]

1.1. Bell's Inequality and the EPR Paradox

The discussion of EPR[1] might have remained merely philosophical if it had not been for the discovery of Bell's inequality in 1964 and the subsequent experimental work carried out by

[1] Nowadays, some authors refer to this as the EPRB paradox in recognition of the fact that Bohm's reformulation of the problem in terms of the spin singlet state of the diatomic molecule was used by Bell to derive his inequality.

IOP Conf. Series: Journal of Physics: Conf. Series **1239** (2019) 012021 doi:10.1088/1742-6596/1239/1/012021

Aspect et al[4]. Bell began his work by interpretating EPR's belief that quantum mechanics was incomplete to mean that the theory "should be supplemented by additional variables" ([2, p. 195]). In other words, he proposed the existence of hidden variables that would conform to Einstein's notion of measurement. He then considered a pair of spin one-half particles in the singlet state moving away from each other in opposite directions. Denoting their spin operators by σ_1 and σ_2 and their respective components of spin in the direction of the unit vector $\mathbf{n}$ by $\sigma_1 \cdot \mathbf{n}$ and $\sigma_2 \cdot \mathbf{n}$, it follows that if $\sigma_1 \cdot \mathbf{n}$ yields a measurement of $+1$, then by definition of a singlet state $\sigma_2 \cdot \mathbf{n}$ will be -1 and vice versa. Quantum mechanics then predicts that the expected value of spin for the singlet state observed in two directions $\mathbf{n_1}$ and $\mathbf{n_2}$ will be given by

$$E(\mathbf{n}_1, \mathbf{n}_2) = \langle \sigma_1 \cdot \mathbf{n}_1 \; \sigma_2 \cdot \mathbf{n}_2 \rangle = -\mathbf{n}_1 \cdot \mathbf{n}_2. \tag{1}$$

Bell proceeded to examine the expected spin measurements in three different directions $\mathbf{n}_1, \mathbf{n}_2, \mathbf{n}_3$ and by means of a triangular inequality established the relationship

$$|E(\mathbf{n}_1, \mathbf{n}_2) - E(\mathbf{n}_1, \mathbf{n}_3)| \leq 1 + E(\mathbf{n}_2, \mathbf{n}_3) \tag{2}$$

If we now choose the three directions $\mathbf{n}_1, \mathbf{n}_2, \mathbf{n}_3$ in the same plane with azimuthal angles 0, $\frac{\pi}{3}$, and $\frac{2\pi}{3}$ respectively then $\mathbf{n}_1.\mathbf{n}_2 = \mathbf{n}_2.\mathbf{n}_3 = \frac{1}{2}$ and $\mathbf{n}_1.\mathbf{n}_3 = -\frac{1}{2}$. Substituting these into (2) gives $1 < \frac{1}{2}$, which is a contradiction. This means that the quantum mechanics method of calculating expectations is not compatible with the expectation derived from Einstein's et al. assumptions previously discussed.

Subsequently in 1982 Alain Aspect and colleagues experimentally measured the spin states of singlet state photons and concluded based on their experiments that nature does not obey Bell's inequalities and that the quantum expectations predictions were in agreement with their experiments. In other words, measurements associated with entangled states exhibited quantum nonlocality, where, in the words of Bell, locality meant that "the result of measurement on one system be unaffected by operations on a distant system with which it has interacted in the past" ([2, p. 195]).

Generalizations of this inequality have been given by Greenberger et al. They moved beyond entangled pairs associated with Bell's inequalities to consider perfectly correlated systems of four particles [5], which means that "knowledge of the outcomes for any three particles enables a prediction with certainty of the outcome for the fourth." They then established the following identity for spin observables of four perfectly correlated particles in the directions $\mathbf{n}_1, \mathbf{n}_2, \mathbf{n}_3, \mathbf{n}_4$:

$$E^{\psi}(\mathbf{n_1}, \mathbf{n_2}, \mathbf{n_3}, \mathbf{n_4}) = \cos\theta_1 \cos\theta_2 \cos\theta_3 \cos\theta_4 - \sin\theta_1 \sin\theta_2 \sin\theta_3 \sin\theta_4 \cos(\phi_1 + \phi_2 - \phi_3 - \phi_4).$$

In the case of the perfect phase correlations given by $\phi_1 + \phi_2 - \phi_3 - \phi_4 = 0$ with $A_\lambda(\phi_1)B_\lambda(\phi_2)C_\lambda(\phi_3)D_\lambda(\phi_4) = -1$ and $\phi_1 + \phi_2 - \phi_3 - \phi_4 = \pi$ with $A_\lambda(\phi_1)B_\lambda(\phi_2)C_\lambda(\phi_3)D_\lambda(\phi_4) = 1$ a mathematical contradiction follows when it is shown that the first set of equations imply $A_\lambda(2\phi) = const$ for all ϕ while the second imply $A_\lambda(\theta + \pi) = -A_\lambda(\theta)$.

1.2. Wigner's Approach

Wigner in 1970 inspired by Bohm's gedanken experiment used a pair of spin-1/2 particles to give a simple reformulation of Bell's inequality based on the properties of the rotation group [6]. He recognized that nonlocality and the mathematical contradiction associated with the inequality followed from the group properties. Indeed, given the importance of the $SU(2)$ group in the subsequent discusion, Wigner's complete derivation is as follows:
Assuming hidden parameters exist, then the set of all possible outcomes of spin values in the arbitrary chosen directions $\mathbf{n}_1, \mathbf{n}_2, \mathbf{n}_3$ are given by

$$\{(+++), (++-), (+-+), (-++), (+--), (-+-), (--+), (---)\} \tag{3}$$

IOP Conf. Series: Journal of Physics: Conf. Series **1239** (2019) 012021 doi:10.1088/1742-6596/1239/1/012021

where + indicates spin up and − indicates spin down. Specifically, if the predetermined values lie in the same plane at orientations $\theta_i,\ \theta_j,\ \theta_k$ with respect to a chosen axis then

$$\{(++-),(+--)\} \subset \{(++-),(+--),(-+-),(+-+)\} \tag{4}$$

implies

$$P\{(++-),(+--)\} \leq P\{(++-),(+--),(-+-),(+-+)\}. \tag{5}$$

Using the properties of the $SU(2)$ group and the probability interpretation associated with superposition, we obtain

$$P(++\,|\theta,\phi) = P(--\,|\theta,\phi) = \frac{1}{2}\sin^2\!\left(\frac{\theta-\phi}{2}\right). \tag{6}$$

It then follows from equation (5) that

$$\frac{1}{2}\sin^2\frac{\theta_{ki}}{2} \leq \frac{1}{2}\sin^2\frac{\theta_{jk}}{2} + \frac{1}{2}\sin^2\frac{\theta_{ij}}{2}\,. \tag{7}$$

This represents another variation on Bell's inequality. Taking $\theta_{ij} = \theta_{jk} = \frac{\pi}{3}$ and $\theta_{ki} = \frac{2\pi}{3}$ gives $\frac{1}{2} \geq \frac{3}{4}$, a contradiction. The spin orientation cannot be an element of reality in the Einstein sense and measurments corresponding to three spin states cannot be predetermined with probability 1, and in the case of entangled states measurements exhibit quantum nonlocality.

It is also worth noting that although Wigner's derivation refers to singlet states and proves that quantum measurements involving such states are necessarily nonlocal, nevertheless equation (3) and the subsequent theory could also have been directly applied to a single particle. In other words, if hidden variables existed in Bell's understanding of the term then it should be possible to specify the states in all three directions in accordance with Einstein's belief. However, Wigner's inequality shows that regardless of entanglment, no such hidden variables exist (at least in the case of spin).

2. The EPR Paradox and Causality

The work of Bell, Wigner and Aspect (et al.) indicate that in contrast to Einstein's belief, it is precisely the hidden variable approach that is mathematically inconsitent while quantum theory is quite complete. Nevertheless, as a vast number of publications in the area of EPR indicate, oftentimes, there is an intuitive gap between our understanding of the quantum model and our notion of local causality. In what follows, I will suggest an alternate intuitive approach, bearing in mind Bohr's observation that the laws of quantum mechanics try to explain an "entirely new situation as regards the description of physical phenomena that the notion of complementarity aims at characterizing." We begin with a more complete analysis of the singlet state.

2.1. The whole is more than the sum of its parts

Unlike the original description of EPR proposed by Einstein, both Bohm's and Bell's approach which relie on the use of the spin singlet state, offer an additional element that can be easily overlooked and which is implicit in Bohr's reference to "this entirely new situation." We note that in the case of the singlet state the new emerging physical laws can be subsumed under the phrase *the whole is more than the sum of its parts*[2] [9]. We now elaborate upon this.

[2] Aristotle, *Metaphysics*, London, Everyman's Library, Book Z, XI, p. 195: *Why the definition of a whole sometimes includes that of its parts and sometimes does not.* p. 218, VI, *The unity of definition: All wholes which are not mere aggregates, i.e. must have a principle of unity*

IOP Conf. Series: Journal of Physics: Conf. Series **1239** (2019) 012021 doi:10.1088/1742-6596/1239/1/012021

Consider the singlet state

$$\Psi = \frac{1}{\sqrt{2}}\begin{pmatrix}1\\0\end{pmatrix}\otimes\begin{pmatrix}0\\1\end{pmatrix}-\begin{pmatrix}0\\1\end{pmatrix}\otimes\begin{pmatrix}1\\0\end{pmatrix} \tag{8}$$

As a member of $C^2 \otimes C^2$, Ψ is invariant under the action of the group elements $g \otimes g/|g|$ where $g \in GL(2,C)$ and $|g|$ is the determinant of g. Invariance means that

$$|g|^{-1}(g \otimes g)\Psi = \Psi$$

In particular, if $|g| = 1$ then, as has already been proven, $g \in SL(2,C)$ [10], which also includes rotational invariance. Moreover, as shown in [10], Ψ is also unique in that it is the only eigenstate common to all elements of $g \in SL(2,C)$.

As a tensor product, Ψ can be embedded in C^4 and written as $\tilde{\Psi} = (0,1,-1,0)^T$ (where T refers to the transpose). Note that $\Psi = \tilde{\Psi}$. However, since $C^2 \otimes C^2 \subset C^4$, we can say that C^4 is of a higher order than $C^2 \otimes C^2$ and use Ψ and $\tilde{\Psi}$ respectively to maintain the distinction between the two spaces. Moreover, if we consider $\tilde{\Psi} \in C^4$ we find that it is invariant under a 24-parameter semigroup, which we denote by $SM(4,C)$, whose elements are given by

$$A = \begin{pmatrix} a_{11} & a_{12} & a_{12} & a_{14} \\ a_{21} & a_{22} & a_{22}-1 & a_{24} \\ a_{31} & a_{33}-1 & a_{33} & a_{34} \\ a_{41} & a_{42} & a_{42} & a_{44} \end{pmatrix}$$

where each $a_{ij} \in C$. Note that $a_{12} = a_{13}$, $a_{42} = a_{43}$, $a_{23} = a_{22} - 1$ and $a_{32} = a_{33} - 1$, which means that A has 12 independent complex parameters. We formulate this as a lemma:

Lemma 1 *$\tilde{\Psi} \in C^4$ is invariant under the action of the semigroup $SM(4,C)$. In other words, for all $A \in SM(4,C)$: $A\tilde{\Psi} = \tilde{\Psi}$.*

The proof follows by direct multiplication. It should also be noted that the determinant of A can be any real number including 0, which also explains why $SM(4,C)$ is not a group. In general, the set of matrices $\{A\}$ are irreducibile, although it does contain the four complex parameter reducible subgroup whose elements are $g \otimes g/|g|$, already mentioned above. For example, in the case of

$$A = \begin{pmatrix} 1 & 0 & 0 & 0 \\ 1 & 1 & 0 & 0 \\ 1 & 0 & 1 & 0 \\ 1 & 1 & 1 & a_{44} \end{pmatrix} \quad \text{where } a_{44} \neq 1,\ A \text{ is irreducible.}$$

Whereas in the case of $a_{44} = 1$,

$$A = \begin{pmatrix} 1 & 0 & 0 & 0 \\ 1 & 1 & 0 & 0 \\ 1 & 0 & 1 & 0 \\ 1 & 1 & 1 & 1 \end{pmatrix} = \begin{pmatrix} 1 & 0 \\ 1 & 1 \end{pmatrix} \otimes \begin{pmatrix} 1 & 0 \\ 1 & 1 \end{pmatrix}.$$

In either case, we can say that the semigroup $SM(4,C)$ represents a new "physical reality" in Bohr's sense associated solely with the singlet state. It does not stand on its own nor can it be associated with the individual states. The irreducibility of $SM(4,C)$ is a mathematical (or a quantum mechanical) way of saying that the whole is more than the sum of its parts. It also demonstrates that Einstein ed al.'s interpretation of reality is reductionist in that it fails to conceive of higher order realities represented by $SM(4,C)$ that cannot be understood by

IOP Conf. Series: Journal of Physics: Conf. Series **1239** (2019) 012021 doi:10.1088/1742-6596/1239/1/012021

analyzing the separate components. The incompleteness suggested by EPR is consequence of overlooking this new "physical reality."

As an application, we can consider ionic and covalent bonds in chemistry. Both types of bonding involve spin singlet states and consequently entanglement is intrisic to our understanding of chemistry. One cannot understand water H_2O by merely studying hydrogen alone or oxygen alone. The chemistry of water is to be understood primarily in terms of the covalent bonds between the hydrogen and oxygen atoms, bonds which in their ground state configuration form spin-singlet states. Indeed, from the perspective of chemistry, the spin-singlet state constitutive of covalent bonding is an essential element of reality. It represents a rotationally invariant pure state that cannot be decomposed into its individual components (collapsed wave function) without destroying the correlation. It's a concrete example of the whole being greater than the sum of its parts.

2.2. Entanglement and the Projection Postulate

Keeping in mind the previous discussion of a"physical reality" that is not reducible to the sum of its parts and characterized by the semigroup $SM(4, C)$, let us consider an alternate approach to the EPR problem. For simplicity, we will work with those elements of reality that are specifically related to the invariant subgroup $g \otimes g/|g|$ where $g \in GL(2, C)$. Moreover, each $g \in GL(2, C)$ can be reparameterized in terms of polar coordinates defined with respect to a non-orthogonal frame:

$$g = \begin{pmatrix} w & y \\ x & z \end{pmatrix} = \begin{pmatrix} r_1 \cos\theta & r_2 \sin\phi \\ -r_1 \sin\theta & r_2 \cos\phi \end{pmatrix}$$

A frame becomes orthogonal only when dealing with proper (and improper) rotations. For convenience, we change notation and write $R(\theta, \phi)$ for g to reflect the dependancy upon polar coordinates. There is also a dependancy upon r_1 and r_2. However, for the discussion that follows, we let $r_1 r_2 = 1$ or equivalently $|R(\theta, \phi)| = \cos(\theta - \phi)$ and so we do not explicitely refer to them in the notation.

Once again, let us consider a singlet state. Three states can be associated with it. The first is given by Ψ (cf. equation (8)). It is an invariant pure eigenstate of the operator $R(\theta, \phi) \otimes R(\theta, \phi)/|R(\theta, \phi)|$. The second and third are reduced states (accessed through a Stern-Gerlach experiment) which come from taking a spin measurement of the singlet state in the direction of $\mathbf{n}$ and are given by $|\mathbf{n}, +\rangle\, |\mathbf{n}, -\rangle$ and $|\mathbf{n}, -\rangle\, |\mathbf{n}, +\rangle$ respectively. In terms of the elements of reality associated with EPR, we can speak of elements of reality prior to spin measurements (the singlet state), and elements of reality after the spin measurements (the two reduced states).

Quantum mechanics is something new. We may say that the original spin singlet state and the reduced SG states are related as coherent and decoherent states of spin. At any one time the system is either in an entangled or a reduced state but not contemporaneously. They are mutually exclusive. We can observe one or the other. In the case of entanglement the correlation is intrinsic to the definition of the state and is not an epiphenomenon. The whole is greater than the sum of the parts, with the rotational invariance of the pure singlet state constituting an additional element that cannot be captured by either of the reduced states alone and independent of any choice of angle.

Specifically, if

$$\mathbf{a} = \begin{pmatrix} a_1 \\ a_2 \end{pmatrix}, \ \mathbf{b} = \begin{pmatrix} b_1 \\ b_2 \end{pmatrix}$$

then the wedge product $\mathbf{a} \wedge \mathbf{b} = \det[\mathbf{a}, \mathbf{b}]\sqrt{2}\Psi$, where Ψ is the singlet state of equation (8) and $\det[\mathbf{a}, \mathbf{b}] = a_1 b_2 - b_1 a_2$. Note that we use $\mathbf{a}$ and $\mathbf{b}$ to represent general vectors and reserve $\mathbf{n}_i$ to designate the direction of the spin components $\sigma.\mathbf{n}_i$ measured in the different directions $\mathbf{n}_i$.

IOP Conf. Series: Journal of Physics: Conf. Series **1239** (2019) 012021 doi:10.1088/1742-6596/1239/1/012021

This use of $\mathbf{n}_i$ is consistent with notation already introduced. Consequently by invariance

$$R(\theta,\phi)\otimes R(\theta,\phi)(\mathbf{a}\wedge\mathbf{b}) \tag{9}$$
$$\equiv R(\theta,\phi)\mathbf{a}\wedge R(\theta,\phi)\mathbf{b} \tag{10}$$
$$= \cos(\theta-\phi)(\mathbf{a}\wedge\mathbf{b})\ . \tag{11}$$

Moreover, although the spin operator

$$\sigma_3 = \begin{pmatrix} 1 & 0 \\ 0 & -1 \end{pmatrix}$$

is such that $[\sigma_3, R(\theta,\phi)] \neq 0$ and $[\sigma_3\otimes\sigma_3, R(\theta,\phi)\otimes R(\theta,\phi)] \neq 0$ for all $\theta \neq 0$ and $\phi \neq 0$, nevertheless in the latter case the two operators share an eigenvector in common, namely $(\mathbf{a}\otimes\mathbf{b}-\mathbf{b}\otimes\mathbf{a})$. It follows that

$$[\sigma_3\otimes\sigma_3, R(\theta,\phi)\otimes R(\theta,\phi)] \neq 0 \text{ but } [\sigma_3\otimes\sigma_3, R(\theta,\phi)\otimes R(\theta,\phi)](\mathbf{a}\wedge\mathbf{b}) = 0$$

and

$$(R(\theta,\phi)\sigma_3\otimes R(\theta,\phi)\sigma_3)(\mathbf{a}\wedge\mathbf{b}) = -\cos(\theta-\phi)(\mathbf{a}\wedge\mathbf{b})$$

In other words, they commute while operating on the subspace

$$sp\{\mathbf{a}\otimes\mathbf{b}-\mathbf{b}\otimes\mathbf{a}\}\ ,$$

spanned by the rotationally invariant state $\mathbf{a}\wedge\mathbf{b}$. It also follows that the expected value of the operator restricted to the above domain is given by

$$\langle R(\theta,\phi)\sigma_3\otimes R(\theta,\phi)\sigma_3\rangle = -\cos(\theta-\phi) = -\mathbf{n}_1.\mathbf{n}_2 \tag{12}$$

which is as expected from quantum mechanics.

This offers a new insight into the nature of coherency and the projection postulate. Spin measurements of the isotropically spin-correlated states in directions of θ and ϕ, or equivalently in the directions $\mathbf{n}_1$ and $\mathbf{n}_2$ respectively, depend only on the difference $\lambda = \theta - \phi$ and this difference is a consequence of its action on $\mathbf{a}\wedge\mathbf{b}$. It reflects the isotropy of the prior correlation. The intrinsic correlation is defined prior to any projections and while in this pure state, the singlet will continue to exhibit this correlation. This also refines our understanding of the projection postulate of quantum mechanics. A single SG measurement will destroy the isotropy and reduce the pure singlet state of the joint wave function into two independent particle states, while at the same time maintaining a polarization in the direction of the original measurements. In other words, causality is still preserved not in a deterministic way but rather in accordance with the rules of (quantum) probabilities. Polarization is a consequence of the SG experiment destroying the isotropic correlation except in the direction of measurement. The polarizion is not induced by the SG magnets or a polarizing lens, rather the SG experiment destroys the entanglement except in the direction of measurement. Collapsing the wave function in this case means collapsing the isotropy.

2.3. Hidden Variables versus Correlations

At the core of Bell's argument lies the fact that if hidden variables were to exist and uniquely determined the spin value of each individual particle then spin correlations would be merely epiphenomenal. This means that once we understood how the hidden parameters function with respect to each particle, we could then correlate as many particles as we wish by reproducing each one in the same way as prescribed by the hidden parameter. For example, to borrow from

IOP Conf. Series: Journal of Physics: Conf. Series **1239** (2019) 012021 doi:10.1088/1742-6596/1239/1/012021

geometry, once we learn to draw an equilateral triangle with a ruler and compass, we can draw an endless number of similar equilateral triangles using the same principle. Each equilateral triangle would then be automatically correlated. However, in this case the correlations would then be a secondary consequence of having made each triangle in a similar way.

Returning to the singlet state, the existence of hidden parameters, in Bell's sense, would mean that it can be implicitly considered to be a product of two reduced states $A(\mathbf{n}_1)B(\mathbf{n}_2)$ associated with the probability equation:

$$E(\mathbf{n}_1, \mathbf{n}_2) = -\int d(\lambda)\rho(\lambda)A(\mathbf{n}_1, \lambda)A(\mathbf{n}_2, \lambda) \tag{13}$$

This leads to a contradiction. Bell's work and the subsequent experiments of Aspect et al. show that in the case of spin systems, no such hidden parameters exist. Yet, there is no doubt that in the case of a singlet state, particles are correlated. If a spin measurement is made on one particle then we immediately know the outcome of the spin measurement on the second particle, measured in the same direction. The only alternative for Bell is to seek an explanation in terms of his understanding of the projection postulate of QM. But this too is unsatisfactory since it would mean that "the results of one magnet now depend on the setting of the distant magnet, which is just what we would wish to avoid" ([2], p.197).

He fails to recognize the significance of the rotational invariance of the state $\mathbf{a}\wedge\mathbf{b}$ prior to any measurements having been taken, and consequently that this can also be interpreted to mean that the pre-established correlations are real and dependent upon a correlation parameter $\theta - \phi$, with the direction of the magnetic fields measuring this intrinsic correlation without any need of action at a distance. In other words, the correlation $\lambda = \theta - \phi$ can itself be considered a new hidden variable but not in Bell's (nor Christian's [7]) sense, which was strictly meant to be "a hidden variable account of spin measurements on a single particle"([2, p. 196]). Instead, here it is intrinsic to the *pairing* associated with isotropically spin correlated particles. This can be seen in equation (11). The $\lambda = \theta - \phi$ emerges only from the uniqueness of the wedge product $\mathbf{a}\wedge\mathbf{b}$. It cannot be extended beyond pairs and it cannot be applied to individual particles. In other words, the spin-singlet state is a higher order phenomena that incorporates some new principle. We have already called it the "coupling principle," meaning that *isotropically spin-correlated particles can only occur in pairs*[11]. It is intrinsically related to the notation of rotational invariance or more precisely to $SM(4, \mathcal{C})$ invariance and is strictly a quantum mechanical effect. Its existence requires a new intuitive (not mathematical) understanding of the projection postulate, in that the collapsing of the wave function in this case means the collapsing of the isotropy as noted above. We explore this in greater detail below and offer an alternative interpretation in terms of $SU(2)$ relativity.[3]

3. Reference Frames and the SU(2) Group

The entanglement of the singlet state guarantees that both particles will have equal and opposite values when measured in the same reference frame. However, entanglement does not include the assigning of unique numerical values over the entire plane which are associated not only with the correlation but with the relative perspective of the observer. Although the SG magnets correctly observe the preexisting correlations of singlet states, Wigner's version of Bell's inequality derived from the laws of probability and the properties of the group $SU(2)$ applied to any of the eight elements of equation (3), clearly shows that such states cannot actually be consistently defined. Their existence would require that individual hidden variables be associated with each particle,

[3] I use the words "$SU(2)$ relativity" to reflect the fact that $SL(2, C)$ is homomorphic to the Lorentz group $SO(3, 1)$ and that $SU(2) \subset SL(2, C)$ is homomorphic to $SO(3) \subset SO(3, 1)$. Moreover, as we shall see in sections 3.2 and 3.3, spin values are a consequence not only of the intrisic correlations related to the singlet state but also of the specific frame of reference within which it is measured.

in contrast with Bell's inequality which shows that such hidden variables do not exist. States such as $(+,+,-)$ cannot be coherently defined within this context and attempting to do so hides the fact that the problem is ill-posed.

To better understand the ill posedness, we approach the question from the perspective of what we call SU(2,C) relativity [12, 13]. Essentially this means that although the inherent nature of the isotropically spin correlated states resides in the rotational invariance and is independent of the observer, the observed spin values are dependent upon the SU(2) frame of refence, or equivalently are dependent upon the direction of measurement of the SG instruments. In other words, the same spin state can be interpreted as $+1$ or -1 depending on our point of view. Consequently, as we shall see below, this permits two different values to be assigned the same event, depending on one's SU(2) frame of reference.

An analogy might help. The Earth as viewed from above the North pole can be seen as rotating in an anti-clockwise direction (+1), while the same state of motion can be viewed as being in a clockwise direction (-1), if viewed from the South pole. Denoting the corresponding states by $(1,0)$ and $(0,1)$ respectively, if we attempt to characterize these two states from the perspective of at least two other frames, pointing in different directions, a contradiction will arise if the SU(2) relativity effects are ignored. For example, we cannot consistently characterize these states from a frame on the equator. One cannot consistently define clockwise and counterclockwise unless handedness (trivector) is taken into consideration.

In EPR something similar is happening. Each observer can uniquely describe spin up and spin down from the perspective of his or her reference frame. However, three different observers cannot consistently define spin-up and spin-down relative to each other for a pure quantum state in terms of reduced states. The resulting inconsistency is encapsulated in Bell's inequality. The same reality will take on two different spin values, depending on one's viewpoint. Moreover, since the essence of Bell's inequality involves correlations from three different directions, its a prime candidate for bringing forth inconsistencies, unless this SU(2) relativity effect (which is explained below) is taken into account.

3.1. What is a Pure Quantum State?

In terms of mathematics we define a pure state as an eigenvector in a Hilbert space that corresponds to an operator. The simplest example can be seen by considering

$$\begin{pmatrix} 1 & 0 \\ 0 & -1 \end{pmatrix}\begin{pmatrix} 1 \\ 0 \end{pmatrix} = \begin{pmatrix} 1 \\ 0 \end{pmatrix} \qquad \begin{pmatrix} 1 & 0 \\ 0 & -1 \end{pmatrix}\begin{pmatrix} 0 \\ 1 \end{pmatrix} = -\begin{pmatrix} 0 \\ 1 \end{pmatrix} \tag{14}$$

In this case both $\begin{pmatrix} 1 \\ 0 \end{pmatrix}$ and $\begin{pmatrix} 0 \\ 1 \end{pmatrix}$ are pure states of the operator σ_3. Now consider equation (14) multiplied on the left by a rotation matrix $R(\theta)$ and rewritten in the form

$$R(\theta)\sigma_3 R(-\theta)R(\theta)\begin{pmatrix} 1 \\ 0 \end{pmatrix} = R(\theta)\begin{pmatrix} 1 \\ 0 \end{pmatrix} \tag{15}$$

This simplifies to

$$\begin{pmatrix} \cos 2\theta & \sin 2\theta \\ \sin 2\theta & -\cos 2\theta \end{pmatrix}\begin{pmatrix} \cos\theta \\ \sin\theta \end{pmatrix} = \begin{pmatrix} \cos\theta \\ \sin\theta \end{pmatrix} \tag{16}$$

In other words, the $\begin{pmatrix} 1 \\ 0 \end{pmatrix}$ and $\begin{pmatrix} \cos\theta \\ \sin\theta \end{pmatrix}$ can be viewed as the **same** state but viewed from the perspective of two different reference frames and from the perspective of the different spin operators, σ_3 and $R(\theta)\sigma_3 R(-\theta)$ respectively. In this sense both are pure states. In contrast, from the perspective of $R(\theta)\sigma_3 R(-\theta)$ and σ_3 respectively they are both mixed states. It should also be noted that the spin-singlet state is an eigenvector of all linear operators defined on the tensor product of the two two-dimensional Hilbert spaces $\mathcal{H} \otimes \mathcal{H}$. With this in mind, we now introduce the notion of $SU(2)$ relativity.

3.2. SU(2) Relativity

So, what is SU(2) relativity? To begin, let us erect two orthogonal frames, the laboratory frame (i, j, k) and a spinor frame $(e_i, e_{r(\theta)}, e_\theta)$, where $e_i = i$ represents the direction of motion of a particle with spin, $e_{r(\theta)}$ is related to the direction of the detector orientated at an angle θ with respect to i, and $e_\theta = e_i \times e_r$ is a bivector. Note that relative to the laboratory frame (i, j, k) there are an infinite number of choices for $e_{r(\theta)}$ and consequently e_θ (one for each θ). In general, for a rotation through an angle θ in the $j - k$ plane, we will obtain

$$(e_i, e_r, e_\theta) = (i, j\cos(c\theta) + k\sin(c\theta), k\cos(c\theta) - j\sin(c\theta)) \ , \tag{17}$$

where c is a constant to be determined from the physics of the situation.

For example, in the case of a photon $c = 1$. It follows that if the initial alignment of the polarizer is along the positive j-axis then $(e_i, e_{r(0)}, e_0) = (i, j, k)$ and $(e_i, e_{r(\pi/2)}, e_{\pi/2}) = (i, k, -j)$ if the polarizer is rotated through an angle $\pi/2$ in the $j-k$ plane. This means that if $k = e_0 = |+\rangle$ denotes spin-up in the direction k then $e_{\pi/2} = -j = |-\rangle$ denotes spin-down in the direction $-j$ with respect to the laboratory frame.

In the case of an electron $c = 1/2$, we identify $e_{r(\theta)}$ not with the direction of the SG magnetic field but with the axes of rotation (or reflection) of the original e_0 after it is rotated through an angle θ. This now means that if spin-up is represented by the vector k, then spin-down should be represented by a bivector j with respect to the triad (i, j, k). Moreover, again based on the above, we know that if the original spin is measured as $+1/2$ with respect to the (i, j, k) triad then after a rotation of π in the $j - k$ plane, spin will be measured as $-1/2$ in the same triad. All of this can be encapsulated in the formula,

$$(e_i, e_r, e_\theta) = (i, j\cos(\theta/2) + k\sin(\theta/2), k\cos(\theta/2) - j\sin(\theta/2). \tag{18}$$

It is important to recognize that this formula reflects the physics associated with the SU(2) representation of spin. Note that for $\theta = \pi$, the spin vector e_θ will be spin down and in a direction orthogonal to the original k.

Different frames (e_i, e_r, e_θ) for different θ can be obtained by rotating the original triad $(e_i, e_{r(0)}, e_0)$ about different axes of rotation. However, when three frames are introduced, EPR tells us that contradictions will arise (Bell's inequality) if we identify the results of one frame with those of another. This should not be seen as a flaw in quantum mechanics but rather as a geometrical consequence of introducing three or more frames of the type (e_i, e_r, e_θ) for different angles θ. In particular, consider the three frames $(e_i, e_{r(0)}, e_0)$, $(e_i, e_{r(\theta)}, e_\theta)$ and $(e_i, e_{r(\theta+\phi)}, e_{\theta+\phi})$ with the understanding that the triad $(i, j, k) = (e_i, e_r, e_0)$. It follows from (18) that with respect to the frame (i, j, k),

$$(e_i, e_{r(\theta)}, e_\theta) = (e_i, j\cos(\theta/2) + k\sin(\theta/2), k\cos(\theta/2) - j\sin(\theta/2)) \ ,$$

and

$$\begin{aligned}(e_i, e_{r(\theta+\phi)}, e_{\theta+\phi}) &= \\ &(e_i, j\cos((\theta+\phi)/2) + k\sin((\theta+\phi)/2), k\cos((\theta+\phi)/2) - j\sin((\theta+\phi)/2))\end{aligned}$$

but with respect to the frame $(e_i, e_{r(\theta)}, e_\phi)$

$$(e_i, e_{r(\theta+\phi)}, e_{\theta+\phi}) = (e_i, \cos(\phi/2)e_{r(\theta)} + \sin(\phi/2)e_\theta, \cos(\phi/2)e_\theta - \sin(\phi/2)e_{r(\theta)}) \ .$$

From the perspective of quantum mechanics, we can interpret this last equation to mean that the spin state $e_{\theta+\phi}$ can be identified with the spin state e_θ with probability $\cos^2(\phi/2)$ and with the spin state $e_{r(\theta)}$ with probability $\sin^2(\phi/2)$ with respect to the frame $(e_i, e_{r(\theta)}, e_\theta)$. However, it should be noted that $e_\theta = k\cos(\theta/2) - j\sin(\theta/2)$ and therefore cannot be interpreted as being in the spin up state k nor the spin down state $-j$. Instead, e_θ can be interpreted as been in the spin-up state with respect to the $(e_i, e_{r(\theta)}, e_\theta)$ but not with respect to the basis (i, j, k). To ignore this leads to the EPR paradox. A comparable situation occurs in differential geometry if one attempts to crect an orthogonal frame which uniquely spans the entire manifold. It cannot be done in general.

3.3. Different Representations

The first thing to grasp is that there are two equal but different representations of spin. Specifically, consider the following:

$$\begin{pmatrix} 1 & 0 \\ 0 & -1 \end{pmatrix}\begin{pmatrix} 1 \\ 0 \end{pmatrix} = \begin{pmatrix} -1 & 0 \\ 0 & 1 \end{pmatrix}\begin{pmatrix} -1 \\ 0 \end{pmatrix} \tag{19}$$

which means that

$$\begin{pmatrix} 1 \\ 0 \end{pmatrix} = -\begin{pmatrix} -1 \\ 0 \end{pmatrix}. \tag{20}$$

It is clear that if the ket $|s\rangle$ represents the spin state and $U \in SU(2)$ then $\langle s|U^*U|s\rangle$ is invariant regardless of the representation. This is what we mean by $SU(2)$ spin invariance. We now separately analyze each representation. Returning to the notation of Greenberger et al [5], we can write $|\mathbf{n}, +\rangle$ and $|\mathbf{n}, -\rangle$ to represent spin-up and spin-down respectively along the $\mathbf{n}$ direction. Therefore,

$$|\mathbf{n}_1, +\rangle = (\cos\theta/2)\,|\mathbf{n}_2, +\rangle + (\sin\theta/2)\,|\mathbf{n}_2, -\rangle\,. \tag{21}$$

Now choose $\mathbf{n}_2$ by rotating through $\theta = \pi/2$ (clockwise) with respect to $\mathbf{n}_1$ and choose $\mathbf{n}_3$ by rotating through $\theta = -\pi/2$ (anti-clockwise) with respect to $\mathbf{n}_1$ (Fig. 1). Then equation (21) gives

$$\begin{aligned} |\mathbf{n}_1, +\rangle &= 1/\sqrt{2}\,|\mathbf{n}_2, +\rangle + 1/\sqrt{2}\,|\mathbf{n}_2, -\rangle && (22) \\ &= 1/\sqrt{2}\,|\mathbf{n}_3, +\rangle - 1/\sqrt{2}\,|\mathbf{n}_3, -\rangle\,. && (23) \end{aligned}$$

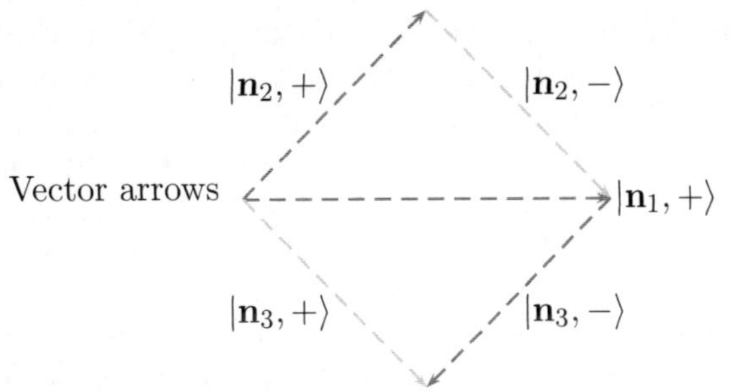

Fig. 1

In particular if $|\mathbf{n}_1, +\rangle = \begin{pmatrix} 1 \\ 0 \end{pmatrix}$, then $|\mathbf{n}_2, +\rangle = \begin{pmatrix} 1/\sqrt{2} \\ 1/\sqrt{2} \end{pmatrix}$ and $|\mathbf{n}_2, -\rangle = \begin{pmatrix} 1/\sqrt{2} \\ -1/\sqrt{2} \end{pmatrix}$, and $|\mathbf{n}_3, +\rangle = \begin{pmatrix} 1/\sqrt{2} \\ -1/\sqrt{2} \end{pmatrix}$ and $|\mathbf{n}_3, -\rangle = \begin{pmatrix} -1/\sqrt{2} \\ -1/\sqrt{2} \end{pmatrix}$. Note, immediately that $\langle \mathbf{n}_2, \pm|\mathbf{n}_3, \pm\rangle = 0$

and $|\mathbf{n}_2, -\rangle = |\mathbf{n}_3, +\rangle$. This creates the ambivalent situation of identifying, from the perspective of $\mathbf{n}_1$, a spin-down state ($|\mathbf{n}_2, -\rangle$) with a spin-up state ($|\mathbf{n}_3, +\rangle$), which forces the obvious question, as to what is the meaning of up or down in this case. In fact from the perspective of $\mathbf{n}_1$ alone there is no consistent meaning; for the value depends not only on the direction $\mathbf{n}_1$ but on the choice of angle θ defined relative to $\mathbf{n}_1$. It is precisely this ambivalence that is at the heart of both the Greenberger, Horne, and Zeilinger (GHZ) inconsistency discussed in section 1.1 and Bell's inequality; inconsistancies that only occur if we assume the existence of hidden variables as a parameter that determines spin values independently of a reference frame.

Finally, one might still be wondering why we cannot deduce spin values in the different reference frames with equations analogous to the Lorentz transformations of special relativity. In fact they are very similar. Once initial values of parameters and variables are specified in a given frame (usually the rest frame), Lorentz transformations allow us to calculate the values of parameters and variables such as mass, length and velocity in all inertial frames. Similarly, in the case of EPR, the $SU(2)$ transformations apply to probability states and once the initial probability state is specified as $(+1, 0)$ or $(0, -1)$ the remaining probability states can be predicted by way of the $SU(2)$ transformations.

4. Conclusion

In this paper we have traced the development of the EPR paradox and have offered a new approach to understanding it based on the fact that $\mathbf{a} \wedge \mathbf{b} = \det[\mathbf{a}, \mathbf{b}]\sqrt{2}\Psi$ is $SL(2, C)$ invariant[10] and that assigned spin values depend on the $SU(2)$ frame of reference.

At the core of the EPR paradox is a new reality that comes from the fact the the whole is more than the sum of its parts, with the *more*, in the case of the singlet state, being encapusulated by the semi group $SM(4, C)$. However, it should be possible to extend this approach, in general, to GHZ states. This becomes even more important given the recent work on entangled qudits [14]. The GHZ states must be invariant under some yet to be determined algebraic structure that cannot be reduced to the sum of its parts. For example, the GHZ state given by

$$\Psi = \frac{1}{\sqrt{2}}\left[\begin{pmatrix} 1 \\ 0 \end{pmatrix} \otimes \begin{pmatrix} 1 \\ 0 \end{pmatrix} \otimes \begin{pmatrix} 0 \\ 1 \end{pmatrix} \otimes \begin{pmatrix} 0 \\ 1 \end{pmatrix} - \begin{pmatrix} 0 \\ 1 \end{pmatrix} \otimes \begin{pmatrix} 0 \\ 1 \end{pmatrix} \otimes \begin{pmatrix} 1 \\ 0 \end{pmatrix} \otimes \begin{pmatrix} 1 \\ 0 \end{pmatrix}\right] \tag{24}$$

can in turn be written as $(0, 0, 0, 1, 0, 0, 0, 0, 0, 0, 0, 0, -1, 0, 0, 0)^T$ and is invariant under the $SM(16, C)$, where $a_{i4} = a_{i(13)}$, $i \neq 4, i \neq 13$ and $a_{44} + a_{4(13)} = 1$ and $a_{(13)4} + a_{(13)(13)} = 1$.

This article serves as a first step in classifying these inherent symmetries associated with perfect correlations and has suggested a general approach to handling the statistics of multparticle systems. Our interpretation is based on a critical realist approach which takes the given data to be objective, without falling into some Newtonian dualistic interpretation of reality. We do not put the observer outside of the world he or she is observing, as is the approach in classical mechanics, nor do we go to the other extreme of giving a singular importance to the observer. We interact with the physical world because we are part of it and indeed it is precisely this interaction that permits us to do physics. The laws of interaction are part of the objective laws of the universe and are there to be discovered. Quantum physics has tried to incorporate those interactions pertaining to physics into its axioms. To the extent that it has succeeded, quantum mechanics gives a more real and objective picture of physical reality, the uncertainty relations being a case in point. However, its success does not preclude that a more intuitive explanation of quantum theory might enrich our understanding of non-locality but in a way that preserves causality and thus avoiding interpretations based on a "spooky action at a distance." The combination of the principle that the whole is more than the sum of the parts, taken together with $SU(2)$ relativity is meant to be a first step in that direction.

IOP Conf. Series: Journal of Physics: Conf. Series **1239** (2019) 012021 doi:10.1088/1742-6596/1239/1/012021

References

[1] Einstein A, Podolsky B, Rosen N 1935 *Phys. Rev.* [**47**] 777
[2] Bell J S 1964 *Physics* [**1**] 195
[3] Bohr N www.marxists.org/reference/subject/philosophy/works/dk/bohr.htm
[4] Aspect A, Dalibard J and Roger G 1982 *Phy. Rev. Lett.* [**49**](25) 1804
[5] Greenberger D, Horne M, Shimony A and Zeilinger A 1990 *Am. J. Physics* [**58**](12) 1139
[6] Wigner E 1970 *Amer. J. Phys.* [**38**](8) 1005
[7] Christian J 2010 *Preprint* arXiv:1005.4932v2[quant-ph]
[8] M. Holman 2009 *Preprint* arXiv:0704.2038v2[quant-ph]
[9] Aristotle 1970 *Metaphysics* (Everyman: London) Book Z XI p.195 Chap VI p 218
[10] O'Hara P 2017 *J. Phys: Conf. Series*[**845**] 012030.
[11] O'Hara P 2003 *Foun.Phy.*, **33**(9) 1349
[12] O'Hara P 2000 *SEFIR*[**1**] 133
[13] O'Hara P 1995 *Ann. NYAS*[**755**] 880
[14] Baumgartner B, Hiesmayr B and Narnhofer H 2007 *J. Phys. A: Math Theor.* **40** 7919

IOP Conf. Series: Journal of Physics: Conf. Series **1239** (2019) 012022 doi:10.1088/1742-6596/1239/1/012022

On the motion of microscopic bodies and observation of position in quantum theory

Alexey A. Kryukov
Department of Mathematics & Natural Sciences, University of Wisconsin-Milwaukee, USA
E-mail: kryukov@uwm.edu

Abstract. Newtonian and Scrödinger dynamics can be formulated in a physically meaningful way within the same Hilbert space framework. The resulting unexpected relation between classical and quantum motions goes beyond the results provided by the Ehrenfest theorem. The normal probability distribution and the Born rule turn out to be related. Here a dynamical mechanism responsible for the latter formula is proposed and applied to measurements of macroscopic and microscopic systems. A new meaning of the wave function collapse is proposed. The results are used to explore the classical behavior of macroscopic bodies in quantum theory.

1. Introduction

In the papers [1] and [2] a new connection between classical and quantum dynamics was derived. The starting point was a realization of classical and quantum mechanics on an equal footing within the same Hilbert space framework, and identification of observables with vector fields on the sphere of normalized states. A physically meaningful interpretation of the components of the velocity of a state was then obtained. Newtonian dynamics was shown to be the Schrödinger dynamics of a system whose state is constrained to the classical phase space submanifold in the Hilbert space of states, made of coherent states. This resulted in a formula relating the normal probability distribution and the Born rule, and interpretation of quantum collapse in terms of diffusion of the state on the projective space of states. Simply put, the classical space and classical phase space of a system of particles are identified with a submanifold of the space of states of the corresponding quantum system. When the system is constrained to the submanifold, it behaves classically. Otherwise, it behaves quantum-mechanically. The velocity of the state at any point of the classical space submanifold can be decomposed into the classical (velocity, acceleration) and non-classical (phase velocity, spreading) components.

In this paper, we continue to explore the implications of the proposed geometric framework. It is shown that the curvature of the sphere of states is determined from the canonical commutation relations. It is also shown that the state of a microscopic particle exposed to the random potential, typically experienced by the pollen grain in a liquid, is equally likely to get displaced in any direction tangent to the projective space of states. The relationship between the normal probability distribution and the Born rule that was established earlier signifies, then, that the probability density for the state to reach a particular point in the space of states is given by the Born rule. A diffusion equation for the motion of the state in these conditions is then obtained and used in explaining the classical behavior of macroscopic bodies in quantum theory.

IOP Conf. Series: Journal of Physics: Conf. Series **1239** (2019) 012022 doi:10.1088/1742-6596/1239/1/012022

Because the results reported in [1] and [2] are essential to this paper, the main derivations of these papers are reproduced here in detail.

2. The classical mechanics in the language of quantum states

Experience shows that macroscopic bodies have a well-defined position in space at any time. In quantum mechanics the state of a spinless particle with a known position $\mathbf{a} \in \mathbb{R}^3$ is described by the Dirac delta function $\delta^3_{\mathbf{a}}(\mathbf{x}) = \delta^3(\mathbf{x} - \mathbf{a})$. The map $\omega : \mathbf{a} \longrightarrow \delta^3_{\mathbf{a}}$ provides a one-to-one correspondence between points $\mathbf{a} \in \mathbb{R}^3$ and state "functions" $\delta^3_{\mathbf{a}}$. This allows us to describe points in $\mathbb{R}^3$ in functional terms and identify the set $\mathbb{R}^3$ with the set M_3 of all delta functions in the space of state functions of the particle.

The common Hilbert space $L_2(\mathbb{R}^3)$ of state functions of a particle does not contain delta functions. By writing the inner product of functions $\varphi, \psi \in L_2(\mathbb{R}^3)$ as

$$(\varphi, \psi)_{L_2} = \int \delta^3(\mathbf{x} - \mathbf{y})\varphi(\mathbf{x})\overline{\psi}(\mathbf{y})d^3\mathbf{x}d^3\mathbf{y} \tag{1}$$

and approximating the kernel $\delta^3(\mathbf{x} - \mathbf{y})$ with a Gaussian function, one obtains a new inner product in $L_2(\mathbb{R}^3)$

$$(\varphi, \psi)_{\mathbf{H}} = \int e^{-\frac{(\mathbf{x}-\mathbf{y})^2}{8\sigma^2}}\varphi(\mathbf{x})\overline{\psi}(\mathbf{y})d^3\mathbf{x}d^3\mathbf{y}. \tag{2}$$

The Hilbert space $\mathbf{H}$ obtained by completing $L_2(\mathbb{R}^3)$ with respect to this inner product contains delta functions and their derivatives. In particular,

$$\int e^{-\frac{(\mathbf{x}-\mathbf{y})^2}{8\sigma^2}}\delta^3(\mathbf{x} - \mathbf{a})\delta^3(\mathbf{y} - \mathbf{a})d^3\mathbf{x}d^3\mathbf{y} = 1. \tag{3}$$

It follows that the set M_3 of all delta functions $\delta^3_{\mathbf{a}}(\mathbf{x})$ with $\mathbf{a} \in \mathbb{R}^3$ form a submanifold of the unit sphere in the Hilbert space $\mathbf{H}$, diffeomorphic to $\mathbb{R}^3$.

The kernel $\delta^3(\mathbf{x} - \mathbf{y})$ of the metric on $L_2(\mathbb{R}^3)$ is analogous to the Kronecker delta δ_{ik}, representing Euclidean metric in orthogonal coordinates. The "skewed" kernel $e^{-\frac{(\mathbf{x}-\mathbf{y})^2}{8\sigma^2}}$ of the metric on $\mathbf{H}$ is then analogous to the Euclidean metric represented in linear coordinates with skewed axes by a constant non-diagonal matrix g_{ik}.

The map $\rho_\sigma : \mathbf{H} \longrightarrow L_2(\mathbb{R}^3)$ that relates L_2 and $\mathbf{H}$-representations is given by the Gaussian kernel

$$\rho_\sigma(\mathbf{x}, \mathbf{y}) = \left(\frac{1}{2\pi\sigma^2}\right)^{3/4} e^{-\frac{(\mathbf{x}-\mathbf{y})^2}{4\sigma^2}}. \tag{4}$$

In fact, multiplying the corresponding operators (integrating the product of kernels) one can see that

$$k(\mathbf{x}, \mathbf{y}) = (\rho^*_\sigma\rho_\sigma)(\mathbf{x}, \mathbf{y}) = e^{-\frac{(\mathbf{x}-\mathbf{y})^2}{8\sigma^2}}, \tag{5}$$

which is consistent with (2). The map ρ_σ transforms delta functions $\delta^3_{\mathbf{a}}$ to Gaussian functions $\widetilde{\delta}^3_{\mathbf{a}} = \rho_\sigma(\delta^3_{\mathbf{a}})$, centered at $\mathbf{a}$. The image M^σ_3 of M_3 under ρ_σ is an embedded submanifold of the unit sphere in $L_2(\mathbb{R}^3)$ made of the functions $\widetilde{\delta}^3_{\mathbf{a}}$. The map $\omega_\sigma = \rho_\sigma \circ \omega : \mathbb{R}^3 \longrightarrow M^\sigma_3$ is a diffeomorphism. Here ω is the same as before. In what follows, the obtained realizations will be used interchangeably.

Let $\mathbf{r} = \mathbf{a}(t)$ be a path with values in $\mathbb{R}^3$ and let $\varphi = \delta^3_{\mathbf{a}(t)}$ be the corresponding path in M_3. It is easy to see that the norm $\left\|\frac{d\varphi}{dt}\right\|^2_H$ of the velocity in the space $\mathbf{H}$ is given by

$$\left\|\frac{d\varphi}{dt}\right\|^2_H = \left.\frac{\partial^2 k(\mathbf{x}, \mathbf{y})}{\partial x^i \partial y^k}\right|_{\mathbf{x}=\mathbf{y}=\mathbf{a}} \frac{d\mathbf{a}^i}{dt}\frac{d\mathbf{a}^k}{dt}. \tag{6}$$

IOP Conf. Series: Journal of Physics: Conf. Series **1239** (2019) 012022 doi:10.1088/1742-6596/1239/1/012022

Here $k(\mathbf{x},\mathbf{y}) = e^{-\frac{(\mathbf{x}-\mathbf{y})^2}{8\sigma^2}}$ as in (5), so that

$$\left.\frac{\partial^2 k(\mathbf{x},\mathbf{y})}{\partial x^i \partial y^k}\right|_{\mathbf{x}=\mathbf{y}=\mathbf{a}} = \frac{1}{4\sigma^2}\delta_{ik}, \tag{7}$$

where δ_{ik} is the Kronecker delta symbol. Assuming now that the distance in $\mathbb{R}^3$ is measured in the units of 2σ, we obtain

$$\left\|\frac{d\varphi}{dt}\right\|_H = \left\|\frac{d\mathbf{a}}{dt}\right\|_{\mathbb{R}^3}. \tag{8}$$

It follows that the map $\omega : \mathbb{R}^3 \longrightarrow \mathbf{H}$ is an isometric embedding. Furthermore, the set M_3 is complete in $\mathbf{H}$ so that there is no vector in $\mathbf{H}$ orthogonal to all of M_3. By defining the operations of addition $\oplus$ and multiplication by a scalar $\lambda\odot$ via $\omega(\mathbf{a})\oplus\omega(\mathbf{b}) = \omega(\mathbf{a}+\mathbf{b})$ and $\lambda\odot\omega(\mathbf{a}) = \omega(\lambda\mathbf{a})$ with ω as before, we obtain M_3 as a vector space isomorphic to the Euclidean space $\mathbb{R}^3$.

The projection of velocity and acceleration of the state $\delta^3_{\mathbf{a}(t)}$ onto the Euclidean space M_3 yields correct Newtonian velocity and acceleration of the classical particle:

$$\left(\frac{d}{dt}\delta^3_{\mathbf{a}}(\mathbf{x}), -\frac{\partial}{\partial x^i}\delta^3_{\mathbf{a}}(\mathbf{x})\right)_{\mathbf{H}} = \frac{da^i}{dt} \tag{9}$$

and

$$\left(\frac{d^2}{dt^2}\delta^3_{\mathbf{a}}(\mathbf{x}), -\frac{\partial}{\partial x^i}\delta^3_{\mathbf{a}}(\mathbf{x})\right)_{\mathbf{H}} = \frac{d^2a^i}{dt^2}. \tag{10}$$

The Newtonian dynamics of the classical particle can be derived from the principle of least action for the action functional S on paths in $\mathbf{H}$, defined by

$$\int k(\mathbf{x},\mathbf{y})\left[\frac{m}{2}\frac{d\varphi_t(\mathbf{x})}{dt}\frac{d\overline{\varphi}_t(\mathbf{y})}{dt} - V(\mathbf{x})\varphi_t(\mathbf{x})\overline{\varphi}_t(\mathbf{y})\right] d^3\mathbf{x}d^3\mathbf{y}dt, \tag{11}$$

where m is the mass of the particle, V is the potential and $k(\mathbf{x},\mathbf{y}) = e^{-\frac{1}{2}(\mathbf{x}-\mathbf{y})^2}$ (same as in (5) with $2\sigma = 1$; see (8)). Namely, under the constraint $\varphi_t(\mathbf{x}) = \delta^3(\mathbf{x}-\mathbf{a}(t))$ the action (11) becomes

$$S = \int \left[\frac{m}{2}\left(\frac{d\mathbf{a}}{dt}\right)^2 - V(\mathbf{a})\right] dt, \tag{12}$$

which is the classical action functional for the particle.

This shows that a classical particle can be considered a constrained dynamical system with the state φ of the particle and the velocity $\frac{d\varphi}{dt}$ of the state as dynamical variables. A similar realization exists for mechanical systems consisting of any number of classical particles. For example, the map $\omega\otimes\omega : \mathbb{R}^3 \times \mathbb{R}^3 \longrightarrow \mathbf{H}\otimes\mathbf{H}$, $\omega\otimes\omega(\mathbf{a},\mathbf{b}) = \delta^3_{\mathbf{a}}\otimes\delta^3_{\mathbf{b}}$ identifies the configuration space $\mathbb{R}^3\times\mathbb{R}^3$ of a two particle system with an embedded submanifold $M_6 = \omega\otimes\omega(\mathbb{R}^3\times\mathbb{R}^3)$ of the Hilbert space $\mathbf{H}\otimes\mathbf{H}$. Consider a path $(\mathbf{a}(t),\mathbf{b}(t))$ in $\mathbb{R}^3\times\mathbb{R}^3$ and the corresponding path $\delta^3_{\mathbf{a}(t)}\otimes\delta^3_{\mathbf{b}(t)}$ with values in M_6. For any t, the vectors $\frac{d}{dt}\delta^3_{\mathbf{a}(t)}\otimes\delta^3_{\mathbf{b}(t)}$ and $\delta^3_{\mathbf{a}(t)}\otimes\frac{d}{dt}\delta^3_{\mathbf{b}(t)}$ are tangent to M_6 at the point $\delta^3_{\mathbf{a}(t)}\otimes\delta^3_{\mathbf{b}(t)}$ and orthogonal in $\mathbf{H}\otimes\mathbf{H}$. The space M_6 with the induced metric is isometric to the direct product $\mathbb{R}^3\times\mathbb{R}^3$ with the natural Euclidean metric. Projection of velocity and acceleration of the state $\varphi(t) = \delta^3_{\mathbf{a}(t)}\otimes\delta^3_{\mathbf{b}(t)}$ onto the basis vectors $\left(-\frac{\partial}{\partial x^i}\delta^3_{\mathbf{a}(t)}\right)\otimes\delta^3_{\mathbf{b}(t)}$ and $\delta^3_{\mathbf{a}(t)}\otimes\left(-\frac{\partial}{\partial x^k}\delta^3_{\mathbf{b}(t)}\right)$ yields the velocity and acceleration of the particles by means of the formulas similar to (9) and (10).

3. Observables as vector fields

Quantum observables can be identified with vector fields on the space of states. Given a self-adjoint operator $\widehat{A}$ on a Hilbert space L_2 of square-integrable functions (it could in particular be the tensor product space of a many body problem) one can introduce the associated linear vector field A_φ on L_2 by

$$A_\varphi = -i\widehat{A}\varphi. \tag{13}$$

If D is the domain of the operator $\widehat{A}$, then A_φ maps D into the vector space L_2. Because $\widehat{A}$ is self-adjoint, the field A_φ, being restricted to the sphere S^{L_2} of unit normalized states, is tangent to the sphere. The commutator of observables and the commutator (Lie bracket) of the corresponding vector fields are related in a simple way:

$$[A_\varphi, B_\varphi] = [\widehat{A}, \widehat{B}]\varphi. \tag{14}$$

Furthermore, a Hilbert metric on the space of states yields a Riemannian metric on the sphere. For this consider the realization $L_{2\mathbb{R}}$ of the Hilbert space L_2, i.e., the real vector space of pairs $X = (Re\psi, Im\psi)$ with ψ in L_2. If ξ, η are vector fields on S^{L_2}, define a Riemannian metric $G_\varphi : T_{\mathbb{R}\varphi}S^{L_2} \times T_{\mathbb{R}\varphi}S^{L_2} \longrightarrow \mathbb{R}$ on the sphere by

$$G_\varphi(X, Y) = \mathrm{Re}(\xi, \eta). \tag{15}$$

Here $X = (\mathrm{Re}\xi, \mathrm{Im}\xi)$, $Y = (\mathrm{Re}\eta, \mathrm{Im}\eta)$ and (ξ, η) denotes the L_2-inner product of ξ, η.

The Riemannian metric on S^{L_2} yields a Riemannian (Fubini-Study) metric on the projective space CP^{L_2}, which is the base of the fibration $\pi : S^{L_2} \longrightarrow CP^{L_2}$. For this an arbitrary tangent vector $X \in T_{R\varphi}S^{L_2}$ is decomposed into two components: tangent and orthogonal to the fibre $\{\varphi\}$ through φ (i.e., to the plane C^1 containing the circle $S^1 = \{\varphi\}$). The differential $d\pi$ of the projection π maps the tangential component to zero-vector. The orthogonal component of X can be then identified with $d\pi(X)$. If two vectors X, Y are orthogonal to the fibre $\{\varphi\}$, the inner product of $d\pi(X)$ and $d\pi(Y)$ in the Fubini-Study metric is equal to the inner product of X and Y in the metric G_φ:

$$(d\pi(X), d\pi(Y))_{FS} = G_\varphi(X, Y). \tag{16}$$

The resulting metrics can be used to find physically meaningful components of vector fields A_φ associated with observables. Since A_φ is tangent to S^{L_2}, it can be decomposed into components tangent and orthogonal to the fibre $\{\varphi\}$ of the fibre bundle $\pi : S^{L_2} \longrightarrow CP^{L_2}$. These components have a simple physical meaning, justifying the use of the projective space CP^{L_2}. From

$$\overline{A} \equiv (\varphi, \widehat{A}\varphi) = (-i\varphi, -i\widehat{A}\varphi), \tag{17}$$

one can see that the expected value of an observable $\widehat{A}$ in a state φ is the projection of the vector $-i\widehat{A}\varphi \in T_\varphi S^{L_2}$ onto the fibre $\{\varphi\}$. Because

$$(\varphi, \widehat{A}^2\varphi) = (\widehat{A}\varphi, \widehat{A}\varphi) = (-i\widehat{A}\varphi, -i\widehat{A}\varphi), \tag{18}$$

the term $(\varphi, \widehat{A}^2\varphi)$ is the norm of the vector $-i\widehat{A}\varphi$ squared. The vector $-i\widehat{A}_\perp\varphi = -i\widehat{A}\varphi - (-i\overline{A}\varphi)$ associated with the operator $\widehat{A} - \overline{A}I$ is orthogonal to the fibre $\{\varphi\}$. Accordingly, the variance

$$\Delta A^2 = (\varphi, (\widehat{A} - \overline{A}I)^2\varphi) = (\varphi, \widehat{A}_\perp^2\varphi) = (-i\widehat{A}_\perp\varphi, -i\widehat{A}_\perp\varphi) \tag{19}$$

is the norm squared of the component $-i\widehat{A}_\perp\varphi$. Recall that the image of this vector under $d\pi$ can be identified with the vector itself. It follows that the norm of $-i\widehat{A}_\perp\varphi$ in the Fubini-Study metric coincides with its norm in the Riemannian metric on S^{L_2} and in the original L_2-metric.

The Schrödinger equation

$$\frac{d\varphi}{dt} = -i\widehat{h}\varphi \tag{20}$$

is an equation for the integral curves of the vector field $-i\widehat{h}\varphi$ on the sphere S^{L_2}. Let's decompose $-i\widehat{h}\varphi$ onto the components parallel and orthogonal to the fibre. The parallel component of $\frac{d\varphi}{dt}$ is numerically

$$\mathrm{Re}(-i\varphi, -i\widehat{h}\varphi) = \overline{E}, \tag{21}$$

i.e., the expected value of the energy. The decomposition of the velocity vector $\frac{d\varphi}{dt}$ into the parallel and orthogonal components is then given by

$$\frac{d\varphi}{dt} = -i\overline{E}\varphi + -i(\widehat{h} - \overline{E})\varphi = -i\overline{E}\varphi - i\widehat{h}_{\perp}\varphi. \tag{22}$$

The orthogonal component of the velocity $\frac{d\varphi}{dt}$ is equal to $-i\widehat{h}_{\perp}\varphi$. From this and equation (19) we conclude that: *The velocity of evolution of the state in the projective space is equal to the uncertainty of energy.* Equation (22) also demonstrates that the physical state is driven by the operator $\widehat{h}_{\perp}$, associated with the uncertainty in energy rather than the energy itself.

The uncertainty relation

$$\Delta A \Delta B \geq \frac{1}{2}\left|\left(\varphi, [\widehat{A}, \widehat{B}]\varphi\right)\right| \tag{23}$$

follows geometrically from the comparison of areas of rectangle $A_{|XY|}$ and parallelogram A_{XY} formed by vectors $X = -i\widehat{A}_{\perp}\varphi$ and $Y = -i\widehat{B}_{\perp}\varphi$:

$$A_{|XY|} \geq A_{XY}. \tag{24}$$

There is also an uncertainty identity, [3]:

$$\Delta A^2 \Delta B^2 = A^2_{XY} + G^2_{\varphi}(X, Y). \tag{25}$$

4. Commutator of observables and curvature of the sphere of states

Commutators of observables are related to the curvature of the sphere of states. To see this, consider first the space C^2 of electron's spin states. The sphere S^3 of unit-normalized states in C^2 can be identified with the group manifold $SU(2)$. For this one identifies the space C^2 of complex vectors $\varphi = \begin{bmatrix} z_1 \\ z_2 \end{bmatrix}$ with the real vector space Mat of 2×2 matrices

$$\widehat{\varphi} = \begin{bmatrix} z_1 & z_2 \\ -\overline{z}_2 & \overline{z}_1 \end{bmatrix} \tag{26}$$

with the usual operations of addition and multiplication by numbers. The map $\omega : \varphi \longrightarrow \widehat{\varphi}$ is an isomorphism of (real) vector spaces C^2 and Mat. The sphere S^3 of unit states in C^2 is identified via ω with the subset of matrices with unit determinant. The latter subset is the group $SU(2)$ under matrix multiplication.

The differential $d\omega$ of the map ω identifies the tangent space $T_{e_1}S^3$ to the sphere S^3 at the point $e_1 = \begin{bmatrix} 1 \\ 0 \end{bmatrix}$ (that is, the hyperplane $\mathrm{Re} z_1 = 1$) with the Lie algebra $su(2)$ of traceless anti-Hermitian matrices

$$\widehat{A} = \begin{bmatrix} ia_2 & a_3 + ia_4 \\ -a_3 + ia_4 & -ia_2 \end{bmatrix}, \tag{27}$$

$a_2, a_3, a_4 \in R$. Under $d\omega$ the basis $e_2 = \begin{bmatrix} i \\ 0 \end{bmatrix}$, $e_3 = \begin{bmatrix} 0 \\ 1 \end{bmatrix}$, $e_4 = \begin{bmatrix} 0 \\ i \end{bmatrix}$ in the tangent space $T_{e_1}S^3 = R^3$ becomes the basis $\{i\widehat{\sigma}_3, i\widehat{\sigma}_2, i\widehat{\sigma}_1\}$ in the Lie algebra $su(2)$. In particular, the real numbers a_2, a_3, a_4 acquire the meaning of coordinates of points on the tangent space $\text{Re}z_1 = 1$ in the basis $\{e_2, e_3, e_4\}$.

The embedding of S^3 into C^2 induces the usual Riemannian metric on the sphere. A direct verification demonstrates that this metric coincides with the Killing metric on $SU(2)$. The latter metric can be defined on the tangent space $T_eSU(2)$ at the identity e (i.e., on the Lie algebra $su(2)$) by $(\widehat{X}, \widehat{Y})_K = \frac{1}{2}Tr\widehat{X}\widehat{Y}^+$ and then extended to the entire $SU(2)$ by the group action. Here $(\widehat{X}, \widehat{Y})_K$ denotes the Killing inner product of tangent vectors and $\widehat{Y}^+$ on the right is the Hermitian conjugate of $\widehat{Y}$. The constant $1/2$ in the Killing metric ensures the equality of the metrics. However, the unit of measurement must be specified. The tangent space $su(2)$ is spanned by the spin operators having the dimension of angular momentum and measured in the units of $\hbar$. Therefore, the Planck system of units will be used. The spin generators $\widehat{s}_1 = \frac{i}{2}\widehat{\sigma}_1, \widehat{s}_2 = \frac{i}{2}\widehat{\sigma}_2, \widehat{s}_3 = \frac{i}{2}\widehat{\sigma}_3$ are orthogonal in the defined metric and have a norm equal to $1/2$ in Planck units.

The integral curves of the left-invariant vector fields $L_{\widehat{X}}(\widehat{\varphi}) = \widehat{\varphi}\widehat{X}$ are geodesics on $SU(2)$. They are given by $\widehat{\varphi}_t = \widehat{\varphi}_0 e^{-i\widehat{X}t}$. In the usual coordinates on C^2 the equation of these geodesics takes the form $\varphi_t = e^{-i\widehat{X}t}\varphi_0$, where $\omega(\varphi_0) = \widehat{\varphi}_0$. The carriers of geodesics are the great circles on the sphere S^3. The commutators of the spin observables are directly related to the sectional curvature of the sphere S^3. This is not surprising as the non-trivial Lie bracket of vector fields whose integral curves are geodesics can only be due to curvature of the underlying space. If $\widehat{X}, \widehat{Y} \in su(2)$ are linearly independent generators and $L_{\widehat{X}}(\widehat{\varphi}), L_{\widehat{Y}}(\widehat{\varphi})$ are the associated left-invariant vector fields, then the sectional curvature $R_\varphi(p)$ of S^3 in the plane p through $L_{\widehat{X}}(\widehat{\varphi}), L_{\widehat{Y}}(\widehat{\varphi})$ is given at any point $\widehat{\varphi}$ by

$$R_\varphi(p) = \frac{1}{4} \frac{\left\| [\widehat{X}, \widehat{Y}] \right\|_K^2}{\left\| \widehat{X} \right\|_K^2 \left\| \widehat{Y} \right\|_K^2 - \left(\widehat{X}, \widehat{Y} \right)_K^2}. \tag{28}$$

In particular, if the generators $\widehat{X}, \widehat{Y}$ are orthonormal in the Killing metric, (28) simplifies to

$$R_\varphi(p) = \frac{1}{4} \left\| [\widehat{X}, \widehat{Y}] \right\|_K^2. \tag{29}$$

Using the formula (28), we obtain the following expression for the sectional curvature $R_\varphi(p)$ in the plane p through orthogonal vectors $L_{\widehat{s}_1}(\widehat{\varphi}), L_{\widehat{s}_2}(\widehat{\varphi})$:

$$R_\varphi(p) = \frac{1}{4} \frac{([\widehat{s}_1, \widehat{s}_2], [\widehat{s}_1, \widehat{s}_2])_K}{(\widehat{s}_1, \widehat{s}_1)_K (\widehat{s}_2, \widehat{s}_2)_K} = 4 (\widehat{s}_3, \widehat{s}_3)_K = 1. \tag{30}$$

This means that the radius of S^3 in Planck units is equal to 1, confirming the isometric nature of the isomorphism ω considered as a map from the unit sphere S^3 in C^2 onto $SU(2)$ with the Killing metric. Note that in an arbitrary system of units the sectional curvature would be equal to $1/\hbar^2$ (i.e., radius=$\hbar$). The dimension of sectional curvature is consistent with the fact that the tangent space $su(2)$ is spanned by the spin operators.

The obtained relationship between commutators of spin observables and radius of the sphere of states can be extended to other observables. In particular, the commutator $[\widehat{p}, \widehat{x}]$ of position and momentum observables of an arbitrary non-relativistic particle whose state is in the space $L_2(\mathbb{R})$ yields similarly the sectional curvature of the unit sphere S^{L_2} in $L_2(\mathbb{R})$. In fact, let's compute the sectional curvature of the sphere S^{L_2} in the plane through the tangent vectors $-i\widehat{p}\varphi$, $-i\widehat{x}\varphi$ at a point $\varphi \in S^{L_2}$. It is convenient to represent the action of operators $\widehat{p}$, $\widehat{x}$ in the basis $\varphi_n(x) = \frac{1}{\sqrt[4]{\pi}2^n n!}H_n(x)e^{-\frac{x^2}{2}}$, $n = 0, 1, 2, ...$ of the quantum harmonic oscillator. Here $H_n(x)$ are the Hermite polynomials. Note that the vectors φ_n are in the domain of the operators $\widehat{p}$, $\widehat{x}$, $\widehat{px}$ and $\widehat{xp}$. The matrices of the operators $\widehat{p}$, $\widehat{x}$ in the basis are given by

$$\widehat{x} = \frac{1}{\sqrt{2}}\begin{bmatrix} 0 & 1 & 0 & 0 & \cdots \\ 1 & 0 & \sqrt{2} & 0 & \cdots \\ 0 & \sqrt{2} & 0 & \sqrt{3} & \cdots \\ 0 & 0 & \sqrt{3} & 0 & \cdots \\ . & . & . & . & \cdots \end{bmatrix} \tag{31}$$

and

$$\widehat{p} = \frac{1}{\sqrt{2}}\begin{bmatrix} 0 & -i & 0 & 0 & \cdots \\ i & 0 & -i\sqrt{2} & 0 & \cdots \\ 0 & i\sqrt{2} & 0 & -i\sqrt{3} & \cdots \\ 0 & 0 & i\sqrt{3} & 0 & \cdots \\ . & . & . & . & \cdots \end{bmatrix}. \tag{32}$$

In light of the unboundedness of the operators, the validity of such a matrix representation requires a discussion. However, for the purpose of computing the sectional curvature it will be sufficient to point out that the matrices (31), (32) correctly reproduce the action of operators on all vectors with finitely many non-vanishing components in the basis $\{\varphi_n\}$.

Let us find the sectional curvature of the sphere S^{L_2} at the "vacuum" point $\varphi_n|_{n=0} = \varphi_0$. For this, consider the subspace $C^2 \subset L_2(\mathbb{R})$ formed by the first two vectors of the basis. Note that up to the coefficient $\frac{1}{\sqrt{2}}$, the sub-matrices formed by the first two rows and columns of matrices (31), (32) coincide with the Pauli matrices $\widehat{\sigma}_x, \widehat{\sigma}_y$ respectively. Let us introduce the bounded operators $\widehat{s}_p, \widehat{s}_x$ on $L_2(\mathbb{R})$ defined by $\widehat{s}_x\varphi = \frac{1}{\sqrt{2}}\widehat{\sigma}_x\varphi$, $\widehat{s}_p\varphi = \frac{1}{\sqrt{2}}\widehat{\sigma}_y\varphi$ for φ in C^2, and by $\widehat{s}_x\varphi = 0$, $\widehat{s}_p\varphi = 0$ for φ in the orthogonal complement of C^2 in $L_2(\mathbb{R})$. Note that the action of operators $\widehat{p}$, $\widehat{x}$ and $[\widehat{p}, \widehat{x}]$ on the point φ_0 is correctly reproduced by the operators $\widehat{s}_p, \widehat{s}_x$:

$$\widehat{x}\varphi_0 = \widehat{s}_x\varphi_0 \tag{33}$$
$$\widehat{p}\varphi_0 = \widehat{s}_p\varphi_0 \tag{34}$$
$$[\widehat{p}, \widehat{x}]\,\varphi_0 = [\widehat{s}_p, \widehat{s}_x]\,\varphi_0. \tag{35}$$

Consider the sphere $S^3 = S^{L_2} \cap C^2$ with the metric induced by the inclusion. As discussed, this metric coincides with the Killing metric on the group $SU(2) = S^3$. The point φ_0 is given in the basis $\{\varphi_0, \varphi_1\}$ in C^2 by the column $\begin{bmatrix} 1 \\ 0 \end{bmatrix}$. The image $\widehat{\varphi}_0$ of the column φ_0 under the isomorphism (26) is the identity e in the group $SU(2)$. Accordingly, one can compute the norms of the right sides of (33), (34) and (35) in the Killing metric. Such a computation verifies that these norms are equal to the norms of the corresponding left sides in the L_2-metric. For example, the norm of the right side of (35) in the Killing metric is given by $\left\|\widehat{\varphi}_0\frac{1}{2}[\widehat{\sigma}_y, \widehat{\sigma}_x]\right\|_K = \|i\widehat{\sigma}_z\|_K = \sqrt{\frac{1}{2}Tr(\widehat{\sigma}_z)^2} = 1$. This coincides with the L_2-norm of the corresponding left side: $\|[\widehat{p}, \widehat{x}]\varphi_0\|_{L_2} = \|\varphi_0\|_{L_2} = 1$.

The sectional curvature of S^{L_2} in the plane through vector fields $-i\widehat{x}\varphi$, $-i\widehat{p}\varphi$ at $\varphi = \varphi_0$ is equal to the sectional curvature $R_{\varphi_0}(p)$ of S^3 in the plane p through the fields $-i\widehat{\sigma}_x\varphi$, $-i\widehat{\sigma}_y\varphi$ at this point. By (28), (33), (34) and (35) this sectional curvature is given in terms of the Lie brakets of these fields, i.e., in terms of the commutator $[\widehat{p}, \widehat{x}]$ evaluated at φ_0 and is equal to 1. Because sphere has a constant sectional curvature, the same result applies to any point.

It follows that the commutator of vector fields associated with the operators of position and momentum has the same geometric interpretation as the commutator of vector fields associated with the operators of spin.

5. Components of the velocity of a state under the Schrödinger evolution

From (22) we know that for any state $\varphi \in S^{L_2}$ the velocity $\frac{d\varphi}{dt}$ under the Schrödinger equation can be decomposed onto the components parallel and orthogonal to the fibre $\{\varphi\}$ of the bundle $\pi : S^{L_2} \longrightarrow CP^{L_2}$:

$$\frac{d\varphi}{dt} = -i\overline{E}\varphi - i\widehat{h}_{\perp}\varphi. \tag{36}$$

The norm of the parallel component $-i\overline{E}\varphi$ is the expected value of energy $\overline{E}$. It represents the phase velocity of the state. The norm of the orthogonal component $-i\widehat{h}_{\perp}\varphi$ is equal to the uncertainty of energy ΔE on the state φ. It represents the velocity of motion of the fibre $\{\varphi\}$. In particular, from (36) it follows that under the Schrödinger evolution the speed of evolution of a state in the projective space is equal to the uncertainty in energy.

The orthogonal component $-i\widehat{h}_{\perp}\varphi$ of the velocity can be further decomposed into physically meaningful components. Suppose that at $t = 0$, a microscopic particle is prepared in the state $\varphi_0 \equiv \varphi_{\mathbf{a},\mathbf{p}}$ given by

$$\varphi_{\mathbf{a},\mathbf{p}}(\mathbf{x}) = \left(\frac{1}{2\pi\sigma^2}\right)^{3/4} e^{-\frac{(\mathbf{x}-\mathbf{a})^2}{4\sigma^2}} e^{i\frac{\mathbf{p}(\mathbf{x}-\mathbf{a})}{\hbar}}, \tag{37}$$

where σ is the same as in (4) and $\mathbf{p} = m\mathbf{v}_0$ with $\mathbf{v}_0$ being the initial group-velocity of the packet. The set $M^{\sigma}_{3,3}$ of all initial states $\varphi_{\mathbf{a},\mathbf{p}}$ given by (37) form a 6-dimensional embedded submanifold in $L_2(\mathbb{R}^3)$. Consider the set of all fibres of the bundle $\pi : S^{L_2} \longrightarrow CP^{L_2}$ through the points of $M^{\sigma}_{3,3}$. The resulting bundle $\pi : M^{\sigma}_{3,3} \times S^1 \longrightarrow M^{\sigma}_{3,3}$ identifies $M^{\sigma}_{3,3}$ with a submanifold of CP^{L_2}, denoted by the same symbol. The map $\Omega : \mathbb{R}^3 \times \mathbb{R}^3 \longrightarrow M^{\sigma}_{3,3}$,

$$\Omega(\mathbf{a},\mathbf{p}) = \left(\frac{1}{2\pi\sigma^2}\right)^{3/4} e^{-\frac{(\mathbf{x}-\mathbf{a})^2}{4\sigma^2}} e^{i\frac{\mathbf{p}(\mathbf{x}-\mathbf{a})}{\hbar}} \tag{38}$$

is a diffeomorphism from the classical phase space of the particle onto $M^{\sigma}_{3,3}$. For $\Omega = re^{i\theta}$, where r is the modulus and θ is the argument of Ω, the vectors $\frac{\partial r}{\partial a^{\alpha}}e^{i\theta}$ and $i\frac{\partial \theta}{\partial p^{\beta}}re^{i\theta}$ are orthogonal to the fibre $\{\varphi_0\}$ at the point φ_0 in $L_2(\mathbb{R}^3)$. These vectors can be then identified with vectors tangent to $M^{\sigma}_{3,3}$, considered as a submanifold of CP^{L_2}. They form a basis in the tangent space $T_{\{\varphi_0\}}M^{\sigma}_{3,3}$. Furthermore, the induced Riemannian metric is the usual Fubini-Study metric on CP^{L_2}, restricted to $M^{\sigma}_{3,3}$.

For any path $\{\varphi\} = \{\varphi_{\tau}\}$ with values in $M^{\sigma}_{3,3} \subset CP^{L_2}$ the norm of velocity vector $\frac{d\{\varphi\}}{d\tau}$ in the Fubini-Study metric is given by

$$\left\|\frac{d\{\varphi\}}{d\tau}\right\|^2_{FS} = \frac{1}{4\sigma^2}\left\|\frac{d\mathbf{a}}{d\tau}\right\|^2_{\mathbb{R}^3} + \frac{\sigma^2}{\hbar^2}\left\|\frac{d\mathbf{p}}{d\tau}\right\|^2_{\mathbb{R}^3}. \tag{39}$$

It follows that under a proper choice of units, the map Ω is an isometry that identifies the Euclidean phase space $\mathbb{R}^3 \times \mathbb{R}^3$ of the particle with the submanifold $M^{\sigma}_{3,3} \subset CP^{L_2}$ furnished with the induced metric. The map Ω is an extension to the phase space of the isometric embedding $\omega_{\sigma} = \rho_{\sigma} \circ \omega$ of the space $\mathbb{R}^3$, introduced in the section 2, entitled "The classical mechanics in the language of quantum states".

The obtained embedding of the classical phase space into the space of quantum states is physically meaningful. To see this let us first project the orthogonal component $-\frac{i}{\hbar}\widehat{h}_{\perp}\varphi$ of the velocity $\frac{d\varphi}{dt}$ onto vectors tangent to the curves of constant values of $\mathbf{p}$ and $\mathbf{a}$ (classical space and momentum space components) in the projective manifold $M^{\sigma}_{3,3}$. Calculation of the projection of the velocity $\frac{d\varphi}{dt}$ onto the unit vector $-\widehat{\frac{\partial r}{\partial a^{\alpha}}e^{i\theta}}$ (i.e., the classical space component of $\frac{d\varphi}{dt}$) for an arbitrary Hamiltonian of the form $\widehat{h} = -\frac{\hbar^2}{2m}\Delta + V(\mathbf{x})$ yields

$$\mathrm{Re}\left(\frac{d\varphi}{dt}, -\widehat{\frac{\partial r}{\partial a^{\alpha}}e^{i\theta}}\right)\Bigg|_{t=0} = \left(\frac{dr}{dt}, -\widehat{\frac{\partial r}{\partial a^{\alpha}}}\right)\Bigg|_{t=0} = \frac{v_0^{\alpha}}{2\sigma}. \tag{40}$$

Calculation of the projection of velocity $\frac{d\varphi}{dt}$ onto the unit vector $i\widehat{\frac{\partial\theta}{\partial p^{\alpha}}\varphi}$ (momentum space component) gives

$$\mathrm{Re}\left(\frac{d\varphi}{dt}, i\widehat{\frac{\partial\theta}{\partial p^{\alpha}}\varphi}\right)\Bigg|_{t=0} = \frac{mw^{\alpha}\sigma}{\hbar}, \tag{41}$$

where

$$mw^{\alpha} = -\left.\frac{\partial V(\mathbf{x})}{\partial x^{\alpha}}\right|_{\mathbf{x}=\mathbf{x}_0} \tag{42}$$

and σ is assumed to be small enough for the linear approximation of $V(\mathbf{x})$ to be valid within intervals of length σ.

The velocity $\frac{d\varphi}{dt}$ also contains a component due to the change in σ (spreading), which is orthogonal to the fibre $\{\varphi\}$ and the phase space $M^{\sigma}_{3,3}$, and is equal to

$$\mathrm{Re}\left(\frac{d\varphi}{dt}, i\widehat{\frac{d\varphi}{d\sigma}}\right)\Bigg|_{t=0} = \frac{\sqrt{2}\hbar}{8\sigma^2 m}. \tag{43}$$

Calculation of the norm of $\frac{d\varphi}{dt} = \frac{i}{\hbar}\widehat{h}\varphi$ at $t = 0$ gives

$$\left\|\frac{d\varphi}{dt}\right\|^2 = \frac{\overline{E}^2}{\hbar^2} + \frac{\mathbf{v}_0^2}{4\sigma^2} + \frac{m^2\mathbf{w}^2\sigma^2}{\hbar^2} + \frac{\hbar^2}{32\sigma^4 m^2}, \tag{44}$$

which is the sum of squares of the found components. This completes a decomposition of the velocity of the state at any point $\varphi_{\mathbf{a},\mathbf{p}} \in M^{\sigma}_{3,3}$.

For a closed system the norm of $\frac{d\varphi}{dt} = \frac{i}{\hbar}\widehat{h}\varphi$ is preserved in time. For a system in a stationary state, this amounts to conservation of energy. In fact, in this case $\varphi_t(\mathbf{x}) = \psi(\mathbf{x})e^{-\frac{iEt}{\hbar}}$, which is a motion along the phase circle, and

$$\left\|\frac{d\varphi}{dt}\right\|^2 = \frac{E^2}{\hbar^2}. \tag{45}$$

For a closed system in any initial state the norm of the phase component (expected energy) and orthogonal component (energy uncertainty) of the velocity $\frac{d\varphi}{dt}$ are both preserved.

Based on (40) and (41) one can verify that the phase space components of the velocity $\frac{d\varphi}{dt}$ assume correct classical values at any point $\varphi_{\mathbf{a},\mathbf{p}} \in M^{\sigma}_{3,3}$. This remains true for the time dependent potentials as well. The immediate consequence of this and the linear nature of the Schrödinger equation is that under the Schrödinger evolution with the Hamiltonian $\widehat{h} = -\frac{\hbar^2}{2m}\Delta + V(\mathbf{x}, t)$, the state constrained to $M^{\sigma}_{3,3} \subset CP^{L_2}$ moves like a point in the phase space representing a particle in Newtonian dynamics. More generally, the Newtonian dynamics of n particles is the Schrödinger dynamics of n-particle quantum system whose state is constrained to the phase-space submanifold $M^{\sigma}_{3n,3n}$ of the projective space for $L_2(\mathbb{R}^3)\otimes\ ...\ \otimes L_2(\mathbb{R}^3)$, formed by tensor product states $\varphi_1 \otimes\ ...\ \otimes \varphi_n$ with φ_k of the form (37). On the contrary, there exists a unique extension of the Newtonian dynamics on the classical phase space submanifold to a unitary dynamics in the full Hilbert space [2].

Note again that the velocity and acceleration terms in (44) are orthogonal to the fibre $\{\varphi_{\mathbf{a},\mathbf{p}}\}$ of the fibration $\pi : S^{L_2} \longrightarrow CP^{L_2}$, showing that these Newtonian variables have to do with the motion in the projective space CP^{L_2}. The velocity of spreading is orthogonal to the fibre and to the phase space submanifold $M^{\sigma}_{3,3}$.

6. The Born rule and the normal probability distribution

The isometric embedding of the classical space M^{σ}_3 into the space of states $L_2(\mathbb{R}^3)$ results in a relationship between distances in $\mathbb{R}^3$ and in the projective space CP^{L_2}. The distance between two points $\mathbf{a}$ and $\mathbf{b}$ in $\mathbb{R}^3$ is $\|\mathbf{a}-\mathbf{b}\|_{\mathbb{R}^3}$. Under the embedding of the classical space into the space of states, the variable $\mathbf{a}$ is represented by the state $\tilde{\delta}^3_{\mathbf{a}}$. The set of all states $\tilde{\delta}^3_{\mathbf{a}}$ form a submanifold M^{σ}_3 in the Hilbert spaces of states $L_2(\mathbb{R}^3)$, which is "twisted" in $L_2(\mathbb{R}^3)$. It belongs to the sphere S^{L_2} and spans all dimensions of $L_2(\mathbb{R}^3)$. The distance between the states $\tilde{\delta}^3_{\mathbf{a}}$, $\tilde{\delta}^3_{\mathbf{b}}$ on the sphere S^{L_2} or in the projective space CP^{L_2} is not equal to $\|\mathbf{a}-\mathbf{b}\|_{\mathbb{R}^3}$. In fact, the former distance measures length of a geodesic between the states while the latter is obtained using the same metric on the space of states, but applied along a geodesic in the twisted manifold M^{σ}_3. The precise relation between the two distances is given by

$$e^{-\frac{(\mathbf{a}-\mathbf{b})^2}{4\sigma^2}} = \cos^2\theta(\tilde{\delta}^3_{\mathbf{a}}, \tilde{\delta}^3_{\mathbf{b}}), \tag{46}$$

where θ is the Fubini-Study distance between the states in CP^{L_2}. The distance θ in the projective space of states CP^{L_2} appears here for a good reason: the fibres of the fibration $\pi : S^{L_2} \longrightarrow CP^{L_2}$ through the points of the classical space M^{σ}_3 are orthogonal to this space. This is why the distance in M^{σ}_3 can be expressed in terms of the distance in CP^{L_2}. Despite the non-trivial geometry contained in (46), the formula itself is easy to derive. The left hand side is the result of integration in $|(\tilde{\delta}^3_{\mathbf{a}}, \tilde{\delta}^3_{\mathbf{b}})|^2$. On the other hand, the same expression is equal to the right side of (46) by definition of the Fubini-Study metric.

The relation (46) has an immediate implication onto the form of probability distributions of random variables over M^{σ}_3 and CP^{L_2}. In particular, consider a random variable over CP^{L_2}. Suppose that the restricted random variable defined over $M^{\sigma}_3 = \mathbb{R}^3$ is distributed normally on $\mathbb{R}^3$. Then the direction-independent probability of transition between arbitrary quantum states must satisfy the Born rule. So, *The normal distribution law on M^{σ}_3 implies the Born rule on CP^{L_2}. Conversely, the Born rule implies the normal distribution law for the states in M^{σ}_3.*

The fact that the Born rule implies the normal distribution on M^{σ}_3 is straightforward. According to the Born rule, the probability density $f_{\mathbf{a},\sigma}(\mathbf{b})$ to find the particle in a state $\tilde{\delta}^3_{\mathbf{a}}$ at a point $\mathbf{b}$ is equal to

$$f_{\mathbf{a},\sigma}(\mathbf{b}) = |\tilde{\delta}^3_{\mathbf{a}}(\mathbf{b})|^2 = |(\tilde{\delta}^3_{\mathbf{a}}, \delta^3_{\mathbf{b}})|^2 = \left(\frac{1}{2\pi\sigma^2}\right)^{3/2} e^{-\frac{(\mathbf{a}-\mathbf{b})^2}{2\sigma^2}}, \tag{47}$$

which is the normal distribution function. It follows that on the elements of M^{σ}_3, the Born rule *is* the rule of normal distribution.

The Born rule on M^{σ}_3 can be also written in term of the probability $P(\tilde{\delta}^3_{\mathbf{a}}, \tilde{\delta}^3_{\mathbf{b}})$ of transition between the states $\tilde{\delta}^3_{\mathbf{a}}, \tilde{\delta}^3_{\mathbf{b}}$ in M^{σ}_3:

$$P(\tilde{\delta}^3_{\mathbf{a}}, \tilde{\delta}^3_{\mathbf{b}}) = |(\tilde{\delta}^3_{\mathbf{a}}, \tilde{\delta}^3_{\mathbf{b}})|^2. \tag{48}$$

Assuming $\widetilde{\delta_{\mathbf{b}}^3}$ is sufficiently sharp, the formulas (47) and (48) mean the same thing. In fact,

$$|(\widetilde{\delta_{\mathbf{a}}^3}, \widetilde{\delta_{\mathbf{b}}^3})|^2 = f_{\mathbf{a},\sqrt{2}\sigma}(\mathbf{b})(\Delta x)^3, \tag{49}$$

where $f_{\mathbf{a},\sqrt{2}\sigma}$ is the normal distribution function with variance $\sqrt{2}\sigma$ and $\Delta x = \sqrt{4\pi\sigma^2}$. This relates the probability in (48) to the normal probability density in (47) and identifies $P(\tilde{\delta_{\mathbf{a}}^3}, \tilde{\delta_{\mathbf{b}}^3})$ with the probability of finding the macroscopic particle near the point $\mathbf{b}$.

Conversely, suppose we have a rule for probability of transition between states in CP^{L_2} which gives the normal distribution law for the states in M_3^{σ} and depends only on the distance between states. Let's show that this must be the Born rule. In fact, the Fubini-Study distance between the states $\tilde{\delta_{\mathbf{a}}^3}$, $\tilde{\delta_{\mathbf{b}}^3}$ takes on all values from 0 to $\pi/2$, which is the largest possible distance between points in CP^{L_2}. By assumption, the probability $P(\varphi, \psi)$ of transition between any states φ and ψ depends only on the Fubini-Study distance $\theta(\pi(\varphi), \pi(\psi))$ between the states. Given arbitrary states $\varphi, \psi \in S^{L_2}$, let then $\tilde{\delta_{\mathbf{a}}^3}$, $\tilde{\delta_{\mathbf{b}}^3}$ be two states in M_{σ}^3, such that

$$\theta(\pi(\varphi), \pi(\psi)) = \theta(\tilde{\delta_{\mathbf{a}}^3}, \tilde{\delta_{\mathbf{b}}^3}). \tag{50}$$

It then follows that

$$P(\varphi, \psi) = P(\tilde{\delta_{\mathbf{a}}^3}, \tilde{\delta_{\mathbf{b}}^3}) = \cos^2\theta(\tilde{\delta_{\mathbf{a}}^3}, \tilde{\delta_{\mathbf{b}}^3}) = \cos^2\theta(\pi(\varphi), \pi(\psi)), \tag{51}$$

which yields the Born rule for arbitrary states. This proves the claim and completes a review of [1] and [2].

7. Measurements on macroscopic and microscopic particles

We are now in a position to compare the process of measurement in the classical and quantum physics. First of all, the classical space and phase space are now submanifolds in the Hilbert space of states. This allows us to use the same language when analyzing both types of measurement. Second, the Newtonian dynamics is now a restriction of the Schrödinger dynamics to the classical phase space submanifold. Conversely, the Schrödinger dynamics is a unique unitary extension of the Newtonian dynamics from the classical phase space to the Hilbert space. This allows us to begin with a model of measurement satisfying Newton laws and extend it to a model consistent with the rules of quantum mechanics. Finally, the normal probability law is the restriction of the Born rule to the classical space submanifold. Conversely, the Born rule is the unique isotropic extension of the normal probability law from the classical space to the space of states. In particular, a classical model of measurement with a normal distribution of the measured quantity should lead us to a model consistent with the Schrödinger dynamics and the Born rule.

Measurements performed on a macroscopic particle satisfy generically the normal distribution law for the measured observable. This is consistent with the central limit theorem and indicates that a specific way in which the observable was measured is not important. To be specific, consider measurements of position of a particle. A common way of finding the position of a macroscopic particle is to expose it to light of sufficiently short wavelength and to observe the scattered photons. Due to the unknown path of the incident photons, multiple scattering events on the particle, random change in position of the particle, etc., the process of observation can be described by the diffusion equation with the observed position of the particle experiencing Brownian motion from an initial point during the time of observation. This results in the normal distribution of observed position of the particle.

The ability to describe measurement of position of a macroscopic particle as a diffusion seems to be a general feature of measurements in the macro-world, independent of a particular measurement set up. The averaging process making the central limit theorem applicable and

IOP Conf. Series: Journal of Physics: Conf. Series **1239** (2019) 012022 doi:10.1088/1742-6596/1239/1/012022

leading to the normal distribution of the position random variable can be seen as the result of random hits experienced by the particle from the surrounding particles participating in the measurement. These random hits are equally likely to come from any direction, independent of the initial position of the particle, leading to Brownian motion and the validity of the diffusion equation for the probability density of the position random variable for the particle.

It is claimed now that at any time t the initial state ψ of a microscopic particle undergoing a position measurement is equally likely to shift in any direction in the tangent space $T_{\{\psi\}}CP^{L_2}$. Suppose, for example, that the particle is exposed to a stream of photons of sufficiently high frequency and number density. The scattered photons are then observed to determine the position of the particle. The field of photons in the experiment will be treated classically, as a fluctuating potential in a region surrounding the source. The assumptions about the potential that will be made will determine to what extent the derived results can be applied to different experiments of measuring position of a microscopic particle.

Recall that the space $M^{\sigma}_{3,3}$ is complete in $L_2(\mathbb{R}^3)$. Consider the subset of $M^{\sigma}_{3,3}$ formed by the states

$$\varphi_{\mathbf{mn}}(\mathbf{x}) = \left(\frac{1}{2\pi\sigma^2}\right)^{3/4} e^{-\frac{(\mathbf{x}-\alpha\mathbf{n})^2}{4\sigma^2}} e^{i\frac{\beta\mathbf{m}\mathbf{x}}{\hbar}}, \tag{52}$$

where $\alpha = \sqrt{2\pi}\sigma$, $\beta = \frac{h}{\sqrt{2\pi}\sigma}$ and $\mathbf{m}, \mathbf{n}$ take values on the lattice $\mathbb{Z}^3 \times \mathbb{Z}^3$ of points with integer coordinates in $\mathbb{R}^3 \times \mathbb{R}^3$. The set of functions (52) is known to be also complete in $L_2(\mathbb{R}^3)$. Any state in $L_2(\mathbb{R}^3)$ can be then represented by a linear combination of the states $\varphi_{\mathbf{mn}}$. (For $\alpha\beta < h$ the system of functions $\varphi_{\mathbf{mn}}$ is called the Gabor or Weil-Heisenberg frame.) In particular, the initial state ψ of the particle can be represented by a sum

$$\psi = \sum_{\mathbf{m},\mathbf{n}} C_{\mathbf{mn}} \varphi_{\mathbf{mn}}. \tag{53}$$

The set M^{σ}_3 is also complete in $L_2(\mathbb{R}^3)$. Here too there exist countable subsets of M^{σ}_3 that are complete in $L_2(\mathbb{R}^3)$. Moreover, an arbitrary initial state ψ in $L_2(\mathbb{R}^3)$ can be approximated as well as necessary by a finite discrete sum

$$\psi \approx \sum_{\mathbf{n}} C_{\mathbf{n}} \widetilde{\delta}^3_{\mathbf{a}-\gamma\mathbf{n}}, \tag{54}$$

where $\mathbf{a}$ is arbitrary, $\mathbf{n} \in \mathbb{Z}^3$, and the value of $\gamma > 0$ together with the number of terms in the sum depend on ψ and the needed approximation. Taking γ sufficiently small, let's partition the space $\mathbb{R}^3$ into the cubes of edge γ centered at the lattice points $\mathbf{a}-\gamma\mathbf{n}$ and consider the indicator function $1_{\mathbf{n}}$ for each cube. The potential $\widehat{V}$ can be written as a sum $\sum_{\mathbf{n}} 1_{\mathbf{n}} \widehat{V}_{\mathbf{n}}$. The components $\widehat{V}_{\mathbf{n}}$ for different $\mathbf{n}$ can be assumed to be independent, identically distributed random variables. In the case of position measurement by scattering photons off the particle, the components $\widehat{V}_{\mathbf{n}}$ can be associated with a single photon at time t.

For simplicity, let's neglect the kinetic energy term in the Hamiltonian $\widehat{h}$. Let us denote the solution of the Shrödinger equation with the initial state ψ by $\chi(t)$ and let us write $\chi(t) = e^{-\frac{i\overline{V}t}{\hbar}}\psi(t)$, where $\overline{V} = (\widehat{V}\psi, \psi)$ and $\psi(0) = \psi$. We then have at $t = 0$:

$$\frac{d\psi}{dt} = -\frac{i}{\hbar}\widehat{V}_{\perp}\psi, \tag{55}$$

where $\widehat{V}_{\perp} = \widehat{V} - \overline{V}$, as before. This equation gives the velocity of the states $\psi(t)$ and $\chi(t)$ in the projective space CP^{L_2} at $t = 0$. To prove that under the action of $\widehat{V}_{\perp}$ all directions of velocity of state in $T_\psi CP^{L_2}$ are equally likely, consider the components of the velocity in the basis $-i\widetilde{\delta}^3_{\mathbf{m}} \equiv -i\widetilde{\delta}^3_{\mathbf{a}-\gamma\mathbf{m}}$

$$\left(\frac{d\psi}{dt}, -i\widetilde{\delta}^3{}_{\mathbf{m}}\right) = \frac{1}{\hbar}(\widehat{V}_\perp\psi, \widetilde{\delta}^3_{\mathbf{m}}). \tag{56}$$

The probability density of the random variables given by (56) is a function of the distance between the points ψ and $\widetilde{\delta}^3_{\mathbf{m}}$ in the Fubini-Study metric and, possibly, of the direction vector $\eta \in T_\psi CP^{L_2}$ of a fixed norm. Let us divide the random variables in (56) by $|(\psi, \widetilde{\delta}^3_{\mathbf{m}})|$:

$$\frac{\frac{1}{\hbar}(\widehat{V}_\perp\psi, \widetilde{\delta}^3_{\mathbf{m}})}{|(\psi, \widetilde{\delta}^3_{\mathbf{m}})|}. \tag{57}$$

Because $|(\psi, \widetilde{\delta}^3_{\mathbf{m}})|$ depends only on the distance between $\{\psi\}$ and $\widetilde{\delta}^3_{\mathbf{m}}$, the probability densities of the random variables given by (56) and (57) are either both dependent or both independent of η. Provided the potential does not change much within each cube, the expression in (57) is equal to

$$\tfrac{1}{\hbar}(V_{\mathbf{m}} - \overline{V}). \tag{58}$$

As discussed, the random variables $V_{\mathbf{m}}$ at different cubes, i.e., for different values of $\mathbf{m}$ can be considered independent and identically distributed. It follows that the probability distributions of the random variables $V_{\mathbf{m}} - \overline{V}$ have a zero mean and are identical for all values of $\mathbf{m}$. With the help of the central limit theorem one can also claim that these distributions are normal. So the random vector given by the components (57) is the standard normal random vector.

The standard deviation ΔV for the distribution of $V_{\mathbf{m}}$ in time satisfies the uncertainty relation $\Delta V\tau > h$. Therefore, the phase $-\frac{1}{\hbar}\overline{V}\tau$ in the phase factor of $\chi(t)$ should be considered uniformly distributed on $[0, 2\pi)$ so that any value of the phase factor is equally likely. It follows that the components of $\frac{d\chi}{dt}$ may have arbitrary constant phase factors at any given time and so the complex random vector made of these components has a normal circularly symmetric distribution. Accordingly, the vector $\eta = \frac{d\chi}{dt}\tau$ is equally likely to point in any direction in $T_\psi CP^{L_2}$ at $t = 0$. To make this result valid at an arbitrary moment of time, it remains to assume that the distribution of potentials is stationary and has independent increments.

Under a measurement of position of a macroscopic particle the observed particle is exposed to a random potential that is responsible for the normal distribution of the position random variable. We now see that the state of a microscopic particle undergoing a similar measurement and exposed to the same potential will experience a random motion on the space of states CP^{L_2} under which any direction η of displacement of the state is equally likely. From the section 6, entitled "The Born Rule and the Normal Probability Distribution" we know that the normal distribution on M^σ_3 and equal probability of any direction of displacement of the state in CP^{L_2} result in the Born rule. That is, under the random potential produced by the measuring device the state ψ of the measured microscopic particle performs a random walk on the sphere of states and the probability for the state of reaching a neighborhood of any point φ on the sphere is given by the Born rule: $P(\varphi, \psi) = |(\varphi, \psi)|^2$.

Given the lack of Lebesgue measure on an infinite-dimensional Hilbert space, one may wonder how the state would have any chance of reaching a neighborhood of a given point in S^{L_2}. However, a realistic measuring device occupies a finite volume in the classical space. So the potential created by it can only affect a region $Q \subset \mathbb{R}^3$ of a finite volume V. The initial state ψ of the particle can be split onto the state $\psi_Q = \psi|_Q$ with support in Q (restriction of ψ to Q) and the leftover state $\chi = \psi - \psi_Q$. The state χ is not going to change under the potential and does not participate in the measurement (the probability for it of reaching a detector in Q is zero). Furthermore, possible group-velocity $\mathbf{v}_g$ of the measured particle in the given potential is also bounded. The corresponding momentum $m\mathbf{v}_g$ of the particle belongs then to a bounded region $P \subset \mathbb{R}^3$. Therefore the state ψ_Q of the particle is limited to a superposition of states in the region $P \times Q$ in the phase space $\mathbb{R}^3 \times \mathbb{R}^3 = M^{\sigma}_{3,3}$. But there are only finitely many elements of the Weil-Heisenberg frame in such a region. Therefore, under the measurement the state ψ_Q evolves in a finite-dimensional subspace $V_{P\times Q}$ of the Hilbert space $L_2(\mathbb{R}^3)$. In particular, the Lebesgue volume of a neighborhood $Q_\mathbf{a}$ of any point $\tilde{\delta}^3_\mathbf{a}$ (a state of particle with known position) in $V_{P\times Q}$ is well defined. Accordingly, the state has a non-vanishing probability of reaching $Q_\mathbf{a}$ and the relative probabilities of reaching neighborhoods of different points are given by the Born rule.

8. The motion of state under measurement

Let's now look into details of the motion of a state under a measurement. Note that in the non-relativistic quantum mechanics the particle, and, therefore, its state in a single particle Hilbert space, cannot disappear or get created. The unitary property of evolution means that the state can only evolve along the unit sphere in the space of states $L_2(\mathbb{R}^3)$. To express this conservation of states in the case of observation of position of the particle, consider the density of states functional $\rho_t[\varphi; \psi_Q]$. This functional measures the number of states that belong to a neighborhood of a point φ on the sphere of states $S^{V_{P\times Q}} \subset V_{P\times Q}$ at time t for an ensemble of particles, prepared in the initial state ψ_Q each. Under the isometric embedding $\omega : \mathbb{R}^3 \longrightarrow M^{\sigma}_3 \subset L_2(\mathbb{R}^3)$ the states in M^{σ}_3 are identified with (positions of) particles. So the density of states functional $\rho_t[\varphi; \psi_Q]$ must be an extension of the density of macroscopic particles $\rho_t(\mathbf{a}; \mathbf{b})$ on $\mathbb{R}^3$. Here $\mathbf{b}$ is the initial position of the particle. In other words, $\rho_t(\mathbf{a}; \mathbf{b}) = \rho_t[\tilde{\delta}^3_\mathbf{a}; \tilde{\delta}^3_\mathbf{b}]$.

In the case of macroscopic particles the conservation of the number of particles is expressed in differential form by the continuity equation. For instance, if $\rho_t(\mathbf{a}; \mathbf{b})$ is the density at a point $\mathbf{a} \in \mathbb{R}^3$ of an ensemble of Brownian particles with initial position $\mathbf{b}$ and $\mathbf{j}_t(\mathbf{a}; \mathbf{b})$ is the current density at $\mathbf{a}$ of the particles, then

$$\frac{\partial \rho_t(\mathbf{a}; \mathbf{b})}{\partial t} + \nabla \mathbf{j}_t(\mathbf{a}; \mathbf{b}) = 0. \tag{59}$$

The conservation of states of an ensemble of microscopic particles is expressed by the continuity equation that follows from the Schrödinger dynamics. This is the same equation (59) with

$$\rho_t = |\psi|^2, \text{ and } \mathbf{j}_t = \frac{i\hbar}{2m}(\psi\nabla\overline{\psi} - \overline{\psi}\nabla\psi). \tag{60}$$

For the states $\psi \in M^{\sigma}_{3,3}$ we obtain

$$\mathbf{j}_t = \frac{\mathbf{p}}{m}|\psi|^2 = \mathbf{v}\rho_t. \tag{61}$$

Because the restriction of Schrödinger evolution to $M^{\sigma}_{3,3}$ is the corresponding Newtonian evolution, the function ρ_t in (61) must be the density of particles. This density was denoted earlier by $\rho_t(\mathbf{a}; \mathbf{b})$. It gives the number of Brownian particles that start at $\mathbf{b}$ and by the time t reach a neighborhood of a point $\mathbf{a}$. The relation $\rho_t(\mathbf{a}; \mathbf{b}) = \rho_t[\tilde{\delta}^3_\mathbf{a}; \tilde{\delta}^3_\mathbf{b}]$ tells us that ρ_t in (60) must be then the density of states $\rho_t[\tilde{\delta}^3_\mathbf{a}; \psi]$. It gives the number of particles in an initial state ψ found under the measurement at time t in a state $\tilde{\delta}^3_\mathbf{a}$ (i.e., on a neighborhood of the point $\mathbf{a}$ in $\mathbb{R}^3$).

In the integral form the conservation of states in $L_2(\mathbb{R}^3)$ can be written in the following form:

$$\rho_{t+\tau}[\varphi; \psi] = \int \rho_t[\varphi + \eta; \psi]\gamma[\eta]D\eta, \tag{62}$$

where $\gamma[\eta]$ is the probability functional of the variation η in the state φ and integration goes over all variations η such that $\varphi+\eta \in S^{L_2}$. When the state of the particle is constrained to $M_3^\sigma = \mathbb{R}^3$ this equation must imply the usual diffusion on $\mathbb{R}^3$. The restriction of (62) to M_3^σ means that $\varphi = \tilde{\delta}^3_{\mathbf{a}}$ and $\eta = \tilde{\delta}^3_{\mathbf{a}+\epsilon} - \tilde{\delta}^3_{\mathbf{a}}$, where ϵ is a displacement vector in $\mathbb{R}^3$. As we already know, the function $\rho_t[\tilde{\delta}^3_{\mathbf{a}}; \tilde{\delta}^3_{\mathbf{b}}] = \rho_t(\mathbf{a}; \mathbf{b})$ is the usual density of particles in space. Let us substitute this into (62), replace $\gamma[\eta]$ with the corresponding probability density function $\gamma(\epsilon) \equiv \gamma[\tilde{\delta}^3_{\mathbf{a}+\epsilon} - \tilde{\delta}^3_{\mathbf{a}}]$ and integrate over the space $\mathbb{R}^3$ of all possible vectors ϵ. As in the Einstein derivation of the Brownian motion, assume that $\gamma(\epsilon)$ is the same for all $\mathbf{a}$ and independent of the direction of ϵ (space symmetry). Therefore, the terms $\int \epsilon^k \gamma(\epsilon) d\epsilon$ and $\int \epsilon^k \epsilon^l \gamma(\epsilon) d\epsilon$ with $k \neq l$ vanish. It follows by the Einstein derivation that

$$\frac{\partial \rho_t(\mathbf{a}; \mathbf{b})}{\partial t} = k \Delta \rho_t(\mathbf{a}; \mathbf{b}), \tag{63}$$

where $k = \frac{1}{2\tau} \int \epsilon^2 \gamma(\epsilon) d\epsilon$ is a constant.

The diffusion equation (63) describes the dynamics of an ensemble of particles in the classical space M_3^σ. If initially all particles in the ensemble are at the origin, then the density of the particles at a point $x \in \mathbb{R}$ (one dimensional case) at time t is given by

$$\rho_t(x; 0) = \frac{1}{\sqrt{4\pi k t}} e^{-\frac{x^2}{4kt}}. \tag{64}$$

The path of a Brownian particle under random hits from the surrounding particles is a particular path of the state ψ in the space of states under the corresponding random potential. Since the distribution of displacement of ψ is isotropic, the distribution (64) can be expressed in terms of the Fubini-Study distance between states. From (46) we have

$$e^{-\frac{x^2}{4\sigma^2}} = \cos^2\theta, \tag{65}$$

where θ is the Fubini-Study distance between the state $\tilde{\delta}_x(u) = \left(\frac{1}{2\pi\sigma^2}\right)^{1/4} e^{-\frac{(u-x)^2}{4\sigma^2}}$ and the like state centered at $x = 0$. Therefore,

$$x^2 = -4\sigma^2 \ln(\cos^2\theta). \tag{66}$$

Equating the probability density for the Brownian particle initially at the origin to be found at time t at the point x with the probability density for transition between the corresponding states (see (49)), we have

$$\frac{1}{\sqrt{4\pi k t}} e^{-\frac{x^2}{4kt}} = \frac{1}{\sqrt{4\pi\sigma^2}} \rho_t(\theta), \tag{67}$$

where by (46) the function $\rho_t(\theta)$ is equal to $\cos^2\theta$ for $t = \frac{\sigma^2}{k}$. Denoting this value by τ and substituting (66) into (67), we have

$$\rho_t(\theta) = \sqrt{\frac{\tau}{t}} \left(\cos^2\theta\right)^{\frac{\tau}{t}}. \tag{68}$$

As t increases, $\left(\cos^2\theta\right)^{\tau/t}$ approaches 1 for all $\theta \in [0, \pi/2)$, while $\rho_t(\pi/2) = 0$. Note that the integral of ρ_t with respect to $dx = \frac{dx}{d\theta} d\theta$ is, of course, 1.

From (66) we have

$$\left(\frac{d\theta}{dx}\right)^2 = -\frac{1}{2\sigma^2} \ln(\cos\theta) \cdot \cot^2\theta \tag{69}$$

and

$$\frac{d^2\theta}{dx^2} = \frac{\cot\theta}{4\sigma^2}\left[1 + 2\ln(\cos\theta)\cdot\csc^2\theta\right]. \tag{70}$$

This yields the second derivative $\frac{d^2}{dx^2}$ in the the form of the following operator Δ_θ:

$$\Delta_\theta = \frac{\cot\theta}{4\sigma^2}\left[1 - 2\ln(\cos\theta)\cdot\frac{d}{d\theta}\left(\cot\theta\frac{d}{d\theta}\right)\right]. \tag{71}$$

The diffusion equation takes the form

$$\frac{\partial\rho_t(\theta)}{\partial t} = k\Delta_\theta\rho_t(\theta), \tag{72}$$

where $\rho_t(\theta) = \rho_t[\varphi;\psi]$ with θ being the Fubini-Study distance between the initial state ψ and the variable point φ. The corresponding fundamental solution is the function in (68).

The parameter $\tau = \frac{\sigma^2}{k}$ should be interpreted as the time interval of observation. It is the time from the beginning of the diffusion process of the initial state ψ to the moment when the state has reached the end point φ (the moment of observation). Of course, this time may vary in different trials. However, the same is true of the time interval of observation of a Brownian particle in a particular experiment. By the central limit theorem the normal distribution on the left side of (67) is correct for the averaged time interval of observation τ. Because the left hand side of (67) is valid, the density (68) is accurate as well, giving the Born rule at $t = \tau$. One can say that a specific measuring device has an associated time parameter τ and diffusion coefficient k, and, therefore, the variance of the normal distribution of the position $\sigma^2 = k\tau$. The value of σ is then used in the isometric embedding $\omega : \mathbf{a} \longrightarrow \widetilde{\delta}^3_{\mathbf{a}}$, giving the relationship (65) and leading to the validity of the Born rule at the time of observation.

Note that if the diffusion of a state takes place but an observation is not made, the state continues its random walk on the sphere of states. Because of this additional (Schrödinger) evolution, the Born rule for any future observation has to be applied to the evolved state, rather than the original state ψ.

9. Collapse of a quantum state

It was shown that under the action of a random potential typically experienced by a Brownian particle, the state of a microscopic particle is equally likely to fluctuate in any direction tangent to the projective space of states. This, together with the normal distribution of position of the Brownian particle, signifies the validity of the Born rule for the motion of a state. The motion satisfies the diffusion equation (72), whose fundamental solution is (68). It was shown that the state under the random potential evolves in the finite-dimensional subspace of the space of states. As a result, there is a non-vanishing probability for the initial state ψ of reaching a neighborhood of any non-orthogonal state φ.

The presented process of transition between states is very different from what is usually understood by the collapse. The fact that a noise may lead to random fluctuation of a state is rather simple and goes against of what one normally tries to achieve when explaining collapse under a measurement. The collapse models utilize various ad hoc additions to the Schrödinger equation with the goal of explaining why the state under the resulting stochastic process "concentrates" to an eigenstate of the measured observable (usually, position or energy) [6]-[17]. Instead, it is argued here that under a generic measurement an ensemble of states in the same initial position ψ "diffuses" isotropically into the space of states. Whenever a particular state in the ensemble reaches a neighborhood of an eigenstate of the measured observable, we say that the "collapse" has occurred. In this case the measuring device can record the value of the measured physical quantity.

IOP Conf. Series: Journal of Physics: Conf. Series **1239** (2019) 012022 doi:10.1088/1742-6596/1239/1/012022

The role of the measuring device in this scenario is reduced to initiating the diffusion(creating a "noise") and to registering a particular location of the diffused state. For instance, the "noise" in the position measuring device under consideration is due to the stream of photons. The device then registers the state reaching a point in M_3^g. In a similar way, a momentum measuring device registers the states that reach under the diffusion the eigen-manifold of the momentum operator (which is the image of M_3^g under the Fourier transformation). Note also the similarity in the role of measuring devices in quantum and classical mechanics: in both cases the devices are designed to measure a particular physical quantity and inadvertently create a "noise", which results in a distribution of values of the measured quantity.

It follows, in particular, that the measuring device in quantum mechanics is not responsible for creating a basis into which the state is to be expanded. If several measuring devices are present, they are not "fighting" for the basis. When the eigen-manifolds of the corresponding observables don't overlap, only one of them can "click" for the measured particle as the state can reach only one of the eigen-manifolds at a time.

What does it all say about measurement of position of macroscopic and microscopic particles? During the period of observation of position of a macroscopic particle, the position is a random variable with normal distribution. Normally, observation happens during a short enough interval of time and the spread of the probability density is sufficiently small. A particular value of the position variable during the observation is simply a realization of one of the possible outcomes. The change in observed position of the particle can be equivalently thought of as either a stochastic process $\mathbf{b}_t$ with values in $\mathbb{R}^3$ or a process $\tilde{\delta}^3_{\mathbf{b},t}$ with values in M_3^g. The advantage of the latter representation is that the position random variable $\tilde{\delta}^3_{\mathbf{b}}$ gives both, the position of the particle in $M_3^g = \mathbb{R}^3$ and the probability density to find it in a different location $\mathbf{a}$ (in the state $\tilde{\delta}^3_{\mathbf{a}}$), due to uncontrollable interactions with the surroundings under the observation.

Measuring position of a microscopic particle has, in essence, a very similar nature. Under observation the state ψ is a random variable with values in the space of states CP^{L_2}. To measure position is to observe the state on the submanifold M_3^g or $M_{3,3}^g$ in CP^{L_2}. In this case, the random variable ψ assumes one of the values $\tilde{\delta}^3_{\mathbf{a}}$, with the uniquely defined probability density compatible with the normal density in the space $\mathbb{R}^3$. This probability density is given by the Born rule. Here too the random variable ψ gives both, the position of the state of the particle in CP^{L_2} and the probability density to find the particle in a different state $\tilde{\delta}^3_{\mathbf{a}}$.

So the difference between the measurements is two-fold. First, under a measurement the state ψ of a microscopic particle is a random variable over the entire space of states CP^{L_2} and not just over the submanifold M_3^g. Second, unless ψ is already constrained to M_3^g (the case which would mimic the measurement of position of a macroscopic particle), to measure position is to observe the state that "diffused" enough to reach the submanifold M_3^g. To put it differently, the measuring device is not where the initial state was. Assuming the state has reached M_3^g, the probability density of reaching a particular point in M_3^g is given, as we saw, by the Born rule.

We don't use the term collapse of position random variable when measuring position of a macroscopic particle. Likewise, there seems to be no physics in the term collapse of the state of a microscopic particle. Instead, due to the diffusion of state, there is a probability density to find the particle in various locations on CP^{L_2}. In particular, the state may reach the space manifold $M_3^g = \mathbb{R}^3$. If that happens and we have detectors spread over the space, then one of them clicks. If the detector at a point $\mathbf{a} \in \mathbb{R}^3$ clicks, that means the state is at the point $\tilde{\delta}^3_{\mathbf{a}} \in CP^{L_2}$ (that is, the state *is* $\tilde{\delta}^3_{\mathbf{a}}$). The number of clicks at different points $\mathbf{a}$ when experiment is repeated is given by the Born rule. The state is not a "cloud" in $\mathbb{R}^3$ that shrinks to a point under observation. Rather, the state is a point in CP^{L_2} which may or may not be on $\mathbb{R}^3 = M_3^g$. When the detector clicks we know that the state is on M_3^g.

Note once again that there is no need in any new mechanism of "collapse" in the model. An observation is not about a "concentration" of the state and the stochastic process initiated by

IOP Conf. Series: Journal of Physics: Conf. Series **1239** (2019) 012022 doi:10.1088/1742-6596/1239/1/012022

the observation is in agreement with the conventional Schrödinger equation with a randomly fluctuating potential ("noise"). The origin of the potential depends on the type of measuring device or properties of the environment, capable of "measuring" the system. Fluctuation of the potential can be traced back to thermal motion of molecules, atomic vibrations in solids, vibrational and rotational molecular motion, and the surrounding fields.

10. Generalities of the classical behavior of macroscopic bodies

It was demonstrated that the Schrödinger evolution of a state constrained to the classical phase space $M^{\sigma}_{3,3}$ results in the Newtonian motion of the particle. A similar result holds true for systems of particles. To reconcile the laws of quantum and classical physics, one must also explain the nature of this constraint. Why would microscopic particles be free to leave the classical space, while macroscopic particles be bound to it?

Suppose first that the macroscopic particle under consideration is a crystalline solid. The position of one cell in the solid defines the position of the entire solid. If one of the cells was observed at a certain point at rest, the state of the solid immediately after the observation (in one dimension) is the product

$$\varphi = \tilde{\delta}_a \otimes \tilde{\delta}_{a+\Delta} \otimes \ ... \ \otimes \tilde{\delta}_{a+n\Delta}, \tag{73}$$

where Δ is the lattice length parameter. The general quantum-mechanical state of the solid is then a superposition of states (73) for different values of a in space:

$$\varphi = \sum_a C_a \tilde{\delta}_a \otimes \tilde{\delta}_{a+\Delta} \otimes \ ... \ \otimes \tilde{\delta}_{a+n\Delta}. \tag{74}$$

Why would non-trivial superpositions of this sort be absent in nature?

The classical phase space $M^{\sigma}_{3n,3n}$ of a n-particle system consists of all tensor products $\varphi_1 \otimes ... \otimes \varphi_n$, defined up to a phase factor, with the state φ_k of each particle given by (37). As demonstrated in [1] and discussed in this paper, the Schrödinger dynamics of n-particle system constrained to $M^{\sigma}_{3n,3n}$ is the Newtonian dynamics of the system. Furthermore, there exists a unique unitary evolution whose restriction to $M^{\sigma}_{3n,3n}$ reproduces the motion of n particles in Newtonian dynamics [2]. This evolution is governed by the usual Hamiltonian $\sum_k \frac{\hat{p}_k^2}{2m_k} + V(\mathbf{x}_1, ..., \mathbf{x}_n)$. Note also that the isomorphism $\omega_n : \mathbb{R}^3 \times ... \times \mathbb{R}^3 \longrightarrow M^{\sigma}_{3n}$, $\omega_n(\mathbf{a}_1, ..., \mathbf{a}_n) = \tilde{\delta}^3_{\mathbf{a}_1} \otimes ... \otimes \tilde{\delta}^3_{\mathbf{a}_n}$ allows us to interpret n-particle states in M^{σ}_{3n} as positions of n particles in the classical space $\mathbb{R}^3$. A similar map identifies the submanifold $M^{\sigma}_{3n,3n}$ with the classical phase space of n particles. These maps together with the discovered relationships of the classical and quantum dynamics allow us to think of M^{σ}_{3n} and $M^{\sigma}_{3n,3n}$ as the classical space and phase space with n particles.

To understand the dynamics of macroscopic bodies under measurement, consider the Brownian motion of a crystalline solid. The motion of any solid can be represented by the motion of its center of mass under the total force acting on the body and a rotational motion about the center of mass. The motion of the center of mass is the motion of a material point under the random force term, which is the sum of forces acting from the surrounding particles on each cell. Suppose for simplicity that the solid is one-dimensional and consists of n-cells. Let $\rho[\varphi; \psi]$ be the density of states functional on the space CP^{L_2}, where $L_2 = L_2(\mathbb{R}) \otimes ... \otimes L_2(\mathbb{R})$ is the tensor product of n single particle Hilbert spaces. The conservation of states for the system reads as before

$$\rho_{t+\tau}[\varphi] = \int \rho_t[\varphi + \eta; \psi]\gamma[\eta] D\eta, \tag{75}$$

where the meaning of terms is clear from (62). Define $\tilde{\delta}^{\otimes}_a = \tilde{\delta}_{a+\Delta_1} \otimes \tilde{\delta}_{a+\Delta_2} \otimes ... \otimes \tilde{\delta}_{a+\Delta_n} \in M^{\sigma}_{3n}$ and consider the functions

$$\rho_t(a; b) = \rho_t[\tilde{\delta}^{\otimes}_a; \tilde{\delta}^{\otimes}_b], \tag{76}$$

and

$$\rho_t(a;\psi) = \rho_t[\tilde{\delta}_a^{\otimes};\psi], \tag{77}$$

where a, b denote the center of mass and Δ_k describe the positions of each cell relative to the center of mass. Applying the results of the sections titled "Measurements on macroscopic and microscopic particles" and "The motion of state under measurement" we conclude that the state of the solid will experience a random motion on CP^{L_2}, such that any direction of displacement of the state in $T_{\{\psi\}}CP^{L_2}$ at any time t is equally likely. In particular, if ψ is constrained to M^{σ}_{3n}, then (75) leads to the usual diffusion equation for the material point positioned at the center of mass of the solid.

It is a well established and experimentally confirmed fact that macroscopic bodies experience an unavoidable interaction with the surroundings. Their "cells" are pushed in all possible directions by the surrounding particles. For instance, a typical Brownian particle of radius between $10^{-9}m$ and $10^{-7}m$ experiences about 10^{12} random collisions per second with surrounding atoms in a liquid. The number of collisions of a solid of radius $10^{-3}m$ in the same environment is then about 10^{19} per second. Collisions with photons and other surrounding particles must be also added. Even empty space has on average about 450 photons per cm^3 of space.

Now, suppose the state of a macroscopic body (in one dimension) is initially given by $\psi = \tilde{\delta}_{b+\Delta_1} \otimes ... \otimes \tilde{\delta}_{b+\Delta_n}$. Recall that this means that the initial distribution of position random variable is Gaussian with the center of mass at b. Under interaction with the surroundings the state ψ undergoes a random walk on the space of states CP^{L_2}. Consider the spatial (i.e., restricted to M^{σ}_{3n}) component of the walk near $t = 0$. As we know, the mean position of the center of mass will remain equal to b. Also, macroscopic bodies are distinguished by a large number of "cells". The sum of forces from the collisions of cells with the surrounding particles at any time is almost exactly zero. In addition, the mass of the body increases with the number of cells, so the displacements generated by the total force become even smaller and for sufficiently large massive bodies can be safely disregarded. This means that the diffusion coefficient k in (63) vanishes so that the diffusion in space is trivial. But we know that the probability density of states is direction-independent: if the state does not diffuse in the space M^{σ}_{3n}, then it does not diffuse in the space of states either! Accordingly, the probability distribution remains constant in time. In the absence of additional potentials acting on the macroscopic body it will maintain its original state ψ.

The situation is surprisingly similar to that of pollen grains and a ship initially at rest in still water. While under the kicks from the molecules of water the pollen grains experience a Brownian motion, the ship in still water will not move at all. Because of the established relation of Newtonian and Schrödinger dynamics, this is more than an analogy. In fact, when the state is constrained to the classical phase space submanifold, the "pushes" experienced by the state become the classical kicks in the space that could lead to the Brownian motion of the system.

Suppose now an external potential V is applied to the macroscopic system. According to (44), this will "push" the state that belongs to the classical phase space submanifold in the direction tangent to the submanifold. Therefore, the external potential applied to a macroscopic body will not affect the motion of the state in the directions orthogonal to the classical phase space submanifold. That means that the state will remain constrained to the submanifold. On the other hand, as we know from the same section, the constrained state will evolve in accord with Newtonian dynamics in the total potential $V + V_S$, where V_S is the potential created by the surroundings. However, since at any time t the total force $-\nabla V_S$ exerted on the macroscopic body by the particles of the surroundings is vanishingly small, the body will evolve according to Newtonian equations with the force term $-\nabla V$. To be sure, the particles of the surroundings are responsible for the friction. In the Hamiltonian description of interaction of the body with the surroundings (as in the Ullersma model [4]) the friction comes from a contribution to the total potential in the Hamiltonian. However, whenever the friction can be neglected, the dynamics of the solid is determined by the force $-\nabla V$.

This qualitative analysis must be complemented by an accurate calculation. An estimate of fluctuations of the state about the mean state ψ in the classical space submanifold M^{σ}_{3n} requires a selection of the parameters σ and k and will be done elsewhere. This estimate should also give us an idea about where the boundary between the classical and the quantum world should be placed.

In many cases the state of the measured particle, whose position is measured, and the measuring device plus the environment cannot be described independently. The state of the particle and the surroundings is then a linear combination of the terms

$$\tilde{\delta}^3_{\mathbf{a}} \otimes E_{\mathbf{a}}, \tag{78}$$

where $E_{\mathbf{a}}$ represents the state of the apparatus and the environment when the particle is in the state $\tilde{\delta}^3_{\mathbf{a}}$. At the same time, the result of measurement is always a single term like (78). How could it be?

This situation is, however, similar to the one already considered. Under the measurement, the initial superposition ψ of states (78) undergoes a random walk on the projective space CP^{L_2}, where L_2 is the tensor product of the Hilbert spaces of the particle and the surroundings. At any time during this walk, the total state is equally likely to shift in any direction in the tangent space $T_{\{\psi\}}CP^{L_2}$. The position of the particle in the classical space becomes defined when the state reaches the product manifold $M^{\sigma}_3 \otimes H$ that consists of all states of the form $\tilde{\delta}^3_{\mathbf{a}} \otimes E_{\mathbf{a}}$. As was verified, the probability of reaching a particular state in $M^{\sigma}_3 \otimes H$ is given by the Born rule.

11. Summary

The dynamics of a classical n-particle mechanical system can be identified with the Schrödinger dynamics with the state of the system constrained to the classical phase space submanifold $M^{\sigma}_{3n,3n}$. Conversely, as explained in [2], there is basically a unique unitary time evolution on the space of states of a quantum system that yields Newtonian dynamics when constrained to the classical phase space submanifold. This results in a tight, geometry-based relationship between classical and quantum physics. In particular, the normal distribution of position of a macroscopic particle extends to the Born rule for transitions between arbitrary quantum states. The central limit theorem ensures that the outcomes of a measurement of a classical system satisfy generically the normal distribution law. It follows that the Born rule in measurements of a quantum system is as generic as the normal distribution law in classical measurements.

This seems to indicate that the physical laws that govern the behavior of macroscopic and microscopic bodies are fundamentally the same. For instance, the Brownian motion can be used to model the process of measurement on a macroscopic particle. The state of a microscopic particle, exposed to the same conditions, is equally likely to be displaced in any direction in the space of states CP^{L_2}. The state performs then a random walk on the space of states and the probability of reaching a particular point of CP^{L_2} is automatically given by the Born rule.

The role of the measuring device is reduced to creating a "noise" that triggers the diffusion in CP^{L_2} and in recording the states that reach a particular region in CP^{L_2}. It follows that the collapse of the wave function in the framework is not an instantaneous "concentration" of the state near an eigenstate of the measured observable. Instead, it is a random walk of the state on the space of states under interaction with the measuring device and the environment. The "collapse" to an eigenstate of an observable happens when the state reaches the eigen-manifold of that observable. In the case of position measurements, the state must reach the classical space or phase space submanifolds in CP^{L_2}. Due to the enormous amount of collisions between a macroscopic body and the particles in the surroundings, position of the body is constantly measured. As a result, the diffusion process for macroscopic bodies can trivialize, which may explain why they remain in the classical space and, therefore, have a definite position.

Acknowledgments

I would like to express my deep gratitude to Larry Horwitz for inviting me to give a talk at Bar-Ilan and Ariel universities and for numerous fruitful discussions, and to Martin Land for organizing my talk at IARD 2018 conference and for his sincere interest and great comments.

References

[1] Kryukov A 2017 *J. Math. Phys.* **52** 082103
[2] Kryukov A 2018 *J. Math. Phys.* **59** 052103
[3] Kryukov A 2007 *Phys. Lett.* A **370** 419
[4] Sanz A and Miret-Artés S 2012 *Dynamics of Open Classical Systems* in: Lecture Notes in Physics vol 850 (Berlin: Springer) p 47
[5] Percival I 1998 *Quantum State Diffusion* (Cambridge: Cambridge University Press)
[6] Pearle P 1976 *Phys. Rev.* D **13** 857
[7] Pearle P 1999 *Collapse Models* in: Lecture Notes in Physics vol 526 (Berlin: Springer) p 195
[8] Ghirardi G, Rimini A and Weber T 1986 *Phys. Rev.* D **34** 470
[9] Ghirardi G, Pearle P and Rimini A 1990 *Phys. Rev.* A **42** 78
[10] Diósi L 1989 *Phys. Rev.* D **40** 1165
[11] Adler S 2002 *J. Phys.* A **35** 841
[12] Adler S 2004 *Quantum theory as an emergent phenomenon* (Cambridge: Cambridge University Press)
[13] Adler S, Brody D, Brun T and Hughston L 2001 *J. Phys.* A **34** 8795
[14] Adler S and Brun T 2001 *J. Phys.* A **34** 4797
[15] Adler S and Horwitz L 2000 *J. Math. Phys.* **41** 2485
[16] Hughston L 1996 *Proc. Roy. Soc. London* A 953
[17] Bassi A, Lochan K, Satin S, Singh T and Ulbricht H 2013 *Rev. Mod. Phys.* **85** 471
[18] Joos E, Giulini D, Kiefer C, Kupsch J and Stamatescu I 2003 *Decoherence and the appearance of a classical world in quantum theory* (Berlin: Springer)